Large Power Transformers

STUDIES IN ELECTRICAL AND ELECTRONIC ENGINEERING

STUDIES IN ELECTRICAL AND
ELECTRONIC ENGINEERING 25

Large Power Transformers

K. KARSAI D. Sc.
Institute for Electrical Power Research
Budapest, Hungary

D. KERÉNYI D. Sc.
GANZ Electric
Budapest, Hungary

L. KISS D. Sc.
GANZ Electric
Budapest, Hungary

ELSEVIER
Amsterdam – Oxford – New York – Tokyo 1987

This book is the revised English version of
Nagytranszformátorok
published by Műszaki Könyvkiadó, Budapest

Translated by
Z. Náday

Joint edition published by
Elsevier Science Publishers, Amsterdam, The Netherlands and
Akadémiai Kiadó, Budapest, Hungary

The distribution of this book is being handled by the following publishers

for the U.S.A. and Canada
Elsevier Science Publishing Company, Inc.
52 Vanderbilt Avenue
New York, NY 10017, U.S.A.

*for the East European countries, Democratic People's Republic of Korea,
People's Republic of Mongolia, Republic of Cuba and the Socialist Republic of Vietnam*
Kultura, Hungarian Foreign Trading Company
P. O. Box 149, H-1389 Budapest 62, Hungary

for all remaining areas
Elsevier Science Publishers
25 Sara Burgerhartstraat
P.O.Box 211, 1000 AE Amsterdam, The Netherlands

Library of Congress Cataloging-in-Publication Data
Karsai, K. (Károly)
    Large power transformers.
    (Studies in electrical and electronic engineering ;
25)
    Rev. English version of: Nagytranszformátorok.
    Bibliography: p.
    Includes index.
    1. Electric transformers. I. Kerényi, D. (Dénes)
II. Kiss, László. III. Title. IV. Series.
TK2792.K3413 1986   621.31'4   86-4596

ISBN 0-444-99511-0 (Vol. 25)
ISBN 0-444-41713-3 (Series)

Printed in Hungary

PREFACE

The Hungarian version of this book was published more than ten years ago. Its main aim was to fill the gap in the literature due to the appearance of new magnetic and insulating materials and the substantial increase of rating and voltage limits. Since then several new requirements have been added, such as the losses arising in stray magnetic field, resonance phenomena due to oscillating switching overvoltages, the survey of mechanical design, the review of main accessories, etc. By arranging for publication of this English edition, Elsevier Science Publishers and Akadémiai Kiadó have made it possible for us to include these requirements in the text. In particular, we appreciate their understanding of the problems related to the inclusion of the latest R&D results in transformer engineering.

In 1985, the 100th anniversary of the invention of the transformer and its parallel operation was celebrated by almost all the leading manufacturers of the world at an international meeting in Budapest.

The authors dedicate this book to the memory of those experts from all over the world who, through 100 years of continuous development, contributed to the present state-of-the-art of transformer engineering.

Chapters 1–3 were written by Dr. K. Karsai, Chapters 4 and 5 by Dr. D. Kerényi, and Chapters 6 and 7 by Dr. L. Kiss.

K. Karsai
D. Kerényi
L. Kiss

CONTENTS

8

1. THE PHYSICAL BACKGROUND
OF TRANSFORMER OPERATION

1.1 Principal factors in the early development
of transformers

As is often the case with new technical ideas, different authors writing about the transformer are surprisingly inconsistent in their views of the priority of its invention. Undoubtedly, Faraday's experiments and Maxwell's theoretical results made the theory of electromagnetic fields a public property on which the operating principle of the transformer is based. Development of the transformer may be characterized by the important steps below.

In 1831, Faraday induced a voltage pulse across the secondary terminals of his experimental apparatus (Fig. 1.1) by interrupting the flow of direct current. In 1836, G. J. Page repeated the experiment in his open-core device that would today be called an auto-transformer. In 1838, N. J. Callon again repeated the experiments of Faraday.

At the GANZ Factory in Budapest (originally an iron foundry established by the successful Swiss iron-founder, Abraham Ganz in 1844), the first closed-core transformer was built on 16th September, 1884 under the guidance of M. Déry, O. Bláthy and K. Zipernovsky. The capacity of this transformer was 1400 VA, its frequency 40 Hz and its voltage ratio 120/72 V (Fig. 1.2). The GANZ inventors were the first to use the term "transformer" in their patent application, a name created from the Latin. They had their invention patented in some countries of major commercial importance at that time. With a few exceptions, their applications for patent were successful. The file numbers of some of the more important patents were 42.793-35/2.446 in Austro–Hungarian Monarchy, 68.583 in Belgium, XXXVI 154 in Italy, 5.201 in the United Kingdom and 352.105 in the USA.

M. O. Dolivo–Dobrowolsky, an engineer working with AEG in Germany, invented the three-phase transformer in 1890 and in the same year the first oil-cooled, oil-insulated transformer was constructed by Brown. The oil–cellulose two-phase insulation system opened the way to higher capacities and voltage levels. At present, transformer designers are aiming at 2000 MVA capacity, and units of 1500 kV voltage rating are now at the stage of experimental operation.

In the field of transformer development and production the upper voltage limit achievable appears to be somewhere between 1500 and 2000 kV, with capacity limits lying around 5000 MVA. With growing capacities, an ever increasing reliability of products in service is demanded of power transmission systems. The design of transformers is based on the expected sustained and short-term power-frequency voltage stresses originating on the network and on those caused by lightning and switching pulses. Beside the growing geometrical dimensions, disposal of heat losses and the determination of local hot spots resulting from stray magnetic fields are factors of increasing concern in the design.

International standardization is trying to establish a basis for proving compliance with the requirements of service reliability by specifying up-to-date test methods, but progress in this field is by no means complete. Thus the partial discharge tests specified for judging the state of insulation give insufficient information for either the location of internal discharges or the degree of hazard associated with them; no measuring or computing methods are yet known for locating hot spots, and only very few short-circuit laboratories are available which are capable of performing the required tests on units of very high ratings.

Difficult transport problems have to be faced by some designers. At present, the road and rail transport of 500-tons transformers is already solved, but development

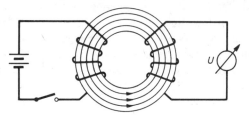

Fig. 1.1. Faraday's experimental apparatus

Fig. 1.2. The first GANZ transformer

16

engineers still concern themselves with the problem of transporting larger transformers as integral units.

Technological advances in recent decades have accustomed the development engineer to rapid changes, a phenomenon which will have to be taken into account when considering the future of transformers. For ten to twenty years, the transformer will surely remain an important element of power systems. It is not easy to predict what new requirements will be raised in the years to come. In any case, the present state of the art is surely an important link in the chain leading into the future.

1.2 The principle of transformer operation

The simplest transformer is a two-winding construction consisting of three active elements: a closed iron core, and two windings inductively coupled with the core. The windings of a two-winding transformer are distinguished either by their service condition of by their voltage levels. In terms of their service condition, the supplied winding is called the primary, and the loaded winding the secondary. In terms of their voltage levels, the terms high-voltage winding and low-voltage winding are used. Apart from a few very individual applications (generator-transformer, metallurgical furnace transformer, etc.), either winding may act as primary or secondary depending on the system condition, whereas the roles of the high- and low-voltage windings obviously cannot be interchanged.

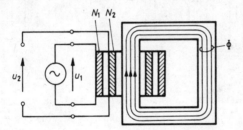

Fig. 1.3. The transformer in no-load conditions

The principle of operation of the transformer can best be understood, to a first approximation, by considering the operation of a loss-free transformer. The loss-free transformer is one in which the resistance of the windings, and the hysteresis and eddy-current losses of the iron core, can be neglected. In the no-load condition of such a transformer, i.e. when no loading impedance is connected across its secondary terminals, there is only a magnetic flux in the iron core (of peak value Φ) and all turns (N_1 and N_2) of the primary and secondary windings are linked with that flux.

In this ideal case (Fig. 1.3) the instantaneous value of primary voltage is

$$u_1 = N_1 \frac{d\Phi}{dt}$$

and the instantaneous value of secondary voltage is

$$u_2 = N_2 \frac{d\Phi}{dt}.$$

Indices 1 and 2 indicate primary and secondary quantities, respectively. In the no-load condition, the law of induction applies and the voltage ratio (being equal to the turn ratio) will be

$$\frac{u_1}{u_2} = \frac{N_1}{N_2}.$$

In the short-circuit condition the secondary winding of the transformer is short-circuited. The primary voltage is then

$$u_1 = N_1 \frac{d\Phi}{dt} \quad \text{and} \quad u_2 = 0.$$

To a first approximation, the flux distribution is represented in Fig. 1.4. The flux distribution depends, basically, on whether the outer or inner winding is short-circuited (shorted). A common feature of the two cases is that there is no resultant

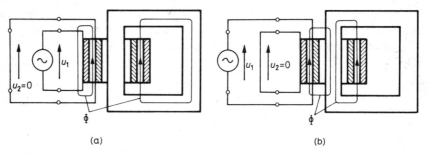

(a)   (b)

Fig. 1.4. Short-circuited transformer (a) with outer winding supplied; (b) with inner winding supplied

flux linkage of the short-circuited winding. This is because there is no flux present in the core when the winding adjacent to the core is short-circuited, whereas if the outer winding is shorted, the winding is linked with an equal number of positive and negative flux lines. In the short-circuited condition the transformer excitations are in equilibrium:

$$i_1 N_1 = i_2 N_2.$$

The instantaneous values, and also the r.m.s. values in the steady-state condition, of the current are inversely proportional to the numbers of turns. For the current ratio the notation

$$\frac{i_1}{i_2} = \frac{N_2}{N_1}$$

is commonly used.

18

In the no-load condition the flux develops practically in the core only, so that both windings are linked with the same flux. Under load, and even more so in the short-circuit condition, the turns of the windings are linked with fluxes of different magnitude. That is why the concept of resultant flux linkage is introduced, obtained by adding the fluxes linked with the turns of the winding:

$$\psi_1 = \sum_{k=1}^{n} N_{1k}\Phi_{1k}, \quad \psi_2 = \sum_{k=1}^{m} N_{2k}\Phi_{2k},$$

where N_{1k} and N_{2k} are the numbers of turns of the respective windings linked with the flux components Φ_{1k} and Φ_{2k}.

The values of the resultant flux linkages in the short circuit condition are

$$\psi_1 = \sum_{k=1}^{n} N_{1k}\Phi_{1k} > 0$$

and

$$\psi_2 = \sum_{k=1}^{m} N_{2k}\Phi_{2k} = 0.$$

The flux linkages of the two windings are different, resulting in a stray flux of

$$\psi_s = \psi_1 - \psi_2 = \sum_{k=1}^{n} N_{1k}\Phi_{1k} - \sum_{k=1}^{m} N_{2k}\Phi_{2k}.$$

The loaded condition lies between the short-circuit and no-load conditions. Thus, because of the presence of the stray field, the turns ratio will no longer be equal to the voltage ratio.

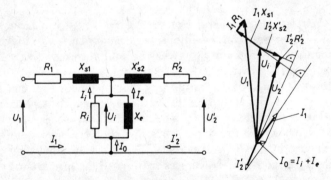

Fig. 1.5. Equivalent circuit and vector diagram of the transformer

For the representation of any load condition of a real transformer, the equivalent circuit is introduced (Fig. 1.5). In this circuit diagram, the stray reactances X_{s1} and X_{s2} calculated from the stray fluxes of the primary and secondary windings are

$$X_{s1} + X'_{s2} = X_s = \omega L_s$$

and

$$L_s = \frac{\psi_1 - \psi'_2}{i_1}.$$

2*

In the above relations $\omega = 2\pi f$ is the angular frequency f is the frequency and L_s is the leakage inductance calculated from the stray field. ψ_1 is the primary resultant flux linkage, and ψ'_2 the secondary resultant flux linkage reduced to the primary. Reduction to the primary can be performed on the basis of the formula

$$\psi'_2 = \frac{N_1}{N_2}\psi_2.$$

The circuit diagrams can be constructed, by way of speculation, from the no-load and short-circuit measurements, but can also be derived by network transformation, as shown later. In the circuit diagram, all quantities represent values reduced either to the primary or to the secondary side, i.e. the respective quantities of one side of the transformer are recalculated corresponding to the number of turns of the other side so as to keep the losses and reactive powers unchanged. The calculations reflect the number of turns $N'_2 = N_1$ when reducing to the primary side and the number $N'_1 = N_2$ when reducing to the secondary side. The secondary quantities reduced to the primary side are

$$N'_2 = \frac{N_1}{N_2}N_2, \quad I'_2 = \frac{N_2}{N_1}I_2, \quad u'_2 = \frac{N_1}{N_2}u_2, \quad \psi'_2 = \frac{N_1}{N_2}\psi_2.$$

Similarly, the reduction to the secondary side is

$$N'_1 = \frac{N_2}{N_1}N_1, \quad I'_1 = \frac{N_1}{N_2}I_1, \quad u'_1 = \frac{N_2}{N_1}u_1, \quad \psi'_1 = \frac{N_2}{N_1}\psi_1.$$

For a loss-free transformer the value of X_e may be taken as very large, and that of I_e as very low. Under practical conditions, the primary current is higher by three orders of magnitude than the exciting current, and correspondingly X_s is three orders of magnitude lower than X_e.

In the equivalent circuit of the transformer the resistance losses and core losses are taken into account, together with the exciting current. The resistance losses of windings can be divided into primary-side and secondary-side values. The stray reactance can also be split into two parts by calculation but often, quite arbitrarily, it is simply divided into two equal parts $X_{s1} = X'_{s2}$. In the equivalent circuit, the values of resistances and stray reactances reduced to the primary are

$$R'_2 = \left(\frac{N_1}{N_2}\right)^2 R_2 \quad \text{and} \quad X'_{a2} = \left(\frac{N_1}{N_2}\right)^2 X_{s2}.$$

The values of R_1 and X_{s1} reduced to the secondary are $R'_1 = \left(\frac{N_2}{N_1}\right)^2 R_1$ and X'_{s1} $= \left(\frac{N_2}{N_1}\right)^2 X_{s1}$. The losses arising in the core, termed core losses, are represented by

20

resistance R_i, and the excitation reactance by X_e. The current flowing through resistance R_i is X_i, and that through reactance X_e is $\bar{I}_e$. The resultant of currents $\bar{I}_i$ and $\bar{I}_e$ is the no-load current to flowing across the primary terminals in the no-load condition of the transformer.

Due to the non-linear excitation characteristic of the core, the instantaneous value of current I_0 will be a non-sinusoidal function of time in the case of a sinusoidal supply voltage.

The r.m.s. value of current I_0 is

$$I_0 = \sqrt{I_{01}^2 + I_{02}^2 + \ldots I_{0k}^2},$$

where I_{01}, I_{02} and I_{0k} are the r.m.s. values of the 1st, 2nd and kth harmonics of the no-load current. In Fig. 1.5 the voltage and current vector diagram pertaining to the load current I_2' of the transformer is represented.

For transformers of 1 MVA up to several hundred MVA ratings the following approximate information can be given on the quantities appearing in the equivalent circuit. The resistances fairly closely follow the relation $R_1 = R_2'$, whereas the stray reactances may be a few times higher than the resistance in smaller transformers and ten to fifty times higher in large units. In smaller units the no-load current is one to two per cent of the rated current, but in large transformers it is usually as low as one-tenth of one per cent. The active component of the no-load current is 1/3 to 1/10 of the reactive component. From these data it can be seen that the two parallel network elements X_e and R_i in the equivalent circuit are in order about three to four times higher than the series elements $(R_1, R_2, X_{s1}$ and $X_{s2})$. In the vector diagram of Fig. 1.5, the relative magnitudes of I_0, $I_1 X_{s1}$, $I_2 X_{s2}$, $I_1 R_1$ and $I_2' R_2'$ are somewhat exaggerated for clarity. For analysing the loaded condition, the parallel network elements of the equivalent circuit may usually be neglected. In the short-circuit condition of the transformer, this neglect is even more justifiable. In such a case, assuming the supply voltage remains unchanged, the short-circuit current may be about 20 times higher (with larger transformers about 6 to 12 times higher) than the rated current.

A transformer is designed for a definite rated voltage U_n and a definite rated current I_n. The product of these two values determines the capacity and dimensions of the transformer. The rating of a single-phase transformer is $P_n = I_n U_n$. At its rated voltage the losses arising in a large tansformer loaded with its rated current are negligible compared to the power transmitted.

1.2.1 Equivalent circuit of the transformer

In addition to the T-type equivalent circuit, the transformer can also be represented by other four-terminal networks. Originally the T-network was proposed [7], but the π-network was also mentioned in literature [11] as another equally suitable possibility. From either of the two equivalent circuits, closely similar calculation results can be obtained. Using the notation of Fig. 1.6, either of the two four-terminal networks can be derived from the other with the help of relations

$$\bar{Z}_1 + \bar{Z}_2 = Z_{1,2} \quad \text{and} \quad \bar{Z}_a = \frac{\bar{Z}_{a1} + \bar{Z}_{a2}}{\bar{Z}_{a1} + \bar{Z}_{a2}}.$$

Of the two circuits the T-network is the one more commonly used. In practical network calculations it is all the same which model is used for substituting the transformer, because the quantities that can be measured at the transformer terminals are practically identical in both cases.

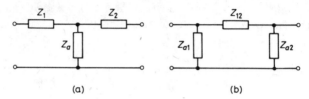

(a)   (b)

Fig. 1.6. T and π equivalent circuits of the transformer

Short circuit tests performed on large transformers at rated supply voltage have shown that, in the case of concentrically arranged windings, the flux in the core limb following the application of the short-circuit, will depend considerably on whether the inner winding adjacent to the core is supplied and the outer winding short-circuited, or vice-versa. When the outer winding is supplied, the flux in the core limb may drop to 5 to 15% of its no-load value. On the other hand, in the case of supplying the inner winding the flux may rise to 105 to 115% of the no-load value.

Such a phenomenon can only be explained by an equivalent circuit—be it a T or π-network—in which a negative series element is also present.

An equivalent circuit correctly representing the magnetic conditions can be derived as suggested by Cauer [19], Cherry [20] and Edelmann [28]. According to these authors, the magnetic network components can be transformed in practical cases into networks whose terminals are terminals of the electric circuit linked with the magnetic circuit and which contain only electric network elements.

In Fig. 1.7, two alternatively excited magnetic circuits are represented. Each is composed of four branches, all branches being connected in parallel in case (b), and in series in case (a). Assuming in each of the four branches an equal number of turns $N = N_1 = N_2 = N_3 = N_4$, the magnetic node-point equation for the parallel magnetic circuits may be written as

$$\Phi_1 + \Phi_2 + \Phi_3 + \Phi_4 = 0.$$

Multiplying this equation with the number of turns and differentiating the product with respect to time, equation

$$u_1 + u_2 + u_3 + u_4 = 0$$

containing electric voltages is obtained. This equation corresponds to a network consisting of four inductive reactances with their voltages connected in series (Fig.

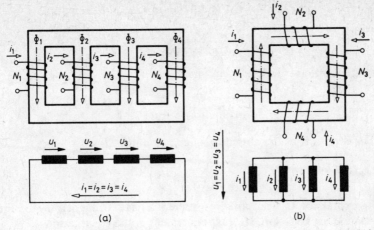

Fig. 1.7. Duality between magnetic and electric quantities in circuits connected (a) in series; (b) in parallel

1.7(a)). In the parallel magnetic branches the magnetomotive forces must be in equilibrium, hence

$$i_1 = i_2 = i_3 = i_4 .$$

It can be seen that the equivalent of the parallel magnetic circuits is the series electric circuit. Based on a similar consideration, for the series magnetic circuit of Fig. 1.7(b), the equation

$$i_1 + i_2 + i_3 + i_4 = 0$$

can be written, and also the constancy of the magnetic flux is satisfied:

$$\Phi_1 = \Phi_2 = \Phi_3 = \Phi_4 .$$

Multiplying this equation also by the number of turns and differentiating gives

$$u_1 = u_2 = u_3 = u_4 .$$

From the relations $\Sigma i = 0$ and $u_1 = u_2 = u_3 = u_4$ it becomes clear that the equivalent of the magnetic circuit of Fig. 1.7(b) is the parallel electric circuit of Fig. 1.7(a). The duality relations by means of which the required network transformation can be performed are as follows:

(a) the relation between magnetic flux (Φ), permeance (Λ) and magnetomotive force (Θ) is the magnetic Ohm's law:

$$\Phi = \Lambda \Theta$$

and its dual equivalent is the induction law of the electric circuit linked with the magnetic circuit:

$$u = L \frac{di}{dt} .$$

23

(b) The dual equivalent of the nodal law of magnetic circuits:

$$\sum_{v=1}^{n} \Phi_v = 0$$

is the loop law of electric circuits:

$$\sum_{v=1}^{n} N \frac{d\Phi_v}{dt} = \sum_{v=1}^{n} U_v = 0.$$

(c) The loop law of the closed magnetic circuit is

$$\sum_{\mu=1}^{m} U_\mu = 0.$$

In this expression U_μ may equally represent a passive magnetomotive force or an active magnetic excitation. Its dual equivalent is the electric nodal law:

$$\sum_{\mu=1}^{m} \frac{1}{N} U_\mu = \sum_{\mu=1}^{m} i_\mu = 0.$$

By combining the duality relations shown above, the rules of network transformation between electric circuits and their dual magnetic circuits are obtained:

1. the electric node corresponds to the magnetic loop;
2. the electric loop corresponds to the magnetic node;
3. the electric induced voltage corresponds to the magnetic flux linked with exciting turns;
4. the electric currents correspond to the magnetomotive force;
5. the permeance corresponds to the induction coefficient.

The described duality relations are suitable for the determination of equivalent electric circuits linked with more complicated magnetic circuits. When judging the practicability of results arising from this determination, consideration should always be given to the division of the sections of the spatially distributed magnetic field into sections to be transformed, because each of these sections will appear in the equivalent electric circuit as a lumped network parameter. This fact should be considered not only in applying the Cauer–Cherry–Edelmann duality relations, but in all cases when calculations are performed with the aid of transformer equivalent circuits, because in such cases the spatially distributed field is also always substituted by a network element of lumped parameters.

With the aid of the duality relations described, the equivalent circuit can be constructed from the magnetic circuit of the transformer shown in Fig. 1.8. The two windings of the transformer shown are in loose magnetic coupling, and the stray fields of the two windings can be clearly distinguished from each other. In the full-line drawing of Fig. 1.8(b) the magnetic circuit of the transformer built up of lumped elements is shown. Θ_1 and Θ_2 are the magnetomotive forces represented by the two windings, Λ_{s1} and Λ_{s2} are the permeances of stray fields linked with the two windings, and Λ is the permeance of the flux linked with both windings. The dual

24

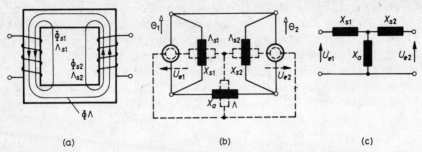

Fig. 1.8. Equivalent circuit of a two-winding transformer with loose magnetic coupling: (a) magnetic circuit; (b) equivalent electric circuit; (c) the dual circuit of (b)

configuration should be constructed so that one node is placed in each loop and the nodes are then connected with the intersections of the magnetic circuit elements to be transformed $\Theta_1, \Theta_2, \Lambda_{s1}, \Lambda_{s2}$. Where the element intersected is a permeance, the transformed configuration is an inductance (or reactance), and where a magnetomotive force is represented by the element intersected, the transformed network element is an induced voltage. The result of the construction described is shown by dashed lines in the diagram. The configuration obtained by this construction appears separately in Fig. 1.8(c), which is the well known T-network.

The other example demonstrated is the case of the transformer with concentric windings. The simplified flux pattern on which the design is based is shown in Fig. 1.9. The two different diagrams appearing there are required because the flux

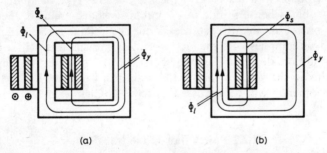

Fig. 1.9. Simplified flux distribution of a transformer with concentric windings: (a) with outer winding supplied; (b) with inner winding supplied

distribution with the outer winding supplied and the inner loaded (case (a)) differs from that where the inner winding is supplied (case (b)). In the diagrams Φ_l is the flux in the limb, Φ_y that in the yoke and Φ_s the stray flux.

The relation between the fluxes in (a) is

$$\Phi_y = \Phi_l + \Phi_s$$

and in case (b):

$$\Phi_l = \Phi_s + \Phi_y.$$

25

In large transformers the effect of the resistance on the flux pattern resistance can be neglected. Thus in the short-circuited condition, in case (a) the flux reduces to a very low value ($\Phi_l \cong 0$), and a large portion of the stray flux passes through the yoke.

In Fig. 1.9(b) the inner winding is supplied and the outer loaded. In this case the magnitude of the limb flux is practically equal to the stray flux, and the yoke flux is very low. The connections of magnetic circuits corresponding to the cases indicated in Fig. 1.9 are shown by the circuits of Fig. 1.10(a), drawn with continuous lines. The

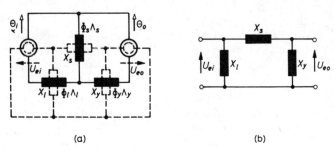

(a)   (b)

Fig. 1.10. Derivation of π equivalent circuit

magnetic circuit of the core consists of series-connected sections of the magnetomotive force in the limb Λ_l and in the yoke Λ_y. This circuit is excited by the inner winding Θ_i and by the outer winding Θ_0. The broken-line circuit in variant (a) of Fig. 1.10 shows the dual circuit of the magnetic configuration drawn with continuous lines. The transformed circuit is repeatedly shown in diagram (b). This circuit corresponds to a network of the π type. This type of model satisfies the requirement that, when one side of the circuit is short-circuited, the full voltage appears on one of the shunt reactances, while the voltage zero on the other shunt reactance. This complies with the fact proved by measurements that in the short-circuited condition, depending on whether the outer or inner winding is shorted, either the flux in the yoke or that in the limb drops to zero.

This phenomenon will be discussed in detail in Chapter 3.

1.2.2 Power flow in the transformer

Transformation of electric power within a transformer takes place, in such a way that the power fed into the primary winding flows into the secondary winding. The power flowing through a unit area can be described by the Poynting vector of the electromagnetic field. The vector itself can be calculated from the electric and magnetic components of the electromagnetic field. Figure 1.11 shows a cross-sectional detail of a transformer, where the numbers of turns of the primary and secondary windings are N_1 and N_2, respectively, and the instantaneous respective currents are i_1 and i_2. Between the two windings, the instantaneous magnetic field strength is approximately

$$H = \frac{i_1 N_1}{l_s} = \frac{i_2 N_2}{l_s},$$

26

where l_s is the axial length of the winding indicated in the drawing. For the purpose of more accurate calculations, l_s is usually taken to be longer than the geometrical axial length of the winding, but shorter than the height of the limb. The determination of its exact value is dealt with in the description of the Rogowsky

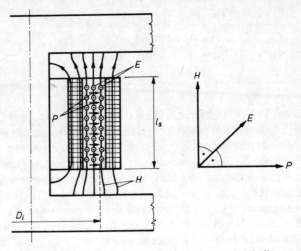

Fig. 1.11. Power flow of a transformer with concentric windings

factor. The electric field strength in the gap between the two windings is induced by the alternating flux in the core. This electric field is tangential in direction and its instantaneous strength at a distance of $r_i = D_i/2$ from the centre line of the core is

$$E = \frac{1}{D_i \pi} \frac{d\Phi}{dt} = \frac{u_w}{D_i \pi},$$

where Φ is the instantaneous value of the flux linked with the windings, t is the time, u_w is the turn voltage, and D_i is the diameter of the circle of radius r_i. The instantaneous value of the vector of power flow (Poynting) is

$$\mathbf{p} = \mathbf{E} \times \mathbf{H}.$$

Substituting the values of $\mathbf{E}$ and $\mathbf{H}$ into this formula and assuming, with good approximation, the angle between $\mathbf{E}$ and $\mathbf{H}$ to be $90°$, the instantaneous value of the Poynting vector will be

$$p = \frac{U_w i_1 N_1}{D_i \pi l_s}.$$

This relation can be transformed into a form to give the mean value of power flow (P) between the two windings per unit surface area of the cylinder of diameter D_i:

$$P = \frac{U_1 I_i \cos \varphi}{D_i \pi l_s}.$$

27

Table 1.1

Poynting vector (power density) and electromagnetic field strengths in
electrical equipment

	Turbo-generator	Trans-former	Trans-mission line	Cable
E, kV m^{-1}	0.3	0.3	2000	8000
H, A m^{-1}	250×10^3	100×10^3	30×10^3	100×10^3
P, MW m^{-2}	0.075	0.03	60	800

In these expressions, U_w is the r.m.s. value of the turn voltage, and U_1 is N_1 times the turn voltage. The density of power flow is a measure of utilization of the transformer. In the design of very large transformers, increase of the Poynting vector is a general objective. In Table 1.1 the magnitudes of the Poynting vector and electromagnetic field strengths presently attained are given for a few power transmission facilities. It can be seen that the values achievable in transformers and turbogenerators still fall far below the respective values for cables and overhead lines. This deficit is mainly due to the low level of the values of **E**. A drastic increase in the induced electric field strength, both in transformers and turbogenerators, is possible through increasing the induction **B** in the magnetic flux inducing the voltage, or by increasing the magnetic field strength (i.e. the current density). Both aims can be achieved by intensifying the cooling.

2. THE CORE

2.1 No-load characteristics

One of the active parts of the transformer is the laminated iron core. This core carries the magnetic flux linked with the transformer windings, and it is the size and magnetic stress of this core that determine the capacity and no-load characteristics of the transformer. These no-load characteristics are the no-load current, no-load loss, inrush current and magnetic noise of the transformer.

The no-load current of smaller transformers is a few per cent of the rated current, and for higher rated capacities it is even lower. The no-load current consists of two parts, being the reactive and the active components. The variation of the reactive component as a function of time, assuming a sinusoidal supply voltage, shows several harmonics in addition to the fundamental harmonic. The harmonic content of the active component, again with a sinusoidal supply voltage, is very low. The reactive component is considerably larger than the active component, so the absolute value of the no-load current of the transformer is very close to the value of the reactive component. The reactive component is a multi-wave current, but in various calculations (provided that approximations are permissible) only the fundamental harmonic or the equivalent sinusoidal wave of that multi-wave current is taken into account.

The no-load losses consist of the iron losses arising in the course of successive reversals of magnetization of the iron core. In addition, an insignificant core loss also occurs due to the no-load current flowing in the primary winding, and similarly negligible losses are produced in the inactive metallic parts by the flux leaving the core in no-load condition.

Due to the successively reversed magnetization, eddy-current and hysteresis losses are produced in the core. The eddy-current losses are caused by the eddy currents induced in the laminations of the core, their magnitude being

$$P_e = \frac{1.65 f^2 B_m^2 t^2}{\rho d} m = \sigma_e (f B_m t)^2 m,$$

where P_e is the eddy-current loss in W, f is the frequency in Hz, B_m is the peak value of the flux density in T, t is the thickness of the individual steel lamination sheets (minimum dimension perpendicular to the magnetic flux lines) in m, ρ is the specific resistance of the sheet material in Ωm, d is its density in kg m^{-3} and m is the mass of the iron core in kg. In the second form of the equation the values of σ_e lie in the range of 200 to 1000 m^2 Ω^{-1} kg^{-1} for up-to-date core materials.

The hysteresis loss is

$$P_h = \sigma_h f B_m^n m,$$

where P_h is the hysteresis loss in W, f is the frequency in Hz, B_m is the peak value of flux density in T, n is the Steinmetz exponent, having numerical values of 1.6 to 2 for hot-rolled laminations and increasing beyond 2 for cold-rolled materials, m is the mass of the iron core in kg and the values of σ_h vary in the range of 3×10^{-3} to 20×10^{-3} [39].

The magnitude of core loss is influenced by the shape of core flux variation (i.e. that of the supply voltage) as a function of time. This is of significance mainly in laboratory testing of transformers of larger ratings and single-phase transformers [11, 109], when the available voltage source is of insufficient capacity to prevent excessive distortion of the voltage–time curve under the effect of no-load current of the transformer to be tested. Distortion of the voltage wave shape supplied by a laboratory voltage source may be even more pronounced in the case of single-phase loads when additional distortion is caused by the armature reaction of the supplying synchronous generator. Results of no-load loss measurements performed with a non-sinusoidal supply voltage can be reduced to the sine-wave basis in the following way. When a voltmeter responsive to the mean value is used, the supply voltage of the unloaded transformer is adjusted to the voltage at which the loss is to be measured. The reading of this mean-value voltmeter will be proportional to the peak value of flux density developing in the core. For equal peak values the hysteresis losses are also equal, independently of the wave shape. As the eddy-current loss is approximately proportional to the square of the r.m.s. value of voltage, higher eddy-current losses will be measured than the eddy-current losses pertaining to the sinusoidal wave shape, in proportion to the square of the quotient of the r.m.s. voltmeter reading and the r.m.s. value of the sinusoidal voltage. Thus, the value of loss reduced to the sinusoidal wave will be [110].

$$P = \frac{P_m}{p_h + k p_e},$$

where P_m is the measured power, p_h is the hysteresis loss referred to the full core loss, p_e is the eddy-current loss related to the full core loss and k is the square of the ratio between the reading of a voltmeter responsive to the mean value and the reading of a voltmeter responsive to the r.m.s. value. For cold-rolled sheets $p_h = 0.5$ and $p_e = 0.5$ are good approximations, while the respective values for hot-rolled sheets are $p_h = 0.7$ and $p_e = 0.3$.

The magnetizing inrush current of a transformer unloaded on its secondary side belongs to the no-load characteristics. The phenomenon takes place as described in the following. At the instant of switching on, the magnetic flux linked with the primary winding is forced to change suddenly in proportion to the voltage appearing at the primary terminals of the transformer. The magnitudes of flux developing as a result of this change may be very different. For example, if the core has been magnetized to some constant remanent flux density before switch-on and the phase of the voltage is such as to cause the flux linked with the winding to increase further, a flux as high as twice the peak of rated flux may add to the remanent flux and, due to saturation, the peak value of the magnetizing current pertaining to this increased flux may considerably exceed even the peak value of the rated current. High inrush currents may have a disturbing effect on the automatic protection of the transformer and network, and may give rise to significant mechanical stresses within the transformer. Inrush currents of relatively long decay times may result in further difficulties when causing the protection device to trip. The overvoltages associated with such trippings may exceed the value of switching surges normally expected in the network, and may jeopardize the transformer.

Transformer operation is inherently associated with noise of magnetic origin. The magnetic noise is caused by the phenomenon of magnetostriction. Because dimensional changes due to magnetostriction are proportional to the square of flux density, their fundamental frequency is twice that of the transformer supply voltage. Beside the basic magnetic noise (100 or 120 Hz) usually the 2nd, 3rd and 4th harmonics also appear. In the overall noise of a well-designed large transformer construction, in spite of the measures taken to suppress them, noises of magnetic origin predominate because they increase proportionally to the logarithm of transformer rating. The noise of cooling oil pumps (if any) may usually be neglected but the noise of fans associated with forced air cooling may be more disturbing, with their frequency spectrum lying in the range of 300 to 3000 Hz.

2.2 Magnetic steel characteristics

Transformer steels belong to the group of ferromagnetic materials. The kind of laminations applicable to transformers was developed by Headfield [43] as early as 1889, and it was he who recognized the possibility of reducing the hysteresis and

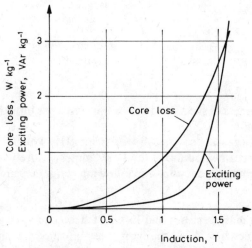

Fig. 2.1. Specific loss and magnetizing power of hot-rolled transformer steels with 0.9 W kg^{-1} loss figure at $f = 50$ Hz as a function of flux density

eddy-current losses in 1 mm and thinner steel sheets by alloying with silicon, at the same time maintaining their magnetization properties within acceptable limits. The development of hot-rolled sheets was carried on until the 1950s, their essential properties being as follows:

— Their magnetization properties and losses, with the magnetic field acting in the direction of rolling or in any other direction, are essentially identical. The attainable minimum loss of a 0.35 mm thick sheet is 0.9 W kg^{-1} at a flux density of 1 T of 50 Hz frequency. Under the above conditions the reactive magnetizing power is 5 VAr kg^{-1}. In Fig. 2.1, the losses and magnetizing power of hot-rolled steel are shown as a function of peak flux density.

— Hot-rolled sheets saturate at a peak flux density as low as 1.9 T. Thus the limit of their excitation does not exceed the value of 1.5 T.

— The magnetic properties of sheets are stable under the effect of mechanical stresses (impacts, bends, vibration). Even after protracted service no signs of aging or deterioration have been observed.

— Sheets are delivered as plates or in coils. The surface of sheets should be provided, at least on one side, with an insulating layer in order to reduce eddy-current losses (Table 2.1).

Table 2.1

Thickness of steel insulation applied
to one side of sheets

Material	Thickness, mm
Paper	0.03
Lacquer	0.02
Sodium silicate	0.01–0.015
Phosphate coating	0.005
Magnesit	0.005
Oxide layer	0.005
Carlite	0.001–0.003

— Surface unevenness of sheets and insulating layers has limited the space factor of cores to a maximum of 92%.

The cold-rolled steel was invented by Goss in 1935. This type of transformer lamination rapidly became popular in the 1950s, almost entirely ousting hot-rolled sheets from transformer production. The following properties are characteristic of cold-rolled sheets:

— The magnetic characteristics and losses of steels are strongly dependent on the direction of magnetizing. A 30° change in direction may cause more than double losses, and in the case of a 60° deviation the magnetizing power may be 50 times higher than normal (Fig. 2.2).

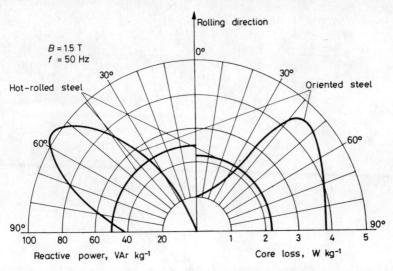

Fig. 2.2. Directional dependence of core losses and magnetizing powers of hot-rolled and cold-rolled transformer steels at 1.5 T flux density and $f = 50$ Hz

— In the rolling direction, the figures for core loss and magnetizing power are considerably lower than those of hot-rolled sheets. Losses at 1 T and 50 Hz are 0.3 to 0.5 W kg^{-1}, and the magnetizing power is in the range of 0.4 to 0.6 VA kg^{-1}. The losses and magnetizing powers of the various cold-rolled sheets in the rolling direction are shown in Figs 2.3 and 2.4.

— In the rolling direction, saturation takes place at 2.03 T. The d.c. hysteresis loop at 1.5 T and the d.c. magnetization curve are shown in Fig. 2.5.

— The sheets are placed on the market in coils, with widths of 700 to 800 mm, lengths of 500 to 1000 m and thicknesses of 0.35 mm or less. Inner diameters are about 800 mm. The surfaces of coiled sheets are extremely smooth, and even with a carlite insulating coating they fill out 97% of the available core volume. As the carlite coating is heat resistant up to 820 °C, the insulation withstands the effect of heat treatment subsequent to cutting.

— An unfavourable property of cold-rolled steels is their instability. Mechanical stresses due to impacts, cuts, vibration, punching or bending increase the loss in the sheets by 5 to 30%, and the no-load current may exceed even these values. Aging of cold-rolled sheets built into a transformer does not grow beyond 1 to 3% even after decades of operation.

— The deterioration due to stresses caused by mechanical processing can be corrected by subsequent annealing at 800 to 820 °C, whereby the original magnetic properties of the sheets can be fully restored.

In a core manufactured of cold-rolled sheets, magnetic stresses about 20% higher may be permitted than with hot-rolled sheets, and even with these increased flux densities the losses remain below those arising in core structures made of hot-rolled

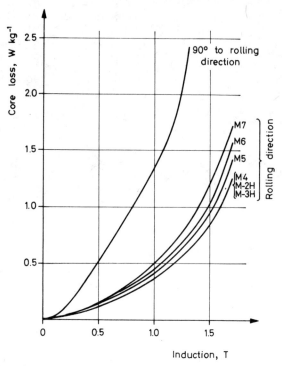

Fig. 2.3. Specific loss of 0.35 mm thick cold-rolled steels at $f = 50$ Hz as a function of flux density

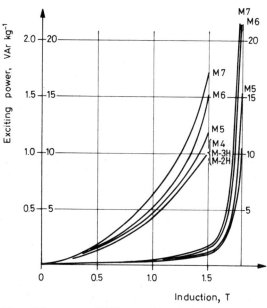

Fig. 2.4. Magnetizing power of 0.35 mm thick cold-rolled steels at $f = 50$ Hz

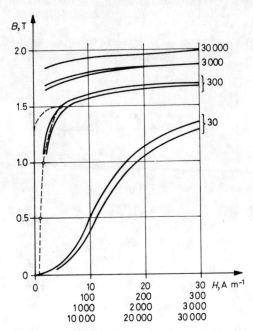

Fig. 2.5. Initial magnetizing and hysteresis curves of cold-rolled steels

material and designed for lower stresses. A transformer may be smaller and have lower loss figures if built with a core made of cold-rolled laminations than its hot-rolled sheet counterpart of identical rating. In Table 2.2 the main physical and technical properties of cold-rolled sheets are shown. The use of hot-rolled sheets in transformer manufacture has declined considerably.

The advent of cold-rolled sheets in transformer production has stimulated designers to strive for better matching between construction and physical properties, by keeping the flux in line with the grain orientation. With this intention cores with mitred joints and equal yoke and limb cross-sections have been introduced, and efforts have been made to accommodate the structural elements required for fixing and clamping of cores outside the core, with the advantage of either omitting or considerably reducing the number of holes through the laminations. An expedient core arrangement is offered by the uncut wound core whose use has already been attempted in the design of larger transformers.

Further improvements of cold-rolled sheets are still in progress. Latest results [118] have led to the appearance of transformer sheets exhibiting at 1.7 T a loss figure of 1.7 W/kg not exceeding those obtainable by the best sheets at 1.5 T a decade ago. The new sheets are characterized by a slightly lower saturation level (about 1.9 T at $H = 800$ A m^{-1}). Very high permeability and low losses have been achieved by producing a relatively rough crystal structure in the direction of rolling and by applying a constant pulling force of 5 MPa to the sheet. In producing this force, the insulating layer plays an important part [57].

Table 2.2

Physical and technical properties of transformer steel sheets

Physical property	Unit	Cold-rolled sheet					Hot-rolled sheet
		in direction of rolling				in trans-verse direction	0.9N kg^{-1}
		M4	M5	M6	M7		
Loss figure at 1 T, 50 Hz	W kg^{-1}	0.36	0.41	0.48	0.51	1.37	0.9
Loss figure at 1.5 T, 50 Hz	W kg^{-1}	0.86	0.95	1.1	1.2	3.35	2.2
Magnetic flux density at 1000 A m^{-1}	T	1.84	1.83	1.82	1.81	1.39	1.29
Magnetic flux density at 2500 A m^{-1}	T	2.00	1.90	1.89	1.88	1.47	1.44
Magnetic flux density at 10 000 A m^{-1}	T	2.01	2.00	1.99	1.95	1.90	1.9
Magnetic flux density at 30 000 A m^{-1}	T	2.01	2.01	2.01	2.01	1.99	1.9
Saturation flux density	T	2.03	2.03	2.03	2.03	1.90	1.9
Magnetizing power at 1 T, 50 Hz	VA kg^{-1}	0.43	0.48	0.52	0.63	2.75	4
Coercitive field strength up to 1.5–1.8 T magnetization	A m^{-1}	6–7	7–8	7–8	11	28	30
Remanent field strength at 1.7 peak	T	1.53	1.49	1.48	1.45	0.25	0.7
Max. a.c. permeability	—	78 000	72 000	70 000	70 000	5000	8300
Curie temperature	°C	855	855	855	855	855	
Resistivity	μΩ cm^{-1}	48	48	48	48	48	60
Density	kg m^3	7650	7650	7650	7650	7650	7600
Hardness, H_{v5}	N mm^{-2}	1800	1800	1800	1800	1800	2600
Limit of elasticity	N mm^{-2}	300	300	300	300	300	350
Tensile strength	N mm^{-2}	340	340	340	340	370	420
Modulus of elasticity in rolling direction	N mm^{-2}	135 000	135 000	135 000	135 000	205 000	172 000
Space factor with insulation	%	97	97	97	97	97	92

2.3 The exciting current

The main dimensions of the core are determined by economic considerations once the required can be satisfied, and effect the mass and electromagnetic properties of the transformer. The voltage in the transformer winding is induced by the flux alternating in the core. The r.m.s. value of the voltage U_i induced in a winding of N turns surrounding the core is

$$U_i = 4k_f f N B_{lm} A_l,$$

where k_f is the form factor (for a magnetic flux varying sinusoidally with time, the variation of voltage is also sinusoidal, thus the form factor is $k_f = \pi/2 \sqrt{2} = 1.11$, f is the frequency in Hz, B_m is the actual peak value of flux density in T and A_1 is the cross-sectional area of the limb in m². The peak value of flux in the limb is $B_{lm} A = \Phi_{lm}$, the flux in the limb and in the yoke being the same in the no-load condition, i.e. $\Phi_{lm} = \Phi_{ym}$. The relation between the current inducing the flux and the required excitation (θ) is

$$\theta = \oint H dl = iN,$$

where θ is the excitation, H is the magnetic field strength in A m^{-1}, dl is the elementary path length in m and i is the current flowing in the conductor of the N-turn winding in A. The law of excitation is valid both for the instantaneous values of current and excitation and for their r.m.s. values. When using the above formula there is no need to apply the method of integration, because in the transformer core

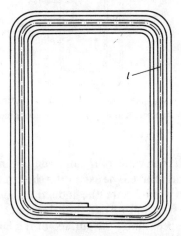

Fig. 2.6. Diagram of a single-phase uncut wound core

a closed induction line can be selected along which, in successive sections, the magnetic induction can be taken as constant to a good approximation, Figures 2.6 and 2.7 show single-phase iron cores where the flux density and field strength can be considered constant along line l represented in the first diagram and along each successive section in the second. The excitation required by an uncut wound core

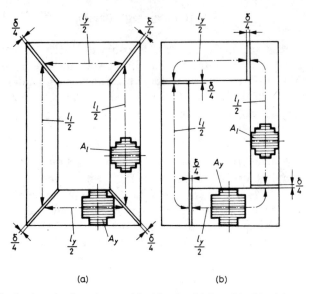

Fig. 2.7. Single-phase laminated cores: (a) with mitred joints; (b) with plain overlap joints

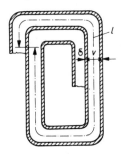

Fig. 2.8. Dimensions of a turn in a wound core: l is the mean length of line of force of the turn, v is the thickness of the sheet and δ is the thickness of sheet insulation

shown in Fig. 2.6 is $\theta = Hl$. The value of H can be determined from the flux densities by means of the magnetizing curve. The excitation thus calculated is an approximate value only, since in the wound core the induction lines do not remain in the lamination while passing from one turn to the other, but continuously re-enter their own turn in the wound sheet along the surface of the lamination so as to form a closed loop. In Fig. 2.8, along the line marked with length l, the magnetomotive force is $V_m = H_i l$, where H_i is the field strength in the sheet along length l. In each turn of the wound core the flux is

$$\Phi' = B_i vb,$$

where b is the core width, v is the sheet thickness and B_i is the flux density in the sheet. The re-entering flux along the full length of a turn re-enters along a surface area of lb. The flux which leaves the turn re-enters through the insulating material, whose permeability equals that of the air, so the magnitude of flux in the air can be determined from the equality of the turn flux in that medium and of the flux

$$B_\delta = B_i \frac{vb}{lb} = B_i \frac{v}{l}.$$

In the field leaving the turn the magnetic field strength is $H_\delta = B\delta/\mu_0$ and the magnetomotive force is $V_\delta = H\delta$.

Example 2.1. With flux lines of mean length $l = 5$ m and flux density of $B_i = 1.6$ T, the magnitude of field strength in a given lamination is $H_i = 100$ Am^{-1}. The required excitation in the steel sheet is $\theta_i = H_i$ $l = 100 \times 5 = 500$ A. With a sheet thickness of $v = 35 \times 10^{-5}$ m, flux density in the space between two adjacent sheets is

$$B_\delta = 1.6 \frac{35 \times 10^{-5}}{5} = 0.112 \times 10^{-3} \text{ T}.$$

The magnetic field strength pertaining to this flux density is $H = 0.112 \times 10^{-3}/1.256 \times 10^{-6} = 89.172$ A m^{-1}. The flux lines between two adjacent turns are perpendicular to the plane of sheets, because the magnetic permeability of sheets is much higher than that of the air. Assuming a space factor of 0.97, the length of flux lines between the two sheets is $35 \times 10^{-5} \times 0.03 = 1.05 \times 10^{-5}$ m, and the magnetomotive force acting along this distance is $U_m = 89.172 \times 1.05 \times 10^{-5} = 9.363 \times 10^{-4}$ A, which is negligible beside the 500 A excitation pertaining to the core.

The required excitation of single-phase laminated cores with mitred joints in the corners as shown in Fig. 2.7(a) or joined at 90° as in Fig. 2.7(b) is

$$\theta = H_{il} l_l + H_{iy} l_y + H_\delta l_\delta,$$

where H_{il} is the magnetic field strength pertaining to flux density B_{il} in the limb and H_{iy} is that for flux density B_{iy} in the yoke. In the air gaps at the joints the magnetic field strength is $H_\delta = B_\delta/\mu_0$. Flux densities in the limb, the yoke and the air can be computed from flux continuity:

$$B_{il} A_{il} = B_{iy} A_{iy} = B_\delta A_\delta.$$

The formulae given above for determining the required excitation apply equally to the instantaneous and mean values. In practice, the calculation of mean values is preferred, because they lend themselves to determination by electrical measuring instruments. The magnitude of mean values depends on how an electric or magnetic quantity varies as a function of time. As the magnetizing curve of iron is non-linear, a simple calculation method requires compromises. In practice, the approximation of assuming a sinusoidal supply voltage may be followed, whereby the flux variation will also have a sine-wave shape. Figure 2.9 shows a construction method by which the curve shape of excitation (field strength) from a sinusoidal flux (magnetic flux density) can be determined. It can be seen there that a high harmonic content is present in the field strength curve consequently the same harmonics also appear in the curve of excitation current. In a core made of cold-rolled sheets the harmonics of current magnetizing the iron are of the following specific magnitudes: fundamental harmonic 1, 3rd harmonic 0.4 to 0.5, 5th harmonic 0.1 to 0.25, 7th harmonic 0.05 to

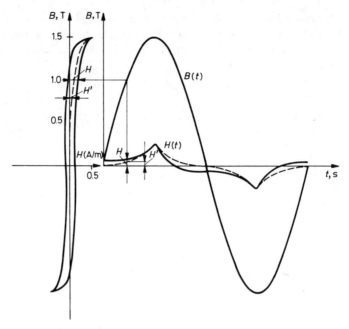

Fig. 2.9. Construction of magnetic field strength for sinusoidally alternating magnetic induction

0.1. When considering also the magnetization of air gaps, the harmonic contents of magnetizing current will be lower than those indicated above. The no-load apparent power of the transformer is

$$P_{b0} = U\sqrt{I_1^2 + I_3^2 + \dots},$$

where U is the r.m.s. value of the sinusoidal supply voltage and I_1, I_3, ... are the r.m.s. values of the fundamental and higher harmonic components of excitation current. In normal practice of transformer dimensioning, the simpler method below is followed.

The values of P_{b0} are given as a function of the peak value of flux density, in the form of specific exciting power as supplied by the manufacturers or obtained by measurements on a test piece, as shown in Fig. 2.4. The specific exciting power is calculated by dividing the core into sections, along each of which the flux density may be considered constant. From the given curve (Fig. 2.4), the specific exciting power for each section is read, and by multiplying the mass of each core section by the specific exciting power, the exciting power of each individual section is obtained. After adding up the exciting powers of all sections the exciting power falling to the air gaps is determined.

For a sinusoidally varying flux density, the specific energy of the magnetic field in the air gap (e, Vm^{-3}) varies as

$$e = \frac{1}{2}\frac{B_m^2}{\mu_0}\sin^2 \omega t,$$

40

where B_m is the peak value of flux density in T, and $\omega = 2\pi f$ is the angular frequency. The power ($p_{b\delta}$, VAr) is the differential quotient taken with respect to time:

$$p_{b\delta} = \frac{\pi}{\mu_0} f B_m^2 \sin 2\omega t$$

and the exciting power of an air gap with volume V, cm³ is

$$p_{b\delta} = \frac{\pi}{\mu_0} f B_m^2 V.$$

When applying the calculation described above, the peak value of flux density B_m in the air gap of normal overlap ($\alpha = 90°$) laminations is equal to the flux density in the sheets, whereas in the case of mitred joints ($\alpha = 30°$ to $60°$) it is lower by a factor $\sin \alpha$, than that present in the core, since the flux lines in the air gap assume a position perpendicular to the joining edges of sheets. From this it follows that the excitation requirement of the air gaps of mitred sheets is lower, since beside the square of flux density in the above formula, $\sin^2\alpha$ appears as a multiplying factor, whereas the volume (assuming an equal gap length) has to be divided by $\sin \alpha$ only. Thus for air gaps of equal length, the resultant excitation requirement of a core with mitred joints is $\sin \alpha$ times that with plain ($\alpha = 90°$) overlap. The volume of an air gap can be determined by estimation, based on the fairly good approximating assumption that from within the air gap the flux lines do not tend to expend into a cross-sectional area exceeding that of the steel, i.e. with the cutting angle taken into account, the cross-sectional area of the air gap is $A_i/\sin \alpha$. The length of the air gap can be estimated from the dimensional tolerance of the sheets constituting the core. It can be assumed that at the joints, the air gap length appearing between sheets cut to size and those produced to positive tolerance may be neglected, while for sheets of negative tolerance, the largest air gap is equal to the maximum permissible negative tolerance. In cutting operations the statistical distribution of lengths corresponds to

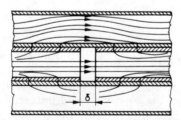

Fig. 2.10. Magnetic induction lines by-passing a gap in the joints of sheets

the normal distribution, so the calculation may be considered as being shifted in the safe direction if the air gap length is taken as half of the negative tolerance. Similarly, there is also a margin of safety if the flux in the lamination in the above considerations is assumed to be identical to that present in the air gap. As shown in Fig. 2.10, some of the induction lines pass round the air gap of the joint and flow in the adjacent sheets. Thus the losses will increase in consequence of the higher flux density developing in the lamination, but because of the lower flux density in the air gap the excitation requirement of air gaps will be less. The calculation of contributions to the magnetizing reactive power for both the lamination and air gaps contains considerable uncertainties. Anisotropy of the sheet itself and dimensions of the developing air gaps may both deviate from estimated conditions, so the calculated values may be in error by $\pm 30\%$. That is why a $+30\%$ deviation in the magnitude of no-load current is generally permitted by the standard specifications, most of which covers the excitation [109].

Example 2.2. In an obliquely joined ($\alpha = 45°$) core with equal limb and yoke cross-sections the peak value of flux density is $1.6 \times$ T and the mass of lamination built into the core is 20×10^3 kg. The air gap volume in the core is 100×10^{-6} m^3. The specific magnetizing power of the sheets is 2.4 VAr kg^{-1}, that of the entire core is $20 \times 10^3 \times 2.4 = 48\,000$ VAr. The magnetizing power of the air gap is

$$P_{l\delta} = \frac{\pi}{\mu_0} 50 \times 1.6^2 \times 100 \times 10^{-6} \sin 45° = 22\,639 \text{ VAr.}$$

The entire magnetizing power is 70 639 VAr.

2.4 Calculation of core losses

The eddy-current losses arising in the core of a transformer, provided the insulation of the sheets is satisfactory, reliably follow a quadratic relation with flux density. The behaviour of the hysteresis loss is not at all so regular. In cold-rolled sheets the reduced core loss has been achieved mainly by lowering the hysteresis component. At 50 Hz, in medium-quality cold-rolled sheets, the hysteresis and eddy-current losses are closely equal. It is not customary to assume a quadratic dependence of the core losses of cold-rolled sheets for different flux densities, except for rough estimations. For calculation purposes, either the curves giving peak values of flux density vs. specific loss supplied by the manufacturer or similar curves based on personal measurements are used.

2.4.1 Losses of cores built up from hot-rolled strips

Hot-rolled strips are used for building plain overlap ($\alpha = 90°$) cores. For the calculation of core losses, the core is divided into sections within which the flux density may be considered constant. From the knowledge of specific losses for the various sections and of their masses, the core loss is obtained by summing up the losses obtained for the individual sections. Nowadays, cores of large transformers are very seldom built up from hot-rolled sheets. Additional losses are taken into account by multiplying the calculated losses by a correction factor τ, which may vary in the range 1.15 to 1.25. This variation depends on what percentage of the full core volume is represented by the overlap cores, the corner areas being shown by shaded lines. For corner areas not above 20% $\tau = 1.15$, for 30% $\tau = 1.2$ and for areas of 40% and above $\tau = 1.25$. The masses of those sections of the cores shown in Fig. 2.11 having identical flux densities are given in Table 2.3, using the notation given in

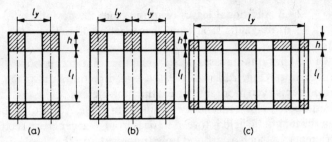

Fig. 2.11. Cores: (a) single-phase two-limb core; (b) three-phase three-limb core; (c) three-phase five-limb core (shading indicates corner areas of the core)

Fig. 2.11. In the table the cross-sectional area of the limb is A_l, that of the yoke A_y and the density of sheet material is $\rho_i (\rho_i = 7.65 \times 10^3 \text{ kg m}^{-3})$. In the five-limb core the cross-sectional area of the two outer limbs and that of the yokes are equal. When calculating the core sections, it should be taken into account that the space filled out by the sheets is smaller than that occupied by the core. Therefore the geometrical

Table 2.3

Calculation of steel masses

	Single-phase two-limb core	Three-phase three-limb core	Three-phase five-limb core
Mass of yoke, m_y	$2l_y A_y \rho_i$	$4l_y A_y \rho_i$	$2(l_y + l_l + h)$ $A_y \rho_i - 3h A_l \rho_i$
Mass of limb, m_l	$2(l_l + h) A_l \rho_i$	$(3l_l + 2h) A_l \rho_i$	$3(l_l + h) A_l \rho_i$

cross-sectional area should be multiplied by the space factor, for which the value of 0.97 may be assumed in the absence of a more accurate information. Using data relating to the masses of the limbs and yokes

$$P_i = \tau(w_l m_l + w_y m_y),$$

where w_l is the specific core loss corresponding to the flux density B_{lm} in the limbs and w_y corresponding to the flux density B_{ym} in the yokes.

2.4.2 Losses in mitred joint cores

The favourable properties of cold-rolled sheets can best be exploited with cores built up from mitred joint laminations. In that case the primary aim is to reduce the additional losses to a minimum. These additional losses result partly from the geometry of the construction (the direction of flux lines deviates from that of rolling), and partly in the machining of sheets.

2.4.3 Additional losses resulting from the deviation
of magnetic field from the direction of rolling

It is known that minimum losses occur when the rolling direction coincides with that of the flux lines (Fig. 2.2), while if the two directions differ the increase of magnetizing power is even more significant than the increase of losses. The magnetic field tends in the core to have an energy of minimum value. Additional losses may be caused when the magnetic field is forced to pass from a larger cross-section through a more restricted area, such as around holes punched into the sheets. In such cases, even relatively far away from the hole, the flux lines remain along their diverted paths (where the losses are higher due to the increased flux density) because the repeated directional change to a larger cross-sectional area of lower loss would considerably increase the magnetizing loss. Thus, the relative permeability for the 1.5 T peak flux density is 21×10^3, but it drops to 10^3 with a $20°$ directional deviation. The losses resulting from the directional deviation are as follows.

2.4.3.1 Additional loss due to inaccurately fitted joints

Such an inaccurate fitting is shown in Fig. 2.12, where the magnetic field becomes compressed towards the smaller air gap, while along the path following the larger air gap the utilization of material is reduced. This additional loss is difficult to calculate,

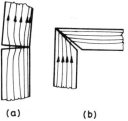

(a)   (b)

Fig. 2.12. Due to inaccurate jointing, magnetic flux lines concentrate at joints fitted with narrower gaps

but should always be reckoned with except in the case of wound cores. With more accurate machining the value of this loss component can be reduced to a minimum.

2.4.3.2 Additional losses resulting from punching

In traditional core constructions the clamping bolts passing through the punched sheets play an important part in ensuring rigidity of the core. In such a core the passage of the magnetic field through the narrow interstices between the holes arranged behind one another is rather restricted (Fig. 2.13). The additional losses caused by this effect may be estimated by calculation [39]. In the unpunched sections of the core, with flux densities of 1.6 to 1.7 T, the loss is

$$P = w_B B_m^{2.8} m,$$

where w_B is the specific loss for flux density B_m and m is the mass of the core section.

In Fig. 2.13 the diverting effect of holes on the flux lines can be observed, the width of the sheet will be taken as unity, and the hole diameter x as a value related to unity.

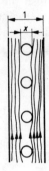

Fig. 2.13. In cold-rolled sheets, the magnetic flux lines do not penetrate into the area shaded by holes

If holes of diameter x are punched into the sheets, the mass of the core to be considered in calculating the loss reduces to $(1-x)$ and at the same time, the flux density increases to $B/(1-x)$. In the punched core this mass reduction and increase of flux density should be considered:

$$P = w_B \left(\frac{B_m}{1-x} \right)^{2.8} m(1-x) = w_B B_m^{2.8} m \left(\frac{1}{1-x} \right)^{1.8}.$$

From the formula it can be seen that the losses have increased by

$$\frac{1}{(1-x)^{1.8}} \cong \frac{1}{1-2x}.$$

The punching loss may be somewhat reduced by decreasing the diameter of holes. Even in the lamination of large transformers the holes are either entirely omitted, or the size and number of holes are drastically reduced. When calculating the punching losses, the effect of the deformation zone developing in the course of punching has to be assessed. Holes of 5 to 50 mm diameter effectively increase by 1 to 15 millimetres, because the permeability of the sheet material decreases in a zone corresponding to these figures, and the flux lines are driven out from it. In modern transformers up to about 16 MVA rating the cores are built up of bonded sheets, thus superseding punched holes. Bonding is performed by spreading or spraying a suitable adhesive over the edges of sheets of the stacked core. The adhesive used is a cross-linked epoxy resin of high capillarity, intruding between the laminations to a depth of about 20 to 30 mm measured from the edges of sheets and assuming its cross-linked texture between 20 and 50 °C within a period of 12 to 48 hours. The strength of the bond thus obtained is satisfactorily reducing the acoustical noise of the core and diminishing the increase of core loss associated with higher voltage stresses.

Investigations have shown that overvoltage stresses have often been followed by a permanent increase of a few per cent in the no-load loss of the transformer. This increase is due to partial damage of sheet insulation, the breakdowns being traceable to the edges of sheets or the vicinity thereof. Since just these parts are reinforced by the applied adhesive, no increase can be detected in the no-load loss following an overvoltage stress in transformers with bonded cores.

Efforts are also being made to banish the holes from the cores of the largest transformers by applying the bonding method to such units, but the limbs and yokes are first compressed by glass-fibre reinforced plastic bandages.

2.4.3.3 Losses occurring in the overlap corner area

In Fig. 2.14 the advantages of the mitred joint (b) are shown, as opposed to the plain overlap joint (a). In the plain overlap joint, the direction of magnetization differs from that of rolling in the entire corner area. Also the additional losses are proportional to the corner volume. The limited diversion of flux lines in mitred

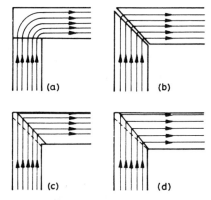

Fig. 2.14. Distribution of flux lines in cold-rolled sheets at plain overlap joints and at mitred joints

joints is due to the fact that, the permeability of iron being finite, a smaller directional deviation occurs in the vicinity of the air gap of the joint. A further additional loss arises because in the part of the magnetic circuit adjacent to the window of the core the length of the flux lines is shorter, hence the values of magnetic field strength and flux density are higher. This uneven flux distribution causes increased losses. Again, a further reason for the additional losses in the corner overlap area is that, for the sake of a more rigid core structure and a better transition of flux lines, overlapping have found wide application for mitred joints: either the 5° cutting lines are alternately displaced with respect to each other as shown in Fig. 2.14(c), or the sheets forming the joints are cut at angles complementing each other to 90° (e.g. 40° + 50°) and stacked alternately, so that the straight lines of the joints are correspondingly displaced in successive layers (Fig. 2.14(d)). When either the arrangement (c) or (d) is adopted, a few per cent increase of core loss will have to be reckoned with, due to the corner effect.

2.4.3.4 Additional loss resulting from changes in the cross-sectional area of the core

In the attempt to accommodate the core within the innermost winding with the best utilization of space, the core is given a stepped cross-section as shown in Fig. 2.15(b). To provide better axial support for the windings, the bearing surfaces of

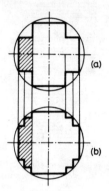

Fig. 2.15. Cross-sections of yoke (a) and limb (b) of a core

yokes are made as large as possible. With this aim, the yoke cross-section is formed to contain a smaller number of steps (Fig. 2.15(a)).

When the number of steps in the limb differs from that in the yoke, the cross-sections of the adjoining steps of the limb and yoke are different. Hence, the flux lines when passing over from the limb into the yoke tend to step from one step into the other in the direction perpendicular to the plane of the sheets. At such points considerable eddy-current losses occur.

In the most unfavourable case (a high proportion of overlap corner area) the additional losses due to cross-sectional changes may reach 5% of the entire core loss. These losses can be avoided by making the steps of the limb and yoke congruent either by having equal cross-sectional areas for both the limb and yoke or by increasing the yoke cross-section by 5 to 15%.

2.4.3.5 Losses influenced by the lamination pattern

The sheets are stacked in succession, according to the patterns defined by the core plate design. The reason for alternating the pattern is to make the joints between adjacent sheets in the same layers overlap each other, in order to obtain a lower no-load current and higher mechanical stability. Part of the flux lines developing in the sheet pass around the air gap of the joint by stepping over into the adjacent sheets. This will result in a reduced reactive component of the no-load current relative to what would have been developed with the flux lines flowing in the plane of the sheet, but due to eddy-current losses caused by the flux lines leaving the plane of the sheets, the active component of no-load current increases. The losses resulting from the lamination are lowest, when the core plate patterns alternate with each other in every successive layer (Fig. 2.16(a)). In the construction of larger-size cores the large sheets are flexible and tend to suffer permanent deformation, leading to additional losses. To prevent the occurrence of deformations, a recurring series of identical plate patterns is used after 2, 3, or 4 successive layers. Examples of such overlapping

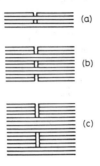

Fig. 2.16. Cross-section of core overlaps in the case of sheet patterns (a) alternating from layer to layer; (b) alternating after two layers; (c) alternating after four layers

joints are given by drawings (b) and (c) of Fig. 2.16. The extent of the migration of flux lines into adjacent layers in the vicinity of overlapping joints is the greater, and the losses are higher, the more identical core plate patterns are placed above each other.

Measurements show that relative to the case of changing the core plate pattern from layer to layer, the losses are 5 to 10% higher when the plate pattern is changed after every other layer; 10 to 15% higher for changing the pattern after every three layers, and 15 to 20% higher when the pattern is changed after every four layers.

2.4.4 Losses due to mechanical processing of sheets

The magnetic properties of the sheets arriving from the rolling mill, under satisfactory packing and transport conditions, usually show good compliance with the values established in manufacturing. The various processing and mounting methods used in transformer factories inevitably lead to additional losses and growth of no-load currents even if optimum conditions are ensured. Losses resulting from manufacturing are listed below.

(a) Losses due to cutting and punching

During cutting and punching, the sheets suffer deformations in the vicinity of cut or punched edges, and mechanical stresses arise in the material which cause the

48

magnetic permeability of the affected section to drop. The losses increase in the deformed material and even more considerable losses result from the increase of flux density in the vicinity of deformed sheet sections which divert the flux lines. The losses due to mechanical processing depend on two factors: the fineness of processing, and the sheet width. The fineness of processing determines the width of the strip where deformation is brought about by cutting or punching tools. The finer the processing, the narrower is the strip within which the magnetic properties are affected. The width of such strips may be as large as 5 to 8 mm when improperly shaped tools or badly maintained machines are used for processing the sheets, whereas the strip width remains within about a tenth of a millimetre with a suitable tool of proper material and a good machine.

The width of sheets influences the additional losses incurred by mechanical processing, because a deformation strip of any particular width affects the losses arising in wider sheets by a lower percentage than in the case of narrower ones. For example, a deformation strip of 2×0.5 mm width along the line of cut in a 30 mm wide sheet may cause a 3% increase in the losses, whereas the additional loss due to a 2×8 mm deformation strip will amount to 50%. In the case of similar deformation strips, the additional losses due to mechanical processing in a 300 mm wide sheet will only be 0.3 and 5%, respectively.

In calculating the punching losses (see Fig. 2.13) the processing loss is taken into account by notionally increasing the actual hole diameter by 1 to 16 mm. Processing losses may be considerably reduced by heat treatment, but they cannot be removed entirely. In the course of heat treatment the original crystalline structure pertaining to the lower loss is restored, but the deformation remains. The forces compressing the core tend to press the deformed parts back into the plane of the sheet. The mechanical stresses thus arising again cause losses. Also corrugated sheets become sources of losses for similar reasons.

(b) During transport and reloading, due to careless handling, the sheets may be bent through radii below 300 mm. Such deficiencies can be eliminated by organizational measures taken in the production process.

(c) Losses due to defective insulation

The majority of cold-rolled sheets are coated on both sides with an insulating layer resistent to heat up to 820 °C. Injured sheet insulation within the core increases the losses caused by eddy currents.

(d) Eddy-current losses of cutting edges

The edges occurring in the course of cutting may give rise to metallic contact between sheets of adjacent layers. Such edges should be flattened down between finishing rollers or ground off with care, as injured insulation may lead to eddy-current losses. These eddy currents developing along cut edges may, in adverse cases, start growing slowly, and this process may even lead eventually to a fault between adjacent sheets of the lamination. As proved by experiments, no faults between sheets can be produced with voltages below 20 V turn voltage even under extremely unfavourable circumstances (uninsulated sheet edges or compression of sheets with excessive force). At turn voltages exceeding 20 V the hazard of inter-sheet faults should be reckoned with, and corresponding counter-measures are necessary. Such protective measures may consist of dividing the core into stacks of sheets isolated from each other by a paper or pressboard layer of at least 0.3 mm thickness, to prevent the voltages induced along the circumference of each stack of sheets from exceeding 20 V.

2.4.5 Heat treatment

The minimum temperature required for heat treatment is 790 °C, and the maximum temperature permissible for the heat-resistant sheet insulation is 820 °C. These two temperature limits define the regulation range of the furnace used for the heat treatment. To restore the insulation of sheets injured in the course of mechanical processing some manufacturers apply a 1 to 2 μm thick synthetic resin coating following the heat treatment. Other manufacturers use adhesives. Epoxy resins of high capillary penetrating capability used for bonding the sheets not only contribute to the mechanical stability of the core, but by acting also as additional insulation, prevent the few per cent increase of core losses associated with overvoltage stresses.

2.4.6 Estimation of additional losses

Generally, the additional losses listed above cannot be accurately calculated, but the methods mentioned are suitable for restricting them. Nevertheless, designers try to find calculation methods to approximate the magnitude of losses with some reliability. As shown by experience, a fair approximation can be obtained by a calculation based on the assumption that the magnitude of additional losses is related to with the corner area of the core. The course of such a calculation is as follows. First, the net yoke and limb masses falling outside the corner area are determined. The mass formula is $m_l = 3\rho_i A_l l_l$, for the limbs, where ρ_i is the density of sheet material, A_l is the cross-sectional area of steel in the limb and l_l is the height of the window. The mass for yokes in a three-limb core is $m_y = 4\rho_i A_y l_y$, where A_y is the cross-sectional area of iron of the yoke and l_y is the width of the window.

In plain overlap cores the corner areas of the widest sheets are larger than those of the narrower sheets. This is because the narrower sheets protrude into the corner areas both along the limbs and the yokes, as shown in Fig. 2.17(a). It can be seen that the surfaces proportional to the shaded corner areas are of different magnitude for each sheet width. In calculating the limb and yoke volumes this is taken into account by increasing the mass of each limb and each yoke obtained above by $m_a = 0.15 \, \rho_i A \, h$. In the formula $h = \dfrac{A_y}{A_l} D_l$, where D_l is the limb diameter. Hence, in a three-limbed core the masses of limbs and yokes are obtained as

$$m_l = m_l' + 3m_a \quad \text{and} \quad m_y = m_y' + 4m_a.$$

The mass of the corner area in a plain overlap core is $m = 0.85 \, \rho_i A \, h$, where the value of h is that given above. In the case of Fig. 2.17(a) altogether six such corner areas have to be considered.

50

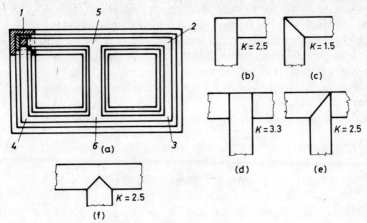

Fig. 2.17. Consideration of additional losses occurring in the corner areas of a three-phase core-type transformer

The calculation has been based on the assumption that the corner areas are geometrical figures resulting from intersecting cylinders. The two cylinders each of diameter D and intersecting each other at an angle of 90° have a common volume for both bodies equal to $0.85\, D^3\pi/4$, with a volume of $0.075\, D^3\pi/4$ protruding into the corner area at each intersection of the cylinders.

The next step is to determine the losses arising in the limbs and those in the yokes by using the specific loss figures pertaining to the flux density in the limbs and in the yokes, respectively. In calculating the losses for the corners the kind of overlapping and the corner position has to be taken into consideration. In case (b) and (c) of Fig. 2.17 the loss factors pertaining to the flux density in the limbs have to be multiplied by $\kappa = 2.5$ and $\kappa = 1.5$, respectively. In cases (d), (e) and (f) of Fig. 2.17 the loss factors

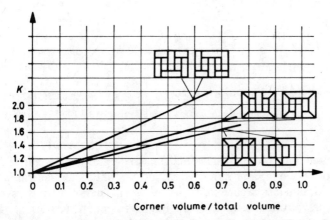

Fig. 2.18. Consideration of additional losses occurring in a core with equal limb and yoke sections

pertaining to the flux density in the yokes have to be multiplied by $\kappa = 3.3$ or $\kappa = 2.5$. The sum of losses thus obtained is the expected core loss.

The case is simpler when the cross-sectional areas of the limb and yoke are equal. Then the loss can be determined with the help of the straight lines of Fig. 2.18. First, the iron mass of the entire core and that of the corner area have to be computed. The iron mass multiplied by the loss factor for the flux density prevailing in the limbs will be the basic core loss. From this result, multiplied by factor κ of Fig. 2.18 the entire core loss including additional losses is obtained.

2.5 Inrush current phenomena

Normally, a transformer is energized by connecting it directly to the network voltage. This direct switching-in may occasionally give rise to an annoying transient phenomenon associated with a current surge, similar to or above the rated current of the transformer, that may cause the automatic protection to operate. The expected maximum instantaneous current can be approximated by the following method.

The differential equation describing the phenomenon, neglecting the attenuating effect of losses occurring in the core, is

$$u(t) = \frac{d\psi(i, t)}{di(t)} \frac{di(t)}{dt} + Rit,$$

where $u(t)$ is a time function of voltage at the transformer terminals, $\psi(i, t)$ is the current and time function of magnetic flux linked with the switched-in winding and R is the resistance of the latter winding. The absolute value of voltage appearing across the resistance of the winding is low, yet it is important from the point of view of the phenomenon, for the attenuation of the inrush current is determined by this resistance, together with the core loss of the transformer and with the self-inductance of the winding. In Fig. 2.19 an oscillogram of the inrush current is shown.

Fig. 2.19. Inrush current of a 400 kV transformer of 3×120 MVA rating

It is to be noted that the inrush current drops to a fraction of its initial value after a few tenths of a second, and its full decay occurs only after several seconds. The expected maximum inrush current, occurring when the transformer is switched-in at the most unfavourable instant, is an important characteristic of the no-load performance of a transformer. In Fig. 2.20(a), the variations of associated voltage

52

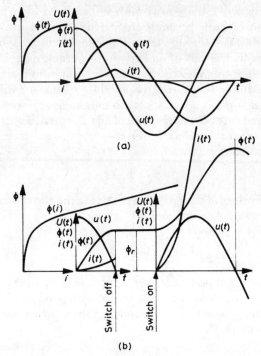

Fig. 2.20. Inrush phenomenon (a) in the most favourable case; (b) in the most unfavourable case

and flux are represented under the conditions of steady-state operation. In large transformers, where the attenuating effect of the winding and core is small, the maximum peak flux Φ_m occurs at zero transitions of voltage. From the point of view of inrush current, the most unfavourable conditions arise when switching-in takes place at a zero voltage transition and at this instant the value of remanent flux is maximum and has the same sign as the tangent of the voltage curve at zero transition. Generally, the sign and magnitude of the remanent flux cannot be influenced, since they are determined by the conditions prevailing at the instant of previous disconnection of the transformer. The most unfavourable conditions are represented in Fig. 2.20(b). The flux, starting from the remanent flux, has to change so as to make its derivative vary as the imposed network voltage variation as a function of time. This is only possible if the flux, and with it the exciting current, increases during the period of the first half cycle following the instant of switching-in. The maximum possible value Φ_e of the flux is the sum of the change of flux $2\Phi_m$ developing in the steady-state condition and of the remanent flux Φ_r:

$$\Phi_e = 2\Phi_m + \Phi_r = A_l(B_r + 2B_m),$$

where A_l is the cross-sectional area of the core, B_r is the remanent induction pertaining to flux density B, and B is the peak value of flux density in the core in the steady-state condition.

Transformer designers usually work with values of 1.5 to 1.75 T selected for B, and the remanent flux density pertaining to this induction may reach values as high as 1.3 to 1.7 T. A switching-in operation occurring at an unfavourable instant, if coinciding with a remanent flux of similarly unfavourable magnitude and polarity, will cause magnetization of the core beyond the saturation limit, and will make a considerable proportion of the flux Φ_e required for inducing a voltage maintaining equilibrium with the supply voltage appear in the air gap (of cross-sectional area A_δ) between winding and core. The magnitude of flux Φ_δ situated outside the core is the difference between the required flux Φ_e and core flux Φ_s determined by the induction B_s in the saturation range:

$$\Phi_\delta = \Phi_e - \Phi_s = B_\delta A_\delta .$$

Substituting the values of $\Phi_s = A B_s$ and of Φ_e into the above formula, the average value of flux density outside the core is

$$B_\delta = \frac{A_l}{A_\delta} (2B + B_r - B_s) . \tag{2.1}$$

Since the tangential component of the magnetic field strength passes continuously across the boundary layers of adjacent media, with good approximation the field strength can be taken as identical in the core and in the space between the core and the switched-in winding, thus

$$B_\delta = \mu_0 H_\delta .$$

If the switched-in winding is of N turns and the length of the induction lines in air is l_0, then the peak value of the inrush current is

$$i_m = \frac{H_\delta l_0}{N} = \frac{B_\delta l_0}{\mu_0 N} .$$

Substituting from equation (2.1) the value of B_δ

$$i_m = \frac{1}{\mu_0} \frac{A_l}{A_\delta} \frac{l_0}{N} (2B + B_r - B_s) . \tag{2.2}$$

Several conclusions can be drawn from this relation concerning the inrush phenomenon and size of the transformer:

— The maximum magnetizing inrush current is influenced by the cross-sectional area between core and winding. Therefore it is expedient, where and when possible, to switch in first the terminals belonging to the winding of larger diameter.
— Another way of reducing the magnetizing inrush current is to increase the resistance of the switched-in circuit. For this purpose, special switchgear is required to bypass the inserted resistor after decay of the inrush phenomenon.
— The peak value of the magnetizing inrush current may exceed the rated current of the winding and may impose considerable electrodynamic stresses on the transformer, and cause the transformer protection to trip. This latter may jeopardize the transformation insulation, because the interruption of magnetizing

currents of such magnitudes may give rise to overvoltages exceeding the switching surges normally occurring in networks.

Example 2.3. Assume a 120 kV transformer of 25 mVA rating and voltage ratio of 120/6.6 kV with its H.V. (high-voltage) winding connected in star formation and its L.V. winding connected in delta. Rated current of the H.V. winding is 120 A (this value being that of the line voltage as well), rated current of the L.V. winding is 1260 A (line current 2180 A). Cross-sectional area of the core limb is $A_l = 1817 \times 10^{-4}$ m², mean diameter of the L.V. winding is 0.626 m, its mean cross-sectional area linked with the flux is $A_{WL} = 3060 \times 10^{-4}$ m², mean diameter of the H.V. winding is 0.862 m, and its mean cross-sectional area is $A_{WH} = 5820 \times 10^{-4}$ m². The number of turns in the H.V. winding is $N_H = 1000$ and in the L.V. winding is $N_L = 95$. In the steady-state condition the peak value of induction in the core (with U_{Hf} being the phase-to-neutral voltage of the H.V. winding and f the frequency) is

$$B = \frac{U_{Hf}}{4.44 f N_H A_l} = \frac{69282}{4.44 \times 50 \times 10^3 \times 0.1817} = 1.717 \text{ T}.$$

The remanent flux density pertaining to the peak value of induction is $B_r = 1.45$ T. In the saturation range the induction is $B_s = 2.03$ T.

When the transformer is switched in from the H.V. side, substitute in Eqn. (2.2) the value $A_{\delta H} = 4003 \times 10^{-4}$ m², instead of A_δ, whereas when switched in from the L.V. side, substitute $A_{\delta L} = A_{WL} - A_1 = (3060 - 1817) \times 10^{-4} = 1243 \times 10^{-4}$.

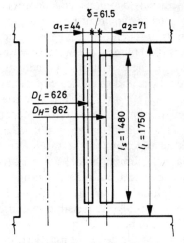

Fig. 2.21. Winding and core dimensions of the transformer dealt with in Example 2.3

In Fig. 2.21 the window section of the transformer is represented. For the sake of increased accuracy, the effect of saturation is taken into account by a calculation using the mean value of the limb length and winding height, instead of the limb length of the core. Thus, the corrected value of l_l will be

$$l_l' = \frac{l_l + l_s}{2} = \frac{1750 + 1480}{2} = 1615 \text{ mm}.$$

In the most unfavourable case of switching-in from the H.V. side, the maximum peak value of magnetizing inrush current will be

$$i_m'' = \frac{1}{1.256 \times 10^{-6}} \frac{1817 \times 10^{-4}}{4003 \times 10^{-4}} \frac{1.615}{1000} \times (2 \times 1.717 + 1.45 - 2.03) = 1665.7 \text{ A}.$$

This current is 10 times the peak value of rated current.

The situation is still worse when the transformer is switched-in from the L.V. side:

$$i''_m = \frac{1}{1.256 \times 10^{-6}} \frac{1817 \times 10^{-4}}{1243 \times 10^{-4}} \frac{1.615}{95} (2 \times 1.717 + 1.45 - 2.03) = 56467.3 \text{ A.}$$

This 56467.3 A is 31.69 times higher than the rated current of the L.V. winding.

2.6 Transformer noises

Designers of transformer stations have to consider matching the transformers into the environment. Two important items are the aesthetic and the acoustic conditions:

The external shape of a transformer is governed by technical considerations, whereas its colour may be influenced by aesthetics as well as the primary aim of improving the conditions of heat radiation. While the shape of the transformer is purely determined by technical and economic motives, in deciding on its surface finish there are some ways open to the designer by which appearance and the matching to the environment may be improved. A competent model designer may help by proposing a better arrangement of reinforcing ribs and baffle plates.

Transformer operation is inseperably linked with magnetic noise, while in addition, the noise of fans and oil pumps will increase the annoying acoustic effects.

Magnetic noise is due to magnetostriction. Magnetostriction is proportional to the square of magnetic induction, and so causes vibration of a frequency twice that of the fundamental harmonic (100 or 120 Hz). The magnitude of dimensional changes due to the phenomenon of magnetostriction depends on the magnetic flux density in the iron sheet. The audible noise is not directly proportional to the magnitude of dimensional change, but to the rate of change. The even and odd upper harmonics of the magnetic fundamental wave are also present in the resultant noise. The fan noise frequency spectrum is in the range of 350 to 3000 Hz, being almost intolerable with peripheral fan speeds exceeding 25 m s^{-1}. The noise of oil pumps is less troublesome, as it is always possible to find quiet-running oil pumps for the required output. More can be found about the noise of fans and oil pumps in Chapter 7 but the rudiments and the noises of magnetic origin will be dealt with in the following paragraphs.

2.6.1 Basics

Noises originating from the core increase proportionally to the logarithm of transformer rating, so the problems associated with them are of special importance in large transformers. In industrially developed countries it is, generally, difficult to have large transformers installed far away from industrial or densely populated areas, hence in most cases reduced-noise transformer designs are preferred.

Audible noise is a pressure variation of the air taking place in the frequency range of 16 Hz to 10 kHz. The sound intensity (i) of audible noise is the energy flow passing across a unit area per unit time, expressed as the ratio of a standard power per unit surface area. The range of audible sounds extends over several orders of magnitude of sound intensity, therefore is expedient to use a logarithmic scale for the representation of sound intensities. For the purpose of comparing different noises the term acoustical noise level has been introduced:

$$L = 10 \log \frac{i}{i_s} \text{ dB},$$

where i_s is, by agreement, 10^{-20} Wm^{-2} at a frequency of 1000 Hz. Since sound intensity (i) is proportional to the square of sound pressure (p), the noise level can also be written

$$L = 20 \log \frac{p}{p_0}.$$

In the above formula p_0 is the sound pressure pertaining to the threshold of audibility at 1000 Hz, having the value of 2×10^{-10} bar. The human auditory sensation is frequency dependent, being more sensitive to sound in the range 1000 to 7000 Hz. Therefore a valuation system has been introduced, giving below 1000 Hz and above 7000 Hz a value lower than that effectively measured. This valuation matched to the character of human hearing is termed "valuation according to the weighted sound level curve e" in international literature. In the range between 1000 Hz and 7000 Hz the effect of this weighted sound level A is insignificant. Valuation may be done with the use of a curve or algorithm, but a suitable filter inserted into the measuring instrument can also perform the required modification.

On moving away from a point source of sound with no acoustic absorbent medium in between, the sound intensity decays inversely as the square of distance whereas the sound pressure reduces inversely as the distance.

In the case of several sources of sound, the sound intensities add up, giving in the presence of sound intensities $i_1 + i_2 + \ldots + i_i$ an overall noise level of

$$L = 10 \log \frac{i_1 + i_2 + \ldots + i_i}{i_s},$$

i.e. with double sound intensity the noise level increases by 3 dB.

2.6.2 Noises originating from the transformer

The basic cause of core noise is magnetostriction or, in other words, the change in dimension of sheets forming the core, which takes place at the fundamental frequency. The dimensional variation caused by magnetostriction associated with the sinusoidally varying magnetic induction is non-sinusoidal, therefore acoustic noise is generated by the even and odd higher harmonics, the noise containing frequencies up to about 600 and 720 Hz (in 50 and 60 Hz networks respectively).

The noise level is proportional to the 8th power of magnetic induction (B) and the logarithm of the 1.6th power of mass (G) [13], [112]:

$$L = 10 \lg \left[\left(\frac{B_2}{B_1} \right)^8 \left(\frac{G_2}{G_1} \right)^{1.6} \right].$$

57

An effective means of reducing the noise may be the selection of a lower magnetic induction. However, this will lead to a larger core mass for a transformer of given capacity, although (as apparent from the above formula) the influence of core mass is less than that of magnetic induction. Reduced magnetic induction is the most effective measure, since a decrease of magnetic induction by about 0.1 T reduces the noise level of a given transformer by as much as 2 dB. To obtain a reduction of flux density the dimensions of the transformer have to be increased, with consequent rise of costs, but this method is economically justified in cases when the reduction of no-load losses (diminishing with flux densities) is important. Nowadays, transformers up to 40 MVA are often installed in densely populated residential areas, and if economically feasible in such cases, it may occur that figures as low as 1.2 T are selected for flux densities in the limbs. Above 40 MVA a substantial reduction of induction levels leads to difficulties in remaining within the loading gauge limitations. With a given flux density, the noise level can be considerably influenced by manufacturing methods and design features. An important role is played in manufacture by the maintenance of dimensional accuracy, beside which the effect of bonding the sheets in the core brings further benefits. Among constructional features the use of mitred overlapped joints and proper fitting of damping elements placed between core and tank are of importance. Such elements may be made of wood or a synthetic material with damping properties.

The noise-generating effect of the Hi–B (High Induction) type sheets is lower. For steels of that kind the value of $\Delta L/\Delta B$ is 22 to 23 dB T^{-1}, which means that selecting a Hi–B sheet can yield a similar result to reducing the induction by $\Delta B = 0.1$ T in a core built up of some other grade of sheet material.

For cooling of transformers up to a rating of about 40 MVA, in view of specifications concerning permissible noise levels, natural oil cooling is often combined with natural air cooling (ONAN). In such cases the noises are of magnetic origin only, for the reduction of which, beside the measures mentioned, further possibilities are provided by selecting suitable tank structures. This question is dealt with in Chapter 7.8, and it should only be noted here that, as with the core, a reduction of the tank surface (and thereby also its dimensions) will have a damping effect, mainly on the lower harmonics.

2.7 Influence of the core and the connection of windings on the electromagnetic properties of the transformer

The calculations of no-load reactive power and loss already described give correct results in the first approximation only. One reason for deviations and some inherent uncertainties in the calculation of supplementary losses has been indicated. The structural design of the core and the connection of windings introduce further deviations. The construction of the core, and the additional losses resulting from it and from the connection of windings will be discussed below together with the phenomena associated with them.

2.7.1 Core structure

The parts of the core covered by the windings are limbs. Other parts of the core with the task of closing the path of the flux, but not surrounded by windings, are the yokes. The space confined by the limbs and yokes is the window. The size of the window is related to the cross-sectional area of the winding. The transformer may either be of the core-type or of the shell-type. Characteristic of the shell-type is that the winding cross-section located in the plane of the core is surrounded from all directions by limbs and yokes. If this condition is not fulfilled, the transformer is of the core type. In Fig. 2.22 sketches of core-type and shell-type transformer cores are shown: (a) is a single-phase two-limb core-type, (b) is a single-phase single-limb shell-type, (c) is a single-phase two-limb shell-type, (d) is a three-phase core-type, (e) is a five-limb shell-type and (f) is a three-phase shell-type lamination.

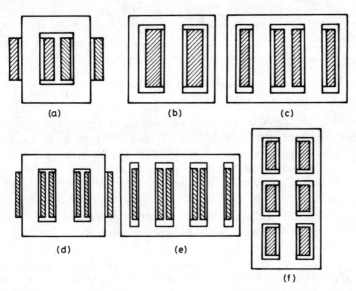

Fig. 2.22. Various core types: (a) single-phase two-limb core-type: (b) single-phase single-limb shell-type; (c) single-phase two-limb shell-type; (d) three-phase core-type; (e) three-phase five-limb core; (f) three-phase shell-type core

In early days, the accepted power distribution system in the USA was based on single-phase transformers. Also the large transformers were composed of single-phase units with their windings connected in star or delta format. With regard to their transport and keeping spare units in stock this system offers obvious advantages. The transformer built with a three-phase core and constituting a single unit has been preferred by the European power distribution systems. The economic advantage of the so-called European system lies in the fact that a three-phase unit is about 20% cheaper than a three-phase bank of identical overall capacity, built up of three-single-phase units. Obviously, the European system is somewhat handicapped by the higher cost of the necessary spare unit. Very high-capacity transformers often consist of three single-phase units also in Europe, in order to cope with the limitations imposed by the railway transport facilities.

Regarding the magnetic field linked with the individual limbs, three-phase cores may be either symmetrical (when the magnetic fields of all the three limbs are congruent) or asymmetrical. In Fig. 2.23 a variant of the symmetrical three-phase core is represented, whereas in Fig. 2.22 (d) the widest-spread three-phase core-type

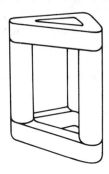

Fig. 2.23. Space symmetrical core

arrangement is shown, which is an asymmetrical core. The advantage of the symmetrical core lies in the minimum mass of lamination, tank size and oil volume, resulting from the spatial symmetry of the arrangement. Increased complexity of production is a drawback of this design.

2.7.2 Influence of the connection of windings on core losses and no-load current

In three-phase transformers the connection of transformer windings, as well as the structural design of the core affects both the losses and the no-load current. In the case of a three-phase core, the electrical connection of windings (star, delta, zig–zag) or the magnetic connection of flux (that may also be star, delta or shell-type) determines whether all upper harmonics of the current exciting the flux variation pertaining to a voltage variation can be drawn by the transformer from the network, or the flux variation corresponding to the voltage change imposed on the transformer can develop. If the transformer can draw the required upper harmonics of the exciting current, or the flux corresponding to the imposed voltage wave shape can develop, the transformer is in the state of free magnetization. If, however, the conditions of free magnetization do not apply, upper harmonic components of the flux appear, and a part of them will leave the core. Both the upper harmonics of the flux and the flux lines leaving the core and entering the surrounding metallic parts will become sources of local temperature rises and of additional losses.

As regards no-load losses it is sufficient to investigate the star and delta connections. In all the following considerations it will be assumed that the supply

60

voltage wave shape is sinusoidal and, in the case of three-phase transformers, that the three-phase supply voltages are approximately symmetrical.

(a) In a single-phase core, in the no-load condition, the flux is of a wave shape whose derivative curve with respect to time is identical with the shape of the voltage curve, so in a single-phase core a sinusoidal flux curve will belong to the sinusoidal supply voltage. In the transformer lamination the relation between flux density and magnetic field strength is non-linear, hence neither will the relation between flux and exciting current be linear. This will have the consequence that from the network supplying a sinusoidal voltage the transformer, in no-load conditions, will draw a current containing several upper harmonic components. Generally the no-load current contains all odd harmonics, corresponding to the wave shape of the magnetizing curve. From the hysteresis phenomenon it follows that there are also even harmonic components present in the magnetizing current.

The area of the hysteresis loop of sheet material used in transformer cores is small compared with the size of the magnetizing curve, so even harmonics are generally of secondary importance, whereas the upper harmonics with ordinal numbers divisible by three deserve special attention.

(b) Upper harmonics with ordinal number divisible by three have identical phase positions in the different phase windings of a three-phase system, hence their behaviour is identical to that of the zero-sequence current varying with network frequency. The same applies to the zero-sequence flux and its components having ordinal numbers divisible by three. Thus, the conclusions drawn on the harmonics of ordinal numbers divisible by three and on their elimination apply also to the zero-sequence currents appearing in asymmetrical loads and to the zero-sequence flux associated with them.

(c) In a single-phase core the sinusoidal flux pertaining to the sinusoidal voltage always develops, and the transformer winding will be capable of drawing from the supply network all required harmonic components of magnetizing current, so the core is in the state of free magnetization.

(d) In a transformer with its primary winding star connected in star format, zero-sequence currents can only flow on the primary side, if the neutral terminal of the transformer is connected with the neutral point of the three-phase supply network. In this case, the transformer is in the state of free magnetization.

(e) In a winding operated with isolated neutral terminal, no current harmonics of ordinal numbers divisible by three can flow, therefore the odd harmonics divisible by three will appear in the flux. The flux harmonics of ordinal numbers divisible by three will induce in each phase of the winding voltage harmonics of ordinal numbers divisible by three. The phase-to-phase voltage being equal to the difference of the phase-to-neutral voltages, in the phase-to-phase voltage no voltage components of ordinal numbers divisible by three appear, and a voltage equilibrium can develop, although the condition of free magnetization is not satisfied. The phenomenon is harmful, because the harmonic components of the flux cause the eddy-current component of the core loss to increase. The magnitude of hysteresis loss is influenced by the harmonics to the extent that they modify the peak value of the flux containing harmonic components. A further increase of losses is caused by the fluxes

61

of ordinal numbers divisible by three and of identical phase positions appearing in the three limbs. These flux components cannot close through the yokes, but they are forced to leave the core, causing thereby further additional losses in surrounding metallic parts (most harmful are the component parts having ferromagnetic properties, such as tank walls, frame structures, etc.). The magnitude of additional losses produced in the core in structural parts outside the core by flux components of ordinal numbers divisible by three may be as high as 30% of the no-load losses. A similar physical phenomenon can be observed in asymmetrically loaded, star/star-connected transformers. If no neutral terminal is brought out on the primary side of a star/star transformer, the primary winding cannot take up the zero-sequence current that would be needed to keep the state of equilibrium with the zero-sequence excitation of the secondary side. The resultant zero-sequence excitation gives rise to a zero-sequence flux of network frequency in the core, closing outside the core. This zero-sequence flux closing through the structural parts causes an additional loss dependent on the magnitude of the asymmetrical load.

(f) In the case of a delta-connected primary winding, every phase winding is in the state of free magnetization, since each phase winding takes up from the terminals of each phase of the line the current harmonics required by the sinusoidal flux curve. The current harmonics of ordinal numbers divisible by three are always present in the phase windings, while the line currents contain harmonic components of ordinal numbers divisible by three only if the core is asymmetrical. In a transformer with its primary winding connected in delta and its secondary winding connected in star format, under asymmetrical load, the primary current necessary for producing the counter-excitation required by the secondary-side load will be taken up independently by each winding wound around different limbs of the core, without forcing the zero-sequence flux to develop.

(g) A transformer connected in star format, on its primary side and operated with isolated neutral, and connected in delta on its secondary side, cannot take up from the network current harmonics of ordinal numbers divisible by three. The flux components of ordinal numbers divisible by three developing in the core induce voltages of identical phase positions in the delta winding. These induced voltages give rise to currents which tend to demagnetize the upper harmonic fluxes. The magnitude of current thus produced (and the measure of demagnetization of flux components of ordinal numbers divisible by three) will depend on the impedance of the delta winding acting against the harmonic currents of ordinal numbers divisible by three. The values of impedance determining the harmonic currents are generally rather low, so the value of flux of ordinal numbers divisible by three is insignificant. The upper harmonic currents flowing in the delta winding cause additional losses in the windings. Such additional losses are produced in the core and in structural parts also by flux components of ordinal numbers divisible by three. Therefore, when the secondary windings are connected in delta, the no-load losses will be somewhat higher than the losses occurring in the case of free magnetization. The difference is not significant and is generally lower than the uncertainties in the measurement of the three-phase core loss.

(h) In an asymmetrical core, the core loss and no-load current of the middle limb are lower than the respective losses in the two outer limbs, since the middle limb is directly connected to the two magnetic nodes of the core, whereas each outer limb is connected to the respective node, through an upper and lower yoke section. With a delta-connected primary winding, the asymmetry of the core manifests itself in the no-load currents and in the input power, without causing inconvenient consequences.

The middle phase of the winding having a neutral terminal brought out also takes up a lower current and power, so in the conductor connected to the neutral terminal not only third, ninth, etc. order current harmonics will flow but also the fundamental harmonic current will appear.

The primary winding, its neutral being isolated, cannot take up fundamental harmonic zero-sequence currents resulting from the asymmetrical load nor upper harmonic currents of ordinal numbers divisible by three. If there is no delta winding on the transformer, then in all the three cores a flux alternating with fundamental frequency and with the frequency of upper harmonics divisible by three will develop which will leave the core and, for the reasons described further above, an additional no-load loss will be caused. The delta-connected secondary winding will also considerably reduce the magnitude of zero-sequence and harmonic flux components in that case.

In order to eliminate the harmful effects of flux components of ordinal numbers divisible by three and of fluxes leaving the core, at least one of the windings of the transformer is connected in delta. Should the application of the delta-connected winding cause difficulties, as a less economical arrangement, a supplementary stabilizing turn is accommodated on each of the upper and lower yokes of the transformer. Such a supplementary stabilizing turn is shown in Fig. 2.24. It should

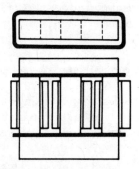

Fig. 2.24. Arrangement of yoke closing turn

be noted that it is less effective than a delta-connected winding, therefore it is applied to transformers not exceeding capacities of 2.5 MVA. For larger transformers, for the compensation of zero-sequence fluxes only, a supplementary delta-connected stabilizing winding is provided. This stabilizing (tertiary) winding is not identical with the third winding of a multi-winding transformer. The terminals of the stabilizing winding serving for the compensation of zero-sequence flux are either not brought out above the top cover of the transformer, or only one or two of its terminals are brought out for the purpose of earthing, or for providing the possibility of disconnecting the delta connection. Thus, no load may be connected across the terminals of the stabilizing winding brought out. The third winding (or further windings) of a multi-winding transformer is not necessarily connected in delta and takes part in power transmission, consequently the three or four terminals of the third winding are brought out to the cover of the transformer. Some star/star connected transformers are produced without supplementary stabilizing turns or supplementary delta-connected stabilizing windings.

The upper harmonics of the no-load currents of transformers can disturb the proper functioning of automatic network protection and can interfere with the operation of telecommunication equipment. Several attempts were made in earlier days to design transformers having no-load currents free of upper harmonics. The problem can however, be solved both theoretically and practically [49], with the wide application of cores made of cold-rolled sheets, the magnitudes of no-load currents and their upper harmonics have been reduced to such an extent that the problem has lost its importance. The specially built transformers of Hueter and Buch, applying cores and compensating windings of increased complexity ensure that with sinusoidal supply voltage the no-load currents will also be sinusoidal. The 5th, 7th, 17th and 19th harmonics can be eliminated from the network by connecting the primary windings of three-phase transformers of identical ratings and magnetic properties in delta and star formats in succession. In this case, the upper harmonic currents are shifted by 180° with respect to each other, and the transformers will not draw these upper harmonics from the network, but they will feed them into each other [83]. This solution is now of minor interest because the application of cold-rolled sheets has reduced the magnitude of harmonics, and in any case the installation of several smaller transformers of equal capacity connected in turn in delta and star, instead of transformers of larger unit ratings, is surely a much more expensive solution.

2.7.3 Influence of core structure on no-load loss and current

Single-phase cores. The simplest shapes of single-phase cores are shown in Fig. 2.25. In drawing (a) chain-loop type, in (b) two-limb core-type and in drawing (c) single-limb-type lamination are represented. Variants (a) and (b) differ in the

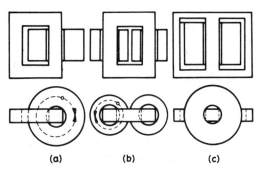

(a)   (b)   (c)

Fig. 2.25. Single-phase cores: (a) chain-loop type; (b) two-limb core-type; (c) single-limb shell-type

arrangement of windings. With type (a) only one limb carries a winding whereas each limb of type (b) is surrounded by a winding. The core-type arrangement is economically more favourable, because the windings around the two limbs utilize the available winding section with a shorter mean turn length (see the top views in Fig. 2.25 (a) and (b)). The single-phase single-limb shell-type lamination represented in Fig. 2.25 (c) is less economical than the core-type, nevertheless its application is recommended when for some reason the height of the lamination has to be kept within a certain limit (e.g. that imposed by the loading gauge). The core-type lamination is used only in small transformers.

The magnetic field developing in the single-phase cores mentioned above is very simple. The structure of the magnetic field is more complicated in a further arrangement, in the single-phase two-limb lamination represented in Fig. 2.26. This

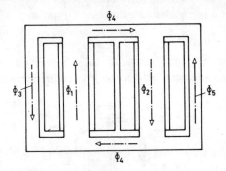

Fig. 2.26. Flux distribution in a two-limb shell-type core

type of core is primarily adopted for the largest capacity transformer banks consisting of three single-phase units. The windings are carried by the two inner limbs of the core, and the two outer limbs without windings are termed yoke closing legs. The windings of the core are linked with fluxes Φ_1 and Φ_2, respectively, as shown in Fig. 2.26. According to the law of magnetic nodes the following two equations may be written:

$$\Phi_1 = \Phi_3 + \Phi_4 \quad \text{and} \quad \Phi_2 = \Phi_4 + \Phi_5,$$

further, for symmetry reasons:

$$\Phi_1 = \Phi_2 \quad \text{and} \quad \Phi_3 = \Phi_5.$$

It is remarkable that, generally, $\Phi_3 \neq \Phi_4$ and $\Phi_4 \neq \Phi_5$. The two inequalities find their explanation in the different magnetic conductances of the parallel branches 3 and 4, and 4 and 5, of the magnetic circuit, with the cross-sectional areas of all yokes being equal and the magnitudes of magnetic conductances being $\Lambda_3 = \Lambda_5 < \Lambda_4$. In the yoke-closing legs the peak value of flux is lower than in the yoke connecting the two inner limbs. An additional loss arises in this core structure because, with a sinusoidal supply voltage, the variation of flux will be sinusoidal in the limbs only. In the yokes and yoke-closing limbs the shape of the flux curve is distorted, the fluxes in the two different parts of the core (connected in parallel, yet of different magnetic conductance) being shared by the said parts in a ratio varying from instant to instant because of the non-linear properties of the core. Due to the unequal flux distribution resulting from the magnetic asymmetry of the yoke and yoke-closing leg and to the non-linear character of the magnetic circuit, harmonic components appear in the flux of the yokes. In earlier designs, in order to compensate for the additional losses caused by these harmonic flux components, the cross-sectional areas of yokes and yoke-closing legs were somewhat increased, from the 50% of the cross-sectional area of a limb (as would be required by half of the flux) to about 55 to 60% of the limb section. A shell-type lamination, if the cross-sectional area of the yoke is 50% of that of the limb, reduces the height of the core just by the height of yoke of a core-type lamination of identical capacity. With 60% yoke area, this reduction, as compared to a core-type lamination, will be only 80% of the yoke height of the core-type lamination. The advantage of the two-limb shell-type core is that the corner area giving rise to additional losses reduces to almost half that of the core-type lamination.

An axonometric drawing of a core with radially stacked lamination is given in Fig. 2.27. With this arrangement of sheets, a circular cross-sectional area can be

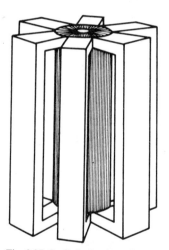

Fig. 2.27. Radially laminated core

filled out with a similar space factor to that of a conventional core of stepped cross-sectional area. The two advantages of the radially stacked lamination can be fully utilized in the design of large transformers. One of its advantageous properties is a further reduction obtainable in the core height. By placing 4, 6, etc. yoke-closing legs along the periphery, the core height can be reduced to a larger extent than in the case of conventional core-type laminations. The other advantage is the reduction of supplementary short-circuit losses. In larger transformers, as it follows from dimensional analysis, higher flux densities are reached in the stray field of windings than in the case of transformers of smaller ratings. Part of the stray field is closing through the core, yoke and limb. Where the stray flux lines enter the core perpendicularly to the plane of sheets, additional losses occur. The additional short-circuit losses mentioned are avoided in radially laminated cores, since they follow the cylindrically symmetrical arrangement of the short-circuited winding system with a geometry closely approximating to cylindrical symmetry. The radially laminated core is advantageous due to its relatively small overlap corner area.

Built up on essentially similar principles, but representing a simpler construction from the point of view of manufacturing, is the cross core having for yoke-closing legs displaced by 90° with respect to each other. The disadvantage of both the radially stacked cores and cross cores is their more complicated manufacture.

The three limbs of the three-phase core arranged symmetrically in space are fully identical as regards their no-load characteristics. The design meets all the requirements that a core manufactured of cold-rolled sheets is expected to satisfy: its overlap corner area is small, and its symmetrical arrangement brings about economy in sheet and tank material. The three limbs are connected either in delta, as shown in Fig. 2.23, or in star, as represented in Fig. 2.28. The yoke cross-sectional

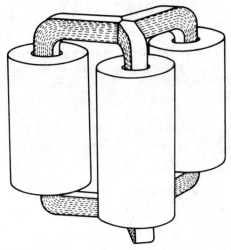

Fig. 2.28. Space-symmetrical star-connected core

area of the symmetrical core connected in delta is selected to be at least $\sqrt{3}$ times the cross-sectional area of the limb. The yoke area of the symmetrical core connected in star may be equal to the limb area,—it has no yokes. Stacking of symmetrical cores can be mechanized, but the associated equipment costs are only justified when manufacturing large series. This type of core is seldom used in transformers of ratings above 2500 kVA.

A three-phase shell-type core is shown in Fig. 2.29 (a). Successive windings on the limbs are wound in alternating senses, giving a flux distribution represented by the vector diagram of Fig. 2.29 (b). Using the notation of the diagram, the following

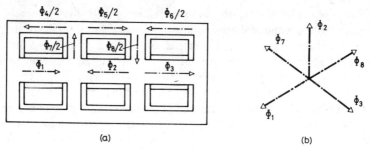

(a) (b)

Fig. 2.29. Flux distribution and the vector diagram of fluxes for a three-phase shell-type core

relations hold:

$$\Phi_1 + \Phi_2 = \Phi_7 \quad \text{and} \quad \Phi_2 + \Phi_3 = \Phi_8,$$

also

$$\Phi_5 = \Phi_4 = \Phi_1 \quad \text{and} \quad \Phi_6 = \Phi_3.$$

The core section may be very simple, if the winding is of the disc type.

In this case, the cross-section of the core may be rectangular, since the electrodynamic forces imposed on a disc-type winding during a short-circuit are in a direction which cannot bring about deformation in the rectangular winding.

The three-phase core-type lamination is shown in Fig. 2.22 (d). This type is the most commonly used arrangement. The core is asymmetrical and, thus, the no-load current is unequally distributed between the three phases. This results in no considerable drawbacks. Preferably, the core should be provided with equal number of steps in the limbs and yokes. The cross-sectional area of the yoke is equal

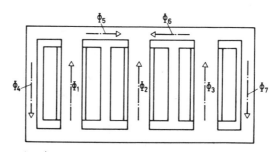

Fig. 2.30. Three-phase five-limb core

to, or somewhat larger than that of the limb. It is unnecessary to increase the cross-sectional area of the yoke by more than 10%. Mitred joints are used in the core.

The three-phase, five-limb core is the arrangement usually adopted for the largest three-phase transformers. The yoke closing limbs allow a reduction of the height of lamination. The magnetic conditions are similar to those of the four-limb single-phase core. With the notation of Fig. 2.30, the conductivity of magnetic channels of yoke fluxes $\Phi_4 = \Phi_7$ is lower than that of fluxes $\Phi_5 = \Phi_6$ in the yoke closing limbs. On the basis of the law of magnetic nodes it may be written:

$$\Phi_1 = \Phi_4 + \Phi_5, \quad -\Phi_2 = \Phi_5 + \Phi_6$$

and

$$\Phi_3 = \Phi_6 + \Phi_7.$$

Because of inequalities in the conductances of the parallel-connected magnetic circuits:

$$\Phi_5 > \Phi_4 \quad \text{and} \quad \Phi_6 > \Phi_7.$$

Due to the non-linear properties of the core already described in the case of four-limbed laminations, the variation of yoke fluxes Φ_4, Φ_5, Φ_6 and Φ_7 with time will be non-sinusoidal even if fluxes Φ_1, Φ_2, and Φ_3 are sinusoidal. The additional losses caused by the uneven flux distribution and non-sinusoidal shape of the flux curves are compensated by adopting yoke sections equal to 55 to 60% of the limb sections. As regards the overlap corner area, the three-phase five-limbed transformer is more favourable.

The cores of split-limb and split-yoke types have advantages, resulting from their smaller overlap corner areas. In Fig. 2.31 it can be seen that the corner area reduces

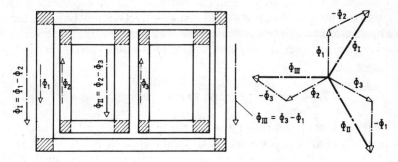

Fig. 2.31. Framed-type core and its vector diagram and corner areas

to half and in Fig. 2.32 to one third of that usual in a core of conventional design. Split limbs and yokes can be applied both in single-phase and three-phase cores. In cores of larger ratings, splitting can be utilized for better cooling of the core. More splitting offers considerable advantages, since heat dissipation of a steel surface in the direction perpendicular to the sheet is about ten times higher than that in the plane of the sheets. The clamping elements compressing the sheets are accommodated in the splittings between the core frames, requiring the provision of oil ducts of 4 to 15 mm width for this purpose. If the oil flow in the ducts is ensured, the oil ducts running parallel to the plane of the sheets may be omitted from cores of smaller ratings and their number may be reduced in larger-size cores. An additional advantage of split cores is the simpler assembly of laminations, the width of strips in the upper yoke being only half of that of corresponding strips used in conventional cores. This reduced dimension is useful in the case of larger cores where the removal and replacement of yoke strips wider than 400 mm would be rather cumbersome. Assuming a very high reluctance between splittings, the flux distribution in a split core is given by the following relations, with the notation of Fig. 2.31. $\Phi_I = \Phi_1 - \Phi_2$, $\Phi_{II} = \Phi_2 - \Phi_3$ and $\Phi_{III} = \Phi_3 - \Phi_1$.

The vector diagram of fluxes is given in Fig. 2.31. As is apparent there, the peak value of flux density in the split core would be $(2/\sqrt{3})$ times that in a three-leg core-

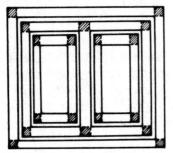

Fig. 2.32. Three-framed core and its corner areas

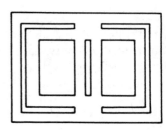

Fig. 2.33. Semi-framed core

type lamination, if there were no upper harmonic flux components present. In reality, the reluctance of the usual 4 to 15 mm wide ducts cannot be neglected and, as shown by measurements, there is an increase of flux of up to 5%. The excess losses of the framed core can be reduced by applying the semi-framed core shown in Fig. 2.33.

2.7.4 Choice of the core, stresses arising in the core

Before the advent of cold-rolled transformer sheets, cores had been manufactured with plain overlap laminations, with yoke areas increased by 5 to 15% with respect to the limbs and with clamping bolts passing through the holes punched into the limb and yoke sheets. The cold-rolled core structures, on the other hand are characterized by mitred joint areas, nearly or fully equal cross-sectional areas in the limbs and yokes and almost total omission of punched holes from the core lamination.

For different working capacities, the following core designs are widely applied.

In the range of 25 to 2500 kVA the three-phase core-type lamination with no holes punched into the sheets, or built up of wound core sections, is normal. In the case of stacked cores, sheets in the outer limbs are joined with mitred corners, whereas sheets of the middle limb are plain overlapped or also mitred. The limbs are clamped together by the windings, and the sheets of yokes are compressed by yoke clamping structures. Resin glass bonded bandages may also be used for mechanical clamping of the core. Bonding of carlite insulated sheets is also a successful method. As adhesive a cross-linked synthetic resin of high capillarity is used. This adhesive is sprayed or applied to the cut edges of the stacked lamination. The adhesive penetrates 25 to 30 mm deep between the sheets. As the next step, sheets are clamped

70

together and allowed to rest for a period of time required by the cross-linked material to set. Such bonded cores have been in service for several decades, and have not only proved their reliability in service, but the adhesives also have a beneficial effect on magnetic noise levels as well. Below 2500 kVA the use of single-phase two-limbed cores has become common practice in the USA. The cores are wound of steel strips, then annealed in an oven in order to improve loss characteristics. Special coiling machines are used for preparing the windings. Spatially symmetrical cores are also applied up to 2500 kVA.

Within the capacity range 2500 to 25 000 kVA the three-phase core-type lamination is the arrangement most often adopted. The sheets of transformers falling in that range do not have punched holes. The use of mitred joints offers advantages in every case. For compressing the sheets, bandages are used to reinforce cementing.

For 25 000 to 125 000 kVA units some of the arrangements used in transformers of lower-ratings are applied. For the elimination of holes the use of split holes has proved practical, the limbs being compressed by bandages and the yokes by clamping elements passed through slots. An arrangement similar to the framed core is the core with split limbs or in which both the limbs and yokes are split. In large transformers the purpose of core bonding is not so much that of reducing mechanical vibrations but rather to restrict magnetic noise, and of applying additional insulation to the cut edges. A similarly common practice followed in this range of capacities is the application of three-phase shell-type lamination (see Fig. 2.29) with mitred joints and without punched holes.

Problems of transport constitute the decisive factor in the selection of core structures for transformer ratings beyond 125 000 kVA. The five-limbed three-phase and three-phase shell-type cores are often adopted. To avoid transportation difficulties the manufacture of single-phase units is often required. In such cases single-phase four-limbed, radially laminated or cross-core types are chosen. Some manufacturers successfully employ cores having no through holes for the largest units. Others are still using clamping bolts passing through holes of reduced cross-sectional area provided in the cores of the largest size units. Laminations of the mitred-joint type are also in general use in this capacity range.

Stresses in the core

Magnetic stresses permitted in the core are governed primarily by economic considerations, namely what significance is given to the value of no-load loss. Beside the economic point of view, the problem of noise level may also be of importance. Magnetic stresses in the core of oil insulated transformers made of hot-rolled sheets used to be 1.4 to 1.5 T. As cold-rolled steels were introduced, the aim of designers was to keep specific losses at earlier levels, and for oil transformers flux densities up to 1.9 T were permitted. The temperature rise conditions of high magnetic stress cores remained at the level usual in hot-rolled cores. Higher stresses allow savings of about 20% in weight and about the same percentage of loss. In large oil transformers, the generally accepted values of flux density fell in the range 1.5 to 1.7 T.

71

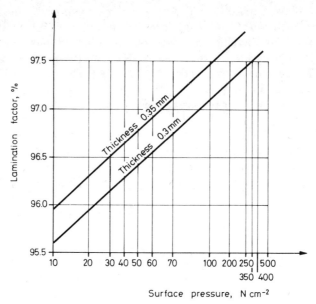

Fig. 2.34. Stacking factor of 0.3 mm and 0.35 mm thick cold-rolled sheets as a function of pressure acting on the sheets

With these stresses, 45 to 50% reduction of core losses and a weight reduction of 10 to 15% were achieved relative to the levels usual with cores made of hot-rolled sheets. Obviously, both the production and transportation costs are considerably influenced by the weight of the transformer.

The mechanical compressive stress prevailing in the core results from the force with which the core is clamped together by windings and bandages or by clamping bolts and other structures. Application of a due compressive force is required to act against shocks occurring during transport of the core, against the loosening effect of forces of magnetostriction alternating with double the system frequency, and, finally, to reduce the electrodynamic effects of short-circuits and to lower the noise level. In a core compressed with an excessive force, high local pressure rises are liable to occur, and sheet insulation may suffer damage at such spots. Damaged sheet insulation may first cause a 10 to 20% increase in the losses, then with further development of deterioration, the defect may lead to an inter-sheet short-circuit. In view of these circumstances no surface pressures exceeding a limit of 4×10^5 Pa are applied when compressing the lamination sheets.

The relation between surface pressure and lamination factor is represented in Fig. 2.34, and that between the same pressure and resistivity of carlite insulating surfaces is shown in Fig. 2.35. Maximum pressure compressing the sheets should not exceed the limit of 10×10^5 Pa, to avoid excessive reduction of resistivity of sheet insulation. Since, however, local increases in pressures up to a ratio of 1:2 have to be reckoned with, due to inhomogenities of sheet material, the peak value of the compressive force should not exceed 4×10^5 Pa in the core. With a lowest pressure level of

2×10^5 Pa and a highest level 4×10 Pa, the ratio between the largest and smallest widths of sheets should not exceed the ratio of $1:2$. The pressure distribution may be made more uniform by applying bandages. A further requirement to be satisfied by sheet insulation is that it should not break down at $100\,°C$ and 10×10^5 Pa pressure. In order to provide increased safety against intersheet short-circuits, the carlite insulation of cold-rolled sheets is coated with a 2 to 3 µm thick additional insulation, or sheets of 0.1 to 0.5 mm thick pressboard or soda cellulose paper are laid between stacks of a certain number of lamination at spacings corresponding to the turn voltage of about 20 V mentioned earlier. The same can be achieved by core bonding.

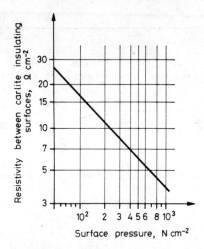

Fig. 2.35. Resistivity between carlite sheet surfaces as a function of pressure acting on the sheets

In order to increase reliability, any two of the three given alternatives may be used in combination.

Suitable measures have to be taken in the design to carry off the heat developing in the core. The temperature of the hottest spot must not exceed the temperature of the surrounding oil by more than $20\,°C$. Should the relative temperature rise be higher, aging of the oil getting between the transformer sheets will set in, and the products of aging, coming into contact with other parts of the oil and with other insulating materials in the transformer, will cause deterioration of the electrical properties of the oil and of the cellulose–base insulation.

For example, if the temperature of the hottest spot of the core rises by $20\,°C$ above that of the surrounding oil, the ambient temperature is $40\,°C$, and the temperature rise of the oil above its surroundings is $50\,°C$, the temperature of the hottest point of the core will rise to $110\,°C$. Determination of the temperature distribution in the core is a rather complex task. In the calculation the main problem is caused by the difference between heat conduction in the plane of the sheets and in the direction perpendicular to that plane. The longitudinal thermal conductivity (i.e. in the direction of the sheet plane) of oil-immersed hot-rolled sheets provided with varnish or sodium silicate insulation is $\lambda_l = 19\ \mathrm{W\ m^{-1}\,°C^{-1}}$, and their transverse conductivity (perpendicular to the sheets) is $\lambda_t = 2.8$ W m$^{-1\circ}$ C^{-1}. In cores composed of cold-rolled sheets, the longitudinal thermal conductivity is $\lambda_l = 21$ W m$^{-1\circ}$ C^{-1}, and the transverse conductivity $\lambda_t = 3.3$ W m$^{-1\circ}$ C^{-1} [83]. An approximate calculation method for assessing the maximum core temperature is given by the investigation of temperature rise conditions in the lamination stack of cross-sectional area represented in Fig. 2.36. For the purpose of investigation it is assumed that in the direction perpendicular to its cross-sectional area the model is a very long laminated limb with uniform flux distribution and, hence, that the loss caused by the

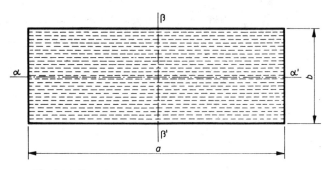

Fig. 2.36. Sheet lamination of rectangular cross-section

alternating flux is equal in each unit volume. Its longitudinal length (i.e. its length in the direction of sheets) is a and, its transverse dimension (perpendicular to the sheets) is b. As an approximation, the heat flow may be conceived so that one part, P_l, of the entire amount of loss P is dissipated in the longitudinal and the other part, P_t, in the transverse direction. In the approximation described in the following, the longitudinal and transverse heat flows are investigated independently of each other. In the longitudinal direction the limb will reach its maximum temperature along axis $\beta - \beta'$, half of the loss P_l being carried off in the direction α and the other half in direction α' by the heat flow. The heat flux enters the oil, or the air in the case of a dry transformer, through an area of width b and surface A_l. In the case investigated, heat flow takes place in a medium where each part of the volume is itself also a heat source. In such a case, the temperature distribution as a function of location assumes the shape of a parabola. The equation of heat flow in the longitudinal direction is [41]

$$\frac{P_l}{A_l} = \frac{\Delta\vartheta_{max}}{\dfrac{a}{4\lambda_l} + \dfrac{1}{\alpha_t}},$$

where $\Delta\vartheta_{max}$ in °C is the highest temperature along the line $\beta - \beta'$, a is the limb width in m and α is the heat-transformer coefficient between core surface and surrounding oil [41]: $\alpha = 110$ W m^{-2} °C^{-1}; P_l is the loss flowing in the longitudinal direction in W and A_l is the dissipation area of longitudinal heat flow in m^2.

74

Based on similar considerations, the equation of transverse heat flow is

$$\frac{P_t}{A_t} = \frac{\Delta\vartheta_{max}}{\dfrac{b}{4\lambda_t} + \dfrac{1}{\alpha}}.$$

The calculated temperature rise of the core is checked by substituting into the two formulae the highest permissible temperature, the respective geometrical dimensions, and the values of heat-transfer coefficients, and as long as the combined quantities of heat carried off in longitudinal and transverse directions are at least equal to the overall loss arising in the core, the core temperature will not exceed the highest permissible temperature. The calculation will be somewhat simpler if instead of the overall losses arising in the core, the computation is carried out with the specific loss p and with the portions of the specific loss carried off in the longitudinal direction, p_l, and the transverse direction p_t:

$$p_l = \frac{\Delta\vartheta_{max}}{\dfrac{a^2}{8\lambda_l} + \dfrac{a}{2\alpha_l}} \frac{10^{-3}}{k_s d},$$

$$p_t = \frac{\Delta\vartheta_{max}}{\dfrac{b^2}{8\lambda_t} + \dfrac{b}{2\lambda_t}} \frac{10^{-3}}{k_s d}.$$

The temperature of the core will remain below the permissible temperature rise, $\Delta\vartheta_{max}$, if $p = p_l + p_t$. In the above formulae d is the density of steel in kg dm^{-3}, and k_s is the lamination space factor, its value being 0.88 to 0.92 for hot-rolled sheets and

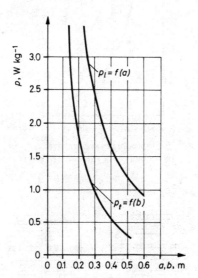

Fig. 2.37. Specific heat flow escaping longitudinally and transversally from a hot-rolled sheet lamination as a function of stack dimensions

0.94 to 0.97 for cold-rolled sheets. The portions of loss carried off in longitudinal and transverse directions are shown as functions of the longitudinal and transverse dimension of the core in Fig. 2.37 for hot-rolled sheets and Fig. 2.38 for cold-rolled sheets [83].

Design and checking of the core for its permissible temperature rise consist briefly of the following steps. The core is first split up, by oil ducts, into quadrangular sections of shapes which ensure, by estimation, a uniform temperature distribution. After arranging the oil ducts along the abscissae of Fig. 2.37 or Fig. 2.38, the values of a and b are marked out, and the magnitudes of the respective components of the carried-off loss p_l and p_t are read. If, during checking, a lower specific loss were obtained than that actually occurring in the core, then the core must be split up into smaller cross-sectional areas. In Fig. 2.39 the cross-section of such a core split up longitudinally and transversely by cooling ducts is shown. It should be pointed out that the calculation described above is only approximate.

The oil ducts are included by cementing pressboard or wooden strips to the surfaces of sheets sorrounding the ducts. When forming the duct, an oil duct of at

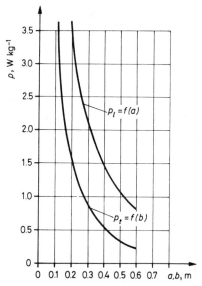

Fig. 2.38. Specific heat flow escaping longitudinally and transversal from a cold-rolled sheet lamination, as a function of stack dimensions

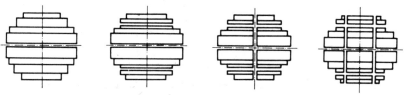

Fig. 2.39. Examples of arrangement of cooling ducts

least 4 mm thickness should be provided at the spacers. At the surfaces of the wooden or pressboard strips the specific compressive load resulting from clamping of the core must not exceed 10×10^5 Pa. If the spacers of oil ducts are made of pressboard or wood, the part covered by them should be subtracted from dimensions a and b when checking the temperature rise conditions.

Usually, there is no need to provide cooling ducts in the cores of small oil transformers. For checking the temperature rise, it is sufficient to calculate the temperature rise of the quadrangle drawn around the cross-section of the core. The following formula may be used for the calculations [83]:

$$p = \frac{\Delta\vartheta_{max}}{\dfrac{R}{2\alpha} + \dfrac{R^2\left(1 + \dfrac{R\alpha_t}{\lambda_t} + \dfrac{\lambda_t}{\lambda_l}\right)}{8\lambda_t + 2R\alpha\left(1 + \dfrac{\lambda_t}{\lambda_l}\right)}} \cdot \frac{10^{-3}}{k_s d}.$$

In the above formula R is the radius of the circle drawn around the core in m, while all other values should be substituted using the same units as in previous formulae.

Example 2.4. The core of the 6300 kVA transformer shown in Fig. 2.40 is built up of M6 grade sheets. The peak value of flux density in the core limb is $B_0 = 1.57$ T. The surfaces and shapes of the limbs and yokes of the core are different, the cross-sectional area of the yoke being larger and the number of its steps less. From the structural point of view this arrangement is favourable, because although its additional

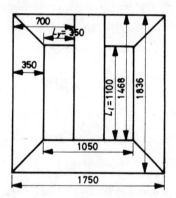

Fig. 2.40. Core of a 6300 kVA transformer

losses are higher it permits better bracing of windings with respect to mechanical short-circuit stresses, and offers better utilization of space in the limb direction. According to the dimensions of Fig. 2.41, the so-called geometric cross-section of the limb, in which the space factor is not yet considered, is $A_{gl} = 925$ cm^2. Assuming an average compressive force of 2×10^5 Pa in the core the value of the space factor, as read from Fig. 2.34 is $k = 0.963$. The limb section filled with sheet material is

$$A_l = A_{gl}k = 925 \times 0.963 = 890 \text{ cm}^2 .$$

77

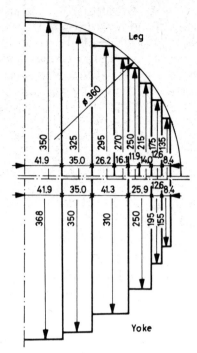

Fig. 2.41. Limb and yoke cross-section of core shown in Fig. 2.40

The geometric yoke section is $A_{gy} = 995 \text{ cm}^2$. The yoke section filled with sheet material is

$$A_y = 995 \times 0.963 = 960 \text{ cm}^2 .$$

With the notation of Figs 2.40 and 2.41, the calculation of core mass proceeds as follows:

In the example

$$k = \frac{A_y}{A_l} = \frac{960}{890} = 1.08$$

and

$$h = kD_l = 1.08 \times 0.36 = 0.388 \text{ m} .$$

Mass of corner area

$$m_c = 0.85 d_r A_l h = 0.85 \times 7650 \times 0.089 \times 0.388 = 224 \text{ kg} .$$

Mass of the six corner areas

$$6 m_c = 1344 \text{ kg} .$$

Mass of limbs

$$3 d A_l l_l = 3.7650 \times 0.089 \times 1.1 = 2250 \text{ kg} .$$

Mass of yokes

$$4 d A_y l_y = 4.7650 \times 0.095 \times 0.35 = 1030 \text{ kg} .$$

Increased mass of limbs

$$3 \times 0.075 d A_l h = 3 \times 0.075 \times 7650 \times 0.089 \times 0.388 = 59 \text{ kg} .$$

Increased mass of yokes

$$4 \times 0.075 d A_l h = 4 \times 0.075 \times 7650 \times 0.089 \times 0.388 = 79 \text{ kg} .$$

Mass of yokes

$$m_y = 1109 \text{ kg} .$$

Mass of limbs

$$m_l = 2309 \text{ kg} .$$

Calculation of core losses:
For the flux density in limb $B_l = 1.57$ T, the loss figure as read from Fig. 2.3 is

$$v_l = 1.22 \text{ W kg}^{-1}.$$

The core loss occurring in the limb is

$$v_l m_l = 1.22 \times 2309 = 2810 \text{ W}.$$

The flux density in yoke is

$$B_y = B_l \frac{A_l}{A_y} = 1.57 \frac{890}{960} = 1.46 \text{ T}.$$

The corresponding loss figure is 0.98 W kg^{-1} and the loss in the yoke $v_y m_y = 0.98 \times 1109 = 1088$ W. Assuming for the corner areas a loss figure equal to that for the limbs, and considering a multiplication factor of 1.5 in calculating the additional loss in the corner areas of outer limbs (Fig. 2.17(c)):

$$4 m_c 1.5 v_l = 4.224 \times 1.5 \times 1.22 = 1640 \text{ W}.$$

Considering a multiplication factor of 3.3 in calculating the additional loss in the corner areas of the middle limb (Fig. 2.17(d)):

$$2 m_c 3.3 v_y = 2 \times 224 \times 33 \times 0.98 = 1450 \text{ W}.$$

The overall loss arising in the core is thus 6988 W.
Now checking the core for temperature rise.
As a first variant, let the temperature rise of the core without cooling ducts be examined. Assume 20 °C for $\Delta\vartheta_{max}$, that the radius of the circle drawn around the core is 0.18 m, the heat transfer coefficient between core surface and oil is $\alpha = 110$ W m^{-2} °C^{-1} and the transverse and longitudinal thermal conductivities are $\lambda_t = 3.3$ W m^{-1} °C^{-1} and $\lambda_l = 21$ W m^{-1} °C^{-1}. The permissible loss figure is

$$p \frac{\Delta\vartheta_{max}}{\dfrac{R}{2\alpha} + \dfrac{R^2\left(1 + \dfrac{R_\alpha}{\lambda_t} + \dfrac{\lambda_t}{\lambda_l}\right)}{8\lambda_t + 2R\alpha\left(1 + \dfrac{\lambda_t}{\lambda_l}\right)}} \frac{10^{-3}}{k_t d} = \frac{20}{\dfrac{0.18}{2 \times 110} + \dfrac{0.18^2\left(1 + \dfrac{0.18 \times 110}{21} + \dfrac{3.3}{21}\right)}{8 \times 3.3 + 2 \times 0.18 \times 110\left(1 + \dfrac{3.3}{21}\right)}} \frac{10^{-3}}{0.96 \times 7.65} = 1.55 \text{ W kg}.$$

The loss figure is lower than the above value in the limb of the core.
Calculation of reactive no-load loss.
From Fig. 2.4, the specific reactive loss for flux density 1.57 T in the limbs is 2.1 VA kg^{-1}, and that for flux density 1.46 T in the yokes is 1.35 VA kg^{-1}.
Reactive power in the limbs is $2309 \times 2.1 = 4850$ VA.
Reactive power in the yokes is $1109 \times 1.35 = 1500$ VA.
In the corner areas the reactive power is much higher, due to the changes in direction. Therefore, as shown by experience, it is expedient to include a factor of 5 in calculating the specific reactive power arising in the corner areas: $1344.5 \times 5 \times 2.1$ VA $= 14200$ VA.
The reactive power requirement of the air gaps of the core is determined as follows. As illustrated in Fig. 2.40, the air gaps are located in the core along the four mitred joints and along the two vertical joints. The approximate cross-sectional area of each mitred joint is

$$A_\delta = \frac{A_l}{\cos 45°}$$

and that of the vertical joint is A_l. In the oblique air gaps the flux density is approximately $B_l \cos 45°$, that in the vertical air gaps is B_y and in the horizontal air gaps B_l. The volumes of air gaps along the joints are calculated from the cross-sectional areas given above and from the average width of air gaps taken in the direction of the flux lines. The average width of air gaps is, with good approximation, equal to half of the fitting tolerance of the joints. The fitting tolerance is taken as 0.3 mm in the oblique planes and 0.05 mm in the horizontal and vertical planes. In the four oblique planes the magnetizing power is

$$4\frac{\pi}{\mu_0} f B_i^2 V = 4\frac{\pi}{1.256 \times 10^{-6}} 50 \times 1.57^2 \times \cos^2 45° \times$$

$$\times \frac{8.92}{\cos 45°} 10^{-2} \times 3 \times 10^{-4} = 23.300 \text{ VAr}$$

and in the two vertical plane it is

$$2\frac{\pi}{1.256 \times 10^{-6}} 50 \times 1.46^2 \times 9.60 \times 10^{-2} \times 5 \times 10^{-5} = 2545 \text{ VAr}.$$

In the two horizontal planes the magnetizing power is

$$\frac{\pi}{1.256 \times 10^{-6}} 50 \times 1.57^2 \times 8.92 \times 10^{-2} \times 5 \times 10^{-5} = 1375 \text{ VAr}.$$

The magnetizing power of the air gaps in the core is 27220 VAr, which is 33% above that required for magnetizing the steel. The overall magnetizing power requirement of the core is 47670 VAr. The no-load power (or no-load current) of the transformer will, thus, be

$$100\frac{47.67}{6300} = 0.75\%$$

of its rated capacity.

3. THE WINDINGS

The windings of a transformer are strictly related to its impedance characteristics, viz. the impedance voltage, the short-circuit current and the short-circuit forces.

3.1 Impedance voltage

Impedance voltage determines the voltage variation of a transformer under load and the magnitude of currents during short-circuits, and it is also of importance under conditions of parallel operation.

The impedance voltage of a transformer is measured after short-circuiting its terminals on one side, and increasing the voltage on its other side until the input current attains its rated value. In this condition, rated current is flowing in both windings. The voltage at which rated currents are flowing in the transformer is termed the impedance voltage. The active component of apparent short-circuit power taken up by the transformer during short-circuit measurement is the nominal load loss or nominal impedance loss of the transformer, which is identical with the loss converted into heat in the transformer windings under rated service conditions.

The impedance voltage is the product of the short-circuit impedance and rated current of the transformer. The short-circuit impedance, comprising active and reactive component, can be expressed in ohms just like any other impedance. The short-circuit impedance of a transformer, however, may vary over several orders of magnitude, depending on the capacity and rated voltage of the transformer, and is therefore less familiar. From the points of view of short-circuit limiting effect, voltage regulation, and parallel operation of a transformer the per-unit (p.u.) or percentage values can be dealt with more easily. The ratio of rated voltage of a transformer to its rated current is termed its rated impedance. The ratio of the short-circuit impedance reduced to one side to the rated impedance of the transformer is the short-circuit impedance referred to the rated data of the transformer. One hundred times this ratio is termed the percentage impedance. The short-circuit impedance referred to the rated data of the transformer may also be obtained as the following ratios: ratio of impedance voltage to rated voltage, ratio of apparent input power measured in the short-circuit test to nominal power. Usual values of the short-circuit impedance referred to nominal values with rated capacities of 10 kVA, 2500 kVA and 400 000 kVA are approximately 0.03, 0.06 and 0.15.

The active component of impedance voltage is the product of the a.c. resistance and current of the winding, and it determines the losses and temperature rise of the transformer. Generally the reactive component of the impedance voltage is higher than the active component, and predominant in large transformers. In designing a transformer, it is the magnitude of this reactive component which accurately determines the desired value of the short-circuit impedance.

3.1.1 Active component of the impedance voltage

The active component of the impedance voltage is the product of a.c. resistance reduced to the primary or secondary side and the respective rated current. Because of the eddy-current losses of the a.c. magnetic field, the resistances of a.c. circuits are higher than the resistances measured with d.c. For the sake of distinction, the concepts of a.c. resistance and d.c. resistance have been introduced. The active component of impedance voltage is given, like the impedance voltage itself, either in terms of p.u. quantities or as a percentage value. The p.u. quantities can be obtained as results of the following quotients: as the ratio of the active component of the short-circuit impedance to rated impedance, of the active component of the impedance voltage to rated voltage, or of the short-circuit loss to rated capacity. The values of active components of the impedance voltage for transformers of 10, 2500 and 400 000 kVA are approximately 0.02, 0.01 and 0.002, respectively.

When calculating the active component of impedance voltage, the d.c. resistances (R_1, R_2) of the primary and secondary windings are found first, then the d.c. losses $(I_1^2 R_1 + I_2^2 R_2)$ pertaining to the respective rated currents and the current-dependent supplementary losses are determined. The sum of d.c. losses and supplementary short-circuit losses is the short-circuit or winding loss (P_s) of the transformer. This loss is the product of the rated transformer current and active component of impedance voltage. The p.u. value of impedance voltage (i.e. its value related to the rated voltage) is denoted by ε_z. The symbol of the active component is ε_R, its value being

$$\varepsilon_R = \frac{P_s}{P_n} = \frac{I_n U_s}{I_n U_n}.$$

The resistance of a wire of length l and cross-sectional area A, made of a material of resistivity ρ, is $R = \rho l / A$. The magnitude of current flowing in that conductor, with a current density J, is $I = JA$. The so-called d.c. loss of the winding, without the supplementary losses arising with an a.c. current, is

$$P_{dc} = \rho \frac{l}{A} J^2 A^2.$$

Considering that the mass m of a winding made of some conductive material of density d is $m = dAl$, the d.c. loss is

$$P_{dc} = \frac{\rho}{d} J^2 m. \tag{3.1}$$

According to the specifications of international and national standards, the losses in oil-immersed transformers should be determined at 75 °C winding temperature. At this temperature the resistivity of copper is $\rho_{Cu\,75}=0.0216\,\Omega m^2\,m^{-1}$, and that of aluminium is $\rho_{Al\,75}=0.036\,\Omega m^2 m^{-1}$. The density of copper is $d_{Cu}=8.9\,kg\,dm^{-3}$ and that of aluminium $d_{Al}=2.7\,kg\,dm^{-3}$. With these data the loss occurring in the unit mass of a copper winding at 75 °C is

$$P_{dc}/m=2.42\,J^2. \tag{3.2}$$

The specific loss of an aluminium winding at 75 °C is

$$P_{dc}/m=13.3\,J^2. \tag{3.3}$$

In the above formulae the specific loss is obtained in terms of $W\,kg^{-1}$, if the value of current density is substituted in $A\,mm^{-2}$.

For calculating the mass, the formula

$$m=m_f ND\pi Ad\,10^{-6} \tag{3.4}$$

is suitable, from which the mass of the winding is obtained in kg, m_f is the number of phases, N is the number of turns, D is the mean winding diameter in mm, A is the cross-sectional area of conductor, and d is the density in $kg\,dm^{-3}$.

In the range of temperatures expected in a transformer, the d.c. losses vary with temperature. The following relation exists between the d.c. loss P_{dcc} in a cold condition and the d.c. loss P_{dcw} in a warm condition of the windings

$$P_{dcw}=P_{dcc}\frac{235+t_w}{235+t_c}, \tag{3.5}$$

where t_w and t_c are temperatures in °C of the warm and cold winding, respectively.

The a.c. losses arising in transformer windings and their surroundings are higher, by the supplementary losses, than the d.c. losses. This difference is due to the eddy-current losses caused in the windings and surrounding metallic parts by the stray flux linked with the windings. The winding losses can be determined by measurement performed in the short-circuited condition of the transformer. This is why the losses mentioned before are called supplementary short-circuit losses.

For calculating the supplementary losses occurring in the windings the model shown in Fig. 3.1(a) may be used. This model may be applied to the calculation of supplementary losses arising both in concentric and in disc windings. When using the formulae described in the following, only the direction of stray flux lines is to be taken into consideration, depending on whether the losses of a concentric or a disc winding are calculated. Physical reality is described only approximately by the model, since its magnetic field is built up as if dimension l_s of the winding, i.e. the dimension in the direction of flux lines, were very long and the flux lines had no components perpendicular to the direction of l_s. It is further assumed by the model that the outside of the windings shown is embedded from all directions, except for the leakage duct, into a medium of infinite permeability. Both assumptions are only approximations. In winding arrangements practically applied, the flux density within the leakage duct at the centre line of the winding is nearly twice as high as at

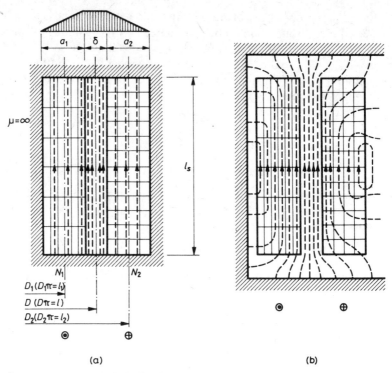

Fig. 3.1. Stray magnetic field of a short-circuited winding: (a) idealized; (b) real condition

the beginning or end of the duct. The cause of this discrepancy is that some flux lines leave the leakage channel while crossing the turns of windings (Fig. 3.1(b)), so the flux components perpendicular to the direction of the leakage channel (which are of radial direction in core-type transformers and of axial direction in shell-type transformers) cannot be neglected. In spite of the approximate nature of the assumption, the formulae deduced are in good agreement with measured results.

In the diagram of trapezoid shape appearing over the cross-sectional drawing of the winding, the distribution of leakage flux components lying in the direction of the leakage duct is shown for different points of the winding and within the leakage channel. According to the assumptions of the diagram, the values of flux density show a linear increase from the edge of the winding up to the leakage channel, remain constant within the channel, then decline linearly along the other winding towards its edge. The stray magnetic field is induced by the currents flowing in the primary and secondary windings. The peak value of flux density in the leakage duct is

$$B = \mu_0 \frac{I_1 N_1 \sqrt{2}}{l_s} R = \mu_0 \frac{I_2 N_2 \sqrt{2}}{l_s} R,\qquad(3.6)$$

where $\mu_0 = 1.256 \times 10^{-6}\ \mathrm{H\,m^{-1}}$, I_1 and I_2 respectively, are the r.m.s. values of primary and secondary current in A, N_1 and N_2 respectively, are the turn numbers of the primary and secondary windings, l_s is the length of the leakage channel in m, R is the Rogowski coefficient (see Eqn. (3.13)) and the flux density B is in T. Eddy-currents are induced in the conductors of the winding by the magnetic field decaying from the leakage channel to the edge of the windings. The loss arising per unit mass of the copper conductor located between flux lines of a sinusoidally varying homogeneous flux density of peak value B is

$$\frac{P}{m} = 1.65 \frac{f^2 B^2 h^2}{\rho d}, \tag{3.7}$$

where P/m is the specified loss in W kg^{-1}, f is the frequency in Hz, B is the flux density in T, h is the size of conductor in the direction perpendicular to the flux lines (see Fig. 3.2(a)) in m, ρ is the resistivity in m and d is the density in kg m^{-3} [86]. The stray flux lines induce eddy-currents in the plane perpendicular to their direction. The highest losses arise where the product $B^2 h^2$ assumes its maximum value. Figure 3.2 illustrates the leakage field distribution for a few winding arrangements. The highest losses arise in the vicinity of the leakage channels. In Fig. 3.3(a) an interpretation of the "size of conductor in the direction perpendicular to the flux lines" is given. In Fig. 3.3(b) the cross-section of conductor located at the end and beginning of the winding, as well as the flux lines linked with the conductor are illustrated. In such cases the products $(B'h')^2$ and $(B''h'')^2$ are characteristic of the magnitude of eddy-current losses, and the value is higher than at the turns adjacent to the leakage channel at the inner parts of the winding, and the conductors are liable to suffer excessive temperature rises.

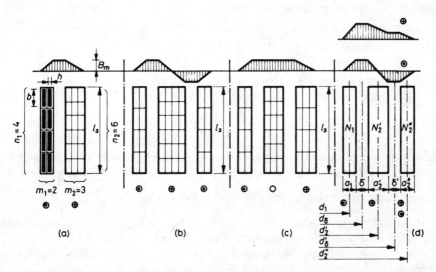

Fig. 3.2. Various stray flux distributions for impedance voltage calculations: (a) one primary and one secondary winding; (b) one primary and two loaded windings; (c) one primary and one loaded winding with middle winding unloaded; (d) one primary, one secondary winding and the regulating winding in two different tap positions

By means of Eqn. (3.7) the eddy-current loss can be determined, in all special cases, if the magnitude and direction of the flux lines are known. It is common practice to calculate, for winding arrangements of general use, average values of supplementary losses in terms of quantities characteristic of the eddy-current losses.

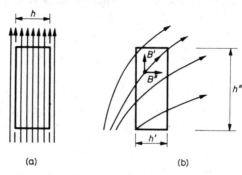

Fig. 3.3. Interpretation of the dimension perpendicular to the flux lines

The ratio of a.c. losses P_{ac} (increased by eddy-current losses) to d.c. losses P_{dc} is [86]

$$\frac{P_{ac}}{P_{dc}} = \varphi + \frac{m^2 - 1}{3},$$

where m is the number of conductors perpendicular to the leakage duct (e.g. in Fig. 3.2(a): $m_1 = 2$ and $m_2 = 3$). The value of φ is

$$\varphi = 2\xi \frac{\mathrm{sh}\, 2\xi + \sin 2\xi}{\mathrm{ch}\, 2\xi - \cos 2\xi} \quad \text{and} \quad \psi = 2\xi \frac{\mathrm{sh}\, \xi - \sin \xi}{\mathrm{ch}\, \xi + \cos \xi}$$

Finally, ξ is the reduced conductor width (a dimensionless number):

$$\xi = h \sqrt{\frac{nb}{l_s} R \frac{\mu_0 f}{\rho}},$$

where h is the conductor size perpendicular to the flux lines in m, n is the number of conductors in the direction of the flux lines (e.g. in Fig. 3.2(a) $n_1 = 4$ and $n_2 = 6$), b is the conductor size in the direction of the flux lines in m, $\mu_0 = \frac{4\pi}{10} 10^{-6}\, \mathrm{H\,m^{-1}}$, f is the frequency in Hz, l_s is the length of leakage duct in m, ρ is the resistivity of the conductor in Ωm and R is the Rogowsky coefficient (see Eqn. (3.13)).

In the cases most frequently occurring in practice $\xi \leq 1$, and results of adequate accuracy can be arrived at by use of the following approximate formula:

$$\varphi = 1 + \frac{4}{45} \xi^4 \quad \text{and} \quad \psi = \frac{1}{3} \xi^4$$

With approximate values of φ and thus obtained, the supplementary loss is usually expressed by the formulae most common in practice as a p.u. quantity referred to the d.c. loss, or expressed as percentage of the latter. The p.u. value of supplementary losses occurring in a winding built up of conductors of rectangular cross-section is

$$k_\square = \frac{m^2 - 0.2}{9} \xi^4$$

and in a winding made of a conductor of circular cross-section

$$k_0 = \frac{m^2 - 0.2}{15.25} \xi^4 .$$

With further approximation, the additional loss in the case of copper of a winding consisting of rectangular-section conductors, expressed in p.u. values is

$$k_\square = \frac{m^2 - 0.2}{9} h^4 \times 10^8 \tag{3.8}$$

and for windings made of a wire of circular section

$$k_0 = \frac{m^2 - 0.2}{15.25} d^4 \times 10^8 . \tag{3.9}$$

In the formulae the size of h and wire diameter d is substituted in m. Both h and d refer to the conductor size and not to the size of the insulated wire. The above approximation is acceptable only as long as $\sqrt{nb/l_s} \cong 1$ is valid.

As observed in the above formula for ξ the supplementary losses vary inversely proportionally the square root of resistivity, therefore the supplementary losses of aluminium windings are relatively lower.

In Fig. 3.2 (b) windings of two magnetically balanced groups are represented. In such a case, the calculation is performed for each group separately. As shown by the diagram, the first magnetically balanced group terminates after the second layer of the middle winding, therefore the supplementary losses for the half of the middle winding are calculated with $m = 2$.

In multi-winding transformers, when a third winding takes its place between the two outer windings, additional losses occur in the middle winding, even if the latter is unloaded (Fig. 3.2 (c)). In the middle winding the value of flux density is the same throughout. The supplementary losses of this winding can be calculated by means of formula (3.7). The peak flux density appearing in that formula can be obtained from formula (3.6).

The distribution of supplementary losses in the windings is not uniform, the highest eddy-current losses occurring in the layer adjacent to the leakage duct. In this layer, the value of supplementary losses related to the d.c. losses is [83].

$$k = \frac{m^2 - m + 0.27}{3} \xi^4 .$$

It is to be noted that the relative values of losses arising in the layer adjacent to the leakage channel are very large. Thus in the case of two layers the losses are 1.75 times higher, with five layers 2.44 times higher and with ten layers 2.71 times higher than the average eddy-current loss calculated for the entire winding. For this reason it is advisable, in the case of large transformers, to check separately the temperature rise expected to occur in the conductor layer lying toward the channel.

An important case encountered in practice is that of the transformer with a regulating winding (Fig. 3.2 (d)). The losses of such transformers are calculated in the following way. The supplementary losses of a regulating winding having N_2'' turns are determined as those of a two-winding transformer. The supplementary loss arising in the so-called main winding, expressed as p.u. value related to the d.c. loss taken as unity is [77]

$$k = \frac{m^2 \left[1 + 3 \frac{N_2''}{N_2'} \left(\frac{N_2''}{N_2'} + \cos \varphi \right) \right] \cdot [-2]}{9} \xi^4 .$$

In the above formula, m is the number of layers perpendicular to the leakage channel in the winding of N_2' turns, N_2' is the number of turns in the inner winding and N_2'' that in the outer winding, φ is the phase angle between currents flowing in the windings of turns N_2' and N_2''. The value of $\cos \varphi$ is $+1$, if the excitations of the two windings are of the same direction (see upper excitation diagram of Fig. 3.2 (d)), or -1 if they are of opposed direction (lower excitation diagram). The formula is of general validity, and is applicable to any arbitrary vector group, such as to zig-zag and rectifier transformers.

Supplementary losses vary inversely with temperature, the relation between loss P_c for a cold temperature t_c and loss P_w for a warm temperature t_w being

$$P_w = P_c \frac{235 + t_c}{235 + t_w} . \tag{3.10}$$

As also mentioned above, the flux lines perpendicular to the leakage channel direction cause a higher temperature rise in the turns located at the beginning and at the end of each winding. In order to reduce this effect, the size of conductors in the leakage channel direction, is generally chosen so as not to exceed 10 mm with copper and 15 mm with aluminium.

In transformers with a rating above 100 MVA special care should be devoted to investigation of supplementary losses occurring in conductors located at the winding ends and the temperature rise conditions of conductors, and, if necessary the size of conductors, at least at the winding ends, should be reduced by dividing their cross-section into several parallel branches. The formulae which serve for determining supplementary losses are only valid as long as the approximating assumptions described hold good. In the case of supplementary losses occurring in short windings, or when calculating losses of conductors over-dimensioned for mechanical reasons, the uncertainties of the calculation method described above may be expected to increase.

Transformer design strives to develop constructions by which the supplementary losses can be reduced. One method of loss reduction is to reduce the conductor dimension perpendicular to the flux lines. In a transformer designed to meet predetermined specification data this aim can be attained by dividing the winding into several parallel branches. In a winding split up into a number of parallel conductors, it may be seen from Eqns (3.8) and (3.9), that the eddy-current losses will be lower, but equalizing current may flow between parallel branches, and the additional losses caused thereby may be considerable. The equalizing currents are due to the different stray flux linkages of the various parallel branches. The equalizing currents can be reduced by transposing the positions of parallel branches in the winding with respect to the leakage channel to have each parallel branch linked, as far as possible, with an equal number of flux lines.

Parallel branches may be transposed within a winding by applying suitable change-overs and crossing, or more recently the continuously transposed conductors have gained growing popularity.

The stray magnetic field of transformer windings gives rise also to losses in the metallic parts outside the windings. In this respect most trouble is caused by the structural parts made of ferromagnetic materials and located in the vicinity of magnetic fields: clamps, frame structure, core and, in larger transformers, the tank. In smaller transformers the magnitude of losses arising in the parts mentioned is generally less significant. An exception is represented by transformers manufactured without a delta winding. In such transformers, under asymmetrical load conditions, the unbalanced excitation causes a substantial additional loss in the metallic parts of the transformer. In transformers with capacities exceeding 40 MVA, the losses dealt with above are to be reckoned with, because their magnitude may reach as much as 10 to 40% of the losses arising in the windings. Known methods for calculating the losses occurring in stray fields are dealt with in Chapter 4.

Example 3.1. Data of L.V. copper winding of a 6300 KVA, three-phase transformer (Fig. 3.4): number of turns $N_1 = 708$, phase current $I_1 = 95.5$ A, current density $J = 3.02$ A mm^{-2}, cross-sectional area of conductor $I_1/J_1 = 31.623$ mm^2. A copper conductor of cross-section 11.6×2.7 mm^2 is chosen. Mean winding diameter is $D_1 = 437$ mm. Mass of the three L.V. phase windings:

$$m_1 N_1 D\pi A_1 d10^{-6} = 2 \times 708 \times 437 \times \pi \times 31.32 \times 8.9 \times 10^{-6} = 813 \text{ kg}.$$

Mass of conductors (of cross-sectional area equal to those of the windings) used for connecting the L.V. windings with each other and to the output terminals is 27 kg (by assessment). Thus, the overall mass of L.V. windings and connections is 840 kg. The d.c. winding loss reduced to 75 °C is

$$P_{dc} = 2.42 \ J^2 m = 2.42 \times 3.02^2 \times 840 = 18 \ 540 \text{ W}.$$

The arrangement of the windig is represented by the cross-sectional drawing of Fig. 3.4. The conductor size perpendicular to the leakage channel is 2.7 mm, the turns being arranged in twelve layers in that direction. With the Rogowski coefficient of

$$R = 1 - \frac{a_1 + a_2 + \delta}{\pi l_s} = 1 - \frac{40.7 + 42.9 + 26.7}{\pi 979} = 0.964$$

and with a reduced conductor width of

$$\xi_1 = h_1 \sqrt{\frac{n_1 b_1}{l_s} R \frac{\pi \mu_0 f}{\rho}} = 2.7 \times 10^{-3} \sqrt{\frac{60 \times 11.6}{979} 0.964 \frac{\pi \mu_0 \times 50}{0.217 \times 10^{-7}}} = 0.213$$

$$k_\square = \frac{m^2 - 0.2}{9} \xi_1^4 = \frac{12^2 - 0.2}{9} 0.213^4 = 0.0329,$$

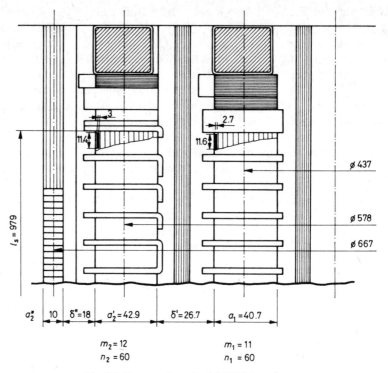

Fig. 3.4. Cross-section of a 6.3 MVA transformer

the a.c. loss of the L.V. winding will be

$$P_{ac} = P_{dc}(1 + k_{1\square}) = 18\,540(1 + 0.0329) = 19\,150 \text{ W}.$$

The number of turns in the H.V. winding is $N_2 = 650$, mean winding diameter is 578 mm, current density is $J_2 = 3.06$ A mm^{-2}, phase current of conductor is 104 A, and the cross-sectional area of the wire is $A_2 = \dfrac{I_2}{J^2} = 34$ mm^2 which can be closely approached by a copper cross-section of 3×11.4 mm^2.

The number of turns in the H.V. winding is higher, because the transformer of the example is a star/delta 35 000/22 000 V unit. Mass of the H.V. winding is

$$m_2 N_2 D_2 \pi A_2 d 10^{-6} = 3 \times 650 \times 578\pi 34 \times 8.9 \times 10^{-6} = 1071 \text{ kg}.$$

Mass of conductors of unchanged cross-sectional area required for the connections of the H.V. winding is assessed as 29 kg, thus the overall mass of conductors in the H.V. side is 1100 kg. The loss refers to the nominal transformation ratio. The transformer of the example is provided with a tap changer permitting a voltage regulation of $\pm 15\%$. The design of the tap changer is such as to permit the selection of 9 tap positions in either direction, in addition to the position pertaining to the nominal ratio. In the latter position only the L.V. and H.V. windings of the transformer are connected. Regulation is effected by the tap-changer by connecting the turns or some of the turns of the so-called regulating winding in series with the H.V. winding (Fig. 3.5).

The nominal d.c. loss of the H.V. winding is

$$P_{dc} = 2.42 \times J^2 m = 2.42 \times 3.06^2 \times 110 = 24\,926 \text{ W}.$$

90

The winding arrangement is shown in Fig. 3.4, and conductor size is 3 mm in the direction perpendicular to the leakage channel, with $m_2 = 11$ conductors being located in that direction. With the same value of Rogowski coefficient as before, the reduced width of conductor is

$$\zeta_2 = h_2 \sqrt{\frac{n_2 b_2}{l_s} R \frac{\pi \mu_0 f}{\rho}} = 3 \times 10^{-3} \sqrt{\frac{60 \times 11.4}{979} 0.964 \frac{\pi \mu_0 50}{0.0217 \times 10^{-6}}} = 0.235 \,.$$

The p.u. value of supplementary loss is,

$$k_{2\square} = \frac{m_2^2 - 0.2}{9} \zeta^4 = \frac{11^2 - 0.2}{9} 0.234^4 = 0.041 \,.$$

The a.c. loss of the H.V. winding, with no regulating turns connected, is

$$P_{ac2} = P_{dc2}(1 + k_{2\square}) = 24\,926(1 + 0.041) = 25\,948 \text{ W} \,.$$

The short-circuit loss pertaining to the nominal voltage ratio of the transformer is obtained by summing the losses of the two windings

$$\varepsilon_r = \frac{P}{R_n} = \frac{45\,098}{6300\,000} = 0.0072 \,.$$

The cross-sectional area of the regulating winding is represented in Fig. 3.4, with the related connection diagram shown in Fig. 3.5. This winding consists of two parallel branches. The beginning of the winding is at its middle, from where it is wound in the two opposed directions toward its ends.

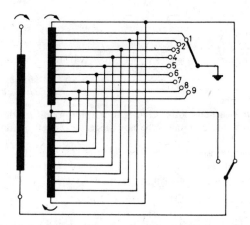

Fig. 3.5. Connection diagram of one phase of the transformer

Regulation is effected by the tap changer in 19 tap positions, by connecting 0 to 15% of the regulating turns in series with the H.V. winding in the same sense as the H.V. turns are wound or in the sense opposed to them. Due to the increased mechanical strength of the regulating winding, the current density in the latter is usually lower than that of the main winding. In our present case, the current density is $J = 1.86$ A mm^{-2} resulting in a conductor cross-sectional area of

$$\frac{I_2}{J} = \frac{104}{1.86} = 56 \text{ mm}^2 \,.$$

This cross-sectional area can be approximated by employing a copper area of 9×3.1 mm^2 in the branch wound in two directions. The required 19 tap positions (middle position ± 9 tap positions) are obtained by providing 10 taps on the regulating winding, because an additional position is necessary to permit the connection of the regulating winding in positive or negative sense. Thus, there are altogether 10 sections in the regulator winding, each section consisting of 11 turns. In the extreme positions of the tap-changer nine sections are connected with altogether $9 \times 11 = 99$ turns. Mean diameter of the regulating winding is 667 mm.

$$mN_r D\pi \, Ad \, 10^{-6} = 3 \times 110 \times 667 \times \pi \times 56 \times 8.9 \times 10^{-6} = 348 \text{ kg}.$$

With the full regulating range inserted, only 9/10 of the mass of the winding is loaded, hence the d.c. loss is

$$P_{dcr} = 2.42 \times 1.86^2 \times \frac{9}{10} \times 348 = 2622 \text{ W}.$$

As there are no oil ducts in the axial direction between the turns of this single-layer winding and so $\xi \approx h$, its supplementary losses can be calculated, with good approximation, substituting instead of the reduced conductor width the true width

$$k_{r\square} = \frac{m^2 - 0.2}{9} h_s^4 = \frac{1^2 - 0.2}{9} 0.9^4 = 0.058.$$

The a.c. losses with rated current flowing in the regulating winding are

$$P_{acr} = P_{dcr}(1 + k) = 2622(1 + 0.058) = 2774 \text{ W}.$$

The additional losses arising in the H.V. winding are influenced by the load of the regulating winding. In the two extreme tap-changer positions the p.u. values of supplementary losses in the H.V. winding are

$$k_2' = \frac{m^2 \left[1 + 3 \dfrac{N_3}{N_2} \left(\dfrac{N_3}{N_2} \pm 1 \right) \right] - 0.2}{9} \xi^4 =$$

$$= \frac{11^2 \left[1 + 3 \dfrac{99}{650} \left(\dfrac{99}{650} \pm 1 \right) \right] - 0.2}{9} 0.235^4 = 0.0625$$

and 0.025, respectively.

Compared to the supplementary losses of 4.4% in the H.V. winding with the regulating winding disconnected, in the two extreme positions of the tap-changer the respective additional losses will be 6.25 and 2.5%.

Finally, the supplementary losses arising in the conductors of H.V. and L.V. windings adjacent to the leakage channel are checked. Taking the respective d.c. loss taken as unity, the additional loss in the outermost conductors of the L.V. winding is

$$k_1 = \frac{m_1^2 - m_1 + 0.27}{3} \, \xi_1^4 = \frac{12^2 - 12 + 0.27}{3} 0.213^4 = 0.0907$$

and that in the innermost conductors of the H.V. winding is

$$k_2' = \frac{m_2^2 - m_2 - 0.27}{3} \, \xi_2^4 = \frac{11^2 - 11 + 0.27}{3} 0.235^4 = 0.112.$$

3.1.2 Inductive component of the impedance voltage

The inductive component of the impedance voltage is the product of stray reactance and rated current of the transformer. The stray reactance referred to the rated data and its percentage value may be calculated from the following ratios: the ratio of the inductive component of the short-circuit impedance to the rated impedance, ratio of the inductive component of the impedance voltage to the rated voltage and ratio of rated short-circuit reactive power to the rated apparent power output. Its value varies in the range of 0.02 to 0.15, the higher values occurring in transformers having ratings of several hundred MVA.

A simplified physical picture of stray reactance is given in Fig. 1.9. In case (a) the primary winding is on the outer side and the secondary winding is arranged inside. The primary winding is linked with flux $\Phi_y = \Phi_l + \Phi_s$, the secondary winding with flux Φ_l, and Φ_s is linked with the primary winding only. In case (b) the supplied (primary) winding is located inside and the secondary outside. In that case the inner winding is linked with flux Φ_l and the outer winding with $\Phi_y = \Phi_l - \Phi_s$. Either the outer or the inner winding is supplied, and the common role of flux denoted by Φ_s in Fig. 1.9. manifests itself, its value being the difference of the fluxes linked with the primary and secondary windings, i.e. $\Phi_s = \Phi_y - \Phi_l$ in case (a) and $\Phi_s = \Phi_l - \Phi_y$ in case (b).

Flux Φ_s is termed the stray flux. Real conditions differ from those illustrated in Fig. 1.9, as not all turns of the windings are linked with the same flux, so the above flux equations have to be given for the winding flux: thus, the stray flux can be calculated from $\psi_s = \psi_y - \psi_l$ for case (a) and from $\psi_s = \psi_l - \psi_y$ for case (b).

Now, consider again the phenomenon mentioned already in Chapter I and manifesting itself in the increase of flux in the limb in short-circuited condition. With the inner winding supplied, the phenomenon of growth of flux in the limb, after the outer winding is short-circuited, may be considered in terms of inserting a reactance of negative sign into the equivalent circuit. The concept of negative reactance can be

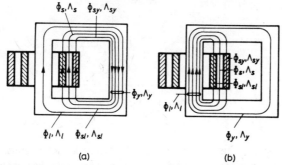

Fig. 3.6. Model considering fluxes partially linked with the windings

found in literature dealing with equivalent circuits of the transformer [12]. Boyajian and Timascheff, in interpreting the equivalent circuit of the three-circuit transformer have pointed out that one of the reactances may be of negative sign, because of the partial flux linkage of the windings. It seems, therefore, appropriate to consider separately the portion of stray flux partially linked with the turns of the windings when deriving the equivalent circuit for the two-circuit transformer. Figure 3.6 illustrates flux linkages when the outer or inner windings are supplied. The partially linked portion of stray flux is taken into account by assuming that the excitations of the outer and inner windings, i.e. Θ_o and Θ_i, bring about a flux also through the respective flux channel characterized by a magnetic conductance reduced according to the partial flux linkage. By the excitation of the inner winding flux Φ_{sl} is produced, the reduced permeance (magnetic conductance) of this flux channel being Λ_{sl}, and by the turns of the outer winding flux Φ_{sy} is excited, the permeance of this flux channel being Λ_{sy}. The magnetic circuit is shown in Fig. 3.7 where the equivalent electric circuit is also represented by dotted lines. The diagram of the constructed dual formation is given in Fig. 3.8. A method for calculating the values of X_{sl}, X_s and X_{sy} is described in the section dealing with three-circuit transformers.

The sign of X_{sl} appearing in the equivalent circuit of Fig. 3.8 is negative, indicating that the increase of flux is caused by this term. The negative sign can be explained on

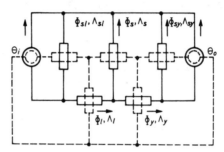

Fig. 3.7. Construction of composite π circuit obtained from the magnetic connection by dual-type representation

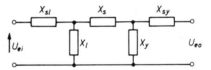

Fig. 3.8. Composite π equivalent circuit

the basis of Fig. 3.6(b). This shows the flux distribution developing when the inner winding is supplied or short-circuited. The turns of the inner winding are also linked with the flux lines opposed to the flux lines in the limb, because the transformer will only be capable of maintaining equilibrium with the voltage imposed on the inner winding, if the flux in the limb increases with respect to the no-load state. Justification of the negative sign is dealt with in the next section, where the three-winding transformer is discussed.

Under rated conditions, depending on the size of the transformer, the stray flux is about 2 to 15% of the flux linked with the primary winding. Depending on which winding is the primary or secondary in these transformers the stray flux leaves at approximately the same part of the core, but its path closes in different directions. As shown by experience, this phenomenon causes no substantial changes in the magnitude of supplementary short-circuit loss, but may give rise to differences in the distribution of local hot spots.

The excitation requirement for the stray flux is the sum of magnetomotive forces of the two sections, one section being the part of the path of stray flux located between the two windings (outside the core) and within the windings themselves along a line of force, and the other section being within the core. The magnetomotive force of the section of stray flux closing through the core is lower than that closing outside the core by about three orders of magnitude. The approximation shown in Fig. 3.1(a) is therefore considered reliable showing the stray flux excited by the two windings and passing along a section of length l_s and permeability μ_0. The coefficient of stray self-induction is

$$L_s = \mu_0 \frac{N^2}{l_s} \left(l\delta + \frac{a_1 l_1 + a_2 l_2}{3} \right), \tag{3.11}$$

where μ_0 is the permeability of vacuum, N is the number of turns of the primary or secondary winding (depending on whether the coefficient of stray self-induction reduced to the primary or to the secondary is being calculated), l_s is the length of the leakage channel and $l = D\pi$ is its mean circumference, in m, $l_1 = D_1 \pi$ and $l_2 = D_2 \pi$ are the respective mean turn lengths of the two windings, in m and a_1 and a_2 are the sizes of the two windings in the direction perpendicular to the stray flux lines, in m (Fig. 3.1). In the above relation the flux passing through surfaces $a_1 l_1$ and $a_2 l_2$ are only partially linked with the respective turns N_1 and N_2; since the flux density changes linearly in the radial direction of the winding, therefore in the expression of the coefficient of self-induction (due to integration) only one-third of the surfaces appear. The expression appearing in brackets in the formula may be written in the following approximate form:

$$l \left(\delta + \frac{a_1 + a_2}{3} \right) = l\delta',$$

where δ' is the width of the reduced leakage duct. The formula is equally valid for transformers having concentric or disc-type winding arrangements. The notation for the pancake arrangement of windings of a shell-type transformer are explained in Fig. 3.9.

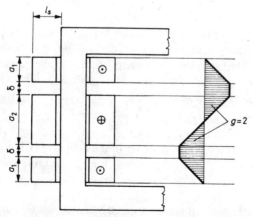

Fig. 3.9. Simplified flux distribution of pancake (disc) arrangement

95

The stray reactance may be calculated from the formula $X_s = 2\pi f L_s$. It is more common to calculate with the p.u. value of the inductive component of the impedance voltage referred to the rated voltage. After multiplying relation (3.11) by $2\pi f$ and by the rated current and dividing it by the rated voltage of the transformer, and performing simple rearrangements, the formula thus obtained, gives the p.u. value of stray voltage (referred to rated data) as

$$\varepsilon_s = 2\pi f \mu_0 \frac{P_{nl}}{U_w^2 l_s g}\left(l\delta + \frac{a_1 l_1 + a_2 l_2}{3}\right), \tag{3.12}$$

where f is the frequency in Hz, P_{n1} is the nominal capacity of one limb in VA, U_w is the turn voltage in V, and g is the number of coil groups of balanced ampere-turns on each limb of the transformer. The concept coil groups of balanced ampere-turns

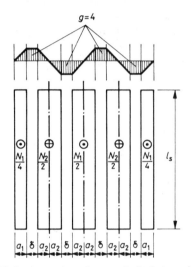

Fig. 3.10. Interpretation of magnetically balanced groups

is explained in Fig. 3.10 (e.g. in Fig. 3.10 the number of groups of balanced ampere-turns is $g=4$, whereas in Fig. 3.1 the respective number is $g=1$). It is to be noted in formula (3.12) that the p.u. value of reactance voltage is directly proportional to rated capacity, and that it is inversely proportional to the square of turn voltage.

The reactance voltage of auto-transformers is calculated from formula (3.12) and the result is multiplied by factor

$$1 - \frac{U_L}{U_H},$$

where U_L and U_H are rated phase voltages of the L.V. and H.V. sides. By expression (3.12) the reduction of number of turns (excitation) in an auto-transformer compared to that of a two-winding transformer of equal capacity is taken into account.

As opposed to the simplified basic model of calculating formulae (3.11) and (3.12), the windings are not embedded in a medium of infinite permeability and the leakage duct is not excessively long relative to the diameter of windings. In fact, since the two conditions mentioned are not satisfied, there are also components of stray flux perpendicular to the leakage channel. Due to the reluctance of flux path outside the windings, fewer flux lines straying in the radial direction are linked with the turns of windings than supposed by the approximating assumption and, for this reason, the measured values of impedance voltage are generally lower than those obtained by calculation. This effect is taken into account by the Rogowski coefficient having the value

$$R = 1 - \frac{a_1 + a_2 + \delta}{\pi l_s} \left(1 - e^{-\frac{ls}{a_1 + a_2 + \delta}} \right). \tag{3.13}$$

The notation appearing in the above formula is given in Fig. 3.1(a), Formula (3.13) is applicable, if

$$\delta < a_1 + a_2 \quad \text{and} \quad \frac{a_1 + a_2 + \delta}{\pi l_s} < 2 .$$

The exponential term of the expression written in brackets may be neglected, if

$$\frac{a_1 + a_2 + \delta}{\pi l_s} < 0.3 .$$

The numerical value of the Rogowski coefficient is invariably lower than unity, and it serves to multiply the values of L_s, X_s and ε_s. In the case of shell-type transformers, the value of coefficient R can be calculated by using the notation of Fig. 3.9.

In larger transformers the layer-windings are subdivided by oil ducts for better cooling. From the point of view of stray reactance, this arrangement is identical with that where a winding, e.g. the H.V. winding, is composed of several concentric elements. Such an arrangement is shown in Fig. 3.11, for which the p.u. value of

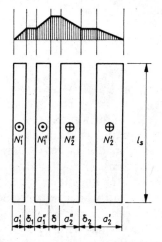

Fig. 3.11. Axial stray field of a transformer of construction subdivided in the axial direction by cooling ducts

impedance voltage is given by

$$\varepsilon_s = 2\pi f \mu_0 \frac{P_{n1} l}{U^2 l_s}\left[\frac{a_1' + a_1'' + a_2' + a_2''}{3} + \delta_1\left(\frac{N_1'}{N_1}\right)^2 + \right.$$

$$\left. + \delta_2\left(\frac{N_2'}{N_2}\right)^2 + \delta\right].$$ (3.14a)

The notation is partly identical with that appearing in formula (3.12) and partly as used in Fig. 3.11, N_1', N_1'', N_2' and N_2'' being the numbers of turns in coil sections with $N_1 = N_1' + N_1''$ and $N_2 = N_2' + N_2''$, and l being the mean length of the duct of width δ. When using the formula, there is no need to calculate with reduced quantities, as only the ratios between turn members are required. The impedance voltage of a winding system subdivided by even more ducts can be calculated by applying the formula (3.14(a)) accordingly.

In order to reduce the short-circuit forces or the reactance voltage, it is a common practice to divide both the primary and secondary windings into several equal parts and to arrange them alternately. Such an arrangement is illustrated in Fig. 3.10. In the upper part of the diagram the distribution of stray flux lines is shown. Each part of the winding system along which the stray induction lines increase from zero to a certain value and decrease again to zero, constitutes a group of balanced ampere-turns. In Fig. 3.10 the number of such groups is $g = 4$. In this case the reactance voltage is computed from formula (3.12) using the notation of Fig. 3.10 and substituting symbol g appearing in the denominator of the formula by the number of groups of balanced ampere-turns. In Fig. 3.10 it can be seen that along the boundary between each two adjacent groups the value of magnetic induction lines is zero. Thus, when locating the oil duct along the boundary line between two magnetically balanced groups, the size of the duct will have no influence on the magnitude of reactance voltage.

The case of the regulating transformer is illustrated in Fig. 3.2(d).

The value of its reactance voltage is [77]

$$\varepsilon_s = 2\pi f \mu_0 \frac{P_{nl}}{U_w^2 l_s}\left\{\frac{a_1}{3}l_1 + \delta l_\delta + \right.$$

$$\left. + \frac{a_2'}{3}\left[1 \pm \frac{N_2''}{N_2} + \left(\frac{N_2''}{N_2}\right)^2\right]l_2 + \delta' l_\delta\left(\frac{N_2''}{N_2}\right)^2 + \frac{a_2''}{3}l_2''\left(\frac{N_2''}{N_2}\right)^2\right\},$$ (3.14b)

where l_1 is the mean turn length of the winding of N_1 turns, l_δ is the mean length of the duct of width δ, l_2'' is the mean turn length of the winding of N_2'' turns, by letter N with the indices shown in the figure the respective number of turns is indicated and $N_2 = N_2' + N_2''$. Information concerning further notation is given in Fig. 3.2(d). In the formula the symbol $\pm$ refers to the directions of excitation.

Partly for design reasons and partly because of excluding one part of the turns of the tapped winding, asymmetries arise in the windings. The effects of asymmetries are calculated by superimposing symmetrical winding arrangements to obtain the required asymmetrical structure. The asymmetrical arrangement of Fig. 3.12(a) may be arrived at by superimposing the windings of Fig. 3.12(b). Winding 1 of (b) is

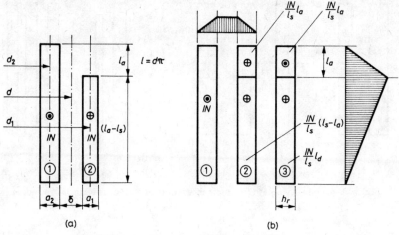

Fig. 3.12. Asymmetry at one winding edge

identical with winding 1 of (a). By superimposing the excitations of windings 2 and 3 of (b) the excitation of winding 2 of (a) is obtained. According to this interpretation, the p.u. reactance voltage of arrangement (b) consists of two parts. One part is the reactance voltage between windings 1 and 2:

$$\varepsilon_s' = 2\pi f \mu_0 \frac{P_{n1} l}{U_w^2 l_s} \left(\delta + \frac{a_1 + a_2}{3} \right)$$

and the other part is the reactance voltage of winding 3 representing the effect of asymmetry:

$$\varepsilon_s'' = 2\pi f \mu_0 \frac{P_{n1} l_1}{U_w^2 3 a_1} \frac{l_s}{3} \left(\frac{l_a}{l_s} \right)^2.$$

The resultant reactance voltage is the sum of the two reactance voltages: $\varepsilon_s = \varepsilon_s' + \varepsilon_s''$. The formula for ε_s'' is, in effect, the known relation of reactance voltage applied to the pair of windings denoted by 3 in the diagram: the mean winding length is in that case $l_1 = d_1 \pi$ and the estimated length of the leakage channel, considering the very short length of the winding in the direction of the stray line of force, is $3a_1$; the width of the leakage duct is 0; the term $(a_1 + a_2)\frac{1}{3}$ is replaced by $\frac{1}{3}(l_s - l_a + l_a) = l_s/3$ and finally, N is replaced by the term Nl_a/l_s. The reactance voltage reduces proportionally with the square of the ratio of turns, therefore the square of the ratio l_a/l_s appears in the formula. The sign of ε_s is always positive, i.e. asymmetries always bring about an increased reactance voltage. It is also apparent that the length of the leakage flux lines is estimated very generously. The formula will not always give a reliable result, but provides a basis for assessing the effect of asymmetry. The effect of asymmetry may be interpreted in another way, by stating that it causes the reduction of dimension l_s. Since this dimension appears in the denominator of the formula for reactance voltage any asymmetry will always lead to the increase of the

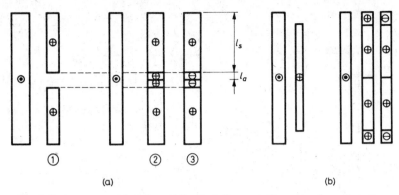

Fig. 3.13. Asymmetry (a) at the middle of winding and (b) at two edges of winding

reactance voltage. Correspondingly, the reactance voltage could be calculated from formula (3.11), also for asymmetrical arrangements, but in such cases, in place of the length of the leakage flux line the term $(l_s - l_a/2)$ has to be substituted.

Figure 3.13 gives an example showing the application of the considerations relating to the effects of asymmetrical winding marked 1 can be obtained by the superposition of windings 2 and 3. Since size l_a of windings 2 and 3 is only half of the size resulting from the superposition when the asymmetry is located at one end of the winding, it can be concluded that the effect of an asymmetry in the interior of the winding (since the square of l_a/l_s appears in the formula) is not as great as that of an asymmetry of identical size at the end section of the winding. In Fig. 3.13(b) a similarly favourable arrangement is presented with the asymmetry uniformly distributed along the two winding ends, instead of being concentrated at its middle.

3.1.3 Impedance voltage of three-circuit transformers

Three-circuit (or three-winding) transformers are widely used, a considerable proportion of power transformers of medium and high power ratings being three-circuit units. Three networks of different voltage levels are coupled through such transformers, mostly in such a way that one winding (e.g. that of 120 kV) acts as primary, while the other two (e.g. of 35 and 20 kV) are in normal service of the secondary windings. Usually, the power ratings of the three windings are different.

A typical application of a three-circuit transformer arises when a high-power transformer is connected through its third winding to a synchronous condenser or a reactor. In effect, two transformers are replaced by a three-circuit transformer, but the investment cost is only about 65 to 75% of that of the two units. The reactance voltage of three-circuit transformers, and also their p.u. copper loss can be calculated from the Boyajian star equivalent circuit shown in Fig. 3.14. In the equivalent circuit diagram one phase of the windings of the three different voltage levels is represented. The beginning and the end of winding 1 are coupled to terminals A_1 and X_1, those of winding 2 to a_2 and x_2 and, finally, those of winding 3

100

to a_3 and x_3. This equivalent circuit has come into use in the literature in a form where the parts pertaining to terminals marked X_1, x_2 and x_3 in the diagram and the branch representing the magnetization of the core are simply omitted. The network elements appearing in the equivalent circuit are calculated by the methods of network theory. The reactance voltage across any two related pairs of terminals, with the third pair of terminals open, should be equal to the reactance voltage

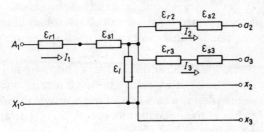

Fig. 3.14. Star equivalent circuit of a three-winding transformer

measured across the corresponding pair of terminals of the transformer. If the calculated or measured voltage components across the related pairs of terminals of the transformer are

$$\varepsilon_{r1-2}, \ \varepsilon_{r1-3}, \ \varepsilon_{r2-3}, \ \varepsilon_{s1-2}, \ \varepsilon_{s1-3}, \ \varepsilon_{s2-3}$$

and the voltage components of the individual branches of the star equivalent circuit are

$$\varepsilon_{r1}, \ \varepsilon_{r2}, \ \varepsilon_{r3}, \ \varepsilon_{s1}, \ \varepsilon_{s2}, \ \varepsilon_{s3}$$

the following relations between the above quantities are valid

$$\left. \begin{aligned} \varepsilon_{r1} &= \frac{1}{2}(\varepsilon_{r1-2}+\varepsilon_{r1-3}-\varepsilon_{r2-3}), \\[2mm] \varepsilon_{r2} &= \frac{1}{2}(\varepsilon_{r1-2}+\varepsilon_{r2-3}-\varepsilon_{r1-3}), \\[2mm] \varepsilon_{r3} &= \frac{1}{2}(\varepsilon_{r1-3}+\varepsilon_{r2-3}-\varepsilon_{r1-2}), \end{aligned} \right\} \tag{3.15}$$

$$\left. \begin{aligned} \varepsilon_{s1} &= \frac{1}{2}(\varepsilon_{s1-2}+\varepsilon_{s1-3}-\varepsilon_{s2-3}), \\[2mm] \varepsilon_{s2} &= \frac{1}{2}(\varepsilon_{s1-2}+\varepsilon_{s2-3}-\varepsilon_{s1-3}), \\[2mm] \varepsilon_{s3} &= \frac{1}{2}(\varepsilon_{s1-3}+\varepsilon_{s2-3}-\varepsilon_{s1-2}). \end{aligned} \right\} \tag{3.16}$$

When applying the above formulae, it should be remembered that, generally, the power ratings of the windings of a three-winding transformer are different. Since the variation of short-circuit loss and impedance voltage referred to the rated data is directly proportional to the rated power, the active and reactive components of impedance voltages have to be reduced to an arbitrarily chosen base in cases when the ratings of the windings are different. For example, if for a three-winding transformer with windings of ratings S_1, S_2 and S_3, S_1 is chosen as the base, the impedance voltage components of winding 2 can be reduced to the base by multiplying its rating by the factor $K_2 = S_2/S_1$ and those of winding 3 by the factor $K_3 = S_3/S_1$. The reactance voltages and, for ratings below about 20 MVA, also the p.u. short-circuit losses obtained from the calculated values of impedance voltage components of three-winding transformers show good agreement with the results of measurements.

In large transformers, where the supplementary short-circuit losses are greater, the results obtained for the latter by calculation and measurement do not always agree. This is because the additional losses occurring outside the windings cannot be pre-assessed accurately and their magnitudes may differ in various load conditions due to the different patterns of flux distribution.

Example 3.2. In the transformer, already dealt with, having a power rating of 6300 kVA (power per limb 2100 kVA), the turn voltage is $U_w = 31$ V, the number of magnetically balanced groups is $g = 1$ and the other data are as given in Fig. 3.4.

The p.u. reactance voltage of the transformer for the nominal voltage ratio is

$$\varepsilon_s = 2\pi f \mu_0 \frac{P_{nl}}{U_w^2 l_s g} \left(l_s'\delta + \frac{a_1 l_1 + a_2 l_2}{3} \right) R =$$

$$= 2\pi \times 50 \times 1.256 \times 10^{-6} \frac{2.1 \times 10^6}{31^2 \times 0.979} \left(0.504\pi \times 2.67 \times 10^{-2} + \right.$$

$$\left. + \frac{4.07 \times 10^{-2} \times 0.437\pi + 4.29 \times 10^{-2} \times 0.578\pi}{3} \right) 0.964 =$$

$$= 7.34 \times 10^{-2}.$$

For calculation of the Rogowski coefficient see page 89. The p.u. values of the reactance voltage in extreme positions of the regulating winding, when $N_s = 99$, $N_2 = 650$, $a_2 = 4.29 \times 10^{-2}$ m, $a_1' = 4.07 \times 10^{-2}$ m, $a_2'' = 1 \times 10^{-2}$ m, $\delta' = 2.67 \times 10^{-2}$ m and $\delta'' = 1.8 \times 10^{-2}$ m are

$$\varepsilon_s' = 2\pi f \mu_0 \frac{P_{nl}}{U_w^2 l_s} \left\{ \frac{a_1}{3} l_1 + \delta' l_s' + \frac{a_2'}{3} \times \left[1 \pm \frac{N_s}{N_2} + \left(\frac{N_s}{N_2} \right)^2 \right] l_2' + \right.$$

$$\left. + \delta' l'\delta \left(\frac{N_s}{N_2} \right)^2 + \frac{a_2''}{3} l_2'' \left(\frac{N_s}{N_2} \right)^2 \right\} R = 2\pi \times 50 \times 1.256 \times$$

$$\times 10^{-6} \frac{2.1 \times 10^6}{31^2 \times 0.979} \left\{ \frac{4.07 \times 10^{-2}}{3} 0.437\pi + 2.67 \times 10^{-2} \times 0.504\pi + \right.$$

$$+ \frac{4.29 \times 10^{-2}}{3} \times \left[1 \pm \frac{99}{650} + \left(\frac{99}{650} \right)^2 \right] 0.578\pi + 2.67 \times$$

$$\left. \times 10^{-2} \times 0.504 \left(\frac{99}{650} \right)^2 + \frac{10^{-2}}{3} 0.667\pi \left(\frac{99}{650} \right)^2 \right\} 0.964 = 7.18 \times 10^{-2}$$

and 7.85×10^{-2}, respectively.

The p.u. value of impedance voltage at the nominal ratio is

$$\varepsilon_z = \sqrt{\varepsilon_r^2 + \varepsilon_s^2} = \sqrt{0.0072^2 + 0.0734^2} = 8.52 \times 10^{-2}.$$

102

In extreme positions the values of impedance voltage, with the data of Example 3.1, are

$$\varepsilon'_z = \sqrt{\varepsilon_r'^2 + \varepsilon_s^2} = \sqrt{0.0072^2 + 0.0714^2} = 7.22 \times 10^{-2}$$

and

$$\sqrt{0.0075^2 + 0.0785^2} = 7.9 \times 10^{-2},$$

respectively.

Example 3.3. The data of a three-winding transformer are rated power of H.V. winding 50 MVA, its rated voltage 120 kV, rated power of intermediate-voltage (I.V.) winding 30 MVA, its rated voltage 36.75 kV, rated power of L.V. winding 20 MVA, its rated voltage 11 kV. Short-circuit data reduced to 75 °C temperature between H.V. and I.V. windings, referred to 30 MVA: $\varepsilon_{r1-2} = 0.0059$, $\varepsilon_{s1-2} = 0.12$; between H.V. and L.V. windings referred to 20 MVA: $\varepsilon_{r1-3} = 0.0045$, $\varepsilon_{s1-3} = 0.0118$; between I.V. and L.V. windings referred to 20 MVA: $\varepsilon_{r2-3} = 0.004$ and $\varepsilon_{s2-3} = 0.031$. Selecting 20 MVA as the base: $\varepsilon_{r1-2} = \dfrac{20}{30} 0.0059 = 0.003\,95$ and $\varepsilon_{s1-2} = \dfrac{20}{30} 0.118 = 0.0787$. The quantities with subscripts 1–3 and 2–3 are left unchanged, since these are already referred to 20 MVA. The p.u. values of active and reactive components of impedance voltages in the branches of the star-connected equivalent circuit are

$$\varepsilon_{r1} = \frac{1}{2}(\varepsilon_{r1-2} + \varepsilon_{r1-3} - \varepsilon_{r2-3}) =$$

$$= \frac{1}{2} 0.003\,95 + 0.0045 - 0.004 = 0.002\,22,$$

$$\varepsilon_{r2} = \frac{1}{2}(\varepsilon_{r2-3} + \varepsilon_{r1-2} - \varepsilon_{r1-3}) =$$

$$= \frac{1}{2}(0.004 + 0.003\,95 - 0.0045) = 0.0071\,72,$$

$$\varepsilon_{r3} = \frac{1}{2}(\varepsilon_{r2-3} + \varepsilon_{r1-3} - \varepsilon_{r1-2}) =$$

$$= \frac{1}{2}(0.004 + 0.0045 - 0.003\,95) = 0.002\,27.$$

The p.u. values of reactance voltages in the branches of the star-connected equivalent circuit are

$$\varepsilon_{s1} = \frac{1}{2}(\varepsilon_{s1-2} + \varepsilon_{s1-3} - \varepsilon_{s2-3}) =$$

$$= \frac{1}{2}(0.007\,87 + 0.118 - 0.031) = 0.0828,$$

$$\varepsilon_{s2} = \frac{1}{2}(\varepsilon_{s1-2} + \varepsilon_{s2-3} - \varepsilon_{s1-3}) =$$

$$= \frac{1}{2}(0.0787 + 0.031 - 0.118) = -0.0041,$$

$$\varepsilon_{s3} = \frac{1}{2}(\varepsilon_{s1-3} + \varepsilon_{s2-3} - \varepsilon_{s1-2}) =$$

$$= \frac{1}{2}(0.118 + 0.031 - 0.0787) = 0.0351.$$

The terms with negative signs appearing among the reactive voltage components indicate that in the winding located between the other two windings, the voltage induced by the latter two is of a polarity opposed to the direction of the main flux linked with the former winding.

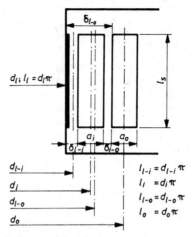

Fig. 3.15. Model of three-winding transformer for calculating the reactances of a two-winding transformer

Example 3.4. In the equivalent circuit of the two-winding transformer of Fig. 3.8 the value of reactance X_{sl} can be determined by means of the relations (3.16) given for the three-winding transformer. The dimensions of Fig. 3.15 marked with subscripts i and o refer to the two-winding transformer, whereas by that with subscript l refers to the winding located on the limb surface and having no radial dimension, in which a voltage proportional to the flux flowing in the limb is induced, when unloaded. The reactances between pairs of the three windings, with N number of turns in each of the windings are

$$X_{l-i} = 2\pi f \mu_0 N^2 \frac{1}{l_s} \left(l_{l-i} \delta_{l-i} + l_{l-i} \frac{a_i}{3} \right),$$

$$X_{i-o} = 2\pi f \mu_0 N^2 \frac{1}{l_s} \left(l_{i-o} \delta_{i-o} + l_{i-o} \frac{a_i + a_o}{3} \right),$$

$$X_{l-o} = 2\pi f \mu_0 N^2 \frac{1}{l_s} \left[l_{l-o} (\delta_{l-i} + \delta_{i-o} + a_i) + l_{l-o} \frac{a_o}{3} \right].$$

The reactance of the inner windings is

$$X_i = \frac{1}{2} \left(X_{l-i} + X_{i-o} - X_{l-o} \right) = \pi f \mu_0 N^2 \frac{1}{l_s} \left[l_{l-i} \left(\delta_{l-i} + \frac{a_i}{3} \right) \right] +$$

$$+ l_{i-o} \left(\delta_{i-o} + \frac{a_i + a_o}{3} \right) - l_{l-o} \left(\delta_{l-i} + \delta_{i-o} + \frac{3a_i + a_o}{3} \right) \right].$$

With the dimensions occurring in practice X_i is negative, which explains the phenomenon that the limb flux increases with increasing load or in the short-circuited state, when the inner winding is supplied. Reactance $X_s + X_{sy}$ is equal to reactance X_0 of the outer winding

$$X_o \frac{1}{2} \left(X_{l-o} + X_{i-o} - X_{l-i} \right) = \pi f \mu_0 N^2 \frac{1}{l_s} \times$$

$$\times \left[l_{l-o} \left(\delta_{l-i} + \delta_{i-o} + \frac{3a_i + a_o}{3} \right) + \right.$$

$$\left. + l_{i-o} \left(\delta_{i-o} + \frac{a_i + a_o}{3} \right) - l_{l-i} \left(\delta_{l-i} + \frac{a_i}{3} \right) \right].$$

104

The sum of inner and outer reactances is

$$X_i + X_o = 2\pi f \mu_o N^2 \frac{l_{i-o}}{l_s} \left(\delta_{i-o} + \frac{a_i + a_o}{3} \right)$$

which is equal to the resultant impedance of the two-winding transformer. The equivalent circuit of Fig. 3.8, with the notation of the above example (neglecting X_y as having lower admittance by several orders

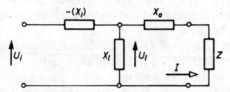

Fig. 3.16. Equivalent circuit of a two-winding transformer composed of windings, marked with subscripts i and o, of the transformer of cross-sectional area shown in Fig. 3.15

of magnitude), is shown in Fig. 3.16. The limb flux is proportional to the voltage $U_l = c\Phi_l$ induced in the winding of subscript l, fitting tightly on the limb. The variation of flux, when the load impedance is Z, is

$$\bar{U}_l = c\Phi_l = U_i \frac{\bar{Z} + \bar{X}_o}{\bar{Z} + \bar{X}_o - \bar{X}_i}.$$

From the above formula it is apparent that in the short-circuited condition $(Z=0)$, the limb flux is higher than in the no-load condition $(Z = \infty)$.

3.2 The voltage drop of transformers

The voltage across the output (secondary) terminals varies according to the magnitude and phase angle of the load. This voltage variation is closely related to the components of the impedance voltage of the transformer.

The p.u. value of voltage variation of a two-winding transformer is

$$\varepsilon_{\varphi 2} = \frac{I_2}{I_n} \Big[\varepsilon_r \cos \varphi_2 \pm \varepsilon_s \sin \varphi_2 +$$

$$+ \frac{1}{2} (\varepsilon_s \cos \varphi_2 \mp \varepsilon_r \sin \varphi_2)^2 \Big], \tag{3.17}$$

where $k = I_2/I_n$ is the ratio of load current to rated current, ε_r and ε_s are the p.u. values of loss voltage and reactance voltage, respectively, and φ_2 is the phase angle between secondary current and voltage. The second, quadratic, term in parentheses can be neglected in most cases. The negative sign refers to the case when the load impedance is of capacitive character, i.e. $\varepsilon_r \cos \varphi_2$ is smaller than $\varepsilon_s \sin \varphi_2$, and a voltage rise will be caused by the load.

The voltage variation of a three-winding transformer is more complicated, because the voltage change induced by the field of the third winding has also to be considered. Generally, a three-winding transformer is operated so that the power fed into one winding is taken up by the loads connected to the second and third windings. The voltage change can be calculated on the basis of the model of Fig. 3.14. It has to be taken into account that each of the three windings may have a different rating. The course of calculation is as follows:

(a) The impedance voltage components given for the various pairs of windings are referred to an arbitrarily chosen base. It is convenient to set out, right at the start, from the impedance voltage values pertaining to the load condition to be examined and to refer these to the base.

(b) From equations (3.15) and (3.16), the impedance voltage components are determined for each branch of the star equivalent circuit.

(c) From the loads of the two windings loaded, the input power of the primary winding is determined. If S_2 and S_3 are the two output powers and S_b is the selected base, then $k_2 = S_2/S_b$ and $k_3 = S_3/S_b$.

The power factor components referred to the power of the primary winding as base are

$$k_1' \cos \varphi_1 = k_2' \cos \varphi_2 + k_3' \cos \varphi_3 ,$$

$$k_1' \sin \varphi_1 = \pm k_2' \sin \varphi_2 \pm k_3' \sin \varphi_3 .$$

From these two equations the magnitude of k_1' can be calculated, this being the ratio of primary input to the base:

$$k_1' = \sqrt{(k_1 \cos \varphi_1)^2 + (k_1 \sin \varphi_1)^2}$$

and

$$\cos \varphi_1 = \frac{k_1' \cos \varphi_1}{k_1'} \quad \text{and} \quad \sin \varphi_1 = \frac{k_1' \sin \varphi_1}{k_1'} .$$

In the preceding formula $\varphi_1, \varphi_2, \varphi_3$ are the phase angles between the currents flowing in the three windings and the respective voltages.

(d) The p.u. values of impedance voltages in branches of the star equivalent circuit are

$$\varepsilon_1 = k_1' \left[(\varepsilon_{r1} \cos \varphi_1 \pm \varepsilon_{sl} \sin \varphi_1) + \right.$$
$$\left. + \frac{1}{2} (\varepsilon_{sl} \cos \varphi_1 \mp \varepsilon_{rl} \sin \varphi_1)^2 \right],$$

$$\varepsilon_2 = k_2' \left[(\varepsilon_{r2} \cos \varphi_2 \pm \varepsilon_{s2} \sin \varphi_2) + \right.$$
$$\left. + \frac{1}{2} (\varepsilon_{s2} \cos \varphi_2 \mp \varepsilon_{r2} \sin \varphi_2)^2 \right],$$

$$\varepsilon_3 = k_3' \left[(\varepsilon_{r3} \cos \varphi_3 \pm \varepsilon_{s3} \sin \varphi_3) + \right.$$
$$\left. + \frac{1}{2} (\varepsilon_{s3} \cos \varphi_3 \mp \varepsilon_{r3} \sin \varphi_3)^2 \right].$$

(e) Considering the selected positive directions shown in Fig. 3.14, the voltage drop between the first and second winding is

$$\varepsilon_{1-2} = \varepsilon_1 + \varepsilon_2,$$

that between the first and third windings is

$$\varepsilon_{1-3} = \varepsilon_1 + \varepsilon_3,$$

and that between the second and third winding is

$$\varepsilon_{2-3} = \varepsilon_3 - \varepsilon_2,$$

or

$$\varepsilon_{3-2} = \varepsilon_2 - \varepsilon_3.$$

In connection with the above calculation the following remarks apply: the quadratic terms appearing in the formulae under (d) may generally be neglected; in the formulae under (e) algebraic summations have to be performed, as all voltage drops given by the formulae under (d) are referred to the same voltage of identical phase.

Example 3.5. In Example 3.3 the impedance voltage was reduced to the 20 MVA base.

Suppose the voltage variations occurring between the different pairs of windings are to be determined, when the load imposed on the intermediate winding is $P_2 = 30$ MVA with power factor $\cos \varphi_2 = 0.8$ lagging, and the load of the low-voltage winding is $P_3 = 20$ MVA with power factor $\cos \varphi_3 = 0.6$ lagging. The steps of the calculation under (a) and (b) have already been dealt with on pages 103. For the intermediate winding $k_2 = P_2/P_a = 30/20 = 1.5$, and for the low-voltage winding $k_3 = P_3/P_a = 20/20 = 1$. The power factors of the primary winding referred to the base are

$$k_1 \cos \varphi_1 = k_2 \cos \varphi_2 + k_3 \cos \varphi_3 = 1.5 \times 0.8 + 1 \times 0.6 = 1.8,$$

$$k_1 \sin \varphi_1 = k_2 \sin \varphi_2 + k_3 \sin \varphi_3 = 1.5 \times 0.6 + 1 \times 0.8 = 1.7,$$

$$k_1 = \sqrt{k_1^2 \cos^2 \varphi_1 + k_1^2 \sin^2 \varphi_1} = \sqrt{1.8^2 + 1.7^2} = 2.48,$$

$$\cos \varphi_1 = k_1 \cos \varphi_1 / k_1 = 1.8/2.48 = 0.725,$$

$$\sin \varphi_1 = k_1 \sin \varphi_1 / k_1 = 1.7/2.48 = 0.685.$$

In branch 1 of the equivalent circuit the p.u. value of the impedance voltage is

$$\varepsilon_1 = k_1 \left[(\varepsilon_{r1} \cos \varphi_1 + \varepsilon_{s1} \sin \varphi_1) + \frac{1}{2} (\varepsilon_{s1} \cos \varphi_1 + \varepsilon_{r1} \sin \varphi_1)^2 \right] =$$

$$= 2.48 \left[(0.002\,22 \times 0.725 + 0.0828 \times 0.685) + \right.$$

$$\left. + \frac{1}{2} (0.0828 \times 0.725 + 0.002\,22 \times 0.685)^2 \right] = 0.145.$$

It is demonstrated by the above calculation that neglect of the quadratic expression in the formula for ε_1, does not cause a large error (in the present case this error is less than 1%). The situation is similar for ε_2 and ε_3 and therefore it is sufficient to calculate with expressions containing first order terms only:

$$\varepsilon_2 = k_2 (\varepsilon_{r2} \cos \varphi_2 - \varepsilon_{s2} \sin \varphi_2) =$$

$$= 1.5 (0.001\,72 \times 0.8 - 0.0041 \times 0.6) = -0.001\,62,$$

$$\varepsilon_3 = k_3 (\varepsilon_{r3} \cos \varphi_3 + \varepsilon_{s3} \sin \varphi_3) =$$

$$= 1 (0.002\,27 \times 0.6 + 0.0351 \times 0.8) = 0.0296.$$

The voltage drops across terminals 1 and 2, 1 and 3, and 2 and 3 are

$$\varepsilon_{1-2}=\varepsilon_1+\varepsilon_2=0.145-0.001\,62\cong0.143,$$

$$\varepsilon_{1-3}=\varepsilon_1+\varepsilon_3=0.145-0.0296=0.175,$$

$$\varepsilon_{2-3}=\varepsilon_3-\varepsilon_2=0.0296-0.001\,62=0.028.$$

3.3 Connection of transformer windings and parallel operation of transformers

3.3.1 Connection of transformer windings

The windings of a single-phase transformer can be connected to the primary and secondary networks in the four different ways shown in Fig. 3.17. It is apparent there that the relative phase positions of primary and secondary voltages are identical for variants (a) and (d) and for variants (b) and (c).

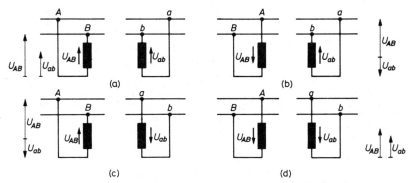

Fig. 3.17. Connection variants of a single-phase transformer

In the case of three-phase transformers the conditions are more complicated:

(a) For connecting the three primary and three secondary line terminals to the respective network the number of possibilities is 3! on the primary side and 3! on the secondary side, resulting in $(3!)^2=36$ variations.

(b) Each of the primary and secondary sides may be connected in star, in delta and in zig-zag (interconnected star) ways, representing a further $3^2=9$ variations.

c) On the primary and secondary side the beginning and the end of windings connected in star, delta or zig-zag can be interchanged, by which 4 variants can be obtained.

The three different groups of possibilities may be varied between themselves, so that the windings of a three-phase two-winding transformer can be connected in 1296 different ways. No such number of variants is needed in practice. So, the connections resulting in an interchanged phase sequence are not considered. The remaining connections based on the phase displacements between voltages of the primary and secondary windings are classified into 12 vector groups. The group symbol consists of two letters and a number. The first letter is a capital and refers to the connection of the primary side. D stands for delta, Y for star and Z for zig–zag. The second letter refers to the connection of the secondary side using the respective small letter, as d for delta, y for star and z for zig–zag. The number attached to the two letters indicates the phase angle between primary and secondary terminals of

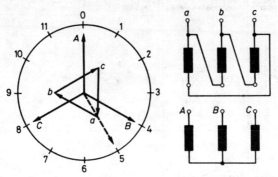

Fig. 3.18. Voltage vectors and connection diagrams of a transformer of vector group Yd5

identical terminal marking. The phase angle is expressed in degrees divided by 30, which is identical with the respective number as appearing on a clock face. In Fig. 3.18 the connection of a $Yd\,5$ transformer is shown, together with the position of voltage vectors. In Table 3.1 the vector groups most often used are represented.

3.3.2 Parallel operation of transformers

Two or more transformers operate in parallel, if they are fed from the same busbars at the input side and feed common busbars at the output side. Parallel operation becomes necessary when the available capacity of one transformer is insufficient for the power transmission between the two connected networks, or when for the purpose of energy saving a transformer is operated in parallel with its standby unit. In the latter case, the coil losses are reduced by half. The load is shared by transformers operated in parallel in proportion to the ratio of their rated capacities, if the following conditions are satisfied:

(a) the phase displacements between the respective input and output voltages are the same, i.e. the transformers are of the same vector group;
(b) the voltage ratios are the same, i.e. the magnitudes of identically marked secondary voltages are equal;
(c) the p.u. impedance voltages are equal;
(d) the ratio between rated outputs of transformers should not exceed 1:3, otherwise the smaller transformer may be overloaded; this requirement results from

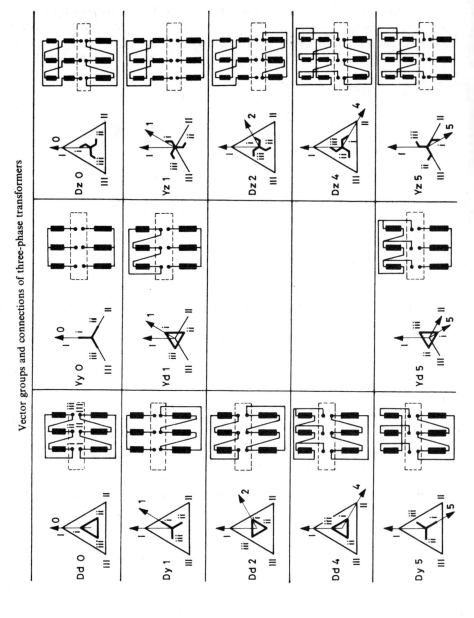

Table 3.1

Vector groups and connections of three-phase transformers

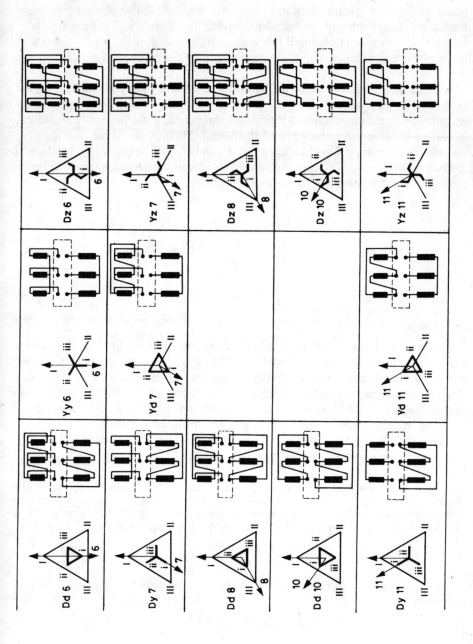

the fact that the ratio of inductive to active components of the short-circuit impedance of a larger transformer is higher than that of the transformer of smaller rating operated in parallel with it. For this reason, the relative phase positions of load currents are different in parallel operated transformers of different ratings, therefore the transferred resultant power of such two transformers, each loaded with its rated current, is lower than the sum of separate ratings of the two transformers.

The first condition is essential for parallel operation. Unless it is satisfied, the act of parallel connection is immediately followed by a catastrophe. Certain tolerances are possible in the observance of the further three conditions. In parallel operation of two transformers of ratings P_I and P_{II}, having p.u. impedance voltages ε_I and ε_{II} and the ratio error $e = (U_I - U_{II})/U_I$ (where U_I and U_{II} are the secondary voltages of the two transformers), a circulating current I will flow whose magnitude relative to the current of transformers I and II, respectively, is

$$\frac{I}{I_I} = \frac{e}{\varepsilon_I + \dfrac{P_I}{P_{II}} \varepsilon_{II}}$$

and

$$\frac{I}{I_{II}} = \frac{\dfrac{P_I}{P_{II}} e}{\varepsilon_I + \dfrac{P_I}{P_{II}} \varepsilon_{II}}.$$

From these equations the value of circulating current flowing in no-load conditions can be determined. Under load, if $\varepsilon_{ZI} \neq \varepsilon_{ZII}$, the input currents of the two transformers will also be unequal. The formula leaves out of consideration that a difference in the phase position of impedance voltages (i.e. the ratios $\varepsilon_r/\varepsilon_s$ of the two transformers being different) may cause further errors. This deviation will be of growing importance with increasingly different transformer ratings.

In the IEC 76 Publication [110] $\pm 0.5\%$ from the rated voltage ratio and a $\pm 10\%$ from the specified impedance voltage are permitted.

Example 3.6. In parallel operation of two transformers, assume the most unfavourable permissible deviations, with $P_{nI}/P_{nII} = 3$, and rated impedance voltage $\varepsilon = 0.1$, from which the actual values deviate by $\pm 10\%$, so that $\varepsilon_I = 0.11$ and $\varepsilon_{II} = 0.09$. Suppose the deviations in voltage ratios are also at their permissible maximum, i.e. $U_I = 1.005\, U$ and $U_{II} = 0.995\, U$ (U is the nominal voltage) resulting in a ratio error of

$$e = \frac{U_I(1.005 - 0.995)}{U_I} = 0.01.$$

The relative values of circulating current referred to the rated currents of transformers I and II, respectively, are

$$\frac{I}{I_I} = \frac{0.01}{0.11 + 3 \times 0.09} = 0.0263,$$

$$\frac{I}{I_{II}} = \frac{3 \times 0.01}{0.11 + 3 \times 0.09} = 0.079.$$

This circulating current imposes a higher stress on the smaller transformer.

112

The conditions to be satisfied when operating two-winding transformers in parallel apply to the case of three-winding transformers only if both transformers connect all the three networks. When a three-winding transformer is operated in parallel with a two-winding unit (this case being, in principle, identical with that when one of two three-winding transformers is connected to all three networks while the other connects only two of the networks), dangerous circulating currents are liable to flow, even if the conditions for parallel operation are fulfilled. This is because the current taken up by the third network influences the impedance voltage between the other two windings.

Parallel operation of a two-winding and a three-winding transformer may be accomplished under more favourable conditions if the primary winding of the three-winding transformer is located between the two secondary windings, and the two-winding transformer in parallel connects the primary system with only one of the secondary networks. In such a case the power flow between the first and second network has hardly any effect on the power exchange of the third network.

3.4 Windings

According to the terminology of the literature dealing with transformers, the winding is the assembly of turns forming an electrical circuit associated with one of the voltages assigned to the transformer.

The material of the windings is usually copper, although aluminium is increasingly being used in smaller transformers.

The physical properties of copper and aluminium are summarized in Table 3.2. Copper and aluminium used for the purpose of windings must not contain inclusions, and no formation of fibres or scales is allowed under the effect of twisting. In the course of manufacturing, electric butt welding or shielded-arc welding should be applied, since torch welding causes hydrogen plague in copper and may lead to the formation of inclusions in aluminium. The copper content of winding material should be at least 99.9%. Tin, aluminium, manganese, chrome, silicon, cobalt, iron or phosphorus content of 0.1% and above causes a deterioration of electrical conductivity by more than 10%. The required minimum Al-content in aluminium is 99.45%. Aluminium is highly sensitive to impurities, a drop in conductivity exceeding 10% being caused by 0.4% magnesium, 0.1% titanium or 0.02% manganese or chrome, whereas relatively high contents (exceeding 0.4%) of copper, iron, silicon, nickel and tin have no significant effect on conductivity. The mechanical properties of copper and aluminium when used as active materials should be within the limits given in Table 3.2. Within those limits, copper and aluminium are far from having the best mechanical properties with regard to short-circuit forces. Still, no efforts are made to use wire materials of higher mechanical strength for two reasons. First, because a winding material of higher strength would be insufficiently ductile for the required tightness of the coils, or the required compactness could only be attained by risking the soundness of wire insulation. On the other hand, after the first few short-circuits inevitably occurring in service, the

Table 3.2

Physical properties of copper and aluminium

Physical property	Unit	Copper (Cu)	Aluminium (Al)
Density at 20 °C	kg dm^3	8.96	2.698
Conductivity hard	mΩ mm^{-2}	55	33
half-hard	mΩ mm^{-2}	56	35
soft	mΩ mm^{-2}	57	37
Temperature coefficient of resistance	°C^{-1}	0.00393	0.00377
Coefficient of linear expansion	°C^{-1}	16.2×10^{-6}	23.9×10^{-6}
Tensile strength (soft/hard)	N mm^{-2}	200/350	70/150
Elongation at rupture (soft/hard)	%	30/20	22/2
Modulus of elasticity	N mm^{-2}	125 000	72 000
Brinell hardness (soft/hard)	N mm^{-2}	35/95	15/25
Thermal conductivity	W m^{-1} °C^{-1}	338 000	231 000
Specific heat	Ws kg^{-1} °C^{-1}	385	920
Fusion point	°C	1 083	659
Boiling point	°C	2 300	2 270
Specific heat of fusion	Ws kg^{-1}	209 000	355 000
Electrochemical potential	V	+0.35	−1.28

winding, while cooling off in oil, is subject to an "annealing treatment" [123] which would inevitably reduce this strength.

Most windings are made of wires having circular cross-sections. Diamaters of such wires used in practice lie in the range of 0.3 to 2.5 mm for copper and of 1 to 2.5 mm for aluminium. Over 5 mm^2, rectangular wire sections are used. The coil is wound so that the smaller dimension of the wire is perpendicular to the stray flux lines. In order to avoid local temperature rises resulting from high additional losses, the dimension perpendicular to the leakage channel is preferably kept below 10 mm for copper and below 15 mm for aluminium, even when there is only a single layer placed perpendicular to the stray flux lines. With a larger number of layers this maximum permissible dimension may be reduced to 2 mm.

At and about the middle of the winding the dimension in the direction of leakage field the conductor may theoretically be of any dimension, but at the beginning and at the end of the coil it should not be wider than 15 mm in the case of copper and 20 mm with aluminium, because of the presence of radial stray flux lines.

The properties of insulation material covering the wire surfaces of the winding are dealt with in Chapter 5.

The voltage of the transformer is determined by the number of turns of the windings and by the flux, while the cross-sectional area is proportional to the current of the transformer. The winding is expected to meet several requirements in short-circuit strength and withstanding voltage and temperature rises. Generally, contemporary types of windings are all suitable for satisfying these requirements, at least at the cost of some compromises. When selecting the type of winding the designer has, in every case, to consider what requirements have to be given preferance in a transformer intended for a given purpose (high short-circuit stresses, high thermal loads, high overvoltages, etc.).

The primary, secondary and possibly any further windings of the transformer are arranged either concentrically as in the units termed core-type transformers, or placing the primary, secondary, etc. as pancakes placed coaxially in an alternating sequence. The latter units are termed shell-type transformers.

3.4.1 Windings of the shell-type transformer

In Fig. 3.9, the shell-type winding arrangement is illustrated. The coil elements of the shell-type winding are shown in Fig. 3.19. In certain cases it may be advantageous if the number of magnetically balanced groups can be varied within wide limits, whereby the magnitude of impedance voltage is reduced, and a further possible advantage consists of lowering the axial short-circuit forces by increasing the number of magnetically balanced groups. Since radial short-circuit stresses arising in the windings are small, it is not necessary that the coils be circular in a shell-type transformer. Its space utilization is very good, and its core can be manufactured without using stepped sections. In such cases the limb cross-section can be made to approach the optimum, if both winding and limb are given a rectangular shape. Economic considerations based on extreme-value calculus show that rectangles with side lengths of $1:2$ ratio are favourable. In shell-type transformers the pancake coils accomodated on the same limb and belonging either to the primary or to the secondary winding may be connected in parallel or in series with each other. Although this system is seldom adopted in the majority of European countries, the arrangement is just as suitable for the construction of highest-voltage transformers as the core-type design. Since the L.V. and H.V. coils are accommodated beside each other, alternating with each other, the roles of turn and pancake insulation are especially important in the manufacture of higher voltage units. With increasing voltage levels, the design of insulation has become

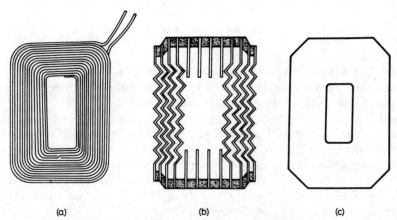

(a)   (b)   (c)

Fig. 3.19. Elements of a pancake winding: (a) pancake coil; (b) spacers; (c) insulating disc

increasingly difficult in the shell-type arrangement. For 400 kV, the designers have succeeded in constructing acceptable transformers with shell-type windings, but for higher voltage levels the transformers have all been designed as core-type units.

With the shell-type arrangement, two types of windings are used. One is built up of single or double discs wound with turns. The cross-section of a double disc is illustrated in Fig. 3.20. In (a) the cross-section of a double disc wound of a single branch is shown. Drawing (b) shows the cross-section of a coil element made of two

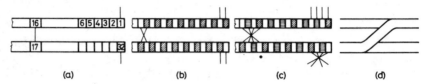

Fig. 3.20. Double pancake windings in the case of: (a) one; (b) two; (c) four parallel branches; (d) transition between two pancake coils

parallel branches, while in drawing (c) the cross-section of a coil wound of four parallel branches is shown. The double discs, constituting the building blocks of the winding, are also termed coil elements. In the discs wound of two or three parallel branches transpositions are made to ensure that every parallel branch be linked with the same number of flux lines, lest the additional losses should increase. Figure 3.20 (d) illustrates the practical execution of transpositions. Of the coil elements described so far, those consisting of a lower number of parallel branches are applied for the H.V. windings, and those with more parallel branches are adopted for the L.V. windings.

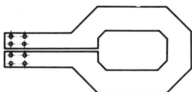

Fig. 3.21. One turn of a solid pancake coil

The other winding design of the shell-type arrangement consists of single turns cut out from a solid conducting plate. A turn of this type is shown in Fig. 3.21. Such elements are usually adopted for high-current windings when the short-circuit strength is of special importance. As regards supplementary losses, this solid turn is unfavourable, because high eddy currents are induced by the stray flux lines perpendicular to the plane of the turn. The shell-type winding arrangement is very favourably employed in combination with the three-phase core-type lamination, and it can often be seen in railway, furnace and short-circuit testing transformers. Generally, shell-type transformers are selected for applications where the short-circuit performance is an outstanding requirement.

116

3.4.2 Windings of core-type transformers

Preference is given by European manufacturers to the most widely used winding arrangement, that of the core-type design. An important feature characteristic of this system is the cylindrical shape of coils for coping with radial short-circuit stresses. In the majority of applications, the number of magnetically balanced groups is just one.

As indicated in Fig. 3.10, provision of several magnetically balanced groups is also possible in core-type designs, but is not at all so self-evident as in the shell-type arrangement. With the core-type system, it is difficult to accomplish very low impedance voltages or moderate radial short-circuit stresses. On the other hand, the electric strength problems can be solved more satisfactorily and by simpler means. The following types of windings are used in the core-type arrangement for the different voltage levels and power ratings.

The layer-type winding can be manufactured for any rating and any voltage level. A few designs of such type of winding are represented in Fig. 3.22. In (a) the so-called

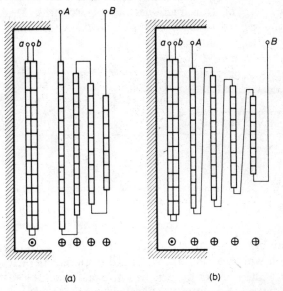

(a)　　　　　(b)

Fig. 3.22. Laminated winding: (a) continuously wound; (b) with uniform field strength between layers

continuously wound layer-type coil is shown where the direction of turns is changed on passing from one layer to the next. Paper insulation is laid between adjacent layers, of a thickness depending on the expected electric stress or, if necessitated by conditions of cooling, oil ducts are provided by means of fillets. In (b) a type of layer winding is shown in which the direction of turns is identical in all layers. Such an arrangement ensures more economical utilization of inter-layer insulation, also its impulse voltage characteristics are better. The minimum permissible thickness of oil ducts provided between adjacent layers is 4 mm. In narrower oil ducts the flow resistance would increase, reducing the flow velocity of the oil and giving rise to excessive local temperature rises. In order for the required oil flow to develop, it is recommended to provide 5 to 6 mm wide oil ducts. Duct widths exceeding this safe limit are not justified for improving the conditions of cooling, because with larger dimensions the so-called inner core of the streaming oil cross-section has no effect on cooling. Electric strength considerations can, of course, justify even larger oil ducts; disc-type windings continuously wound are built up of simple or double discs. The cross-sections of simple and double discs are identical, as shown in Fig. 3.20.

Ducts are provided by spacers placed between the coil elements, permitting radial flow of the oil. The minimum radial oil duct spacing is 4 mm, but wider ducts have been found more efficient in practice. Ring-shaped pressboard or soft paper insulating discs of thickness varying between 0.5 and 2 mm are placed between the double discs. The pancake windings for higher currents are built up from coils made of several parallel branches. For reducing additional losses caused by radial stray lines of force, the method of transposition is used, just as in the case of shell-type windings. Such transpositions are shown in Figs 3.20 and 3.23.

A special case of windings made of several parallel branches is the winding built up of several coils wound to a fractional number of turns. This arrangement is shown by the winding cross-section of Fig. 3.23. The constraints imposed on the

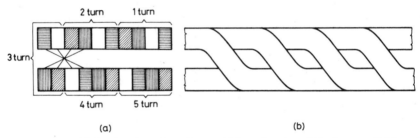

(a) (b)

Fig. 3.23. Winding with fractional number of turns: (a) interchanging of the four parallel branches; (b) transitions between discs

designer may thus be relieved in cases when difficulties are encountered in filling out the space available with turns in axial or radial directions. Characteristic of this kind of winding is that the number of turns is not equal in all parallel branches. For example, in the disc wound with $a = 4$ parallel branches each of the four branches has two turns, but only the first two branches have a third turn. The third and fourth parallel branches have their third turn in the next disc and, in addition, all the four parallel branches have their fourth and fifth turns located here. The winding made up of discs is either manufactured continuously, or the essentially identical winding arrangement is assembled of individually prepared disc elements.

The disc-type winding can be wound from a long continuous wire, whereas with a stacked winding either soldering or welding is required after every other disc. The process of preparing a winding in the first case is more laborious, but the probability of occurrence of defects is lower because of using continuous wires. Manufacturing a stacked winding is simpler, yet its mounting requires more work, because the coil ends have to be jointed by carefully checked soldering or welding. The disc-type windings (either of pancake or stacked design) may be wound or mounted so that the turns situated far apart electrically are geometrically beside each other. By this procedure (inter-leaved windings) the impulse voltage distribution of windings can be improved (for details see Chapter 5).

Spiral coils are used for higher currents. The simplest spiral coils are the single-layer and double-layer coils. Figure 3.24(a) shows a view and cross-section of the single-layer spiral winding, whereas in (b) the double-layer spiral winding is

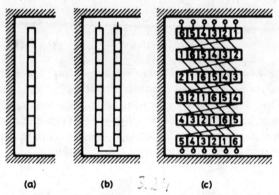

(a)   (b)   (c)

Fig. 3.24. Spiral coils: (a) single-layer; (b) double-layer spiral coil; (c) Wendel coil (spiralled space winding)

represented. For higher currents the single- and double-layer windings are wound of several parallel branches. In such cases the height of winding may considerably increase because of the axially located parallel branches. To avoid both coil ends the windings are "double-threaded", and one half of the parallel branches are brought out from the coil angularly displaced by 180° with respect to the other half. For highest currents the Wendel coil wound of several parallel branches, is used (Fig. 3.24(c)). With this type of coil it must be ensured that the threads of all parallel branches are transposed cyclically as many times as required to make every conductor assume all possible radial directions. With the number of parallel branches equal to a, the minimum number of cyclic transpositions is $(a-1)$. When more than $(a-1)$ cyclic transpositions are made, their number must be an integer multiple of $(a-1)$. There is no constraint between the number of turns and number of transpositions, because it is not necessary to make each transposition after a round number of turns. It is only required to make the transpositions so that they divide the coil into equal sections along the turns of the coil. If N is the number of turns in the coil, and the number of transpositions is $b(a-1)$, transpositions should be made after every $N/b(a-1)$ turns. A variant of the Wendel winding, the continuously transposed Rőbel winding, is also used. A cross-section of this winding is represented in Fig. 3.25. In a Rőbel winding each parallel branch is wound as two bundles each consisting of two $a/2$ parallel conductors. For reducing supplementary losses continuous transpositions are also used in this case. An arrangement of such a cyclic transposition is shown in Fig. 3.25. Also with this type of winding,

119

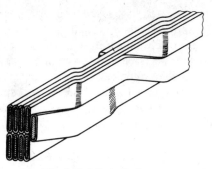

Fig. 3.25. Rőbel winding

transpositions have to be made at least $(a-1)$ times. In practical designs the cyclic transpositions follow each other in a more frequent succession along the coil, say at a spacing of 100 to 300 mm. With cyclic transpositions so frequently performed, the voltages induced in the wires will be equal in spite of changing flux density in the axial direction and in addition, the bundle of conductors will be held together more firmly mechanically by this more frequent transposition.

3.5 Short-circuit forces and phenomena in transformers

In a transformer whose input side is fed from the supply network of rated voltage and whose output side is short-circuited, the magnitude of currents flowing in its primary and secondary windings is determined by the short-circuit impedance of the transformer. This short-circuit current is 8 to 10 times the rated current in larger transformers and 20 to 25 times in smaller units. As a result of interaction between the leakage field of the shorted winding system and the current flowing in the turns, short-circuit forces will act on the turns of the windings. These short-circuit forces being proportional to the square of the short circuit current, the electrodynamic forces arising during a short-circuit may be as high as ten thousand to a million N as compared to a few N associated with the rated load currents. The phase position of voltage prevailing at the instant of the short-circuit and the time constant of the short-circuit loop determine the maximum short-circuit current and force, and the forces acting on the turns and windings after the first current peak keep the windings and their clamping structures in oscillation at double the network frequency until the short-circuit is interrupted. The winding structure should be dimensioned against mechanical strains to prevent the stresses arising in component parts from causing permanent deformations and to limit the movement of the winding structure to a minimum during short-circuits.

In the period between occurrence and clearing of the fault—not exceeding a second even under the most adverse conditions—most of the heat developing in the conductor material does not leave the winding, the thermal time constant being several minutes, but will raise its temperature. For the safety of wire insulation, the

120

winding temperature caused by the short-circuit temperature rise should not exceed that which the insulating materials are capable of withstanding for such short periods of time. The short-circuit temperature rise also causes considerable thermal expansion, imposing a further strain on the winding.

During short-circuits, the mechanical vibration of windings and the thermal expansion of wires take place simultaneously. The relative displacements between wire insulation and other component parts of the winding may injure the insulation, leading to turn faults. Similar damage may result from the displacement of high-current busbar connections or from troubles occurring in the switching mechanism. These hazards can be avoided by proper dimensioning of windings both thermally and dynamically, and by pre-compressing the winding system with a force corresponding to the electrodynamic effects of short-circuits and firm clamping of current-carrying component parts.

A suitable means of safeguarding the insulating materials against short-circuit stresses is to avoid the use of superfluous quantities of insulation. Insulating materials are liable to collapse. A protection measure against collapse of insulation is to subject it to at least two successive forming processes. The first forming consists of compressing the insulating materials with a pressure of 2000 to 5000 N cm^2 at a temperature of 100 to 120 °C for a few minutes. The second forming is performed, after vacuum drying, by compressing the winding in a clamping device.

IEC Publication 76–5 (ability to withstand short circuits) classifies transformers in three categories: Category I covers transformers up to 3150 kVA, Category II those in the range 3150 to 40 000 kVA, Category III those above 40 000 kVA.

3.5.1 Short-circuit current

The symmetrical short-circuit current for a three-phase two-winding transformer is given by

$$\bar{I} = \frac{U}{(Z_t + Z_s)\sqrt{3}}, \tag{3.18}$$

where U is the rated voltage, Z_t the short-circuit impedance of the transformer and Z_s the short-circuit impedance of the system. In the case of transformers of Category I, Z_s can be neglected if it is lower than 0.05 Z_t. In IEC Publication 76–5 the typical impedance voltages of transformers are given as listed in Table 3.3. From the percentage impedance voltage (ε_z), rated voltage (U) and rated power (S) of the transformer, its short-circuit impedance in ohms is given as

$$Z_t = \frac{\varepsilon_z U^2}{100 \, S}.$$

Depending on the connection of transformer windings, the symmetrical short-circuit currents of line-to-line, line-to-earth and double-earth faults of the transformer are not equal in every case to the value obtained from formula (3.18). The short-circuit currents assume the following values in the various cases most often occurring in practice:

(a) In a transformer of star/zig-zag or delta/zig-zag connection, a line-to-earth fault between neutral point and a phase terminal of the zig-zag winding causes the short-circuit current to increase with respect to the symmetrical value by a factor of about 1.5.

(b) In a transformer of star/star connection with neutrals brought out on both sides having a delta connected tertiary or a stabilizing winding, a line-to-earth fault in the secondary circuit will cause a short-circuit current which is of magnitude

$$\bar{I} = \frac{\bar{U}}{\frac{2}{3}\bar{Z}_s + \bar{Z}_d}, \tag{3.19}$$

where U is the rated voltage, $\bar{Z}_s$ is the short-circuit impedance between the two star-connected windings and $\bar{Z}_d$ is the short-circuit impedance between the short-circuited winding and the delta-connected tertiary or stabilizing winding. The line-to-earth fault current is lower than the three-phase fault current if $\bar{Z}_d > Z_s \frac{1}{3}$.

Table 3.3

Impedance voltages of two-winding transformers

Rated power kVA	Impedance voltage %
up to 630	4
631 to 1250	5
1251 to 3150	6.25
3151 to 6300	7.15
6301 to 12 500	8.35
12 501 to 25 000	10
25 001 to 200 000	12.5
200 001 and above	subject to agreement

Generally, this latter condition is not satisfied when the transformer is fed from its winding of largest diameter, the fault occurs in its centrally arranged winding, and its tertiary (or stabilizing) winding is of the smallest diameter. Generally, too, power systems are designed to keep short-circuit levels within reasonable limits, therefore, if fault currents are to be restricted, reactors are inserted in the delta-connected circuits of transformers in order to increase their impedance voltage.

(c) The magnitude of short-circuit currents associated with line-to-line faults on the delta side of star/delta and delta/delta connected transformers is the same as that occurring in the case of a symmetrical fault. In a delta/star transformer (with neutral brought out) and in star/star transformers with neutrals brought out on both sides, the magnitude of single line-to-earth fault currents is closely equal to that of a three-phase fault.

(d) In a star/star connected transformer the short-circuit current of a line-to-line fault is $\sqrt{3}/2 = 0.866$ times the value of the symmetrical fault current. With a fault between two phases on the star-connected side of a delta/star transformer, the secondary-side fault current is $\sqrt{3}/2 = 0.866$ times the symmetrical short-circuit current. Finally, in star/star connection with no neutral brought out on the input side, the fault current associated with a short-circuit between one phase on the secondary side and the neutral terminal is

$$\bar{I} = \frac{\bar{U}}{\frac{2}{3}\bar{Z}_t + \bar{Z}_0}, \tag{3.20}$$

where $\bar{Z}_t$ is the short-circuit impedance pertaining to the symmetrical three-phase fault and Z_0 is the zero-sequence impedance of the transformer. The zero-sequence impedance, with no delta winding being present, is much higher than the positive-sequence impedance (the impedance pertaining to the symmetrical three-phase fault is the positive-sequence impedance), therefore the short-circuit current calculated from the former formula is below what would arise in the symmetrical case.

3.5.2 Thermal capability of withstanding short-circuits

The magnitude of symmetrical short-circuit currents, depending on the actual impedance voltage, may be 8 to 25 times the rated current. As a standard duration of short-circuits, the period of 2 s is specified by IEC Publication 76–5, and within this period the transformer is required to withstand also the thermal effect of a symmetrical fault. Within this specified time the temperature rise in the material of the transformer winding must not exceed a specified limit. This specified limit is an average value of winding temperature such that the properties of the winding material are not endangered, and for a short period (a few seconds) the service life of the wire insulation is not sensibly affected. The maximum permissible value of average temperature specified by IEC Publication 76–5 for the windings of oil-immersed transformers is 250 °C for copper, and 200 °C for aluminium. As regards thermal stability the behaviour of oil-immersed transformers is remarkable, in that—although the liquid and solid components of insulation (oil, paper, pressboard) belong to Class A, because of the rapid heat convection taking place in the oil, the oil-cellulose base—two-component insulations, when exposed to thermal stresses of short duration, are equivalent to Class E materials. The transformer should be so dimensioned, in order to make it capable of withstanding the thermal effect of short-circuit currents, as to prevent the average winding temperature from exceeding 250 °C (in copper) or 200 °C (in aluminium) within the specified 2s duration of short-circuit current, under the effect of the fault current density calculated from the rated current density and short-circuit current density. The thermal time constant of windings in oil-immersed transformers is by at least two orders of magnitude higher than such fault durations. Therefore, when calculating the thermal effect of short-circuits, it is justifiable to neglect the heat flow

123

from the winding into the surrounding oil and to assume that the entire quantity of heat is stored in the winding, thus raising its temperature.

The highest average temperature, Θ_1 of the winding is calculated from the formula

$$\Theta_1 = \Theta_0 + aJ^2t \times 10^{-3}, \tag{3.21}$$

where the value of Θ_1 is obtained in °C, Θ_0 is the initial temperature of the winding in °C (e.g. for an oil-immersed transformer $\Theta_0 = 40 + 65 = 105$ °C), J is the short-circuit current density in A mm^{-2}, t is the duration of the short-circuit current in s and a is a function of $\frac{1}{2}(\Theta_2 - \Theta_0)$ in accordance with Table 3.4, for taking into account the temperature dependence of the resistance of windings. Θ_2 of Table 3.4 is the maximum permissible average winding temperature. The condition to be

Table 3.4

Values of factor a [28]

$\frac{1}{2}(\Theta_2 + \Theta_0)$	$a = $ function of $\frac{1}{2}(\Theta_0 + \Theta_2)$	
°C	for copper windings	for aluminium windings
140	7.41	16.5
160	7.80	17.4
180	8.20	18.3
200	8.59	19.1
220	8.99	—
240	9.38	—
260	9.78	—

satisfied is $\Theta_2 \geq \Theta_1$. Obviously, in formula (3.21) the thermal capacity of wire insulation and the heating effect of eddy currents induced in the alternating field are left out of consideration. By rough estimation it can be verified that the heating effect of eddy currents reduced in their magnitude by increased resistivity pertaining to the higher temperature arising during short circuits is more or less compensated by the heat-absorbing capability of the wire insulation [83]. Therefore, both the heating effect of eddy currents and the insulation heat capacity may be neglected simultaneously. The transient temperature rise of the transformer occurring under short-circuit conditions is checked by substituting into formula (3.21) the current density and duration of the short-circuit. The transformer is considered as satisfying the temperature rise limits, if temperature Θ_1 is lower than 200 °C for aluminium and 250 °C for copper.

3.5.3 Dynamic ability to withstand short-circuits

Generally, after passing through a transient period of a few cycles after the incidence of a fault, sustained short-circuit current of the transformer develops. At any moment, the short-circuit current may be considered as the sum of an a.c. component (i') and a transient component (i''). The a.c. steady-state component is a constant amplitude periodic time function, whereas the transient component shows a relatively rapid decay. Assuming a sinusoidal supply voltage from a system of infinite capacity, and with a voltage/time function of $u = u_m \sin(\omega t)$, the a.c. component (i') and transient component (i'') of the short-circuit current are

$$ i' = \frac{I\sqrt{2}}{\varepsilon_z} \sin(\omega t + \alpha - \varphi) \tag{3.22} $$

and

$$ i'' = \frac{I\sqrt{2}}{\varepsilon_z} \sin(\varphi - \alpha) e^{-\frac{R}{L}t}, \tag{3.23} $$

where I is the r.m.s. value of rated transformer current, ε_z is the p.u. impedance voltage, $\omega = 2\pi f$ is the angular frequency, α is the angle of phase displacement

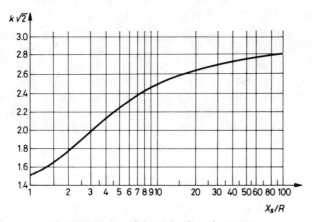

Fig. 3.26. Factor of short-circuit peak current

defining the instant of short-circuiting the transformer and $\varphi = \arctan \dfrac{\omega L_s}{R}$ is the phase angle of short-circuit impedance.

The optimum instant of short-circuiting the transformer is the zero-transition of current ($\varphi = \alpha$), the transient current component being zero in that case. When calculating the short-circuit currents the worst case should, of course, be considered, i.e. assuming the short-circuit to occur at zero transition of voltage ($\alpha = 0$). More precisely, the condition for occurrence of the maximum amplitude of transient short-circuit component is the state when $\alpha = \varphi - \pi/2$. Since for large transformers $\varepsilon_r \ll \varepsilon_s$, in the worst case $(\varphi - \alpha) \approx \dfrac{\pi}{2}$. The neglect acts in the direction of safety, if with the assumption of $\varepsilon_r \ll \varepsilon_s$, the formula

$$\hat{\imath} = \frac{2\sqrt{2}\,I_n}{\varepsilon_z}$$ (3.24)

is used for the calculation.

IEC Publication 76–5 allows dimensioning and testing considering lower current than that resulting from the above formula, the amplitude of the first peak of asymmetrical short-circuit current being

$$\hat{\imath} = Ik\sqrt{2}.$$ (3.25)

The values of $k\sqrt{2}$ appearing in the above formula are shown in the function of ratio X_s/R in the diagram of Fig. 3.26.

3.5.3.1 Dynamic effects of short-circuits on windings with balanced ampere-turns

In a two-winding transformer with windings of uniformly distributed and balanced ampere-turns the arrangement of windings is considered as symmetrical. The electrodynamic short-circuit force acting on a conductor carrying current i placed in a magnetic field of flux density $\bar{B}$ to the elementary conductor lengths $\overline{dl}$, can be computed by summing up the elementary forces $\overline{dF} = i\overline{dl} \times \bar{B}$. These forces oscillating with double the system frequency cause the winding system to vibrate in the same direction. Representing in a diagram the forces acting per unit length of conductors constituting the winding, these forces being termed specific forces, a qualitative survey of stresses arising in the winding is obtained. It is convenient to resolve these forces into axial and radial components. In Fig. 3.27 the distribution of specific radial forces p_r and of specific axial forces p_a is

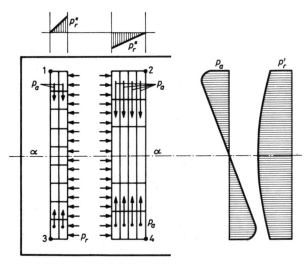

Fig. 3.27. Short-circuit forces acting on symmetrical windings

126

shown. The forces resulting from the interaction of the current flowing in the turns of the winding and the axial components of magnetic flux lines act in the radial direction, whereas the radial components cause axial short-circuit forces. The radial forces arise throughout the entire length of the winding, their magnitude being maximum where the axial flux components are greatest. This point lies on the centre line of the winding, on the leakage channel side of the winding (along the centre line of the outer mantle of the inner winding and along that of the inner mantle of the outer winding, respectively). The radial forces decrease along the turns receding radially from the leakage channel. In Fig. 3.27, the variation of radial forces along the axis is illustrated by the diagram of specific forces p'_r and the variation of the same along the radius by the diagram of specific force p''_r. In the vicinity of lower and upper edges of the windings the stray flux lines turn out of the leakage channel. The largest radial-forces arise along the $\alpha-\alpha$ centre line on the inner part of the leakage channel, the place of the smallest forces being close to points 1, 2, 3 and 4.

The axial forces are represented in the p_a diagram. The biggest forces arise near points 1, 2, 3 and 4. The distribution of axial forces is influenced by the vicinity of the core, so at the beginning and at the end of the inner winding the magnitude of the axial force may be 1.5 to 3 times that acting at the ends of the outer winding. In concentric windings the distribution of short-circuit forces is, to a first approximation, the dual of the shell-type arrangement, i.e. forces which are radial in the core-type concentric arrangement, are axial in the shell-type arrangement. In a shell-type transformer, however, due partly to the radially changing cross-section of the leakage channel and partly to its being relatively shorter, the magnetic field pattern shows a more pronounced deviation from the field pattern of Fig. 3.1 of primary and secondary windings embedded in a medium of infinite permeability. Therefore, the distribution pattern of forces also differs from the dual of Fig. 3.27 mainly quantitatively. The short-circuit forces arising in core-type windings are dealt with in the following.

The inner winding can be regarded as a tube compressed from the outside, and the outer winding as a tube expanded from the inside, by the radial forces. The axial forces tend to compress the coils of a symmetrically arranged winding system, so the forces acting on individual turns have no resultant. In an asymmetrical winding system the axial forces also tend to compress the coil, but the axial forces acting on the turns have a resultant, and the directing forces resulting from the asymmetry is such as to cause the asymmetry to increase further.

The ability of transformer windings to withstand short-circuits is checked as detailed below:

(a) The radial forces should not cause dangerous deformation in the outer winding. This checking is substantially similar to the calculation of mechanical stresses arising in a tube expanded from its inside.

(b) The radial forces should not crush the inner winding, i.e. the winding is checked for deformation as a tube compressed from the outside and braced from the inside at its generatrices along the spacers.

(c) The axial forces should not cause deformation of windings. In the case of balanced ampere-turns, axial forces act on the windings as compressive forces. In layer-windings the compressive stress arises on the surfaces of successive turns, whereas in pancake windings, the axial pressure is transferred through the spacers to the next turns, and the conductor sections between spacers are subject to bending. Since maximum axial forces act on the conductors at the winding ends, the bending stresses are also the highest there. The axial compressive forces add up from both ends toward the centre line of the winding, so that in layer windings the force acting on the conductors, and in pancake windings the compressive forces acting on the spacers, are the biggest. In cases of asymmetry the resultant axial force causes a mechanical stress in the clamping structure as well.

A simpler way of checking mechanical strength uses the method of calculating the resultant short-circuit forces. In this method the resultant radial forces are determined by differentiating the stored magnetic energy in the direction of radial displacement of the winding. Assuming the resistance of the windings to be very small compared to their stray reactance, the expected maximum of the resultant radial force can be calculated from the formula

$$F_{r\,max} = \frac{2P_l}{\pi f \left(\dfrac{a_1 + a_3}{3} + \delta\right)\varepsilon g},$$ (3.26a)

where P_l is the power per limb in VA, f is the system frequency in Hz, ε is the p.u. value of impedance voltage, g is the number of magnetically balanced groups, a_1 and a_2 are the radial dimensions of the respective windings, and δ is the radial dimension of the leakage channel in m, with these units the force is obtained in terms of N. Relation (3.26a) may also be written in the form

$$F_{r\,max} = \frac{\mu_0}{2}(N\hat{\imath})^2 \frac{l}{l_s}\frac{1}{g},$$ (3.26b)

where $\mu_0 = 1.256 \times 10^{-6}$ Hm^{-1}, N is the number of turns, $\hat{\imath}$ is the first peak of short-circuit current in A, l is the mean periphery of the stray channel, l_s is the length of the leakage channel and g is again the number of magnetically balanced groups; the resultant force is obtained in terms of N. It should be noted that the short-circuit force is dependent on the square of ampere-turns and on the dimension of the stray duct: this means that from among transformers designed for the same rating, the developing short-circuit forces will be smaller in those having a higher turn voltage and slimmer windings.

The tensile stress in the outer winding can be determined from the following formula:

$$\sigma = \frac{F_{r\,max}}{2\pi AN},$$ (3.27)

where σ is obtained in terms of N m^{-2} if the force is given in N and the wire cross-sectional area A is expressed in m^2 and N is the number of turns of the winding. The result is the average value of the tensile stress. In the turns adjacent to the leakage channel the stress would be twice as high, if these turns were not leaning against the outer turns of the winding. The outer winding withstands the radial forces, if the short-circuit tensile stress remains below the value causing permanent deformation. Considering unequal distributions of forces, checking will be reassuring if the tensile stress thus obtained is lower than $8000 \, \text{N cm}^{-2}$ for copper windings and $4000 \, \text{N cm}^{-2}$ for aluminium windings.

The inner winding is checked for staving. A winding compressed from the outside is braced from the inside by spacers of number Z. By using a sufficient number of spacers, deformations caused by short-circuit forces can be avoided. In Fig. 3.28 the staving-in of three windings braced with different numbers of spacers is shown.

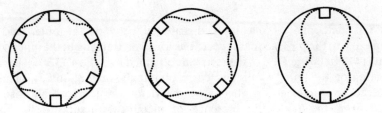

Fig. 3.28. Deformation of internally braced windings

Fischer [31] deduced the following formula for the minimum required number of spacers

$$Z_{\text{min}} = 2 \sqrt{1 + \frac{1.75 F_{r\text{max}}}{ENA} \left(\frac{l}{2\pi b}\right)^2}, \tag{3.28}$$

where $F_{r\text{max}}$ is the short-circuit force in N, A is the conductor area in cm^2, l is the mean length of the inner winding in cm, b is the radial dimension of a turn, N is the number of turns, E is the elastic modulus of conductor material in N cm^{-2}, $E = 12\,500\,000 \, \text{N cm}^{-2}$ for copper and $E = 7\,200\,000 \, \text{N cm}^{-2}$ for aluminium. A winding will withstand the short-circuit forces without suffering permanent deformation, if the number of spacers Z_{min}, obtained from formula (3.28), is lower than that selected by the designer for internal bracing of the winding. It may be that although the winding has been braced against staving by an adequate number of spacers, yet the stresses arising in the wires of the winding cause permanent deformation. In such a case the cross-sectional area of the conductor has to be increased. To check for the hazard of permanent deformation the following relation is used:

$$\sigma_d = \frac{F_{r\text{max}}}{2\pi \, N A}, \tag{3.29}$$

where $\sigma_d = 8000 \, \text{N cm}^{-2}$ for copper and $\sigma_d = 4000 \, \text{N cm}^{-2}$ for aluminium. Should the right-hand side of the above formula be higher than the respective values given above for σ_d, the conductor area has to be increased.

The radial forces arising in the winding are proportional to the value of axial flux densities. Due to the flux lines dispersing radially in the vicinity of the winding ends, the axial flux density tends to decrease towards the ends of windings. Therefore, the radial forces acting on the winding elements (on the discs, or in the case of layer windings on the turns) at the two ends of the windings are only 50 to 60% of that arising at the middle of the winding.

The resultant of axial forces arising in the short-circuited primary and secondary windings is

$$F_{a\,max} = F_{r\,max} \left(\frac{a_1 + a_2}{3} + \delta \right) \frac{1}{l_s} =$$

$$= \frac{\mu_0}{2} (N\hat{\imath})^2 \frac{l}{l_s^2} \left(\frac{a_1 + a_2}{3} + \delta \right). \tag{3.30}$$

This axial component is unequally distributed between the outer and inner windings, due to the presence of the core. The axial force arising in the inner winding amounts to about 65 to 75% of the resultant obtained from Eqn. (3.30), whereas the force acting on the outer winding is, correspondingly, only 35 to 25%. This axial resultant force is influenced not only by the vicinity of the core but also by its distance from the tank, the presence of metallic structural parts and the arrangement of adjacent windings. The unequal distribution of forces is taken into account with a margin of safety acceptable for practical purposes by dimensioning the inner winding to the full value and the outer winding to 50% of $F_{a\,max}$ of formula (3.30). The winding is, then, dimensioned to withstand the axial forces in the following way: the axial short-circuit forces assumed as uniformly distributed along the periphery must not exert on the surface of turns of copper and aluminium windings a specific pressure higher than 4000 N cm^{-2} in the case of spiral winding structures, or higher than 3000 N cm^{-2} in the case of winding parts covered by spacers (in disc-type winding structures). For the surface of turns, the product of mean winding periphery and radial dimension of a turn is to be understood in the case of spiral windings, or the area covered by a spacer and taken along one turn of the disc coil in the case of pancake windings. The method described above is a reliable procedure for checking the short-circuit forces in the case of symmetrical windings. In large transformers or in units expected to operate under very severe short-circuit conditions, the stresses imposed on certain turns of especially critical position within a winding have to be considered separately. In a symmetrical winding higher axial forces act on the turns situated at the end of a winding (or on the discs in extreme positions in a disc-type winding). The magnitude of the axial force imposed on a turn in the extreme position or on a group composed of turns in extreme positions is [139]

$$f_{a\,max} = 0.366 \frac{N'}{N} F_{r\,max} \lg \left[1 + \frac{4 \left(\delta + \frac{a_1 + a_2}{3} \right)^2}{c^2} \right], \tag{3.31}$$

where the force is obtained in N, when the value of $F_{r\,max}$ is substituted in terms of N, N' is the number of turns in extreme positions (the initial disc in a disc-type winding or the initial turn in a spiral winding), N is the number of all the turns constituting the winding, c is the axial dimension (of initial disc or initial turn) and the interpretation of a_1, δ and a_2 has already been given above.

The windings are compressed axially by the clamping structures of the transformer. The axial compression force applied to the winding is chosen to produce a deformation at least equal to that occurring in a short-circuit. By this pre-compression the loosening of transformer windings under the effect of short-circuit forces is prevented.

3.5.3.2 Dynamic effect of short-circuits on windings with unbalanced ampere-turns

The short-circuit forces arising in windings with unbalanced ampere-turns may considerably exceed those occurring in windings of balanced ampere-turns of transformers otherwise identical. The direction of forces occurring in the windings of unbalanced ampere-turns tends to increase the existing asymmetry. In ideally symmetrical windings (with fully balanced ampere-turns) the axial forces tend to compress the winding and are in equilibrium with each other, exerting no force on structural parts outside the windings. The axial force brought about by asymmetry, while tending to increase this asymmetry, have an effect on other structural components of the transformer. No observable increase of radial forces due to the presence of asymmetry is to be expected, so both in the case of symmetry and asymmetry, the formulas (3.26), (3.27), (3.28) and (3.29) are equally suitable for determining the stresses caused by radial forces. It should be noted that a perfectly symmetrical winding system cannot be accomplished, because even in the simplest cases, inaccuracies of manufacture or uneven shrinkage of windings in the course of drying will always introduce some slight asymmetry. Therefore, it is advisable that the clamping structure of windings is dimensioned to withstand the short-circuit force which would result from an asymmetry caused by a relative axial displacement of 1 per cent of the winding length between the centres of ampere-turns of the two windings. Several constructional requirements are to be fulfilled in the design, such as the exclusion of turns of tapped windings, which are liable to give rise to substantial additional forces many times higher than those arising in a symmetrical winding system.

The supplementary magnetic field developing in an asymmetrical winding system can be explained by the method of superposition shown in Fig. 3.12. The two windings placed asymmetrically are shown in Fig. 3.12(a). The field excited by the short winding is the result of the fields excited by windings 2 and 3; asymmetry invariably associated with axially unbalanced ampere-turns, and the radial magnetic field (winding 3 of Fig. 3.12(b)) produced by the resultant excitation will be the consequence of this unbalance. One possible way of calculating the transverse magnetic field is to determine the permeance of the transverse flux duct of unit length in the axial direction. The permanence for radial magnetic field in a section of unit length in the axial direction is

$$\lambda = \frac{l_c}{h_r},$$

Table 3.5

Values for λ

Serial No.	Cross-section of winding	Values of λ for ratio window height dia. of circumscribed circle of limb = 4.2	Values of λ for ratio window height dia. of circumscribed circle of limb	
			> 4.2	< 2.5
	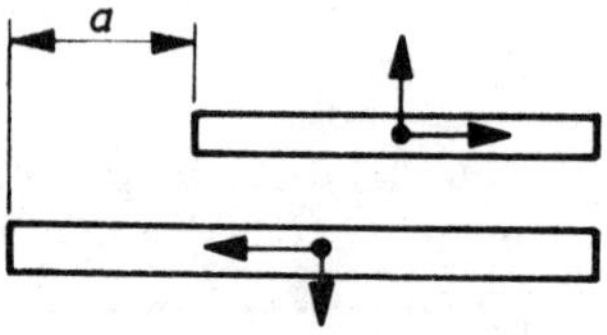	5.5	7.5	
2	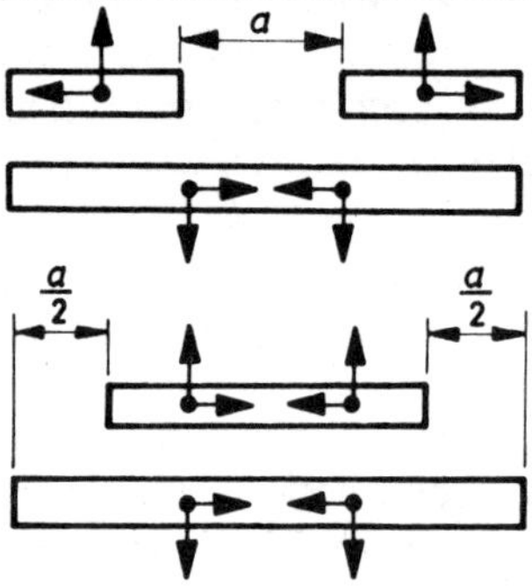	5.8	5.75	7.75
3	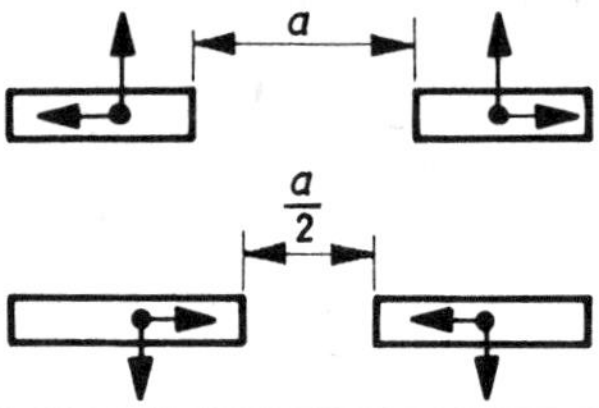	5.8	5.75	7.75
4	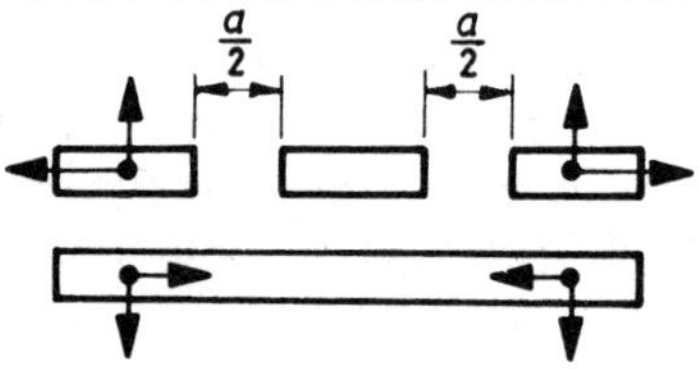	6.0	6	8
5	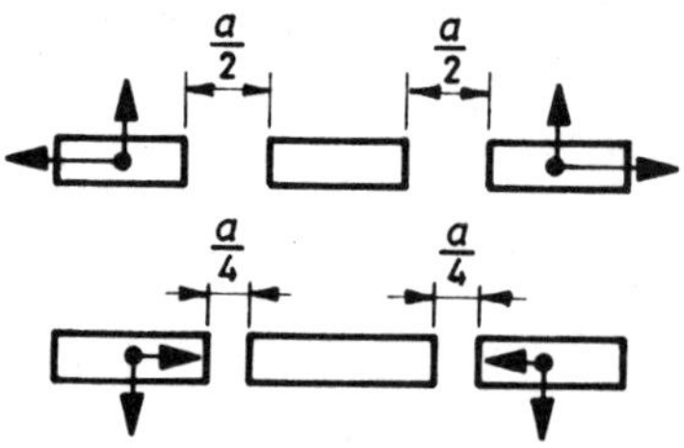	6.0	6	8

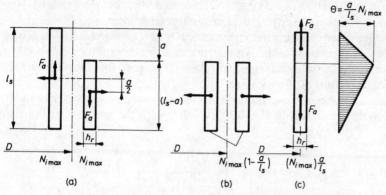

Fig. 3.29. Short-circuit forces acting on asymmetrical windings

where l_c is the mean periphery of the leakage channel and h_r is the length of the radial flux channel. Its magnitude is rather difficult to determine. Waters performed measurements and found that the values of λ depend on the ratio of height of the core window to the diameter of the circumscribed circle of the limb and on the ratio of the latter diameter to the mean winding diameter. Five different cases are compiled in Table 3.5. From among the window height/circle diameter ratios occurring in practice, the values of 2.5 to 4.2 are the most commonly used. Values below 2.5 are encountered in the design of very large transformers of a size approaching the limit of transportability [139].

The case when asymmetry is due to one winding being shorter than the other is shown in the first line of Table 3.5. The related conditions of excitation can be seen in Fig. 3.29. The winding cross-sections are represented by drawing (a), and the conditions of excitation shown can be derived from the superposition of drawings (b) and (c). The peak value of radial flux density represented in drawing (c) is

$$B_{r\max} = \mu_0 \frac{a}{l_s} N\hat{i} \frac{1}{h_r},$$
(3.32)

where dimensions a and l_s are as explained in Fig. 3.29. The radial flux duct length, h_r, can only be estimated, because of the small radial dimension of the space concerned, but its value will not be needed in the following with introduction of the concept of λ. The magnitude of axial repulsing force F_a acting on both windings can be calculated by multiplying the mean magnetic induction $B_{r\max}/2$ by the overall conductor length and by the current flowing in the conductor. In formula (3.33) the repulsing force is obtained in terms of N, if μ_0 is substituted in H/m, current $\hat{i}$ in A and the geometrical dimensions in m:

$$F_a = \frac{\mu_0}{2} \frac{a}{l_s} (N\hat{i})^2 \ \frac{D\pi}{h_r} = \frac{\mu_0}{2} \frac{a}{l_s} (N\hat{i})^2 \lambda .$$
(3.33)

Information about D is given in Fig. 3.29, and the values for λ are to be found in the first line of Table 3.5.

133

Comparing relation (3.30) with (3.33) it can be seen that the axial repulsing force and the symmetrically arranged axial compressive force are of identical order of magnitude. When checking the asymmetrical windings, the calculation according to formula (3.30) serves, as in the symmetrical case, for determining the spacer surface, whereas the force obtained from formula (3.33) is used for dimensioning the axial clamping structure of the winding.

The cross-sections of an axially displaced pair of windings, both having the same axial length, are shown in Fig. 3.30, together with the lateral distribution of flux

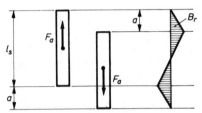

Fig. 3.30. Short-circuit force acting on windings axially displaced with respect to each other, and radial flux distribution

density associated with such an arrangement. The repulsing force is obtained, on the basis of considerations as before, from the following formula:

$$F_a = \mu_0 \frac{a}{l_s}(1 - 0.5a)(N\hat{\imath})^2\lambda . \tag{3.34}$$

The axial repulsing forces are calculated from formula (3.34) using for λ the values given in the first line of Table 3.5.

When examining the formulae for the forces arising in the cases of asymmetry represented in Fig. 3.29 and 3.30, it is evident that they are of considerable magnitude. In order to avoid such great short-circuit forces, attempts are made to eliminate, or at least to reduce, the asymmetry whenever it is feasible. The two cases of asymmetry mentioned above have to be reckoned with, as they can be expected to occur, at least to some lesser extend due to manufacturing inaccuracies or uneven shrinkage of windings in normal service. Should such asymmetry be reckoned with for some reason (the presence of tappings, etc.), the winding arrangements listed in the 2nd, 3rd, 4th and 5th rows of Table 3.5 are selected for ratings below 25 MVA. Beyond 25 MVA, asymmetry should be avoided.

With excluded windings in the middle or at the two ends of either winding (see 2nd row of Table 3.5), the magnitude of the repulsion force is

$$F_a = \frac{\mu_0}{8}\frac{a}{l_s}(N\hat{\imath})^2\lambda . \tag{3.35}$$

In such cases, the additional forces due to asymmetry will result in a distribution of repulsion forces differing considerably from that expected in a symmetrical arrangement. This is because, in contrast to the compressive forces increasing from the winding ends toward the middle, the resultant of forces acting from the middle

134

toward the winding end modifies the distribution, and at some points causes compression forces acting on the spacers to increase. The most critical effects occurring in such cases are calculated by adding to the axial compression force acting on the inner winding (which may be 70% of the force obtained from formula (3.30)) the force resulting from formula (3.35):

$$F_{a\max} = \frac{2\mu_0}{6}(N\hat{i})^2 \frac{l_c}{l_s^2}\left(\delta + \frac{a_1+a_2}{3}\right) + \frac{\mu_0}{8}\frac{a}{l_s}(N\hat{i})^2\lambda =$$

$$= \frac{\mu_0}{2}(N\hat{i})^2 \frac{1}{l_s}\left[\frac{2}{3}\frac{l_c}{l_s}\left(\delta + \frac{a_1+a_2}{3}\right) + \frac{1}{4}a\lambda\right]. \tag{3.36}$$

The compression force in the outer winding is smaller than this value. The highest axial local stress arises in the current-carrying turns lying in the immediate vicinity of the excluded (unbalanced) section of the winding, and this stress will act on the turns as a bending stress. The magnitude of these forces is

$$f_{a\max} = 0.733 \frac{N'}{N} F_{r\max} \lg\left(\frac{2a}{c} + 1\right). \tag{3.37}$$

The force is obtained from the formula in terms of the units in which $F_{r\max}$ is expressed; the interpretation of a is given in Table 3.5 and c is the axial dimension of the turn or disc most stressed.

If excluded turns in the middle of one winding are facing half as many excluded turns in the middle of the other winding, then the thrust is

$$F_t = \frac{\mu_0}{16}\frac{a}{l_s}\frac{(N\hat{i})^2\lambda}{\left(1 - \frac{a}{2l_s}\right)}. \tag{3.38}$$

Computation of the maximum compression force arising in the axial direction is somewhat more complicated, but its value is substantially smaller than that resulting from formula (3.30). Similarly, the value bending force acting on the current-carrying turns in the vicinity of excluded turns is also considerably smaller than that obtained from formula (3.37).

When the excluded turns are located in the first and third quarters of one winding, the thrust (4th row of Table 3.5) is

$$F_a = \frac{\mu_0 a(N\hat{i})^2}{32l_s}. \tag{3.39}$$

The compressive force is of similar magnitude to that acting on a winding of similar dimensions but of symmetrical arrangement. The critical $f_{a\max}$ force acting on the outermost wires can be calculated from formula (3.37) but, corresponding to the notation appearing in the diagram $a/2$ should be substituted instead of a.

The magnitude of the additional force resulting from asymmetry can be further reduced by making the excluded turns in the first and third quarters of one winding face a section of the second winding also containing excluded turns, but whose number is only half as many (fifth row of Table 3.5) as in the first winding. In such a case, the thrust is

$$F_a = \frac{\mu_0 a(N\hat{\imath})^2 \lambda}{64l_s\left(1 - \dfrac{a}{2l_s}\right)}. \tag{3.40}$$

Also in this case, the compressive force is of similar magnitude to that occurring in a symmetrical winding of identical size, and the maximum force acting on the outermost wires can be obtained from formula (3.37), again substituting the value of $a/2$ instead of a.

Example 3.7. Suppose the short-circuit strength of a 25 MVA transformer is to be checked. The vector group is *Yd* 11, voltage ratio 120±15%/11 kV, and line current ratio 120/1310 A, cross-section of windings as shown in Fig. 3.31, and the material of the windings is copper.

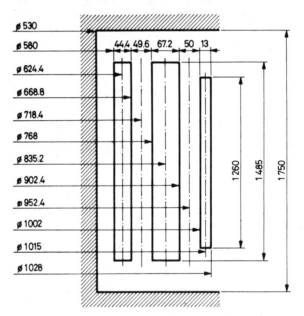

Fig. 3.31. Winding arrangement of a 25 MVA transformer

The H. V. winding comprises 1000 turns, divided into 72 discs in the following sequence: 6 discs (3 at the top and 3 at the bottom) with a wire cross-sectional area of 14.5×2.5 mm^2 (including insulation, 16.3×4.3 mm^2), with 13 turns in each disc. Adjoining these two groups of discs, proceeding toward the inside of the winding, there is one disc at either end, with a wire cross-section of 13×3 mm^2 (with insulation 14.8×4.8 mm^2), also with 13 turns in each disc. The rest of the winding consists of 64 discs, with wire cross-sectional area of 13×3 mm^2 (with insulation 14.8×4.8 mm^2), with 14 turns in each disc. The turns are wound with the longer sides of the wires lying vertically and their shorter sides horizontally.

The L.V. winding comprises 159 turns, the turns being arranged in 106 discs, so that the number of turns per disc is 1.5. This figure is obtained in the following way: in each disc 12 wires are accomodated from the group of 8 parallel wires, hence in addition to the 8 parallel branches of one complete turn, half of the cross-sectional area of the next turn is also located in this disc. The next disc includes 8 parallel

136

wires of the second turn, while the 4 remaining branches of the preceding disc and the 4 remaining branches of the second disc constitute the third turn of the two discs.

The L.V. winding consists of 8 parallel wound wires, each having a cross-sectional area of 9.3×9 mm^2 (with insulation 10×9.7 mm^2), with the smaller conductor dimension arranged in the horizontal and the larger in the vertical direction.

The tapped winding contains 170 turns. Dimensions of the wire are 10×4 mm^2 (with insulation 13×7 mm^2), the larger dimension being horizontal and the smaller vertical.

The p.u. value of impedance voltage is $\varepsilon_z = 0.09$, the resistance voltage is $\varepsilon_r = 0.005$ and the reactance voltage is $\varepsilon_s = 0.09$.

Capability of withstanding thermal effects of short-circuits

Since the winding most exposed to the harmful effects of transient temperature rise is the one in which the current density is highest during a short-circuit, it is sufficient to check the L.V. winding. The symmetrical short-circuit current, considering the delta connection is

$$\left(I_l = \frac{1310}{1.73} = 756 \text{ A} \right),$$

$$I_{ls} = \frac{I_l}{\varepsilon_z} = \frac{756}{0.09} = 8400 \text{ A}.$$

The current density during a short-circuit is

$$J_s = \frac{I_{ls}}{A} = \frac{8400}{8 \times 27.6} = 38 \text{ A mm}^{-2}.$$

The duration of the short-circuit current is taken as $t = 2$ s. At the beginning of the short-circuit the average winding temperature in the most adverse case (i.e. at 40 °C ambient temperature and the 65 °C maximum permissible average temperature rise of the winding at full rated load) is $\Theta_0 = 105$ °C. Selecting from Table 3.4, in advance, for factor a the value 7.41 corresponding to 140 °C and substituting into formula (3.21):

$$\Theta_1 = \Theta_0 + aJ^2 t \, 10^{-3} = 105 + 7.41 \times 38^2 \times 2 \times 10^{-3} \text{ °C} =$$

$$= 126.4 \text{ °C}.$$

This value is below the permissible 250 °C, hence the cross-sectional areas of all windings are sufficient as regards the short-circuit temperature rise.

Dynamic ability to withstand short-circuits

The ratio of stray reactance to resultant resistance of the transformer is

$$\frac{X_s}{R} = \frac{\varepsilon_s}{\varepsilon_r} = \frac{0.09}{0.005} = 18.$$

The factor $k\sqrt{2}$ read from Fig. 3.26 for the above value is 2.6. The peak value of short-circuit current is

$$\hat{i}_h = \frac{k\sqrt{2} I_h}{\varepsilon_z} = \frac{2.6 \times 120}{0.09} = 3467 \text{ A}.$$

Calculation of radial forces

The short-circuit force arising in the H.V. winding (formula (3.26(b)), if the number of magnetically balanced groups is $g=1$) is

$$F_{r\,max\,h} = \frac{\mu_0}{2}(Ni)^2 \frac{l_k}{l_s}\frac{1}{g} = \frac{1.256 \times 10^{-6}}{2}(10^3 \times 3.47 \times 10^3)^2 \frac{0.8352}{1.488} =$$

$$= 13.2 \times 10^6 \text{ N}.$$

The mean tensile stress, with A_h being the cross-sectional area of wire in the H.V. winding (see formula (3.27)) is

$$\sigma_h = \frac{F_{r\,max}}{2\pi A_h N_h} = \frac{13.2 \times 10^6}{2\pi\, 0.352 \times 10^3} = 6000 \text{ N cm}^{-2}.$$

This value is smaller than 8000 N cm², the maximum stress permissible as regards deformation.
The radial short-circuit force arising in the L.V. winding is (because $N_h \hat{i}_h = N_l \hat{i}_l$)

$$F_{r\,max\,l} = \frac{1.256 \times 10^{-6}}{2}(10^3 \times 3.47 \times 10^3)^2 \frac{0.6244}{1.488} =$$

$$= 9.92 \times 10^6 \text{ N}.$$

The mean compression force (with $A = 8.276$ cm², because of the eight parallel branches and 0.276 cm², conductor area per branch) is

$$\sigma_l = \frac{F_{r\,max}}{2\pi A_l N_l} = \frac{9.92 \times 10^6}{2\pi 8 \times 0.276 \times 159} = 4500 \text{ N cm}^{-2}.$$

The L.V. winding should be checked also for buckling. The minimum number of axial spacers (with horizontal wire dimension of 3 mm), by formula (3.28), is

$$Z_{min} = 2\sqrt{1 + \frac{1.75\, F_{r\,max}\, l_h^2}{E N_l A_l (2\pi b)^2}} =$$

$$= 2\sqrt{1 + \frac{1.75 \times 9.92 \times 10^6 \times 62.44^2}{12.5 \times 10^6 \times 159 \times 8 \times 0.276(2\pi\, 0.3)^2}} = 12.$$

24 spacers are applied because, as will be seen later, the forces acting on the bottom and top turns of the winding require closer spacers.

Calculation of axial forces

For calculation of the axial force, the resultant radial force is first calculated from the mean length of stray duct of the transformer. The magnitude of the resultant radial force is

$$F_{r\,max} = \frac{1.256 \times 10^{-6}}{2}(10^3 \times 3.47 \times 10^3)^2 \frac{0.7184 \times \pi}{1.488} =$$

$$= 11.4 \times 10^6 \text{ N}.$$

The resultant axial force (formula (3.30)) is

$$F_{a\,max} = F_{r\,max}\left(\frac{a_1 + a_2}{3} + \delta\right)\frac{1}{l_s} = 11.4 \times 10^6 \left(\frac{0.0444 + 0.0672}{3} + 0.0496\right)\frac{1}{148.8} =$$

$$= 6.85 \times 10^5 \text{ N}.$$

138

The stress imposed on the windings situated closer to the core is higher, therefore it is appropriate to dimension the inner (L.V.) winding to 100% of the resultant axial force ($F_{a\,max}$) and the outer winding to 50% of the latter.

The stressed surface of the L.V. winding is equal to the product of its radial dimension and mean length: $4.44 \times 62 \times 44 = 870$ cm^2. Of this area, 35% ($870 \times 0.35 = 304$ cm^2) is covered by inter-disc spacers. The loaded surface of each spacer, since 24 spacers per disc are used, is $304/24 = 12.7$ cm^2. The thrust acting on the spacers along the centre line of the winding is

$$\frac{685\,000}{304} = 2250 \text{ N cm}^{-2},$$

which lies below the permissible value of 3000 N cm^{-2}.

The stressed surface of the H.V. winding is $6.72 \times 83.52 = 1765$ cm^2. Of this surface, 20% is covered by the area of spacers, i.e. $1765 \times 0.2 = 353$ cm^2. The thrust acting on the spacers in the middle of the winding (it is sufficient to calculate with 50% of the resultant axial force) is

$$\frac{685\,000 \times 0.5}{353} = 970 \text{ N cm}^{-2},$$

which is permissible.

The axial forces acting on the discs located at the top and bottom of the winding should also be checked. In the L.V. winding, the number of turns in the first and in the last disc is 1.5, and the axial dimension of the wire is 9.3 mm (formula (3.31)):

$$f_{amax,\,l} = 0.366 \frac{N'}{N} F_{r\,max} \lg \left[1 + \frac{4\left(\delta + \dfrac{a_1 + a_2}{3}\right)^2}{c^2} \right] =$$

$$= 0.366 \frac{N'}{N} F_{r\,max} \lg \left[1 + \frac{4\left(0.0496 + \dfrac{0.0444 + 0.0672}{3}\right)^2}{0.00932} \right] = 10^5 \text{ N}.$$

This force causes a bending stress in the conductors of the outer discs. The magnitude of this stress can be calculated by considering the wires in the uppermost disc as a beam with uniformly distributed load, clamped at both ends between two adjacent spacers. In that case the bending moment in the cross-section of clamping is

$$M = \frac{f_{amax,\,l} l_s}{12 n_s n_w} = \frac{10^5 \times 8.2}{12 \times 24 \times 12} = 236 \text{ N cm},$$

where l_s is the periphery divided by the number of spacers, n_s is the number of spacers and n_w is the number of wires in the uppermost disc.

The maximum stress arising in the wire of $b \times c$ cross-sectional area is

$$\sigma = \frac{6M}{bc^2} = \frac{6 \times 236}{0.3 \times 0.93^2} = 5460 \text{ N cm}^{-2}.$$

The winding is compressed by the radial forces. At the outer periphery of the winding, the compression stress is about half of that occurring in the middle (2250 N cm^{-2}), thus the compression stress in the conductor is

$$5460 + 2250 = 7710 \text{ N cm}^{-2}.$$

The force acting on the outer disc of the H.V. winding is

$$f_{a\,max\,H} = 0.366 \times \frac{13}{1000} \times 11.4 \times 10^6 \times$$

$$\times \lg \left[1 + \frac{4\left(0.0496 + \dfrac{0.0444 + 0.0672}{3}\right)^2}{0.01452} \right] = 117.2 \times 10^3 \text{ N}.$$

The moment acting on the wire, taken as a beam clamped at both ends ($l_s = 11$ cm, $n_s = 24$ and $n_w = 13$), is

$$M = \frac{11.72 \times 10^4 \times 11}{12 \times 24 \times 13} = 344 \text{ N cm}.$$

The highest tensile stress

$$\sigma = \frac{344 \times 6}{0.25 \times 1.452} = 4100 \text{ N cm}^{-2}.$$

The tensile stress occurring at the edge of the winding under the effect of radial forces is half of the stress arising in the middle of the winding, i.e. its value is 3000 N cm^{-2}. Thus, the maximum tensile stress occurring in a conductor under the effect of axial and radial forces is $4100 + 3000 = 7100 \text{ N cm}^{-2}$, which lies below the permissible value.

Forces acting on a tapped winding

The tapped winding is always the outermost part of the winding system. Only a few flux lines are linked with that winding, therefore the radial forces are small. On the other hand, the axial force may be considerable, especially if (as in the present case) the tapped winding is shorter in the axial direction than the main winding.

Figure 3.32 shows the dimensions of main and tapped windings together with the excitation diagrams to its two extreme tap positions. It should be noted that this distribution of flux lines, mainly because of the different axial lengths of the windings, is an approximation only. It can be seen in Fig. 3.32 that the tapped winding excited in the sense opposed to the main winding, together with a part of the turns of the main winding (the excitation of which part just equals that of the tapped winding), is similar to the case of a two-winding transformer.

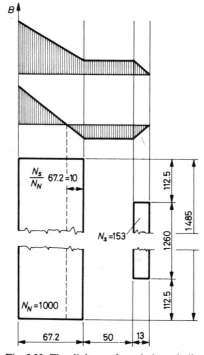

Fig. 3.32. Flux linkage of regulating winding

Based on this approximation, let the axial forces acting on the tapped winding be examined. The asymmetry appearing in this arrangement corresponds to the case of the 2nd row of Table 3.5. The ratio of window height to circumscribed circle of the limb defining the value of λ is $1750/530 = 3.3$. The magnitude of λ pertaining to 3.3 is obtained by interpolation (the values of λ vary in the range 2.5 to 4.2, while the ratio characterizing the core varies from 7.75 down to 5.75)

$$\lambda = 5.8 + \frac{3.3 - 2.5}{4.2 - 2.5}(7.75 - 5.8) = 6.72.$$

The 170 turns of the tapped winding are divided into 10 equal sections. Of these 170 turns $9 \times 17 = 153$ are inserted in extreme positions of the tap changer. (The connection diagram of tap changing is shown in Fig. 3.5)

On the basis of formula (3.35), the thrust acting on one half of the winding, using the notation of Fig. 3.32, is

$$F_t = \frac{\mu_0}{8} \frac{a}{l_s}(N\hat{\imath})^2\lambda = \frac{1.256 \times 10^{-6}}{8}\frac{22.5}{148.8}(153 \times 3.47 \times 10^3)^2 \times 6.72 =$$

$$= 4.5 \times 10^4 \, \text{N}.$$

The surface of the series winding is $1.0 \, \pi \, 101.5 = 318 \, \text{cm}^2$. The pressure acting on the adjacent turns is

$$\frac{4.5 \times 10^4}{318} = 142 \, \text{N cm}^{-2}.$$

This value is much smaller than the permissible limit of $4000 \, \text{N cm}^{-2}$.

The tapped winding, when excited in the same sense as the main winding, increases the radial forces acting on the turns located in the outer layer of the H.V. winding, yet this effect is of no importance due to the low excitation of the tapped winding.

3.5.4 Sophisticated methods

Experience has shown that the calculation methods described on the preceding pages give good overall information on the short-circuit forces, so they have retained their suitability for making quick surveys or comparisons of conditions to be expected. It is, however, obvious that modern computerized methods can furnish more detailed and more accurate data concerning short-circuit forces:

1. The forces acting on the elements of a winding system can be taken into account.

2. The influence of adjacent windings can be computed.

3. The effects of almost any arrangement of active parts and asymmetries can be calculated.

These methods seem to have been fully developed, so the dynamic forces can be calculated by them. The following account is given of one such method, known in literature as Roth's method.

Some important conditions of its application are as follows:

1. The method is a two-dimensional (planar) method based on a solution of the Laplace–Poisson equation. From the density function of ampere-turns δ and equation

$$\Delta^2 A = \mu_0 \delta,$$

the vector potential A is computed, then by Ampere's formula

$$dF = iBdl$$

the forces acting may be computed.
Expressing the above by planar coordinates:

$$\frac{\partial^2 A(x, y)}{\partial x^2} + \frac{\partial^2 A(x, y)}{\partial y^2} = \mu_0 \delta(x, y). \tag{3.41}$$

The ith element of the cross-section of a winding is shown in Fig. 3.33, and within this conveniently chosen element, the ampere-turn density may be considered as

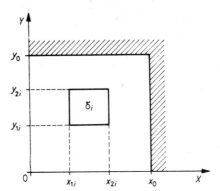

Fig. 3.33. The ith element of the cross section of a winding

being constant. In the case of the ith element the ampere-turn density as a function of position is:

$$\delta_i = \frac{(NI)_i}{(x_{2i} - x_{1i})(y_{2i} - y_{1i})}. \tag{3.42}$$

It is sufficient to approximate function $\delta(x, y)$ either with the first 20 sine and 20 cosine terms of the Fourier series or with 20 exponential terms. The boundary conditions to be considered in solving the equation are as follows:

2. Within the windows, i.e. in the plane laid across the limbs, the flux density is $\frac{\partial A}{\partial x} = 0$ for all values of y where $x = 0$ and $x = X_0$. Further $\partial A/\partial y = 0$ for all values of x where $y = 0$ and $y = y_0$.

3. Outside the windows, e.g. in the plane containing the axis of a limb and lying perpendicular to the plane of the limbs it is convenient to have the boundary conditions $x = 0$, $x = X_0$ and $y = 0$, $y = Y_0$ located sufficiently far (at least 1 metre) away from the windings.

The force in the x direction acting on an element of Fig. 3.33 is

$$|\bar{F}_x|_i = \int\limits_{x_{1i}}^{x_{2i}} \int\limits_{y_{1i}}^{y_{2i}} \delta_i \frac{\partial A}{\partial x} [1_x] \, dx \, dy \tag{3.43}$$

and the force in the y direction is

$$|\bar{F}_y|_i = \int\limits_{x_{1i}}^{x_{2i}} \int\limits_{y_{1i}}^{y_{2i}} \delta_i \frac{\partial A}{\partial y} [1_y]\, dxdy. \tag{3.44}$$

In relations (3.43) and (3.44) 1_x and 1_y are unit vectors in the direction of axes x and y, respectively, A is the vector potential in Vs m^{-1}, dimensions x and y are given in m, $(F_x)_i$ and $(F_y)_i$ are specific forces per unit length in their respective directions in terms of N m^{-1}.

The computation method described is suitable for calculating short-circuit forces both in core- and shell-type transformers, and in a further step, for summing-up the stresses arising in the coil elements.

Computerized methods give more reliable results for dynamic stresses. Results obtained at testing stations indicate that the stresses arising during short-circuits are rather complex. This is proved also by the fact that a transformer defect very seldom occurs during the first 5 to 10 cycles of a short-circuit, i.e. in the period

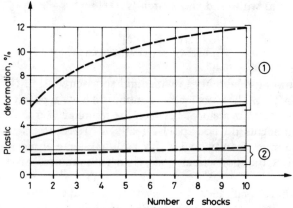

Fig. 3.34. Plastic deformation of pressboard spacers, caused by shocks arising during short-circuits [93]

during which the highest forces are imposed on the transformer windings. During a short-circuit, several simultaneous effects have to be reckoned with, such as vibration of windings, their elongation in longitudinal and transverse directions due to the short-term heating effect of the short-circuit current, rubbing of windings against the spacers and of the turns against one another, gravitational effects acting on the larger winding, inertia of the latter, etc. No method—at least no method of universal validity—has yet been developed for calculation of these effects. The frictional resistance and elastic modulus of insulating materials being dependent on the processing of surfaces (machining, heat treatment, etc.) vary widely from manufacturer to manufacturer, and the modulus of elasticity is far from being linear. In Figs 3.34 and 3.35, for two kinds of pressboard, the plastic deformation and the variation of elastic modulus are shown in terms of the number of shocks. It is clearly demonstrated that the quality of spacers built into the transformer may alone

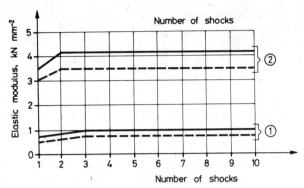

Fig. 3.35. Changes of elastic modulus of pressboard spacers,
due to shocks arising during short-circuits [93]

contribute to a large extent to the improvement of short-circuit stability. The pre-treatment applied in the course of production may also considerably influence the ability of spacers to withstand short circuits [93].

3.5.5 Short-circuit tests

For demonstrating the ability to withstand short-circuits, tests are specified in IEC Publication 75–5. These are neither routine tests, nor type tests, but are termed special tests, subject to agreement between the manufacturer and purchaser. The following requirements do not apply directly to transformers of ratings above 40 MVA having more than two windings, and may serve at most as a guide. For these transformers, the short-circuit tests, if such are required, are specially agreed upon.

Prior to the short-circuit tests the transformer should be subjected to the routine tests, and on units provided with tapped windings, resistance and reactance measurements should be carried out in the tap position used for the short-circuit test. Reproducibility of reactance measurements should be within $\pm 0.2\%$. At the beginning of the test the winding temperature should be in the range of 0 °C to 40 °C. The test should be carried out by applying the asymmetrical short-circuit current calculated from formula (3.18), and the peak value of the same (formula (3.25)).

The values adopted at the testing station should not differ from the result of formula (3.18) by more than 10%, nor from that of formula (3.25) by more than 5%. In the case of transformers of ratings below 3150 KVA, or if the impedance of the system normally supplying the transformer is lower than the transformer impedance, the X_s/R values of Fig. 3.26 should be calculated from the transformer data, whereas in the case of higher system impedances the values of X and R are the sums of the X and R values of the transformer and system.

It is permitted by the test standard to have either the winding nearer to the core or that further from the core-circuited, and the other connected to the supply. It is also

144

permitted either to connect the transformer to the supply in the no-load condition and then to short-circuit the transformer, or to energize the transformer in the short-circuited condition (pre-set short-circuit). In the latter case the winding further from the core should be the one supplied, or else the high magnetizing currents of the core would appear during the first few cycles of the input current. If, however, the secondary terminals of a transformer already energized are short-circuited, the voltage at the primary terminals preceding the short-circuit should not be more than 1.15 times the rated voltage. For transformers with shell-type or multiple-concentric windings, the pre-set short-circuit method should be used only after agreement between the manufacturer and the purchaser. Whenever possible, the tests should be carried out with a three-phase supply. If this is prevented by the limited capability of the testing station, the short-circuit test may be performed using a single-phase supply. For star-connected transformers, the single-phase supply voltage is $2/\sqrt{3}$ times that of the three-phase test voltage whereas in delta connection it is identical with the three-phase test voltage.

From the point of view of the tests, the switchings should not be considered as separate test stresses when the symmetrical current is lower than 70% of the test current. The number of short-circuits applied to transformers below 3150 kVA is three, the duration of each test being 0.5 s with a tolerance of $\pm 10\%$. Usually, the three tests are performed in tap positions corresponding to the lowest, highest and middle voltage ratios. For transformers with ratings higher than 3150 kVA, the test duration, number of switchings and tap position are subject to agreement.

After completion of the short-circuit tests, in order to summarize any defects, the following points may be considered:

1. Whether no defects are indicated by the current and voltage oscillograms taken during the test.

2. Whether the short-circuit reactances do not show a deviation (increase) of more than 2%.

3. Whether no abrupt irregular changes are shown by the oscillograms of the pressure gauge accomodated in the transformer, vibrometer mounted on the tank or acoustimeter.

4. Whether no current between tank and earth is revealed by the oscillographic record.

5. Whether no gases are found in the gas relay, indicative of an internal defect.

6. Whether changes are shown in the oscillograms taken after the test with respect to those taken before the tests, by applying repetitive low-voltage surges (recurrent surge oscillograph method).

7. Whether differences are found between results of repetitive routine tests (dielectric tests, made with 75% of the original test values).

8. No signs of deformation are detected on the transformer, when removed from its tank.

Performance of all the checks listed above is seldom possible. Fault detection methods of major importance are: method of detecting changes in short-circuit reactances (always possible), repetition of routine tests (always possible), and untanking or complete dismantling of the transformer, which is usually carried out only on units delivered in large numbers.

4. STRAY-FIELD LOSSES
IN STRUCTURAL PARTS OF TRANSFORMERS

The problem of stray-field losses becomes increasingly important with growing unit ratings. This can easily be seen when the relation between stray field strength and transformer rating is determined for transformers of different ratings and voltage levels.

At unit ratings of up to about 100 to 150 MVA, transformers of various sizes and rated voltages may be regarded as having cores and windings of geometrically similar shapes. Under such conditions the stray magnetic field strength is

$$H = \frac{IN}{l} = \text{constant} \times \sqrt[4]{P}, \quad \text{A m}^{-1},$$

where I is the current flowing in the H.V. winding in A, N is the number of turns, l is the effective length of the leakage field in m and P is the power of one limb in VA. Both the current and voltage of the transformer increase proportionally with the square root of the unit rating. The number of turns in the H.V. windings are about equal, because both the rated voltage and the turn-to-turn voltage increase proportionally with the square root of rated power (assuming constant flux density in the core and increase of cross-sectional area of the core proportional to the square root of the power rating). The linear dimensions being proportional to the fourth root of power, the field strength will also be proportional to the fourth root of the rated power.

Above unit ratings of about 100–150 MVA the lateral and vertical dimensions of the transformer are limited by the railway loading gauge. Hence, the cross-sectional area and height of the core may be considered approximately constant within the space confined by the railway loading gauge, so the relation between stray field strength and power rating is

$$H = \frac{IN}{l} = \text{constant} \times P.$$

As before, the current is proportional to the square root of the unit rating, and the number of turns (with the cross-sectional area of the core considered unchanged) now also varies in proportion to the square root of rated power. Although this train of thought is not strictly exact, nevertheless it clearly demonstrates that in larger transformers (limited in their size by the loading gauge) the stray field strength increases with growing rating much faster than in smaller transformers that are, more or less, geometrically similar. The stray flux intruding into the structural parts gives rise to eddy currents in them. In parts of more extensive size (such as in tank walls), the eddy-current losses caused by the stray flux may be considerable, thereby increasing the load losses of the transformer and lowering its efficiency. In smaller metallic parts the eddy-current losses are negligible as compared to other losses of the transformer and cannot even be detected by electrical measurements. Still, in such a case the loss density (eddy-current loss per unit of volume) tends to attain excessive levels that may lead to hazardous local temperature rises. Such high temperatures, if occurring in areas of high field strengths, may cause deterioration of the insulation, thereby jeopardizing the service reliability of the transformer. Designers of large transformers are required, on the one hand, to keep the losses caused by stray flux below an acceptable level, and on the other hand, to prevent hot spots from developing in metallic parts located in the stray field.

4.1 Structural parts critical from the point of view of stray losses and temperature rises

In view of potential power losses and temperature rises, the metallic parts located in the stray field of windings, and the stray field of conductors and connections carrying high currents should both be considered.

The stray flux departing radially from the outer surface of windings gives rise to eddy-current losses, in the first place, in the transformer tank walls. Though the stray flux density in the tank wall is rather low (its maximum is in the range of 0.005 to 0.015 T in large transformers), the tank losses may be considerable and represent by far the greater part of the losses caused by the stray flux because of the large size of the tank. Hot spots seldom develop in the tank walls. The heat produced is carried away by the oil in contact with the wall, thereby equalizing differences between surface temperatures. The development of hot spots is further impeded by spreading of the heat due to the good thermal conductivity of the metallic wall.

The stray flux departing radially through the inner surface of windings intrudes into the core and the fittings mounted on it. This flux enters the core partly parallel and partly perpendicular to the latter. Loss, and a consequent temperature rise, is in practice caused only by the flux entering sheets perpendicular to the laminations. Considerable losses may also arise in the flitch plates lying on the outermost packets of the limbs, these plates serving to hold the core laminations together vertically. On the surface of the core flitch plate the flux density may be ten times higher than that developing in the tank surface (its maximum in large transformers is in the range 0.05 to 0.1 T). Although the losses arising in the core and flitch plate are much lower than the tank losses, the local temperature rise may still be considerable, due to the higher loss density figures and poorer conditions of cooling. For holding the core together vertically tie-rods may be used instead of flitch plates. These tie-rods are also located in the stray radial field of the windings, and thus may be dangerous mainly because of their local hot spots.

The stray flux emerging axially from the windings may induce eddy currents in the clamping structures of the windings, in the yokes and in the yoke beams. In the smaller transformers of earlier days, metal clamping rings were used, but for large transformers of modern design they are invariably made of some insulating material (pressboard or wood). The situation is the same with the potential rings bearing on the uppermost discs of a disc-type transformer. These rings are also made of insulating material and only their surfaces are covered with thin metal foil or graphite paint. Although a high loss density may develop in such thin metal layers, their outer surfaces are sufficiently large relative to their thickness to prevent any dangerous temperature rise. In cases when the rings are utilized for mechanical clamping of winding ends, thick metal blocks have to be built into the rings for reinforcement. The losses developing in these blocks represent but a negligible fraction of overall losses of the transformer, not detectable by measurement. Nevertheless, these metallic reinforcements may become critical points of the transformer, by forming local hot spots of elevated temperatures, because they are fully embedded in insulating material and are situated in a high strength electric field.

The losses arising in the yokes and their adjoining structures (yoke beams) are only of concern if they are located close to the windings (as in smaller transformers). In these parts, dangerous hot spots seldom develop.

In large transformers, excessive temperature rise developing in the bolts connecting the flange of the cover to that of the tank may become a hazard. Eddy currents passing through these bolts may cause dangerous hot spots, damaging the sealing between flanges and impairing the insulation of low-voltage wiring and other components that may come into contact with the bolts. All this may even become the source of severe service disturbances.

The electromagnetic field of conductors and internal connections within the transformers causes eddy-current losses in metallic parts situated in their vicinity. Such internal links connect the bushings with the windings, the transformer windings into star or delta, the tappings of the windings to the tap changer, or serve some other similar purposes. The loss caused in the cover by the field of current flowing in the bushing also belongs to this category. A characteristic feature of these losses is that they are distributed along bar conductors and connections, mainly in the cover, in the tank walls, and in the frame clamping structure, i.e. in extensive sheets. Generally, their concentrated development need not be reckoned with, since the field of current-carrying bar conductors is weaker than that of the windings, and the bar conductors being at one potential the isolating clearances represent such long magnetizing paths that restrict the development of higher-strength fields. These losses are, therefore, only of importance from the point of view of overall losses of the transformer (they may reach as much as 10% of the losses in the windings), but they may be ignored as regards local temperature rises.

In addition to the structural parts mentioned which occur in every large transformer, there may be other parts depending on the particular construction (pipes, supporting plates, etc.), which also lie in a stray field and thus may also become sources of loss or hot spots. Their dangerous or harmless character should be decided upon in the knowledge of the construction concerned.

4.2 Permissible temperature rise of structural parts and corresponding magnitude of stray flux density

Values for temperature rise permissible in the various structural parts are not specified in the standards. On the basis of specifications, available for the temperature rise of active parts and of the oil, and from experience gained with transformers, the following considerations should be kept in mind when judging the hazardous character of hot spots:

(a) In the steady state condition, the maximum oil temperature may be 95 °C at the inside of the transformer (assuming 40 °C ambient temperature).

(b) No permissible temperatures are specified by the Standards for the hottest spots of bare or lacquered metallic parts. According to experience, their temperature may be permitted to approach 135–140 °C, provided this temperature arises over a small (few cm²) surface area only, and this area is in contact with a bulk quantity of oil. According to measurements, under these conditions the temperature of the surrounding oil does not even approach the temperature of the hottest spot. (In an experiment, the oil tank of a transformer was heated with a welding torch. The tank wall and the oil adjacent to it were observable from the inside. Although the tank wall discoloured under the effect of intense heating, neither gassing nor tarring was observable in the adjacent oil).

It is not advisable to allow the highest temperature of metallic parts to rise beyond 135 °C over a larger surface area or where only poor flushing is provided, because above this temperature heavy gas formation starts developing in the oil, which would yield misleading information on the condition of transformer insulation. The highest temperature of paper-insulated metallic parts—in a way analogous to the highest temperature permitted for the windings—may be 118 °C.

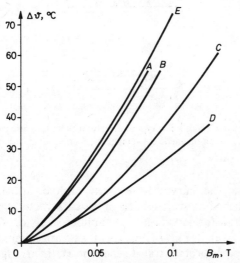

Fig. 4.1. Measured hot spot temperatures as a function of flux density (for arrangements shown in Fig. 4.2)

(In the IEC Publication [84] giving guidelines for conditions of loading 118 °C is permitted as the hottest spot temperature of transformer windings, at 40 °C ambient temperature).

(c) In the case of temperature rises lasting only a few seconds levels as high as 250 °C may be permitted for iron and copper, and 200 °C for aluminium, without harmful consequences. These limits are determined by the metallurgical properties of the materials concerned.

Design work is facilitated by the curves of Fig. 4.1, showing the relation between stray induction and the resulting hottest spot temperature rise for a few simple

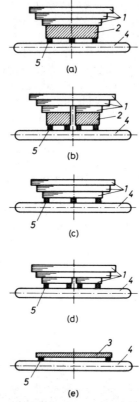

Fig. 4.2. Cross-sections of models of a few typical structural parts located in the radial stray field of windings: *1*—lamination; *2*—non-magnetic clamping plate; *3*—magnetic plate; *4*—field coil; *5*—spacers

structural elements. The curves have been plotted utilizing the results of measurements performed on scale models of typical structural parts located in the radial stray field of transformer windings. Simplified sketches of the models used in the experiments are given in Fig. 4.2. Each of the arrangements (a) to (d) simulates the three outermost packets of the transformer core (1). In arrangements (a) and (b) a 30 mm thick non-magnetic flitch plate (2) is lying on these packets. In arrangement

150

(b) the flitch plate and the packet below it (i.e. the uppermost one) is divided into two parts by a slot. In arrangement (d) the uppermost packet is divided into two parts. The magnetic plate (3) of arrangement (e) simulates the tank wall. In the course of the test, exciting coil (4) produced the flux density distribution of pattern developing in the transformer. The entire equipment was immersed in oil.

It can be seen from the curves of Fig. 4.1 that in the 0.05 to 0.1 T range of flux densities occurring in practice the hot-spot temperature on the core surface without a flitch plate is 10 to 40 °C above that of the surrounding oil, the latter being 95 °C in the worst case. With the use of flitch plates, the temperature rise is approximately doubled. By slotting the flitch plate or uppermost packet, the hot-spot temperature may be reduced by about 10 to 20% (reduction of overall losses is considerably greater). The temperature rise of the tank wall (e), assuming a flux density equal to its usual maximum value of 0.01 T, is negligible.

It should be noted that the curves of Fig. 4.1 apply only to the given arrangement and to generally occurring flux distributions. Therefore they are not of general validity and cannot be immediately used in the practice of transformer design, but may still be considered as valuable in providing a useful starting-point. It can be assessed from the curves whether a hot spot which occurs is dangerous or harmless. In critical cases, however, a more detailed calculation and a knowledge of the distribution of losses will also be required. This subject is dealt with in Section 4.4.

4.3 Thermal stresses in operating conditions, transient overheating

Temperature rises caused by the stray flux may represent a hazard even under normal service conditions, but in two cases they are liable to cause a defect or a permanent deterioration of insulating materials: in the case of overloading and of a prolonged, power-frequency voltage rise. Under overload conditions, due to the saturation characteristics of adjacent ferromagnetic component parts in the leakage channel of the transformer, an increase of flux density causes the flux lines to change their path, and their distribution is also modified. The same symptom occurs with a voltage rise, but the cause is different, this being the saturation of the core, accompanied by a major portion of the main flux leaving the core and, partly, a reduced portion of the leakage flux entering the core.

A sudden increase of load, caused for example by a short-circuit, may also be hazardous. Such abrupt growths of load may lead to a rapid temperature rise of metal parts, jeopardizing the soundness of insulation, even though lasting for a very short time. Transient temperature rises of this kind can be assessed on the basis of the following calculation.

The thermal equilibrium of a metallic body placed in an a.c. magnetic field and heated by eddy currents is given for continuous operating conditions, by the relation

$$p_0 m \, dt = \alpha m (\vartheta_m - \vartheta_s) dt, \tag{4.1}$$

151

where p_0 is the loss arising in unit mass in W kg^{-1}, m is the mass of the body investigated in kg, t is the time in s; α is the quantity of heat in W °C^{-1} kg^{-1} dissipated by conduction, convection and radiation from a unit of mass at a temperature difference of 1 K; ϑ_m is the mean temperature of the body and ϑ_s is the ambient temperature.

When the loss arising in unit mass increases suddenly from p_0 to $p_0 + \Delta p_0$ (where increment Δp_0 may be a multiple of p_0), then the initial conditions are characterized by the relation

$$(p_0 + \Delta p_0) m dt = \alpha m(\vartheta_m - \vartheta_s) dt + cm d\vartheta, \tag{4.2}$$

in which c is the specific heat of the body in Ws kg^{-1} °C^{-1}. From Eqn. (4.2), the following relation is obtained for the initial rate of rise of temperature:

$$\frac{d\vartheta}{dt} = \frac{\Delta p_0}{c}. \tag{4.3}$$

If progressive from the initial condition (ambient temperature) toward the steady-state condition characterized by temperature ϑ_m, then Eqn. (4.2) with the substitutions $\vartheta_m = \vartheta_s$, $d\vartheta = d\vartheta_0$ and $\Delta p_0 = 0$, gives the relation

$$\frac{d\vartheta_0}{dt} = \frac{p_0}{c}. \tag{4.4}$$

The calculation can be simplified if the flux density is used instead of specific loss. It is assumed that flux density B_0 corresponds to loss p_0, i.e. to the steady-state condition, and flux density $B_0 + \Delta B_0$ belongs to the increased loss $p_0 + \Delta p_0$. Since the loss is proportional to the square of the flux density, the temperature rise pertaining to the increased flux density can be calculated from the initial temperature rise corresponding to flux density B_0, following the relation

$$\frac{d\vartheta}{dt} = \frac{\Delta p_0}{c} = \frac{p_0}{c} \left[\left(1 + \frac{\Delta B_0}{B_0} \right)^2 - 1 \right]. \tag{4.5}$$

As it can be seen from relation (4.5), it is sufficient to know the initial temperature rise p_0/c pertaining to steady-state operation, from which the temperature rise resulting from the change of flux density B_0 to $B_0 + \Delta B_0$ can be calculated.

Example 4.1. In a $20 \times 85 \times 140$ mm copper block placed in a magnetic field of flux density $B_0 = 0.04$ T perpendicular to the longest side of the block, a loss of $p_0 = 31.3$ W is induced, whereby its temperature will increase by 22 °C with respect to the surroundings; with 95 °C oil temperature this temperature will be 117 °C. Now, let the load be assumed to increase suddenly, associated with an increase of flux density $\Delta B_0 = 0.08$ T, i.e. the increased flux density will thus be $B = B_0 + \Delta B_0 = 0.12$ T.

For the initial rate of rise of temperature pertaining to flux density B_0 (i.e. to loss p_0), using equation (4.4) and calculating with $c = 377$ J kg^{-1} °C^{-1}, the value of 0.083 °C s^{-1} is obtained. For an increase of flux density ΔB_0, the initial rate of rise of temperature will be 0.67 °C s^{-1}, as given by Eqn. (4.5).

Assume the time constant of the copper block to be 4 minutes, i.e. $T = 240$ s (this value has been found by experiments for bare blocks or for blocks coated with a thin lacquer layer of thickness not exceeding the penetration depth). Then, calculating with the above initial rate of rise of temperature, the temperature rise will be $T d\vartheta/dt = 161$ °C. Adding this to the initial temperature of 117 °C produces 278 °C, which is inadmissible even for a short time. In fact, the temperature will be lower than that in any case, since the rate of rise of the temperature curve decreases as a function of time. Thus, the calculated 278 °C is an upper limit, beyond which the temperature will not rise, yet directs attention to the necessity of somehow reducing this limit, e.g. by slotting the block.

If the flux density increases to $B = 0.26$ T during a short-circuit lasting for 5 s, the temperature rise will be only 23 °C. At the end of this short-circuit the temperature of the block will be only 140 °C, and this is absolutely harmless, since the temperature of copper is permitted to rise to 250 °C for such a short time, as already mentioned.

4.4 Approximate calculation of losses and temperature rises in critical components

4.4.1 Numerical and analytical methods in general

The calculation of losses and temperature rises occurring in structural parts of a transformer is a very complex task. A comprehensive theoretical analysis of the problem is difficult because of the non-linearity associated with the phenomena (effect of saturation of the steel), the three-dimensional field configuration to be calculated, the often complicated shapes of structural parts, and the reaction of induced eddy currents have to be taken into account. It may be expected that with the increasing capability of computers, solution of stray-field problems of any complexity will become possible before long. For the time being, however, there are no generally applicable computation methods available for solving all details without resorting to approximations.

The numerical computation methods known today and applicable in practice are suitable for calculating certain structural elements (tank walls, core, yoke, etc.) only. Most of these methods calculate on the basis of two-dimensional models, and are only capable of approaching the results by introducing simplifications and approximations. In spite of all this, the numerical calculation methods still yield results of the best accuracy obtainable, and further improvement of these methods will lead to the comprehensive and general solution of the problems.

Eddy-current losses arising in metallic parts of rectangular cross-section (bar conductors, blocks, plates) can be calculated also by some analytical methods, which with certain approximations also lend themselves to the calculation of structural parts of less complicated shapes (such as tank and flitch plates) occurring in transformers. Simple formulae are obtained by means of analytical methods, permitting quicker and easier determination of losses, and revealing the role of the various factors having influence on the phenomenon. Their correctness can easily be proved by measurements made on a model or transformer, and their accuracy can be improved by correction factors determined experimentally or from other empirical data. The use of such simple formulae fits well into the present-day practice of transformer calculation. Simple, easy-to-follow approximating formulae clearly showing the role of the various factors affecting the phenomenon are, generally, preferred by designers to computation methods which include all aspects of the problem but are more involved and less understandable.

The role of analytical computation methods and of approximating formulae obtained by them is to put a tool in the hands of the designer to facilitate the assessment of loss arising in a given component part, and to judge whether the temperature rise of that component is to be considered hazardous or harmless. In cases found hazardous, he can immediately decide which factors have to be modified in order to reduce the hazard. In critical cases, the required more accurate determination of figures of loss and temperature rise occurring in the given component part also becomes possible by the use of numerical methods.

Numerical methods are not dealt with here, but reference is made to a few publications of practical importance [18, 23, 26, 89, 140]. In order to facilitate design work an analytical method suitable for approximate calculation of eddy-current losses of rectangular steel plates is described in the following. Its practical application to a few basic structural component parts occurring in transformers is demonstrated by examples.

4.4.2 Analytical method for calculation of losses in plates

The calculation of eddy-current losses of the plate shown in Fig. 4.3 is reduced to the calculation of a two-dimensional model with three layers, as illustrated in Fig. 4.4. The field strength at the edge of the first layer is assumed to vary periodically

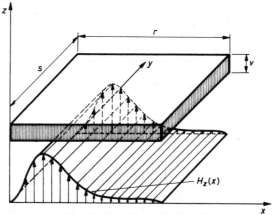

Fig. 4.3. Plate in a magnetic field

with x, along the straight line $z=0$, according to the relation

$$H_{zo} = H_m \sum_{k=1}^{n} b_k \sin k\alpha x \, e^{j\omega t}, \tag{4.6}$$

where $\alpha = \pi/r$ is the angular frequency of the fundamental harmonic of spatial dependence and $\omega = 2\pi f$ is the angular frequency of sinusoidal variation with time, r is the size of the investigated plate or sheet in the x direction in m, and f is the frequency alternating as a function of time in s^{-1}. The Fourier coefficients b_k have to be selected so as to have, over section r, the required distribution. The first and second layers have finite thicknesses, v_1 and v_2 in m, whereas the thickness of the third layer is assumed to be very large ($v_3 \rightarrow \infty$). The three layers are of infinite size in the x direction, and in the y direction perpendicular to the plane of Fig. 4.4, the field strength remaining unchanged in the y direction.

First, the eddy-current loss is determined that arises in a section—of length r in direction x, and of unit length in direction y—of the layers of infinite size (i.e. loss falling under one spatial half-cycle of the field-strength curve). From such calculation, the loss in a sheet of finite length s in direction y is calculated, by applying a multiplication factor. The loss occurring in section $r \times s$ of the middle

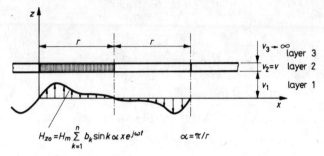

$$H_{zo} = H_m \sum_{k=1}^{n} b_k \sin k\alpha \, x e^{j\omega t} \qquad \alpha = \pi/r$$

Fig. 4.4. Three-layer plane model as basis for the calculation

layer is regarded as the loss of the plate shown in Fig. 4.3. In addition, the following approximations are made:

(a) The field strength developing at the surface of the first layer in the plane $z = 0$ and characterized by relation (4.6) is assumed to remain unaffected by the eddy currents arising in the sheet.

(b) For simplifying the calculation, it is assumed that only the amplitude of harmonic components of the field strength in the z direction (i.e. that perpendicular to the plate) are subject to a change, their wavelength and phase position remaining unaltered. This approximation is permissible as long as $v_1 < 0.25r$.

(c) The permeability of steel plates is considered constant (as being independent of field strength).

(d) The hysteresis loss of steel plates is neglected. This neglect is justified in the case of thick sheets [3]. (It should be noted that about half of the entire loss of best-grade cold-rolled transformer sheets is represented by hysteresis loss. With increasing thickness, however, this proportion drops rapidly)

In our calculations we set out from the Maxwell equations valid for quasi-stationary fields:

$$\Delta \mathbf{H} = \mu \gamma \frac{\partial \mathbf{H}}{\partial t}, \tag{4.7}$$

$$\operatorname{div} \mathbf{H} = 0, \tag{4.8}$$

$$\operatorname{rot} \mathbf{H} = \gamma \mathbf{E}. \tag{4.9}$$

Here $\mathbf{H}$ is the vector of magnetic field strength in A m^{-1}, $\mathbf{E}$ is the vector of electric field strength in V m^{-1}, μ is the magnetic permeability in H m^{-1} and γ is the electric

155

conductivity in S m^{-1}. In our present case, **H** has components in the directions of x and z only ($H_y = 0$), and in the y direction **H** remains constant ($\partial H/\partial y = 0$). From this it follows that a component of current density exists in the y direction only ($i_x = i_z = 0$). It is further assumed that all the three layers have finite electric conductivities.

It can easily be checked that, within our approximating conditions, differential equations (4.7) and (4.8) are satisfied in the case of all the layers by solution

$$H_{zl} = H_m \sum_{k=1}^{n} f_k(x) g_{lk}(z) e^{j\omega t} \tag{4.10}$$

and

$$H_{xl} = H_m \sum_{k=1}^{n} \frac{1}{(k\alpha)^2} f'_k(x) g'_{lk}(z) e^{j\omega t}, \tag{4.11}$$

where

$$f_k(x) = b_k \sin k\alpha x, \tag{4.12}$$

$$g_{lk}(z) = c_{lk1} e^{p_{lk}z} + c_{lk2} e^{-p_{lk}z}, \tag{4.13}$$

$$p_{lk}^2 = (k\alpha)^2 + j\omega\mu_l\gamma_l, \tag{4.14}$$

and where subscript l refers to the layer concerned ($l = 1, 2, 3$). The current density is obtained from the differential equation (4.9), using also (4.10) and (4.11):

$$i_{yl} = j\omega\mu_l\gamma_l H_m \sum_{k=1}^{n} \frac{1}{(k\alpha)^2} f'_k(x) g_{lk}(z) e^{j\omega t}. \tag{4.15}$$

The loss in the elementary prism characterized by coordinates (x, z) and of unit length in the y direction can be calculated from the relation

$$dP_l(x, z) = \frac{1}{2} \frac{1}{\gamma_l} |i_y(x, z)|^2 \, dxdz. \tag{4.16}$$

(The factor 1/2 appears in the formula because $i_{yl}(x, z)$ is the peak value of current). The loss in the respective sheet sections of dimension r in direction x and of unit length in direction y is obtained from Eqn. (4.16) by integration:

$$P_1 = \frac{1}{2} \frac{1}{\gamma_1} \int_0^{v_1} \int_0^{r} |i_{yl}(x, z)|^2 dxdz, \tag{4.17}$$

$$P_2 = \frac{1}{2} \frac{1}{\gamma_2} \int_{v_1}^{v_1+v_2} \int_0^{r} |i_{y2}(x, z)|^2 dxdz, \tag{4.18}$$

$$P_3 = \frac{1}{2} \frac{1}{\gamma_3} \int_{v_1+v_2}^{\infty} \int_0^{r} |i_{y3}(x, z)|^2 dxdz. \tag{4.19}$$

The approximate value of loss in the plate of finite length s in the y direction is obtained by multiplying the above losses by the factor

$$k_s = s \frac{1}{\left(\dfrac{r}{s}\right)^2 + 1},$$
(4.20)

hence

$$P_{sl} = P_l k_s.$$
(4.21)

Integration constants C_{lk1} and C_{lk2} are determined from the boundary conditions.

(a) If $z=0$, then $H_{z1} = H_{z0}$, i.e. from Eqns (4.6), (4.10), (4.12) and (4.13)

$$C_{1k1} + C_{1k2} = 1.$$
(4.22)

(b) If $z \to \infty$, then $H_{z3} \to 0$, hence from Eqns (4.10), (4.12) and (4.13)

$$C_{3k1} = 0.$$
(4.23)

(c) At the boundaries of the layer neither the flux density of direction z, nor the field strength of direction x, changes. Thus, at the boundary between layers 1 and 2, i.e. where $z = v_1$

$$\mu_1 H_{z1} = \mu_2 H_{z2}$$

and

$$H_{x1} = H_{x2},$$

i.e. from Eqns (4.10), (4.11) and (4.13)

$$\mu_1(C_{1k1}e^{p_{1k}v_1} + C_{1k2}e^{-p_{1k}v_1}) = \mu_2(C_{2k1}e^{p_{2k}v_1} + C_{2k2}e^{-p_{2k}v_1})$$
(4.24)

and

$$p_1(C_{1k1}e^{p_{1k}v_1} - C_{1k2}e^{-p_{1k}v_1}) = p_2(C_{2k1}e^{p_{2k}v_1} - C_{2k2}e^{-p_{2k}v_1}).$$
(4.25)

At the boundary between layers 2 and 3, i.e. at $z = v_1 + v_2$:

$$\mu_2 H_{z2} = \mu_3 H_{z3}$$

and

$$H_{x2} = H_{x3},$$

hence

$$\mu_2(C_{2k1}e^{p_{2k}(v_1+v_2)} + C_{2k2}e^{-p_{2k}(v_1+v_2)}) =$$
$$= \mu_3 C_{3k2}e^{-p_{3k}(v_1+v_2)}$$
(4.26)

and

$$p_2(C_{2k1}e^{p_{2k}(v_1+v_2)} - C_{2k2}e^{-p_{2k}(v_1+v_2)}) =$$
$$= -p_3 C_{3k2}e^{-p_{3k}(v_1+v_2)}.$$
(4.27)

By solving the set of Eqns (4.22) and (4.24) to (4.27) the five unknown constants C_{1k1}, C_{1k2}, C_{2k1}, C_{2k2} and C_{3k2} can be found.

The relations dealt with so far have referred to general three-layer arrangements, where any layer may have a finite conductivity. If only the second layer is of metal, with above and below it the presence of air, or some other electrically non-conductive substance, then there will be a loss in the central layer only, whose magnitude can be calculated from the relation

$$P_{s2} = \frac{1}{2}\frac{1}{\gamma_2}s\frac{1}{\left(\dfrac{r}{s}\right)^2+1}\int_{v_1}^{v_1+v_2}\int_0^r |i_{y2}(x,z)|^2 dx dz. \tag{4.28}$$

If the field strength varies spatially over the plate surface according to a simple sinusoidal relation ($v_1 = 0$), i.e. if

$$H_{z0} = H_m \sin \alpha x e^{j\omega t}, \tag{4.29}$$

then

$$P_{s2} = \gamma_2 f^2(\mu_2 H_m)^2 r^3 s \frac{1}{1+\left(\dfrac{r}{s}\right)^2} k_v v_2, \tag{4.30}$$

where the value of factor k_v is represented in Fig. 4.5 as a function of the ratio of plate thickness (v_2) to the depth of penetration ($\delta = 1/\sqrt{\pi f \mu \gamma}$). The product $k_v v_2$ may be considered as the equivalent plate thickness. Formula (4.28) can always be brought to the form (4.30), whatever spatial distribution the field strength H_m has, provided that H_m is multiplied by a factor k_H depending on the field strength distribution. In this way, the relation

$$P_{s2} = \gamma_2 f^2(\mu_2 k_H H_m)^2 r^3 s \frac{1}{\left(\dfrac{r}{s}\right)^2+1} k_v v_2 \tag{4.31}$$

can be written for the losses in every case and from this relation the role of the various factors influencing the phenomenon can immediately be found.

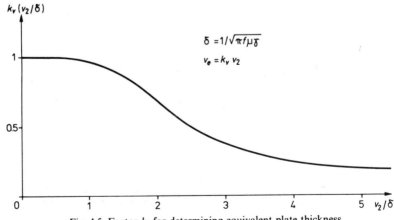

Fig. 4.5. Factor k_v for determining equivalent plate thickness

4.4.3 Eddy current losses in blocks of non-magnetic material

By means of the method presented above, let us determine how the eddy-current losses in a block made of some non-magnetic metal placed in a homogeneous field are influenced by its geometrical dimensions and conductivity. The arrangement investigated is shown in Fig. 4.6. Let a block of width r, lengths s (not shown in the

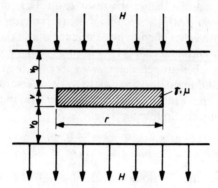

Fig. 4.6. Block-shaped body in a homogeneous field

diagram), thickness v, conductivity γ and permeability μ be placed in a magnetic field of direction perpendicular to side $r \times s$ of the block, having a peak strength of H_m and varying at frequency f. Let the dimensions and units of all quantities be the same as those described for the calculation method. Let the field strength be given for the planes situated at distance v_0 in m from the top and bottom of the block, assuming thereby that the reaction of eddy currents is not effective in the space outside the block beyond a layer of air or oil of thickness v_0. The losses arising in the block can be determined for that case by regarding it as a three-layer arrangement and applying the method described, slightly modified. In our present case, the extension of the third layer is not infinite. So in determining the integration constants condition (b) is modified, namely $H_{z3} = H_m$ pertains to $z = v_0 + v + v_0$. On this basis, the losses occurring in the block are given by the relation

$$P = \frac{\pi^2}{6} f^2 (\mu H_m)^2 r^3 s \frac{1}{1 + \left(\dfrac{r}{s}\right)^2} F(v, v_0, f, \mu, \gamma). \qquad (4.32)$$

Function $F(v, v_0, f, \mu, \gamma)$ of dimension in m appearing in the above relation corresponds to the equivalent thickness encountered in Eqns (4.30) and (4.31), and may therefore be denoted by v_e. Thus,

$$F(v, v_0, f, \mu, \gamma) = v_e. \qquad (4.33)$$

From relation (4.33) the loss is obtained in watts. In the above relation the role of field strength and dimensions r and s are unequivocal. Only the equivalent thickness v_e requires a more detailed investigation.

General conclusions can be drawn if, instead of v_e, its value related to the depth of penetration (v_e/δ) is plotted as a function of the ratio v/δ, and the ratio of overall

159

thickness of air (oil) layers above and below the block to plate thickness, i.e. the ratio $2v_0/v$, is taken as parameter. From Fig. 4.7 plotted in this way it can be seen that, when the field strength is fixed at the surface of the block (viz. $2v_0/v=0$), then the equivalent thickness v_e will attain a maximum in the vicinity of $v/\delta=2$. If in addition a gap is present, it will have two effects. On the one hand, it will reduce the maximum of the curve, and on the other, it will advance this reduced maximum. Both effects are the more pronounced, the higher the ratio $2v_0/v$. The curves of Fig. 4.7 are of general validity and agree with the results of measurements, and can therefore be used for determining the value of equivalent thickness and the locus of maxima for sheets of any thickness and material. When stating the dimensions and material of sheets, care should be taken to get as far away as possible from that maximum.

Suppose the losses and temperature rise are to be determined for a bronze plate serving for clamping the winding end and located in the potential ring of a transformer, considering the following data:

$$r=0.06 \text{ m}, \quad s=0.19 \text{ m}, \quad v=0.02 \text{ m},$$
$$\gamma=8.3\times 10^6 \text{ S/m}, \quad f=50 \text{ Hz}, \quad \delta=24.7\times 10^{-3} \text{ m}.$$

The field strength varies linearly along the side of length r from one edge of the block to the other (cf. field strength variation shown in Fig. 3.1), with the spatial maximum of field strength $H_{\text{r.m.s.}}=0.68\times 10^5 \text{ A m}^{-1}$. The block is coated with 3 mm thick paper.

The block constitutes a part of a large diameter potential ring made of insulating material, and there is no dissipation of heat at the contacting surfaces of the ring. The ring is immersed in oil, and the block can transfer the heat developing in it to the oil through its four sides. (In fact, the situation is, generally, somewhat less favourable, since not all of the surface of the four sides of the block is in direct contact with the oil.)

In view of the triangular shape of field strength distribution, the maximum field strength has to be multiplied by the factor $k_H=0.5$ to make formula (4.32) for homogeneous fields applicable also to this case. The loss occurring in the block is determined for $2v_0/v=2$, for which case Fig. 4.7 shows that, for $v/\delta=0.81$, the value of $v_e/\delta\cong 0.45$. Hence

$$P=\frac{\pi^2}{6}\,50^2\times 8.3\times 10^6 (0.5\times 1.256\times 10^{-6}\sqrt{2}\times 0.68\times 10^5)^2\times$$

$$\times 0.06^3\times 0.19\,\frac{1}{1+\left(\dfrac{0.06}{0.19}\right)^2}\times 0.45\times 24.7\times 10^{-3}=51.5 \text{ W}.$$

As regards temperature rise, the block is considered as a homogeneous body. The temperature difference between the block and the surrounding oil is composed of two parts.

(a) temperature drop across the insulating paper, and
(b) surface heat drop across the boundary layer.

160

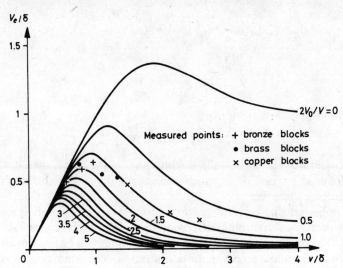

Fig. 4.7. Equivalent relative thickness (v_e/δ) of blocks (related to the depth of penetration) as a function of relative thickness (v/δ) for different relative air gap widths ($2v_0/v$)

(c) The temperature drop across insulating paper is

$$\Delta\vartheta_P = q\frac{\delta}{\lambda_p} \quad °C,$$

where

$q = \dfrac{P}{F}$ is loss dissipated per unit surface in W m^{-2},

P loss induced in the block in W,

F heat dissipating surface in m^2,

δ thickness of paper layer in m and

$\lambda_p = 0.1395$ heat conductivity of oiled paper in Wm^{-1} $°C^{-1}$.

In our case:

$$\delta = 3 \times 10^{-3} \text{ m},$$

$$F = 2s(r+v) = 2 \times 0.19(0.06+0.02) = 0.0304 \text{ m}^2,$$

$$q = \frac{51.5}{0.0304} = 1694 \text{ Wm}^{-2}$$

hence

$$\Delta\vartheta_P = 1694\frac{3 \times 10^{-3}}{0.1395} = 36.4 \text{ °C}.$$

(b) The surface heat drop across the boundary layer between oil and paper:

$$\Delta\vartheta_F = \frac{q}{\alpha},$$

11   **161**

where

$$\alpha = 2.1\sqrt{q} = 2.1\sqrt{1694} = 86.4\ \mathrm{Wm^{-2}\,^{\circ}C^{-1}}.$$

Hence,

$$\Delta\vartheta_F = \frac{1694}{86.4} = 19.6\ ^{\circ}\mathrm{C}.$$

Thus, the entire temperature drop is

$$\Delta\vartheta = \Delta\vartheta_P + \Delta\vartheta_F = 36.4 + 19.6 = 56.0\ ^{\circ}\mathrm{C}$$

If the oil temperature is 75 °C in the surroundings of the block, then it will attain a temperature of $75 + 56 = 131$ °C, which is inadmissible, so some other arrangement will have to be chosen.

Let the potential ring be arranged so that the width of the solid piece of metal serving for mechanical clamping of the winding end should be only a fraction of that of the ring. (The ring made of insulating material should not be interrupted, but only thinned to hold the metal piece). In the case given, select 3.0 mm thick bronze. From Eqn. (4.32), with $v_e = 0.003$ (for small plate thicknesses $v_e = v$), and with all other data remaining unchanged, the loss induced in the block will be

$$P = 13.7\ \mathrm{W}.$$

Now, heat dissipation may be assumed to occur at the top and at the two longer sides of the block. Thus

$$F = (r + 2v)s = (0.06 + 0.006)0.19 = 0.0125\ \mathrm{m^2},$$

$$q = \frac{13.7}{0.0125} = 1096\ \mathrm{Wm^{-2}}.$$

Hence, the temperature drop across the insulating paper is

$$\Delta\vartheta_P = 1096\,\frac{3 \times 10^{-3}}{0.1395} = 23.5\ ^{\circ}\mathrm{C}.$$

The surface temperature drop between paper and oil, from

$$\alpha = 2.1\sqrt{1096} = 69.5\ \mathrm{Wm^{-2}\,^{\circ}C}$$

is

$$\Delta\vartheta_F = \frac{1096}{69.5} = 16\ ^{\circ}\mathrm{C}.$$

With the above, the entire temperature drop is

$$\Delta\vartheta = 23.5 + 16 = 39.5\ ^{\circ}\mathrm{C}$$

i.e., assuming 75 °C oil temperature, the temperature of the block will be $75 + 39.5 = 114.5$, which is below the 118 °C permitted for insulated metallic parts.

4.4.4 Tank losses (the role of the material)

In the next example, let us examine how the losses in the tank are influenced by the tank material. For the sake of simplicity, let the spatial distribution of stray field strength departing from the winding toward the tank in the radial direction be sinusoidal, the maximum field strength be $H_m = 0.68 \times 10^5$ A m^{-1}, and the frequency $f = 50$ Hz. Dimensions of the section of tank wall are $r = 0.4$ m, and $s = 0.46$ m. A sketch of the arrangement is given in Fig. 4.8.

In the following calculation, the first layer is constituted by the oil layer between winding and tank wall ($\gamma_1 = 0$, $\mu_1 = 1.256 \times 10^{-6}$ H m^{-1}), the second layer is the tank

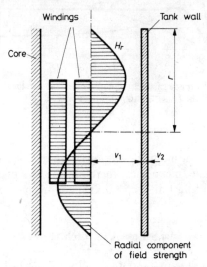

Fig. 4.8. Simplified three-layer model for calculating the plate losses

wall investigated, and the third layer is the air outside the tank wall ($\gamma_3 = 0$, $\mu_3 = 1.256 \times 10^{-6}$ H m^{-1}). Calculations are performed for the cases of conventional magnetic steel ($\gamma_2 = 7 \times 10^6$ S m^{-1}, $\mu_2 = 100 \ \mu_0 = 1.256 \times 10^{-4}$ H m^{-1}), aluminium ($\gamma_2 = 33 \times 10^6$ S m^{-1}, $\mu_2 = \mu_0 = 1.256 \times 10^{-6}$ H m^{-1}) and non-magnetic steel ($\gamma_2 = 1.37 \times 10^6$ S m^{-1}, $\mu_2 = 1.256 \times 10^{-6}$ H m^{-1}). The distance between winding and tank wall has been taken as 100 mm. The loss arising in the plate of size $r \times s$ is plotted in terms of plate thickness in Fig. 4.9.

It can be seen from the plot that the losses in the tank made of magnetic steel attain their maximum at about 4 mm wall thickness, and remain unchanged with increasing wall thicknesses. In the present case, an aluminium tank would only be effective above a plate thickness of 5 mm. A tank made of non-magnetic steel would give a lower loss than that made of magnetic steel in the case of wall thicknesses below about 13 mm.

With increasing thicknesses of the oil layer, the intersection point between the curves drawn for aluminium and magnetic steel is shifted slightly forward, whereas that between the curves of the two kinds of steel remains practically unchanged.

Consequently, for the oil-layer thicknesses of 100 to 500 mm occurring in high-voltage transformers with the usual wall thicknesses, the losses induced in aluminium tank walls are invariably lower than those arising in tank walls made of magnetic steel.

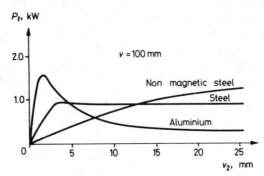

Fig. 4.9. Losses of tank walls made of steel, aluminium and non-magnetic steel as a function of wall thickness for 100 mm thickness of oil layer

Tank walls made of non-magnetic steel give losses lower than those of magnetic steel only for small wall thicknesses, therefore the use of non-magnetic tanks is useful only for substituting thin-wall tanks made of magnetic plates.

4.4.5 Losses in flitch plates

The losses induced in the core and flitch plate by the stray flux departing radially on the inner side of windings can be calculated on the basis of the sketch of Fig. 4.10. Layer 1 is the oil duct between the winding and core (or flitch plate), layer 2 is the flitch plate, and layer 3 is the laminated core. The laminated structure of the core, or the fact that magnetic conductivity in the direction of sheets is many times higher than that perpendicular to it, is not taken into account in the calculation. The core is considered homogeneous, and permeability is taken as equal to the relative permeability pertaining to the saturation flux density ($\mu_1 = 10$ to 50). The permissibility of this approximation has been validated by experiments. Further, the thickness of layer 3 is taken as infinite. This can be done because the thickness of the limb is much greater than the depth of flux penetration (calculated with $f = 50$ Hz, $\gamma = 2 \times 10^6$ S m^{-1} and $\mu = 50\,\mu_0 = 50 \times 1.256 \times 10^{-6}$ H m^{-1}, $\delta = 7.1$ mm is obtained). Losses arising in the flitch plate (layer 2) and in the core (layer 3) are calculated on the basis of relations (4.18) and (4.19). It should be borne in mind that the losses in the sections of core and flitch plate now have to be calculated under both (spatial) half waves of the field strength distribution curve, therefore the losses resulting from the original relations are to be doubled.

The losses caused by the stray flux in a flitch plate made of non-magnetic material and in the core of a 400 kV, 270 MVA three-phase transformer have been

164

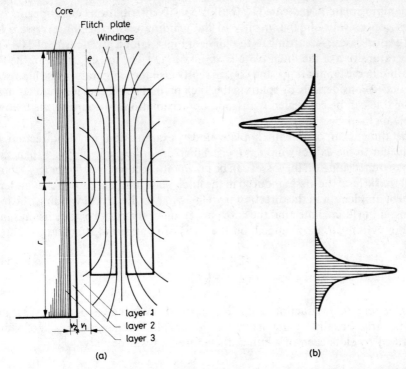

Fig. 4.10. Stray magnetic field of a short-circuited winding (a), and distribution of radial field (B_r) along the inner edge (line e) of the winding (b)

determined by applying the method outlined above. The calculation has been performed with the following data:

$2r = 3.7$ m, $s = 0.36$ m ,
$v_1 = 0.031$ m, $\gamma_1 = 0;\ \mu_1 = 1.256 \times 10^6$ H m^{-1},
$v_2 = 0.025$ m, $\gamma_2 = 1.37 \times 10^6$ S m^{-1}; $\mu_2 = 1.256 \times 10^{-6}$ H m^{-1},
$v_3 \rightarrow \infty$ $\gamma_3 = 2 \times 10^6$ S m^{-1}; $\mu_3 = 6.28 \times 10^{-5}$ H m^{-1}.

When measuring load losses of the transformer (with one winding short-circuited) the distribution of flux density developing at the inner side of the L.V. winding (at the surface of the winding facing the core) is shown in Fig. 4.10. The losses have been calculated for this pattern, with maximum flux density $B_{rm} = 0.08$ T.

Taking these values and our calculation method (relations (4.18) and (4.19)) the following results were obtained. Loss in the flitch plate: $P_2 = 2742$ W; loss in the core packet under the plate: $P_3 = 526$ W; sum of the two losses: $P_s = 3268$ W. In the three-phase transformer the number of flitch plates is 6, making the overall loss

$$P_s = 19\,608 \text{ W} .$$

An additional loss of this size deserves attention even in the case of such a large transformer, yet a still greater hazard is that the major part of the losses occurring in the clamping plate, as indicated by the flux density distribution curve, appears in the plates concentrated in the vicinity of the winding ends, causing excessive local temperature rises. (According to the curve of Fig. 4.1, for a flux density of 0.08 T, the temperature rise of the flitch plate is about 55 °C!) The losses arising in the flitch plate and in the laminations, and consequently also the temperature rise figures, can be reduced considerably by splitting the flitch plate. The effect of loss reduction thus obtainable can be calculated with good approximation by applying the following simple method.

The dimension of the flitch plate under examination in the direction perpendicular to the axis of windings (perpendicular to the plane of Fig. 4.10) is s. The effect of dimension s on the losses can be taken into account by a factor k_s. Denoting by P_1 (basic loss) the losses induced in the flitch plate (and the core packet under the plate) of unit length in the direction perpendicular to the axis of winding, the losses arising in the flitch plate (and the core packet under the plate) of dimension s in the same direction can be obtained, on the basis of relations (4.20) and (4.21), from

$$P_s = P_l \frac{1}{\left(\dfrac{r}{s}\right)^2 + 1} . \tag{4.34}$$

It can be seen that reduction of dimension s will cause the losses to decrease more than linearly. In a flitch plate of overall width s split into n sections each of width s_n separated by slots each of width d, the entire loss will be only

$$nP_{sn} = P_l n s_n \frac{1}{\left(\dfrac{r}{s_n}\right)^2 + 1} . \tag{4.35}$$

From relation (4.35), the overall loss of sections of the flitch plate and core packet under the sections can be calculated. However, the loss caused by the flux directly entering the core packet through the slots should also be taken into account. The approximate magnitude of this additional loss can be calculated from the relation

$$\Delta P = P_l'(n-1)d . \tag{4.36}$$

Here P_l' is the basic loss of the core packet. Factor k_s does not appear in the calculation of ΔP since the width of penetration (dimension d) of flux lines entering directly is much less than the plate width (s), so that the eddy currents flowing in the lateral direction can develop freely. Instead of expressing it as a function of the number of plate sections (n), the corrected value of overall losses induced in the split plate and in the (unsplit) core packet under it may be written as a function of the number of slots ($N = n - 1$) in the following form:

$$P = P_l(N+1)s_n \frac{1}{\left(\dfrac{r}{s_n}\right)^2 + 1} + P_l' N d . \tag{4.37}$$

It is a usual practice to split the core lamination into several sections as well, by providing slots forming extensions of the slots dividing the flitch plate. Additional

166

losses are induced in the core packet under the slots, but because of reduced flux density, the value of P'_l will be lower.

Reverting to the transformer under investigation, let its flitch plate be split up by three 6 mm wide slots into four sections of 85.5 mm width each. With the notation used previously $d = 0.006$ m, $s_n = 0.0855$, $N = n - 1 = 3$. Substituting these values into relation (4.34), the basic loss of the flitch plate is

$$P_l = \frac{P_2}{s\dfrac{1}{\left(\dfrac{r}{s}\right)^2 + 1}} = 208\,759\ \text{W m}^{-1}$$

and the basic loss in the core packets is

$$P'_l = \frac{P_3}{s\dfrac{1}{\left(\dfrac{r}{s}\right)^2 + 1}} = 40\,046\ \text{W m}^{-1}.$$

The overall loss induced in the flitch plate split up into four sections and in the adjacent unsplit core packet, is from (4.37)

$$P = 208\,759 \times 4 \times 0.0855\,\frac{1}{\left(\dfrac{1.85}{0.0855}\right)^2 + 1} + 40\,046 \times 3 \times$$

$$\times 0.006 = 152 + 721 = 873\ \text{W}.$$

By splitting the flitch plates, the overall loss drops to about a quarter of that occurring in the original (unsplit) state of the plates. Correspondingly, the loss that would arise in the flitch plate in unsplit form and would lead to elevated local temperatures also decreases well below any hazardous level.

4.4.6 Three dimensional method for calculation of tank losses

The analytical calculation method is based on a two-dimensional model, and although it usually satisfies practical requirements with adequate accuracy, it is not suitable for computing the distribution of eddy currents within the tank. Thus in a two-dimensional model both geometrical arrangement and material characteristics can only change in the plane containing the axis of windings and lying perpendicular to the tank wall, and eddy currents—in contrast to real conditions—can flow only in the direction perpendicular to this plane. Such a current pattern cannot be used for determining loss distribution, for which purpose knowledge of the real current distribution is required.

The two-dimensional method described, however, can be developed relatively easily into a form suitable for treating three-dimensional arrangements. Here, only a single-layer arrangement is discussed: the losses are to be determined in a part of size

$r \times s$ of a plate of thickness v, and of infinite extension in directions x and y. The distribution of field strength is taken as known over the surface of the plate (see Fig. 4.11) and is characterized by the relation

$$H_{Z0} = H_m \sum_{k=1}^{n} f_k(x) \sum_{l=1}^{m} f_l(y) e^{j\omega t}. \tag{4.38}$$

(This relation now replaces Eqn. (4.6)).
Here,

$$f_k(x) = a_k \cos k\alpha x + b_k \sin k\alpha x \tag{4.39}$$

and

$$f_l(x) = c_l \cos l\beta y + d_l \sin l\beta y. \tag{4.40}$$

The Fourier coefficients a_k, b_k, c_l and d_l have to be selected so as to obtain the required distribution of field strength over a tank surface of size $r \times s$. In expression

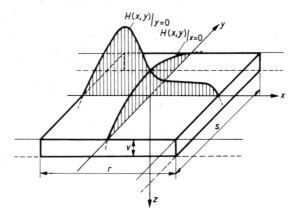

Fig. 4.11. Field distribution in space over a piece of size $r \times s$ of a plate of thickness v and of infinite size in directions x and y

(4.38), the distribution of radial component of the stray field strength, along the $y=0$ surface straight line of the plate, is given by the product $H_m \sum_{k=1}^{n} f_k(x)$. The variation of field strength in the y direction is characterized by the function

$$\sum_{l=1}^{m} f_l(y) \leq 1.$$

Applying the assumptions made in the course of deducing the analytical method to this arrangement, considering the tank dimensions ($r \cong s \cong 2$ m) and tank wall characteristics ($\mu > 100 \, \mu_0 = 1.256 \times 10^{-4}$ H m^{-1}, $\gamma = 7 \times 10^6$ S m^{-1}) usual with large transformers, further introducing some permissible approximations, the following expressions for the currents flowing in the plate are obtained as solution of

168

Eqns (4.7), (4.8) and (4.9):

$$i_x \cong -j\omega\mu\,\gamma H_m \frac{\beta}{\alpha^2+\beta^2} \sum_{k=1}^{n} f_k(x) \times$$

$$\times \sum_{l=1}^{m} \frac{1}{l^2\beta} f'_l(y) e^{-pz} e^{j\omega t}, \tag{4.41}$$

$$i_y \cong j\omega\mu\,\gamma H_m \frac{\alpha}{\alpha^2+\beta^2} \sum_{k=1}^{n} \frac{1}{k^2\alpha} f'_k(x) \times$$

$$\times \sum_{l=1}^{m} f_l(y) e^{-pz} e^{j\omega t} \tag{4.42}$$

and

$$i_z \cong 0.$$

Parameters appearing for the first time in the above expressions are

$$\alpha = \frac{\pi}{r}, \qquad \beta = \frac{\pi}{s} \qquad \text{and} \qquad p^2 \simeq j\omega\mu\gamma,$$

(μ and γ are characteristics of the plate, ω is the angular frequency of variation of field strength with time). The losses induced in an elementary prism of size $\Delta x \times \Delta y \times \Delta z$ defined by coordinates (x, y, z) can be calculated by means of relation

$$P_{xyz} = \frac{1}{2}\frac{1}{\gamma}(|i_x|^2 + |i_y|^2)\Delta x \Delta y \Delta z \tag{4.43}$$

(i_x and i_y are peak values, hence the factor $1/2$).

On the basis of relations (4.41), (4.42) and (4.43) the current and loss distributions within the plates can be found. Further, by integrating P_{xyz} for the entire volume of the plate the overall losses are obtained.

This method has been followed in determining distributions of current density and losses induced in the tank wall by the stray flux of windings wound around two limbs of a 400 kV single-phase transformer, as shown in Fig. 4.12. The calculation has been performed for the following data:

$$r = 1.9\,\text{m}, \qquad s = 2.1\,\text{m}, \qquad f = 50\,\text{Hz}, \qquad \gamma = 7 \times 10^6\,\text{S m}^{-1},$$

$$\mu = 500\,\mu_0 = 6.28 \times 10^{-4}\,\text{H m}^{-1}.$$

The distribution of current density and loss distribution have been calculated for a part of the tank wall of size $r \times s$ (i.e. for the part adjacent to a winding half), representing $1/4$ of the surface on one side of the tank. (From Fig. 4.12 it can be seen that the tank surface may be considered as consisting of four equal parts. It is, therefore, sufficient to calculate current and loss distributions for one quarter of the tank wall, and to deduce the current and loss patterns for the other parts by symmetry). The coordinate system has been chosen so as to have its x-axis ($y = 0$) coincide with the straight line running closest to the winding (this being the axis x_1 (x_2) in Fig. 4.12), and its y-axis at $1/4$ height of the flitch plate (this being at a

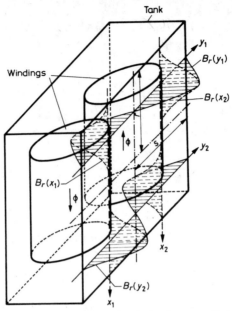

Tank

Windings

$B_r(y_1)$

$B_r(x_2)$

y_2

$B_r(x_1)$

ϕ

ϕ

x_2

$B_r(y_2)$

x_1

Fig. 4.12. Distribution of radial stray flux density of a single-phase, two-limb transformer along four distinctive lines of the tank wall: x_1—vertical line of tank wall, running closest to the left-hand side winding; x_2—vertical line of tank wall, running closest to the right-hand side winding; y_1—horizontal line running at the height of the upper edge of windings; y_2—horizontal line running at the height of the lower edge of windings

distance of $r/2$ from the centre line). With this arrangement, the rectangle of surface $r \times s$ is divided into halves by the two axes, as shown in Fig. 4.11. The flux density distribution along axis x is plotted in Fig. 4.13, and that along axis y is shown in Fig. 4.14. From these distribution curves the following values are obtained for Fourier coefficients a_k, b_k, c_l, d_l:

$$
\begin{array}{lll}
a_1 = 61.69, & b_2 = -40.62, & c_1 = 0.92, \\
a_3 = -1.77, & b_4 = -16.04, & c_3 = 0.068, \\
a_5 = -4.89, & b_6 = 1.77, & c_5 = -0.0015, \\
a_7 = -4.89, & b_8 = -1.67, & c_7 = 0.0085, \\
a_9 = -2.43, & b_{10} = 2.65, & c_9 = 0.003,
\end{array}
$$

$$
\begin{aligned}
a_2 &= a_4 = a_6 = a_8 = a_{10} = 0, \\
b_1 &= b_3 = b_5 = b_7 = b_9 = 0, \\
c_2 &= c_4 = c_6 = c_8 = c_{10} = 0, \\
d_1 &= \ldots = d_{10} = 0.
\end{aligned}
$$

The members of the Fourier series having ordinal numbers above ten have been neglected.

The dimensional unit of Fourier coefficients a_k and b_k is T; c_l are plain numbers. Since the peak value of field strength appears in the example instead of the r.m.s.

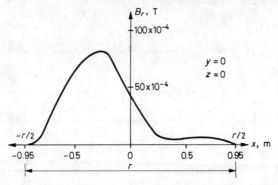

Fig. 4.13. Radial distribution of flux density at the tank wall surface along line $y=0$

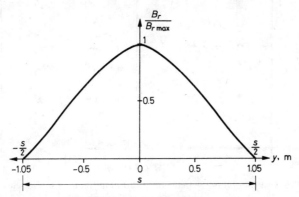

Fig. 4.14. Distribution of relative radial flux density (referred to maximum flux density) at the tank wall surface in the horizontal direction (along line $x=0$)

value of flux density, the coefficients given in T (Tesla) have to be multiplied by $\sqrt{2}/\mu$ $= 0.225 \times 10^4$, where $H_m = 1$.

In our calculation, part of the tank wall of dimension $r \times s$ has been divided into 20 sections in both x and y directions, with 10 divisions spaced in direction z with a pitch of 0.5 mm. (At points more than 5 mm away from the surface, the eddy currents and losses may be neglected.) For the $21 \times 21 \times 11 = 4851$ points thus obtained the current density values, and from these the losses occurring in the elementary prisms each of size $105 \times 95 \times 0.5$ mm pertaining to the division points, have been determined. The absolute values (resultants of real and imaginary components) of surface current densities with their proper directions and magnitudes, are shown in Fig. 4.15. It can be seen there that the maximum current density occurs in the wall section between coils along straight lines $y = -s/2$ and $y = s/2$. It is remarkable that the direction of this maximum current density is parallel to the axis of the coils, a current density of such direction cannot be determined by means of a two-dimensional procedure.

From the losses occurring in the full thickness of the sheet (in the $105 \times 95 \times$ $\times 0.5$ mm prisms) the surface loss densities in the $21 \times 21 = 441$ division points of the

171

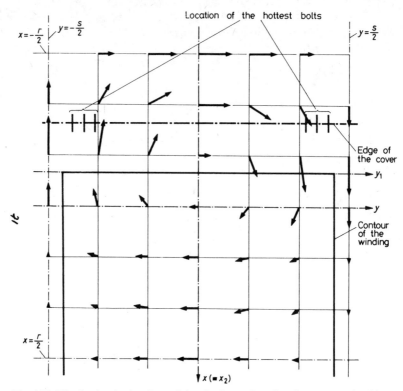

Fig. 4.15. Distribution in the plane of the absolute value of surface current densities
in one quarter of the tank wall

surface have been determined, and with these the contoured sketch of Fig. 4.16 has
been plotted. Each contour line connects the points of equal loss density. It can be
clearly seen that the surface loss density is greatest at the highest winding edges in
the tank wall section between the windings. Relatively large loss densities have also
been obtained for the line lying in the axis of windings at the top and bottom edges of
the tank wall.

The points of maximum loss density do not necessarily coincide with the hot
spots, whose development is largely dependent on the conditions of cooling. Thus,
at the place with maximum loss densities shown in Fig. 4.16 (in the tank wall section
between the windings) considerable oil flow is present which causes the hot spot to
be displaced sideways. On the other hand, it is also apparent from Fig. 4.16 that the
points of the tank wall where the losses are minimum are located just above the
points exhibiting the maximum losses, so the heat generated at the points of
maximum losses will easily be carried off by the wall. As shown by measurements the
temperature of hot spots of the tank wall rises above the average temperature to a
far less extent than the maximum loss density above the average loss density.

As mentioned in Section 4.1, the hot spots of the tank wall seldom represent a risk.
However, excessive temperature rise of bolts connecting the cover with the flange of

the tank may give rise to hazards. The method described above is also suitable for calculating the currents flowing through the cover bolts. Dividing the current uniformly distributed along the flange of the tank and flowing in the x direction by the number of cover bolts the current passing through each bolt is obtained. These currents may reach substantial magnitudes, in the present example the maximum current even exceeds 3000 A.

From knowledge of the magnitude of loss occurring in the bolts, their temperature rise can be assessed by simple calculation. Heat transfer from the cover bolt to the surrounding parts is practically limited to that across the surface of the flange in contact with the bolt head and nut. The temperature rise of the bolt depends on the surface heat drop, which cannot be calculated in a direct way. If, in order to obtain informative data, it is assumed that the difference between the temperatures of the bolt and flange is equal to the heat drop across the piece of steel plate having a surface area equal to that of the bearing surface and having a thickness half that of the 12 mm flange, i.e. 6 mm, then the equation of heat

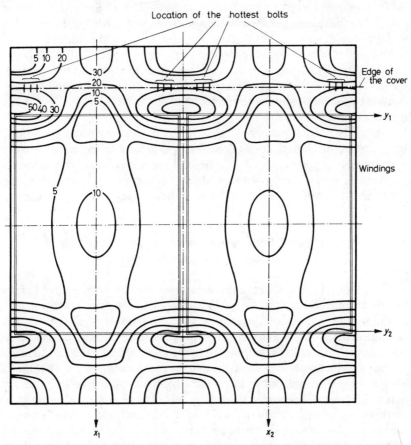

Fig. 4.16. Contour lines connecting the points of equal surface loss densities. The numbers written beside the lines indicate surface loss densities expressed in W dm^{-2} units

173

conduction leads to the following relation between the quantity of heat transferred, thermal conductivity, heat conducting surface, temperature difference and path length of heat flow

$$Q = \lambda A \frac{\Delta t}{l}.$$

From the dimensions and conductivity of the bolts used in the given transformer ($d = 16.9$ mm, $l = 32$ mm, $\gamma = 7 \times 10^6$ Sm^{-1}), and the known magnitude of current flowing through them, the loss generated in them can be determined: calculating with $Q = 250$ W, and substituting $\lambda = 50$ W m^{-1} °C^{-1}, $A = 1.1 \times 10^{-6}$ m^2 and $l = 6 \times 10^{-3}$ m, the temperature difference is obtained as

$$\Delta t = \frac{Ql}{\lambda A} = \frac{250 \times 6 \times 10^{-3}}{50 \times 1.1 \times 10^{-3}} = 27 °C.$$

The temperature of the surrounding parts of the cover may be taken as about equal to the temperature of the oil adjacent to the tank wall, i.e. a maximum of 95 °C. Thus, the bolts in question may attain a temperature of 122 °C. Such a temperature may be harmful to the sealing between flange and cover, and can harm the insulation of low-voltage conductors that may come into contact with the bolts.

The flow of substantial currents can be prevented by isolating the bolts from the cover and tank. Obviously in this case, the cover should be at the same potential as the tank. For this, the said two parts have to be connected through a metal bar or stranded wire, or else one of the cover bolts has to be left uninsulated to establish a metallic connection.

It may be noted here that, according to measurements performed on the transformer, the highest temperature rise figures have not been found to occur in the bolts located at points carrying the highest vertical currents, but in the cover bolts beside these spots (see Figs 4.15 and 4.16). The reason lies in the difference between the location of hot spots and spots of highest losses mentioned before.

4.4.7 Eddy current losses excited by the field of high current carrying conductors

The losses arising in sheets and ducts located in the magnetic field of conductors carrying high currents can easily be calculated by two-dimensional numerical methods, but results of measuring are also readily available. A few typical arrangements are shown in Fig. 4.17 where the losses indicated have been found to occur in the duct of 2.0 m length and in the sheet of 1.0 m × 2.0 m, arranged beside the bar conductor shown. In (a) (b) and (c) the material of the duct and sheet is a normal magnetic steel, whereas the arrangement (d) of Fig. 4.17 refers to a non-magnetic sheet. These and similar diagrams may be utilized directly in the design work; they illustrate clearly the effects of the various factors on the loss. The losses are proportional to the square of current flowing in the bar conductor and to $d^{-0.4}$ approximately when the latter is of magnetic steel, and to $d^{-0.8}$ if the sheet material is non-magnetic [69] where d is the distance between the bar and sheet. The loss

174

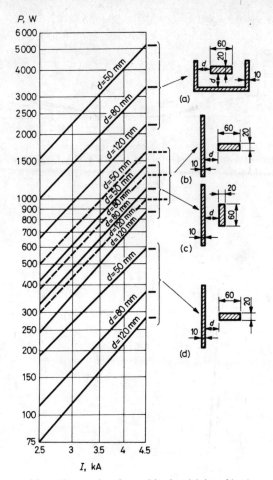

Fig. 4.17. Losses measured in a plate or duct located in the vicinity of leads carrying high current

density is highest in the part of the sheet closest to the bar, and in the given arrangement it decays to zero beyond a distance of about 200 mm (see Fig. 4.18). Obviously, the loss density is highest at the surface, and decreases corresponding to the depth of penetration (see Figs 4.19 and 4.20). Additional losses arise not only in the sheet but also in the conductor, but the latter are negligible compared to the former. The losses occurring in the plate are substantially influenced by the conductivity of the plate material, as shown by the example in the curve of Fig. 4.21. The frequency has a considerable effect as well (see Fig. 4.22), which is to be considered in the case of excitation currents of high harmonic content.

As proved by experience, the eddy-current losses generated in the tank cover by the magnetic field of currents flowing in the bushings can be calculated from a simple formula. A simplified model of the arrangement is given in Fig. 4.23. A current I, kA flowing in the bushing passed through an opening of radius R, m

175

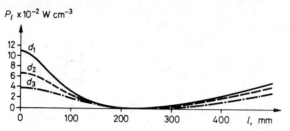

Fig. 4.18. Loss density distribution along the half length in the 0.2 mm thick upper layer of the plate on its side toward the conductor, in arrangement (d) of Fig. 4.17. $I = 4.6\,\text{kA}, f = 50\,\text{Hz}, d_1 = 50\,\text{mm}, d_2 = 80\,\text{mm}, d_3 = 120\,\text{mm}$. Plate thickness: $v = 6\,\text{mm}$ (instead of 10 mm given in the figure); $\sigma_{\text{plate}} = 1.37 \times 10^6\,\text{S m}^{-1}, \mu_r = 1$. (Calculated curves)

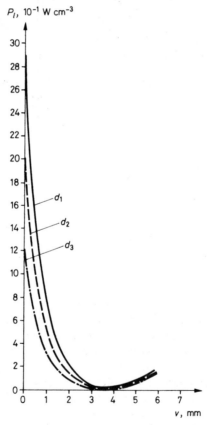

Fig. 4.19. Loss density distribution in the symmetry axis of the plate along the plate thickness, in arrangement (b) of Fig. 4.17. $I = 4.6\,\text{kA}, f = 50\,\text{Hz}, d_1 = 50\,\text{mm}, d_2 = 80\,\text{mm}, d_3 = 120\,\text{mm}$. Plate thickness: $v = 6\,\text{mm}$ (instead of 10 mm given in the figure). $\sigma_{\text{plate}} = 7.8 \times 10^6\,\text{S m}^{-1}, \mu_r = 300$. (Calculated curves)

176

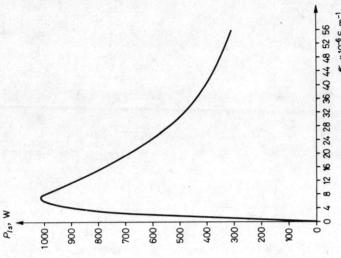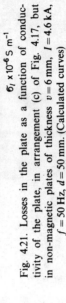

Fig. 4.21. Losses in the plate as a function of conductivity of the plate, in arrangement (c) of Fig. 4.17, but in non-magnetic plates of thickness $v = 6$ mm, $I = 4.6$ kA, $f = 50$ Hz, $d = 50$ mm. (Calculated curves)

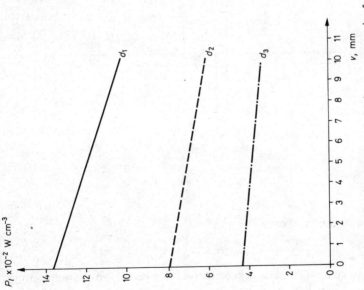

Fig. 4.20. Loss density distribution in the symmetry axis of the plate along the plate thickness, in arrangement (c) of Fig. 4.17, but in the case of a non-magnetic plate ($\sigma_{plate} = 1.37 \times 10^6$ S m^{-1}, $\mu_r = 1$), $I = 4.6$ kA, $f = 50$ Hz, $d_1 = 50$ mm, $d_2 = 80$ mm, $d_3 = 120$ mm. (Calculated curves)

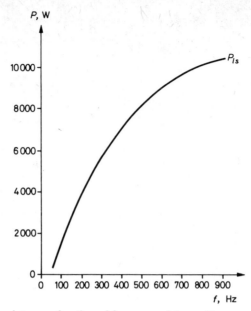

Fig. 4.22. Loss in the plate as a function of frequency of the exciting current, in arrangement (c) of Fig. 4.17, but in a $v=6$ mm thick non-magnetic steel plate (σ plate $=1.37 \times 10^6$ S m^{-1}, $\mu_r=1$), $I=4.6$ kA, $d=50$ mm. (Calculated curve)

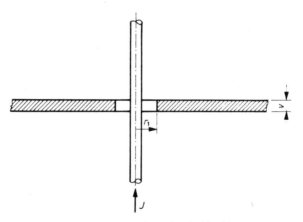

Fig. 4.23. Simplified model of bushing

causes a loss in the tank cover

$$P=405(0.7-\sqrt{R})I^{1.5}. \tag{4.44}$$

This approximate formula has been arrived at by analytic computation applied to the arrangement illustrated, and its correctness has been proven by measurements. The calculation relates to a common magnetic steel ($B_{\text{sat}}=1.4$ T and $\gamma=7 \times 10^6$ S m^{-1}).

178

4.5 Preventive measures against leakage flux effects (magnetic and electromagnetic shielding)

Losses and temperature rises caused by the leakage flux may generally be reduced in the following three ways:

(a) by deflecting the magnetic flux from endangered spots by electromagnetic shields or magnetic shunts;

(b) by proper dimensioning of constructional parts with respect to eddy-current losses;

(c) by proper selection of constructional materials.

For the majority of cases, the effect of dimensions and material exerted on the losses can be judged from the formulas given above from which the losses are calculated. The influence of dimensions has already been illustrated in the preceding examples, and the application of non-magnetic materials will be dealt with briefly at the end of this section. In the following, the problems related to the use of electromagnetic shielding and magnetic shunts will be discussed in detail. The principles of operation of the two procedures are different: the electromagnetic shielding prevents the stray flux from penetrating into the tank by the reacting effect of eddy currents induced in the said shielding, whereas the magnetic shunt (flux diverter) diverts the flux from the endangered location. The problem will, first, be investigated in general.

In Fig. 4.24 the equivalent circuit of the short-circuited transformer is shown, in which the leakage flux outside the windings is also considered. In this circuit X_0

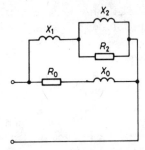

Fig. 4.24. Equivalent circuit of a transformer in short-circuit, with leakage flux outside the winding considered

represents the leakage reactance inside the windings, X_1 and X_2 are the leakage reactances outside the winding, and R_2 is the resistance corresponding to the losses occurring in the space outside the winding. (In the conventional equivalent circuit of short-circuited transformers only X_0 and R_0 appear.) From among the reactances of leakage fluxes outside the winding, X_1 is the reactance corresponding to the part of flux not contributing to the loss, and X_2 is the reactance corresponding to the part of flux which causes the loss. If, from among the losses arising in the structural parts, only the highest in magnitude (i.e. the tank loss) is taken into consideration, neglecting all the others, then X_1 is the reactance corresponding to the part of flux developing between the winding and tank wall; X_2 is the reactance corresponding to the part of flux penetrating into the tank wall, and R_2 is the resistance by which the tank losses are considered.

In order to assess the relative magnitudes of stray losses (U_2^2/R_2) occurring in the structural parts, it is convenient to relate them to the short-circuit reactive power (U_0^2/X_0) of the transformer (U_2 and U_0 are the voltages across reactances X_2 and X_0, respectively). For the relation between the two powers, based on Fig. 4.24 and neglecting R_0, the following formula is obtained

$$v = \frac{\rho}{\rho^2 x^2 + (\rho + \sigma)^2},$$
(4.45)

where

$$\rho = \frac{R_2}{X_0}, \qquad x = \frac{X_1}{X_2} \quad \text{and} \quad \sigma = \frac{X_1}{X_0}.$$

Factor x representing the ratio of flux causing no loss to flux contributing to the loss, depends mainly on the voltage rating of the transformer. Thus in higher-voltage transformers the tank wall is farther from the winding than in transformers of lower voltage ratings, hence the ratio X_1/X_2 will also be greater. Thus, factor x increases with the voltage. The factor σ shows a slight dependence on the power rating in transformers of smaller ratings; where with increasing ratings the limb diameters and, with them, the reactance X_1 also increases; in larger transformers, the value of σ may be taken as constant. Finally, the factor ρ depends on the design and material of the structural parts.

By applying a magnetic shunt, the magnetic resistance of the path of flux causing no loss is reduced, i.e. the magnitude of reactance X_1 is increased. (In a properly designed magnetic shunt the losses may be neglected with respect to the losses occurring in the structural parts, therefore in our present model the magnetic shunt may be regarded as being loss-free). From relation (4.45) it follows that the application of a magnetic shunt will always cause a reduction of losses in the structural parts.

(Care need only be taken to prevent the flux passing through the shunt from causing an excessive temperature rise in some smaller component parts lying in the path of the increased shunted flux). The effect of electromagnetic shielding is not as unequivocal. Such electromagnetic shielding (e.g. an aluminium sheet) can be represented in the model by introducing a resistance in parallel with resistance R_2. The resultant resistance (R_2') will be smaller than the original one $(R_2' < R_2)$, therefore factor ρ will decrease. A reduced factor ρ will have a further effect, in addition to the reduction of resistance: it will reduce the voltage drop U_2 across reactance X_2. Thus, in the quotient U_2^2/R_2 representing the losses both the numerator and denominator will decrease. A reduction of factor ρ leads to a reduced loss only in the case of small values of x and ρ. It may occur that when electromagnetic shielding is used, the flux displacing effect of shielding is weaker than the reduction of resistance due to the increased conductivity (U_2^2 reduces more

180

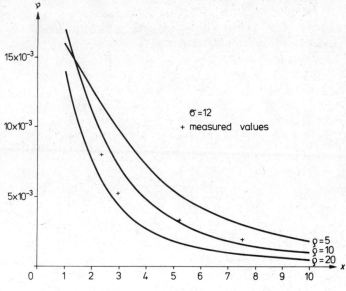

Fig. 4.25. Variation of loss factor v as a function of x

slowly than R_2), in which case the shielding will be ineffective, and will moreover cause an increase of losses.

The role of the various factors is clearly illustrated in Figs 4.25 and 4.26. In Fig. 4.25 the loss factor v has been plotted against x for different values of ρ, whereas in Fig. 4.26 loss factor v is given as a function of ρ for given values of x and σ. (For large transformers the values of factor x lie in the range 1 to 10, those of ρ between 5 and 20, and those of σ between 10 and 15). It is apparent from the diagrams that with increasing x e.g. when a magnetic shunt is used—the loss factor diminishes throughout, whereas curve $v(\rho)$ has a maximum. Reduction of ρ is effective over the section preceding this maximum, with small values of x, i.e. when the distance between the winding and the tank wall is small. In this latter case, i.e. when using an electromagnetic shielding or an aluminium tank care should be taken in the design work to keep the factor ρ of the transformer as far away as possible from the maximum of curve $v(\rho)$.

It is convenient to express the loss factor v relating to all stray losses in kW MVA^{-1}. For older transformers of smaller ratings this figure may be as high as 10, in newer, well-designed transformers it is usually kept between 2 and 4 kW MVA^{-1} or even below these values.

It is common practice to use as a magnetic shunt a core packet laminated parallel to the flux direction. Its arrangement principle is shown in Figs 4.27 and 4.28. The vertically arranged shunt of Fig. 4.27 is employed in three-phase transformers to provide a magnetic path for the leakage flux leaving at the top of the winding to make it re-enter at the bottom. The shunt shown in Fig. 4.28 is a horizontal arrangement generally used in single-phase transformers to form a path for the flux

181

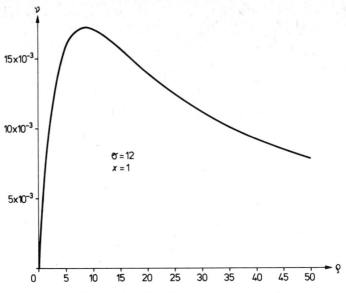

Fig. 4.26. Variation of loss factor v as a function of ρ

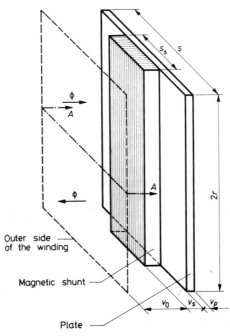

Fig. 4.27. Schematic arrangement drawing of a (vertical) magnetic shunt as employed in three-phase transformers

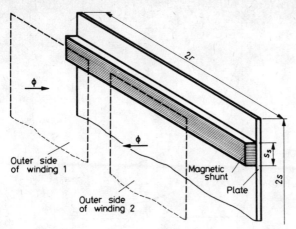

Fig. 4.28. Schematic arrangement drawing of a (horizontal) magnetic shunt for a single-phase transformer

leaving and entering the top parts of adjacent windings excited in opposite senses. The width of the magnetic shunt (s_s) is 0.2 to 0.6 m and its thickness (v_s) is 15 to 20 mm.

When designing a magnetic shunt, the magnitude of leakage flux radially leaving the windings toward the tank and the loss generated by it in the tank wall (without using a shunt) must first be determined. Magnetic shunts are only applied to the tank wall, if the loss is higher than permissible. If so, it must be determined what proportion of the flux entering the tank wall should be diverted by the magnetic shunt in order to reduce the loss to the required level. The shunt is then designed i.e. v_s and s_s are determined. The losses of a properly designed magnetic shunt are negligible relative to the tank losses, amounting to 1 to 2% at the worst.

The required approximate width of the magnetic shunt can be determined with the aid of Fig. 4.29, where the ratio of loss (P_t') in a tank wall section protected by a magnetic shunt to the loss (P_t) in the same wall section without the shunt is shown as a function of relative shunt width. The graphs are based on results of analytical calculations and on experiences gained on large power transformers. By relative shunt width the ratio of dimension s_s of the shunt to the widths of the plate to be protected is meant (for a vertical shunt the latter dimension is equal to the distance between limbs). The loss reduction is highly dependent on the distance (v_0) between the winding and tank wall. In the drawing the values of ratio P_t'/P_t are given for $v_0 = 0.1$ m and $v_0 = 0.35$ m.

Although a magnetic shunt is usually a more effective method of protection against stray flux than electromagnetic shielding, there are still marginal cases where the latter is used (e.g. to avoid the additional weight caused by the magnetic shunt). The loss arising in the electromagnetic shielding and tank wall beneath it can be calculated by means of the analytical method described in Section 4.4. In our three-layer model the first layer is constituted by the oil duct between winding and tank wall, the second layer by the aluminium or copper plate used for shielding, and

183

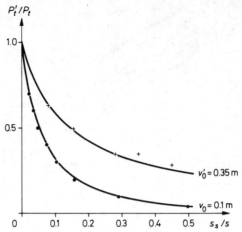

Fig. 4.29. Relative value of losses in a plate protected by a magnetic shunt, as a function of relative shunt width

the third layer by the tank wall. The effectiveness of shielding is demonstrated by the following example. The calculation has been performed with the data listed below:

$$H_m = 0.68 \times 10^{15} \text{ A m}^{-1} \quad \text{(varies sinusoidally in space)},$$
$$f = 50 \text{ Hz}, \quad r = 0.4 \text{ m}, \quad s = 0.46 \text{ m},$$
$$\gamma_1 = 0 \qquad \qquad \mu_1 = 1.256 \times 10^{-6} \text{ H m}^{-1},$$
$$\gamma_2 = 33 \times 10^6 \text{ S m}^{-1}, \quad \mu_2 = 1.256 \times 10^{-6} \text{ H m}^{-1},$$
$$\gamma_3 = 7 \times 10^6 \text{ S m}^{-1}, \quad \mu_3 = 1.256 \times 10^{-4} \text{ H m}^{-1}.$$

The overall loss (P) generated in the tank wall and in the aluminium shield is shown, as a function of thickness v_2 of the aluminium plate, in Fig. 4.30. for an oil duct width of $v_1 = 100$ mm. As it can be seen from the diagram, with increasing plate thickness, first, a rapid growth of overal losses takes place, then, after reaching a maximum, they start to decrease. With increasing distances between winding and

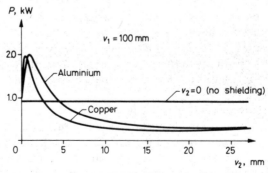

Fig. 4.30. Overall losses (P) in the tank wall and shielding plate made of aluminium and copper respectively, as functions of plate thickness (v_2) with an oil thickness of $v_1 = 100$ mm between tank wall and winding

184

tank wall, this maximum moves toward the smaller plate thicknesses. Shielding is obviously effective only if, by shielding, the amount of overall losses can be reduced below the losses occurring without shielding (i.e. with $v_2 = 0$). This condition is satisfied beyond 5 mm thickness of the shielding plate. In high-voltage transformers the space between winding and tank wall is usually much wider than 100 mm, therefore in practical cases a 8 to 10 mm thick aluminium plate will considerably reduce the tank loss.

Figure 4.30 shows overall losses in a copper shield-plate $\gamma_2 = 56 \times 10^6 \, \mathrm{Sm}^{-1}$, and in the tank wall. The shape of this curve is similar to that of aluminium but the minimum plate thickness representing the limit of effectiveness is smaller with copper than with aluminium, and also the losses are reduced below those obtainable by aluminium shielding.

4.6 The use of magnetic and non-magnetic steel

Finally, it is worth considering in what cases the replacement of conventional structural steel by non-magnetic steel may have advantages, despite being considerably more expensive and less easily machined. The behaviour of magnetic and non-magnetic steels show marked differences as regards losses depending on whether they are placed in a field of constant excitation or of constant flux. In a field of constant excitation (e.g. in the vicinity of busbars carrying high currents), the magnitude of flux entering the component in question may be considerably influenced by modifying the magnetic resistivity of the component material. Therefore, to reduce the losses, in such cases it will be expedient to employ a non-magnetic steel instead of a magnetic steel, e.g. to specify a non-magnetic material for parts of the tank wall adjacent to the conductors normally loaded with high currents (cf. Fig. 4.17). On the other hand, in a field of constant flux (e.g. in the stray field of windings) a non-magnetic material, e.g. a plate, will exhibit a lower loss than a magnetic plate only for very small (3 to 4 mm) thicknesses, so the application of magnetic steels for the thicker plates normally occurring in practice will prove more favourable (cf. Fig. 4.9). For the same reasons it is better to specify a common structural steel to make the flitch plates of windings as opposed to the case of the example presented. Thus, non-magnetic steels should preferably be used as structural material where the component part in question can be expected to have a considerable influence on the distribution of magnetic flux.

4.7 Measuring methods for hot-spot detection

The majority of components lying in the field of stray flux are at earth potential. Thermocouples may be welded to these components or thermistors (heat sensitive resistors) or simple resistance thermometers may be attached to them, by means of which their temperatures can easily be measured. The location of hot spots is unknown in advance, and at best the surroundings of certain points are suspected to be potentially dangerous, so a large number of temperature sensors are installed in a transformer. If feasible, it is appropriate to perform detailed stray flux and temperature distribution measurements as part of the type tests, in different simulated operating conditions of the transformer. For this purpose, several hundred temperature detectors and search coils for measuring flux density should be built into the transformer [129]. In such cases the measuring leads coming from the sensors may cause difficulties. However long the time required for systematization and evaluation of measuring data, the results will be valuable not only for the transformer in question, but by furnishing useful information concerning other similar designs. On conclusion of the measurement the sensing elements, or at least their measuring leads will have to be removed from the transformer.

Measurement of the temperature of electrically live parts is more complicated. In such cases the signal of the temperature sensor is brought out from the inside of the transformer using fibre optics or by means of a small radio transmitter of relatively high power, coded for transmission below the noise level. Obviously, in such a case the application of several hundred measuring points is out of the question, and the temperature of a maximum of one or two critical points can be measured.

The only method for detecting probable hot spots also applicable at present in test-field measurements is the analysis of gases dissolved in the oil of the transformer tested. The method is based on the decomposition of oil under the effect of heat and the formation of gases of different compositions at different temperatures. At normal service temperatures the developing gases consist mainly of hydrogen (H_2) and methane (CH_4). At hot-spot temperatures slightly above normal (e.g. around 150 °C) CH_4 predominates, and with temperatures increasing further ethane (C_2H_6) and etylene (C_2H_4) also appear in a growing proportion. Hence, by analysing gases dissolved in the oil, in the course of the temperature rise measurement, the presence of possible hot spots and their approximate temperatures can be determined. The ratios between the various gases are even more characteristic of the presence of a hot spot and of its temperature. Especially, the following ratios are indicative: C_2H_2/C_2H_4, CH_4/H_2 and C_2H_4/C_2H_6. In the temperature range 150 to 300 °C a considerable increase of the ratio CH_4/H_2 takes place, which may reach the value of 3. With a further temperature rise also of ratio C_2H_4/C_2H_6 increases, approaching the value of ratio CH_4/H_2. Ratio C_2H_2/C_2H_4 remains unaffected by the temperature [17].

In order to obtain a detectable quantity of gases for determining their ratios, a heat run test longer than usual has to be carried out on the transformer (for a period of 12 to 24 hours). During the temperature rise heat run test gas samples are taken at about hourly intervals, and conclusion drawn about possible hot spots from the quantity of critical gases and the trend of changes in their critical ratios. In general practice, it is not necessary to analyse all samples, but sufficient to perform analysis of samples taken after every 4th or 6th hour. The purpose of interim samplings is to make the curve plotted correct so as to detect and discard possible erroneous data and to see the overall tendency.

This method has become a regular practice as a cross-check. Its introduction as a test-field acceptance test can be expected within one or two years.

5. TRANSFORMER INSULATION

The purpose of insulation of a transformer is to isolate parts at different potentials from one another. For designing this insulation the voltage stresses arising in a transformer must be known. The magnitude of these stresses depends on the voltages appearing across the transformer terminals in service and during the dielectric tests. In the following, this question will be discussed first. The stresses arising within the windings can be determined from the voltages appearing at the terminals and knowledge of the voltage distribution. Among the windings, the distribution of power-frequency voltage and overvoltages differing not very much from the system frequency is proportional to the number of turns of the windings, and it is uniform within each individual winding. The distribution of steep-front overvoltages depends on the capacitances and inductances of the windings. In Section 5.2 the calculation of voltage distribution will be dealt with in detail, followed by a discussion of dielectric strength properties of insulating materials and insulation systems, and the fundamental relations required for designing insulations. Finally, the verification of proper designing, i.e. dielectric tests of transformers, will be discussed.

5.1 Voltages appearing across the transformer terminals

The voltages appearing across the transformer terminals may be classified according to their duration. The voltage permanently present in the transformer during its entire time in service is the service voltage. Generally, it is identical to, or only slightly different from, the rated voltage of the transformer. According to the standard specifications, the continuously permissible service voltage should not deviate by more than 5% from the rated voltage, or in the case of a tapped winding, from the voltage pertaining to the adjusted tap.

Voltages of duration shorter by several orders of magnitude than the service voltage which appear across the transformer terminals are termed overvoltages, which may be internal or external depending on their origin. Internal overvoltages arise from a sudden change in some electric property of the system, or when the symmetry of voltages is upset, or as the consequence of a resonance phenomenon. Among them, the temporary overvoltages are mostly undamped or weakly damped periodic overvoltages, of service frequency or upper-harmonic or subharmonic oscillations of relatively long duration (generally, lasting for a few seconds). Such an overvoltage may be caused by one of the most frequent faults, viz. single-phase-to-earth faults, which represent about 70% of all faults occurring in networks [131]. The voltage rise in the sound phases with respect to the conditions prevailing before a single-phase earth fault is given by the earth-fault factor, and depends on the earthing conditions. Networks of very high system voltages are operated with directly earthed neutrals. For such systems the value of the earth-fault factor is below 1.4. For isolated neutral systems or those earthed through arc-suppression coils, the earth-fault factor is taken as 1.73, i.e. the voltages of the terminals belonging to the sound phases rise to the level of line-to-line voltage. Such a service condition represents no hazard in itself for the transformer, and operation under such circumstances may be maintained for several hours without any harmful consequences. The phase-to-phase voltages remain unchanged, and the phase-to-earth insulation of windings not directly earthed (generally having a voltage rating not exceeding 72.5 kV) are capable of withstanding the line-to-line voltage, provided they have been properly designed according to the specified test voltages.

Temporary overvoltages are liable to occur on disconnecting a load at the remote end of a long transmission line. Following a sudden load rejection the end of the open line attains the equilibrium state only after a few seconds of swinging. These overvoltages are highest at the point of switching-off, but their effect cannot be considered negligible at the sending end of the line, i.e. at the transformer terminals. Overvoltages of a similar nature arise when a long transmission line is energized by switching the supplying transformer in on the L.V. side.

A temporary overvoltage may also be caused by phenomena of resonance or ferro-resonance taking place in the network. Such a condition occurs when, in consequence of a switching operation or a sudden change of load, the capacitances and inductances of the system form a combination constituting a resonant circuit tuned to the system frequency or to some of its upper harmonics. In the phenomenon of ferro-resonance the inductive elements of non-linear magnetizing characteristics (such as transformers, and shunt reactors) play a part. Such ferro-resonant overvoltages following energization processes may last through several cycles or even 1 to 2 seconds depending on the transformer inrush current time constant.

As proved by field tests, the service-frequency and higher-harmonic overvoltages lasting a few seconds generally remain below 1.5 times the peak value of the phase-to-neutral voltage. In a few cases they may be higher than that, but do not reach twice the said peak value [38].

Temporary overvoltages are in most cases associated with transient overvoltages lasting a fraction of a second say 200 to 2000 µs. Usually, these voltage surges are referred to as switching overvoltages. Their frequency depends on the natural frequency of the system and they are produced by some switching-in or switching-out operation or by a network fault. Most of these overvoltages are highly damped phenomena of relatively short duration. For transformer insulation, this kind of overvoltage constitutes a hazard.

A transient overvoltage may present itself at the incidence or cessation of an earth fault as mentioned above. A phase to earth fault always gives rise also to a transient

188

process, the overvoltage being highest in the case of an arcing earth fault. Due to these transient phenomena, voltages considerably higher than the rated line-to-line voltage may appear, but according to field measurements, they do not exceed the 2.1 × peak value of the phase-to-neutral voltage. At the cessation of an earth fault, overvoltages lower than this value, of a maximum 1.6 times the said phase-to-neutral voltage may be expected to appear in the sound phases.

For transformers of very high voltage ratings, the transient overvoltages occurring on switching-in of very long transmission lines may be associated with some hazards. When closing unloaded lines, twice the peak value of the switched-on voltage can appear, if the inductance of the supplying voltage source is assumed to be zero and the damping effect of losses are not considered. It follows from the assumption of zero source inductance that the supply voltage can be considered constant during the phenomenon. In reality, however, the voltage source has a finite inductance, therefore the supply-side voltage is not constant and its oscillation is to be reckoned with. Consequently, at the open end of the line, in spite of damping, voltages higher than the double value may appear. This overvoltage is the higher, the higher the supply-side inductance and the longer the transmission line. According to calculation, with 1.4 H supply-side inductivity and 380 km line length, at the open end of the line an overvoltage of 2.8 times the rated voltage, and at the sending end twice the rated voltage, may be expected to appear.

Switching overvoltages due to line energization are influenced, for shorter lines as well, by the mutual effects between phases. In the case of three-phase lines, the closing of the three poles does not take place at exactly the same instant. Immediately on closure, from the line of the phase first connected, a certain amount of charge is transferred capacitively to the second and third phases so that, when the latter are closed, there is residual charge present in their conductors. Consequently, in some sections of the phases closing later, higher overvoltages may arise than in the phase closed first.

Similarly, a higher overvoltage tends to appear in the transmission line in the case of reclosing when the residual charge accumulated on the line may still be present. In such a case, the overvoltage factor may be as high as 3.0 to 3.2. If, however, inductive voltage transformers are installed on the line and the circuit breakers are provided with opening resistors, quick discharging of the line may be expected to take place, whereby the hazard due to such overvoltages is substantially diminished.

If an unloaded transformer is connected to the remote end of the line, the overvoltage arising will be somewhat lower than in the case of a line with its end open. The reclosing of a long transmission line having an unloaded transformer at its remote end may be associated with some hazard, due to oscillations starting at the instant of opening.

In such cases, somewhat higher overvoltages may occur than those associated with the reclosure of unloaded transmission lines.

Dangerous overvoltages may arise on the interruption of small inductive and capacitive currents. When an inductive current is interrupted, high overvoltages may arise if the circuit breaker de-ionizes so rapidly as to force the current prematurely to zero. Again, when interrupting capacitive currents (e.g., when opening an unloaded transmission line, a cable circuit or a capacitor bank), restriking of the circuit breaker may give rise to high overvoltages. Therefore, the circuit breakers used for switching of transformers and associated networks should be of a type free from restriking to avoid this kind of overvoltages.

As proved by field tests, the magnitude of transient (switching) overvoltages of internal origin appearing at the transformer terminals, assuming up-to-date circuit breakers, exceed 2.2 to 2.5 times the peak value of rated phase-to-neutral voltage in exceptional cases only. The wave form of such overvoltages varies, generally taking the form of damped oscillations of frequency below 20 kHz, but, exceptionally, of much higher frequency. In dielectric tests of transformers, switching overvoltages are simulated, in accordance with standard specifications, by aperiodic voltage waves with front times longer than 20 μs and duration to the first zero passage longer than 500 μs. According to recent experiences, however, real (oscillating, damped) switching surges may under favouruable conditions impose higher stresses on transformer windings than an aperiodic voltage wave applied at the test station. This problem will be discussed separately later when describing the calculation of voltage distribution.

The duration of overvoltages of atmospheric (external) origin is even shorter than that of internal transient overvoltages. The former are caused mainly by lightning strikes. Such overvoltages appearing in H.V. networks may either be due to direct strikes hitting the line or to strikes to earth (indirect strikes) very close to the line. In both cases a sudden potential rise takes place on the line, propagating in the form of travelling waves toward the line ends. Obviously, an indirect strike gives rise to a smaller overvoltage than a direct strike. The travelling waves thus produced are unidirectional voltage pulses attaining their peak value very rapidly, within 1 to 2 μs and decaying to zero considerably more slowly, within 100 to 200 μs. The shape of an overvoltage wave caused by a lighting strike can be more readily compared with that of a switching surge if the initial part of the wave, up to its first peak, is substituted by the first quarter cycle (rising section) of a sine wave. In this case, a wave attaining its first peak in 1 μs can be substituted in its initial section by a sine wave of 250 000 Hz. This frequency is by at least one order of magnitude higher than the usual frequency of switching waves. If by the polarity of lightning the polarity of the cloud is meant, then observations show that 60 to 70% of strikes in flat country are of negative polarity. In some regions 80 to 90% are negative strikes. Based on measurements performed in various regions of the world, 95% of lightning strikes are of intensity exceeding 14 kA, 50% are over 30 kA, 4% are over 100 kA, and lightning currents as high as 250 kA have been measured, too.

If a sudden breakdown of voltage occurs under the effect of a wave travelling along a line (e.g. due to a flashover of an insulator), a full wave as characterized above cannot develop, and a chopped wave results. In this case a collapse of voltage occurs, generally after attaining its peak value, but sometimes before that instant. The duration of total voltage collapse is 0.1 to 0.2 μs.

The magnitude of overvoltages produced by a direct or indirect lightning strike is limited by the insulation of the line. The flashover voltage of suspension insulators used in 132 kV networks lies in the 600 to 700 kV range in the case of atmospheric overvoltages, thus no such overvoltages higher than this level can occur on such lines.

For testing the strength of transformer insulation to withstand overvoltages of external origin a test voltage having a wave shape closely similar to that caused by a lightning strike, termed a lightning impulse voltage, is applied.

For limiting the overvoltages appearing at the terminals of a transformer during service, protective devices are installed, first of all surge arresters and spark gaps. For the protection of large transformers, nonlinear resistor type surge arresters are generally mounted at the terminals or in their vicinity. The protective effect of a surge arrester consists in preventing transient or external overvoltages arriving at its terminal from rising beyond its protection level (its switching and impulse sparkover voltages, respectively).

The insulation of a transformer has to be so designed, in view of the voltage stresses expected under service conditions and the protection level of surge arresters employed, as to reduce to an economically and operationally acceptable level the probability of occurrence of voltage stresses that would damage transformer insulation or affect continuity of service. This requirement can be satisfied by proper selection of test voltages, i.e. selection of the insulation level of the transformer [56].

The criterion for the suitability of internal (non-regenerating) insulation is its capability of withstanding the specified test voltage without injury. For judging the quality of external (regenerating) insulation a statistical method is employed: the external insulation of a transformer is properly designed if it withstands with at least 90% probability a switching impulse voltage and a lightning impulse voltage whose peak values are exceeded by the respective kind of impulse voltages occurring in service with a probability not higher than 2%. With the growth of our knowledge concerning properties of insulating materials, this definition will most probably be extended to internal insulations as well, and is indeed already applied in some cases (cf. justification of the number of pulses to be applied in impulse voltage tests discussed in Section 5.4).

The determination of test voltages proceeds as follows. From knowledge of the characteristics of the network concerned, the magnitude of expected overvoltages is first determined. The rated voltage of surge arresters to be employed should be selected to be somewhat higher than the expected temporary overvoltages. The protection level of surge arresters so selected determines the magnitude of test voltages. The switching impulse withstand voltage should be 15 to 25% above, and the lightning impulse withstand voltage by 25 to 35% above the respective protection level of the surge arresters employed.

The insulation levels (power-frequency and impulse-wave withstand voltages) of transformers up to 245 kV voltage rating are shown in Table 5.1. This shows the test voltages, and for defining the respective withstand voltages, the basic voltages (U_b) are also given, these being the peak values of the highest phase-to-phase voltages for the equipment, as well as the peak values of power-frequency withstand voltage and the values of lightning impulse withstand voltage referred to the basic voltage (k_p and k_1). For transformers with voltage ratings below 52 kV two lightning impulse withstand voltages are specified by the Standard. Transformers to be installed in networks less exposed to overvoltages should be tested with the lower test voltage, those for more exposed networks with the higher test voltage given. For transformers of 123 kV and higher voltage ratings two or three power-frequency

Table 5.1

Rated withstand voltages for transformer windings with highest voltage for equipment $U_m < 300$ kV

Highest voltage for equipment U_m, kV	Basic voltage for relating the overvoltages $U_b = U_m \frac{\sqrt{2}}{\sqrt{3}}$, kV	Rated short duration power frequency withstand voltage U_p, kV	$k_p = \frac{\sqrt{2}U_p}{U_b}$	Rated lightning impulse withstand voltage U_l, kV		$k_1 = \frac{U_1}{U_b}$
				Group 1	Group 2	
3.6	2.9	10	4.9	20	40	6.9; 13.8
7.2	5.9	20	4.8	40	60	6.8; 10.2
12	9.8	28	4.0	60	75	6.1; 7.7
17.5	14.3	38	3.8	75	95	5.2; 6.6
24	19.6	50	3.6	95	125	4.8; 6.4
36	29.4	70	3.4	145	170	4.9; 5.8
52	42.5	95	3.2	250		5.9
72.5	59.2	140	3.3	325		5.5
		185	2.6	450		4.5
123	100.4	230	3.2; 2.7; 2.3	550		5.5; 4.6; 4
145	118.4	275	3.3; 2.8	650		5.5; 4.7
170	138.8	325	3.3; 2.3	750		5.4; 3.75
245	200.0	360	2.5	850		4.25
		395	2.8	950		4.75

and lightning impulse withstand voltages are given in the table. Actual test voltages should be selected considering the protection level of surge arresters installed.

From the data compiled in the table it can be seen that the ratios of withstand levels, i.e. the test voltages relative to the basic voltage, steadily decrease with increasing voltage ratings. The explanation lies in the requirement of improving the economy of the design. The larger is the voltage rating of a transformer, the more expensive will be any excess of its insulation, therefore a relatively lower insulation level is to be selected for the transformer. Obviously, this can only be done if there is a way to lower the protection level of protective devices as well. In the case of transformers with voltage ratings of 123 kV and above such lower insulation levels can only be applied if protection is provided by surge arresters of corresponding (lower) protection level.

For transformers with voltage ratings not exceeding 245 kV, no switching impulse, withstand voltage tests are specified by the Standard. Thus, up to this voltage level either the impulse voltage withstand test or the power-frequency voltage withstand test are in themselves suitable for verifying the adequacy of dielectric strength of insulation against internal transient overvoltages. This can be understood from the following.

The switching impulse strength of internal insulation of transformers is about 60 to 90% of that of the lightning impulse strength [48], depending on the material and geometrical arrangement of the insulation and on the way of applying the stress (e.g. on the ratio of length of flashover to that of breakdown). Thus, if the insulation of a transformer withstands a lightning impulse of a magnitude 4.2 times the basic voltage, then assuming a 60% switching impulse strength, it will also be capable of withstanding a switching impulse stress of magnitude 2.5 times that of the basic voltage. This statement would apply fully, if equal stresses were caused by switching surges and lightning impulse waves in definite parts of the insulation. In fact, however, the distribution of switching impulse waves within the windings is generally uniform, as opposed to the distribution of lightning impulse waves which invariably depart from being linear, therefore a transformer properly designed for a lightning impulse withstand test voltage equal to at least 4.2 times the basic voltage will surely be capable of withstanding the switching impulse test voltage equal to 2.5 times the basic voltage. Nor is it necessary to choose a higher switching impulse withstand test voltage, since the internal transient overvoltages occurring in service, as mentioned before, are almost always lower than 2.5 times the peak value of the phase-to-earth voltage.

The dielectric strength of transformers listed in Table 5.1 against switching impulse waves is even more convincingly corroborated by the power-frequency voltage tests. According to the results of experiments [48], the switching surge strength of internal insulation of transformers is 1.4 to 1.5 times their 50 Hz dielectric strength (referred to peak values). Thus, assuming a switching surge strength 1.4 times the power-frequency strength, and assuming within the windings a uniform distribution of switching overvoltages, a transformer having succesfully passed a power-frequency test performed with about 1.8 times the basic voltage will withstand a switching impulse voltage of about 2.5 times the basic voltage. As can be seen in Table 5.1, in every case $k_p > 1.8$.

According to the relevant IEC Specification, there are two alternative methods for the testing of 300, 362 and 420 kV transformers. In Method 1, these transformers are to be subjected to a power-frequency withstand test and a lightning impulse withstand test performed with the test voltages indicated in Table 5.2. Two test voltages apply for each of the three transformer voltage ratings, and selection between the two is made according to the protection level of the surge arresters employed. From the data of the table it can be seen that only the power-frequency test voltage is suitable for verifying the switching impulse strength ($k_p > 1.8$), the value of lightning impulse withstand voltage being too low for this purpose ($k < 4.2$). The question may arise, however, whether it is appropriate to choose such high test voltages merely to verify the switching impulse strength. In particular, if the power-frequency test is regarded as verification of the electric strength of transformer insulation against power-frequency voltage stresses, then this test voltage is exaggeratedly high. A transformer will never be exposed in service to such a high power-frequency voltage stress, or even to one approaching it. In addition, analyses of laboratory tests and figures of probability of occurrence of transformer failures in service have shown that a transformer tested for 1 hour with 50% overvoltage can be expected to operate satisfactorily through 25 years of service with a higher probability than a transformer tested through 30 s with 100% overvoltage [67]. According to experiments performed on oil/paper insulations of transformers,

Table 5.2

Test voltages for line terminals of windings with $U_m \geq 300\,\mathrm{kV}$, specified according to Method 1.

Highest voltage for equipment U_m, kV	Basic voltage for relating the over-voltages $U_b = U_m \dfrac{\sqrt{2}}{\sqrt{3}}$, kV	Rated short duration power frequency withstand voltage U_p, kV	$k_p = \dfrac{\sqrt{2}U_p}{U_b}$	Rated lightning impulse withstand voltage U_l, kV	$k_l = \dfrac{U_l}{U_b}$
300	245	395	2.3	950	3.9
		460	2.7	1050	4.3
362	296	460	2.2	1050	3.5
		510	2.4	1175	4.0
420	343	570	2.35	1300	3.8
		630	2.6	1425	4.2

under corona-free conditions, i.e. if the breakdown is not due to the deterioration of insulation, but only to some deficiency of a statistical character (e.g. engineering design inadequacies, manufacturing deficiencies arising from careless workmanship or improper processing, foreign objects, contamination, etc.), then the relation between breakdown voltages V_1 and V_2 and between times t_1 and t_2 elapsing up to the instant of breakdown is

$$\frac{t_2}{t_1} = \left(\frac{V_1}{V_2}\right)^m,$$

where $m = 15$ to 30. If the voltage that the insulation will sustain for 1 year is taken as 100%, then — with $m = 25$ — the same insulation can be expected to resist a 117% voltage for 1 week, a 145% voltage for 1 day, and a 170% voltage for 1 minute. Longer exposure times have the advantage of constituting a safer basis for extrapolating to a service period of, say, 6 to 12 months, than tests of shorter durations. According to experience gained in practical operation, defects occurring beyond 6 to 12 months of service are such as could not have been disclosed by a factory overvoltage test of one day duration.

With the above in view, for transformers of 300 kV rating and above, the IEC Standards specify (as Method 2) switching and lightning impulse tests omitting the short-duration power frequency test from the rated insulation levels (Table 5.3). Instead, a long-duration power-frequency test of considerably lower voltage than those listed in Table 5.2, combined with a partial discharge measurement are to be performed (see part of the chapter dealing with testing). For transformers beyond 420 kV, only this second method may be used.

The insulation levels of transformers having a voltage rating exceeding 765 kV are not specified by the IEC Standard. For these levels the test voltages recommended in Table 5.4 may be taken as indicative values [22].

194

Table 5.3

Test voltages for live terminals of windings with $U_m \geq 300$ kV, specified according to Method 2

Highest voltage for equipment U_m, kV	Rated switching impulse withstand voltage (phase-to-neutral) (peak) U_s, kV	Rated lightning impulse withstand voltage (peak) U_l, kV
300	750	850
362	850	950
420	950	1050
525	1050	1175
765	1175	1300
	1425	1425
	1550	1550
		1800
		1950

Table 5.4

Suggested test voltages for live terminals of windings with $U_m \geq 765$ kV, specified according to Method 2

Highest voltage for equipment U_m, kV	Rated switching impulse withstand voltage (phase-to-neutral) (peak) U_s, kV	Rated lightning impulse withstand voltage (peak) U_l, kV
	1675	1950–2700
1100	1800	2100–2900
	1950	2250–3100
	1800	2100–2900
1300	1950	2250–3100
	2100	2400–3500
	2100	2400–3500
1500	2250	2250–3700
	2400	2700–3900

Summing up the above, it can be stated that the test voltages impose the highest stresses on the transformers. Thus, transformer insulation has to be designed for these voltages. The standard specifications are based on the assumption that a transformer having once withstood the specified test voltage will with high probability withstand the service voltage without damage during its entire service life as well as the overvoltages of internal and atmospheric origin occurring in service.

5.2 Voltages inside the transformer

As it has been shown in the preceding section, the frequency spectrum of variations of switching impulse voltages extends into the range of several thousand Hz, and that those of the lightning impulse voltages up to several hundred thousand Hz. When calculating phenomena associated with such high frequencies, the capacitances of transformer windings should also be considered, although being of no importance at power frequency voltage levels. There are capacitances present between transformer windings and earthed parts (core, tank, etc.), within each winding between discs, turns and layers, and between individual coils. Owing to these capacitances the voltage distribution of steep-front overvoltage waves within the transformer will not be uniform, i.e. the ratios between such steep-front voltages arising on the various parts of the winding (the voltage stresses appearing across the insulation between the discs, the turns, etc., and between the winding and earthed parts, as well as stresses across that between individual coils) will differ from the ratios between the respective power-frequency voltages. The following sections deal with the calculation of voltage distribution.

5.2.1 Standing waves and travelling waves

The role of capacitances will be demonstrated by the examination of a single homogeneous single-layer transformer winding (Fig. 5.1). This shows the shunt capacitances (C_t) between the turns and the core, and the series capacitances (K_t) between adjacent turns. The simplified mathematical model of such a winding is a ladder network containing the inductances and the shunt and series capacitances (Fig. 5.2). This model differs from the simplified transmission line equivalent circuit in the presence of the series capacitances. A section of the model—two of its elements, each of length dx—is shown in Fig. 5.3. The inductance of a unit length of the winding is denoted by $L(\mathrm{H\ m^{-1}})$, its shunt capacitance by $C(\mathrm{Fm^{-1}})$, and its series capacitance by $K(\mathrm{F\ m})$. The set of differential equations describing the transient process taking place in the winding, using the notations of Fig. 5.3 and based on Kirchhoff's laws, takes the following form:

$$\frac{\partial i_L}{\partial x} + \frac{\partial i_K}{\partial x} = -C\frac{\partial u}{\partial t}, \tag{5.1}$$

$$i_K = -K\frac{\partial^2 u}{\partial x\,\partial t}, \tag{5.2}$$

$$\frac{\partial u}{\partial x} = -L\frac{\partial i_L}{\partial t}. \tag{5.3}$$

The currents and voltages appearing in the above relations are functions of both the space and time.

By eliminating the currents, the set of differential Eqns (5.1), (5.2) and (5.3) can be reduced to a single differential equation of 4th order referring to the voltage:

$$\frac{\partial^2 u}{\partial x^2} - LC\frac{\partial^2 u}{\partial t^2} + LK\frac{\partial^4 u}{\partial x^2\,\partial t^2} = 0. \tag{5.4}$$

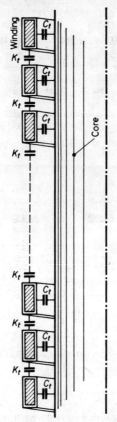

Fig. 5.1. Single-layer transformer winding

Starting from this differential equation, the transient process can be treated both by the standing-wave theory [137], and by the travelling-wave theory [116].

According to the first theory standing waves are produced of frequency $\dfrac{\omega}{2\pi}$ in time and of $\dfrac{\gamma}{2\pi}$ frequency in space. Between angular frequencies ω in time and γ in space the following relations prevail:

$$\omega = \frac{\gamma}{\sqrt{LC\left(1 + \dfrac{K}{C}\gamma^2\right)}} \tag{5.5}$$

and

$$\gamma = \sqrt{\frac{LC\omega^2}{1 - \omega^2 LK}}. \tag{5.6}$$

Thus, to each frequency in space belongs a frequency in time, and with increase of γ the value of ω also increases. With $\gamma \to \infty$ the critical angular frequency of the

197

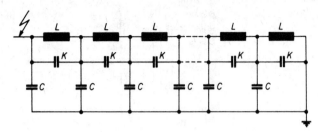

Fig. 5.2. Simplified mathematical model of a single-layer transformer winding

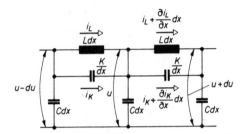

Fig. 5.3. A section of the model shown in Fig. 5.2

winding is obtained:

$$\omega_{cr} = \frac{1}{\sqrt{LK}}.$$
(5.7)

This is the highest angular frequency in time with which the winding is capable of oscillating.

According to the travelling-wave theory, travelling waves of angular frequency ω pass the winding with a velocity of

$$v = \frac{\omega}{\gamma} = \sqrt{\frac{1}{LC} - \frac{K}{C}\omega^2}.$$
(5.8)

It can be seen that the higher the frequency of the wave the lower its propagation velocity. Waves of angular frequencies $\omega \geq \omega_{cr}$ do not propagate in the winding, such waves being unable to penetrate into it. Thus, the winding behaves like a low-pass filter, permitting the passage of waves having angular frequencies of $\omega < \omega_{cr}$.

The determination of voltage distribution within the winding is simpler with use of the standing-wave theory on which our further discussion is therefore based.

5.2.2 Determination of voltage distribution within the winding according to Wagner [137]

A voltage step function U is connected to one end of the simplified transformer winding represented by the model of Fig. 5.2 (i.e. the voltage applied to the terminal is $u(t) = 0$ if $t < 0$, and $u(t) = U$ if $t \geq 0$), and the other end of the winding is earthed. On the appearance of voltage, the initial voltage distribution will immediately develop, i.e. every point of the winding will assume its initial voltage. From this initial value the voltage will tend to attain the final distribution as defined by the inductances (and by the resistances left out of consideration in the model). Since the system contains capacitances and inductances, the transition from the initial to the final state is an oscillating process. In fact, this voltage oscillation is attenuated by the resistances of the system, but this effect will be neglected in the calculation that follows.

5.2.2.1 Initial voltage distribution

If the initial voltage distribution alone is to be investigated, the boundary transition $L \to \infty$ has to be employed in the differential Eqn. (5.4) relating to the voltages.

In that case Eqn. (5.4) will take the form

$$\frac{\partial^2 u(x,0)}{\partial x^2} = \frac{C}{K} u(x,0).$$

(5.9)

For at one end earthed winding of length l — after solving the above differential equation for the boundary conditions $u(0,0) = U$ and $u(l,0) = 0$ — the ratios of voltages arising in various points of the winding with respect to earth related to the terminal voltage U, as a function of ratio x/l, where x is the coordinate of a given point of the winding and l is the full winding length, are obtained from the relation

$$\frac{u\left(\frac{x}{l},0\right)}{U} = \frac{\sinh \alpha \left(1 - \frac{x}{l}\right)}{\sinh \alpha},$$

(5.10)

where

$$\alpha = l \sqrt{\frac{C}{K}} = \sqrt{\frac{C_r}{K_r}},$$

(5.11)

and $C_r = Cl$ is the resultant shunt capacitance and $K_r = K/l$ is the resultant series capacitance, i.e. thus the initial voltage distribution depends on the factor α. The initial voltage distribution curves developing under the effect of the unit step function described by relation (5.10), for different values of the factor α, are shown in Fig. 5.4. From the curves it will be apparent that the higher the factor α, i.e. the lower the resultant series capacitance of the winding compared to the shunt capacitance, the less uniform will be the voltage distribution along the winding. By increasing the series capacitance, i.e. by reducing factor α, the uniform voltage distribution (characterized by $\alpha = 0$) can be more closely approached. The gradient of the voltage distribution curve (i.e. the voltage stress imposed on the winding) is the highest over the initial section:

$$g_{max} = -\frac{\alpha}{l} \coth \alpha.$$

(5.12)

Considering that $g = -1/l$ for uniform voltage distribution, the value of maximum voltage stress related to that arising with uniform voltage distribution can be

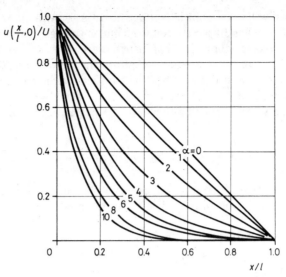

Fig. 5.4. Initial voltage distribution curves for windings earthed at one end

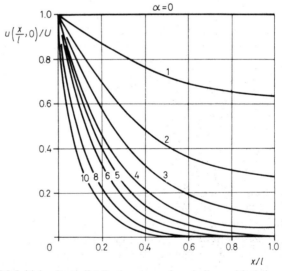

Fig. 5.5. Initial voltage distribution curves for windings with their end open

calculated from the formula

$$\delta = \alpha \coth \alpha .$$ (5.13)

By similar reasoning, the initial voltage arising for an open-end winding with respect to earth, related to the terminal voltage, is given by the ratio

$$\frac{u\left(\dfrac{x}{l}, 0\right)}{U} = \frac{\text{ch}\left(1 - \dfrac{x}{l}\right)}{\text{ch}\ \alpha}$$ (5.14)

(see curves of Fig. 5.5). It will be apparent from a comparison of Figs 5.4 and 5.5 that a substantial difference between the initial voltages arising in an open-end and an earthed-end winding will appear with small values of factor $\alpha(\alpha<5)$. In such cases, the initial voltages arising on the first section of an open-end winding will be lower, and the initial voltages with respect to earth will be higher, than those arising on the corresponding part of an earthed-end winding.

5.2.2.2 Final voltage distribution

For calculating the final voltage distribution we again start from differential equation (5.4). In this case, however, the differential equation with respect to time has to be set equal to zero. Hence

$$\frac{\partial^2 u}{\partial x^2} = 0 .$$
(5.15)

After solving this differential equation and using boundary conditions $(u(0) = U$ and $u(l) = 0)$, the final voltage distribution of the earthed-end winding, again calculating with relative values as in the previous case, will be given by the relation

$$\frac{u}{U} = \left(1 - \frac{x}{l}\right)$$
(5.16)

and that of the open-end winding by the relation

$$\frac{u}{U} = 1 .$$
(5.17)

The final state will, therefore, correspond to the straight line as drawn into Figs 5.4 and 5.5, pertaining to $\alpha = 0$, i.e. to a uniform voltage distribution.

5.2.2.3 Voltage oscillations

The complete oscillation process taking place in the transformer winding is given by the solution of differential equation (5.4) satisfying the boundary conditions. The voltage at an arbitrary point x of an earthed-end winding at instant t using boundary conditions $u(0, t) = U$ and $u(l, t) = 0$, is

$$u(x, t) = U\left[\left(1 - \frac{x}{l}\right) - 2 \sum_{n=1}^{\infty} \frac{A_n}{n\pi} \sin n\pi \frac{x}{l} \cos \omega_n t\right] ,$$
(5.18)

where

$$A_n = \frac{\alpha^2}{\alpha^2 + (n\pi)^2} \quad \text{and} \quad \omega_n = \frac{n\pi}{\sqrt{\alpha^2 + (n\pi)^2}} \omega_{cr} .$$
(5.19)

$n = 1, 2, 3, \ldots$, and α and ω_{cr} can be calculated from formula (5.11) and (5.7), respectively.

For an open-end winding, with initial conditions

$$u(0, t) = U \qquad \text{and} \qquad \frac{\partial u(x, t)}{\partial x}\bigg|_{x=l} = 0 \, ,$$

$$u(x, t) = U \left[1 - 2 \sum_n \frac{A_n^*}{\left(\dfrac{n\pi}{2}\right)} \sin\left(\frac{n\pi}{2}\right) \frac{\pi}{l} \cos \omega_n^* t \right], \tag{5.20}$$

where

$$A_n^* = \frac{\alpha^2}{\alpha^2 + \left(\dfrac{n\pi}{2}\right)^2} \qquad \text{and} \qquad \omega_n^* = \frac{\dfrac{n\pi}{2}}{\alpha^2 + \left(\dfrac{n\pi}{2}\right)^2} \, \omega_{cr} \, . \tag{5.21}$$

Summation has to be performed taking into account only the odd values of n, i.e. for $n = 1, 3, 5, \ldots$, whereas α and ω_{cr} can again be calculated from Eqns (5.11) and (5.7), respectively.

At every point of the winding, the voltage sets out from the initial value developing at instant $t = 0$, and oscillates around the value that corresponds to the final voltage distribution. Plotting as a function of length along the winding the maximum voltages arising during transient voltage oscillations in the various

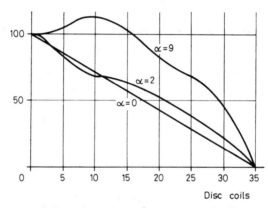

Fig. 5.6. Envelope curves for different values of α

points of the winding, the distribution of maximum stresses to earth i.e. the envelope of the latter is obtained. The amplitude of voltage oscillations is the smaller, the closer the initial voltage is to that prevailing in the final state, i.e. the smaller is the factor α. Thus, by reducing factor α not only is the initial voltage distribution rendered more uniform, but also the amplitude of voltage oscillations is reduced. In Fig. 5.6 envelopes measured on two different transformer windings are shown as examples. One of the windings is of small series capacitance ($\alpha = 9$) and exhibits large oscillations, whereas the other of large series capacitance ($\alpha = 2$) is less liable to oscillate.

202

The primary importance of the Wager model and of the calculation based on it, is the very informative qualitative picture provided by the phenomena taking place in transformer windings under the effect of steep-front overvoltage waves. Also from a practical point of view, the factor calculated from the resultant shunt and series capacitances is of great significance, since conclusions can be drawn from it also as to the uniformity of initial voltage distribution within the winding and to the liability of the latter to oscillations. This method, however, is unsuitable for the purpose of voltage distribution calculations required for the insulation design.

5.2.3 Mathematical models for practical calculations

The infinitesimal homogenous winding model proposed by Wagner, where the series and parallel capacitances and the self-inductances are alone taken into account, idealises the real conditions very much. For one thing, a real transformer winding is composed of a finite number of elements (e.g. discs, pairs of discs), therefore it is more obvious to represent it by a model consisting of a finite number of elements than by an infinitesimal model. Thus in the case of disc type transformer windings, it is convenient to consider each disc, each pair of discs or each group of discs as a separate element. Further, real transformer windings are not homogeneous in every case but generally at the beginning and end of the winding, and sometimes inside it, often contain winding elements (discs) in which the number of turns, geometrical arrangement or shape deviates from the majority of elements constituting the winding. The Wagner model also simplifies real condition by neglecting the mutual inductances within the winding and the damping effect of eddy current losses on voltage oscillations. Finally, in the majority of cases, it is not a single winding that is to be investigated, but a transformer or a winding system consisting of two or more windings.

Moreover, only qualitative information is furnished by the Wagner method of calculating the voltage distribution within the winding, because the result refers to a voltage distribution developing under the effect of voltage step. Real overvoltages are, in fact, either finite front-time (rise-time) and time-of-tail (decay time) aperiodic voltage waves or periodic damped oscillations.

For calculating a voltage distribution, first the impulse voltage characteristics of the winding or winding system should be known with the series and shunt capacitances of winding elements, the coefficients of self- and mutual inductance, and the damping factor (Section 5.2.4). When constructing a finite model, the number of elements in the model should be properly chosen (Section 5.2.5).

In practical calculations it is not always necessary to consider all of the points enumerated above. In the course of design work, often only an assessment of voltage distribution, quick comparison of a number of alternatives or selection of the range of favourable winding capacitances is required. So, for assessing the effect of inhomogeneities present at the beginning and at the end of the winding it is generally sufficient to consider a network containing only capacitances (Section 5.2.6). Similarly, the same capacitive model will furnish sufficient information for the purpose of judging the severity of impulse voltages transferred from one winding to the other and for the determination, on this basis, the winding capacitances favourable from this respect (Section 5.2.7).

In networks consisting of capacitances only, the calculation of voltage distribution is based on assuming a zero front-time and infinite time-of-tail function (step function). In a real system containing inductances as well, the real wave shape should be considered in the calculation of voltage distribution (Section 5.2.8).

The voltage distribution within homogeneous transformer windings can be calculated more accurately than Wagner by taking into account the mutual inductances within the winding (Section 5.2.9). Regulating transformers containing main and regulating windings of high series capacitances may be calculated on the basis of a two-element model (Section 5.2.10). The voltage distribution developing within arbitrarily arranged winding systems under the effect of lightning and switching impulse waves of any arbitrary shapes is rendered possible by a general method utilizing matrix calculus (Section 5.2.12), by which if necessary the damping effect can also be taken into account (Section 5.2.13).

5.2.4 Impulse voltage characteristics of transformer windings

5.2.4.1 Series and shunt capacitances

From among the types of windings dealt with in detail in Chapter 3, only relations for calculating the capacitance of the types most often encountered will be given here. The principle of capacitance calculation will be demonstrated on ordinary disc-type windings. Using the same principle, the capacitance of the types of windings not discussed here can easily be calculated.

Series capacitance of disc coil windings. In Fig. 5.7 the schematic drawing of a disc coil consisting of two sections and its capacitance network is shown. The series

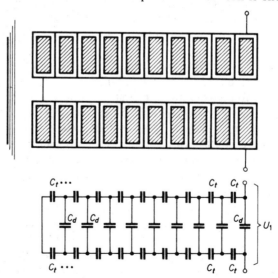

Fig. 5.7. Schematic diagram of a disc coil and its capacitance network

capacitance of this disc coil is composed of two parts, being the resultant of inter-turn capacitances and of the inter-section capacitance. For calculating the capacitance, the voltage is assumed to be evenly distributed within the disc coil. The calculation of the resultant capacitance of the disc coil is based on the principle that the sum of energies accumulated in the part capacitances within the section is equal to the entire energy of the disc coil.

With n turns in each section and with voltage U_1 present on the disc coil, the energy of turn capacitances C_t can be expressed by the following:

$$\frac{1}{2} C_t \left(\frac{U_1}{2n}\right) 2(n-1) = \frac{1}{2} C_{tr} U_1^2 \, ;$$

from which the resultant of turn capacitances is

$$C_{tr} = \frac{C_t}{2n^2} (n-1) \, .$$

The inter-section capacitance, assuming a linear voltage distribution within each section, can be calculated on the basis of Fig. 5.8. Accordingly, the voltage arising in

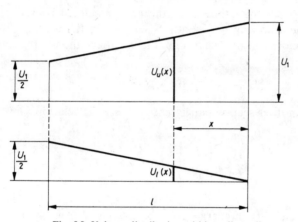

Fig. 5.8. Voltage distribution within a disc coil

point x of the upper section is

$$U_u(x) = U_1 \frac{2l-x}{2l}$$

and in point x of the lower section is

$$U_l(x) = U_1 \frac{x}{2l} \, .$$

The energy of capacitance C_{d1} of a unit length of the winding at point x will thus be

$$\frac{1}{2} C_{d1} [U_u(x) - U_l(x)]^2 \, .$$

The resultant energy of the capacitance between disc-coils, with substituting

$$U_u(x) - U_l(x) = U_1\left(1 - \frac{x}{l}\right)$$

will be

$$\frac{1}{2} C_{dr} U_1^2 = \int_0^l \frac{1}{2} C_{d1} \left[U_1\left(1 - \frac{x}{l}\right)\right]^2 dx.$$

After integration and reduction of the above equation, the following expression is obtained

$$C_{dr} = \frac{C_{d1} l}{3},$$

viz. the resultant capacitances between the sections is one third of the capacitance obtained from the geometrical dimensions.

Thus, the resultant series capacitance of one disc-coil is

$$C_r = C_{tr} + C_{dr}.$$

The series capacitance of the entire winding is calculated from the capacitances of disc-coils in the following way. With N sections in the winding, the calculation performed as before will give the resultant turn-capacitance for the entire winding as

$$\Sigma C_{tr} = \frac{C_t}{Nn^2}(n-1). \tag{5.22}$$

The resultant of inter-section capacitances, using the relation

$$\frac{1}{2} C_{dr}\left(\frac{U}{\dfrac{N}{2}}\right)^2 (N-1) = \frac{1}{2} \Sigma C_{dr} U^2$$

expressing the equality of energies, can be calculated from the formula

$$\Sigma C_{dr} = 4\frac{N-1}{N^2} C_{dr}. \tag{5.23}$$

The resultant series capacitance of the entire winding, with the use of relations (5.22) and (5.23), will be

$$K = \frac{1}{N}\left(\frac{n-1}{n^2} C_t + 4\frac{N-1}{N} C_{dr}\right). \tag{5.24}$$

Since, in most practical cases, $n > 10$ and $N > 30$, relation (5.24) can be reduced to the following simple form:

$$K = \frac{1}{N}\left(\frac{C_t}{n} + 4C_{dr}\right). \tag{5.25}$$

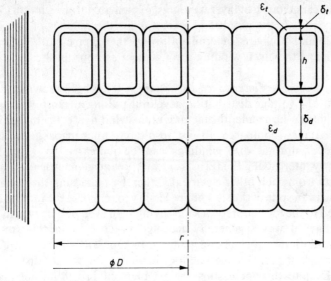

Fig. 5.9. Letter symbols used in the calculation of series capacitance of disc coils

If the mean winding diameter is D, the size of copper conductor used for calculating the turn-capacitance is h, the thickness of inter-turn insulation is δ_t, and its permittivity is ε_t, the radial size of winding is r, the thickness of spacers between discs is δ_d, and the resultant permittivity of (oil + solid) insulation of thickness δ_d is ε_d (notations shown in Fig. 5.9), then the inter-turn capacitance and the inter-disc capacitance can be calculated from the known formula for plane condensers as follows

$$C_t = \varepsilon_t \varepsilon_0 \frac{D\pi(h+2\delta_t)}{2\delta_t} = 27.8D \frac{\varepsilon_t(h+2\delta_t)}{2\delta_t} 10^{-12} \text{ F} \tag{5.26}$$

and

$$C_{dr} = \frac{1}{3}\left[\varepsilon_0 D\pi \frac{r+\delta_d}{\dfrac{2\delta_t}{\varepsilon_t} + \dfrac{\delta_d}{\varepsilon_d}}\right] = 27.8D \frac{1}{3} \frac{r+\delta_d}{\dfrac{2\delta_t}{\varepsilon_t} + \dfrac{\delta_d}{\varepsilon_d}} 10^{-12} \text{ F}. \tag{5.27}$$

When writing the above formulae, the surfaces defining the capacitances have been taken as larger than those of the metal parts facing each other; surfaces twice the thickness of turn insulation have been added to these surfaces for calculating the turn capacitance, and the disc spacing added for calculating the inter-disc capacitance, in order to take into account the effect of stray capacitances.

On the basis of relations (5.25), (5.26) and (5.27), the series capacitance of normal disc windings can be calculated from the equation

$$K = \frac{27.8D}{N}\left[\frac{\varepsilon_t(h+2\delta_t)}{2n\delta_t} + \frac{4}{3}\frac{r+\delta_d}{\dfrac{2\delta_t}{\varepsilon_t} + \dfrac{\delta_d}{\varepsilon_d}}\right]10^{-12} \text{ F}, \tag{5.28}$$

where all geometrical dimensions are to be substituted in metres.

The series capacitance of layer windings is calculated from relation (5.24). In that case, however, N is the number of layers and n the number of turns in a layer, whereas C_t and C_{dr} are determined from the geometrical dimensions and permittivities of the given winding arrangement as appropriate.

Series capacitances of interleaved disc windings. It has been shown in the preceding sections that the voltage distribution developing along a transformer winding will be more uniform, the smaller the factor α of the winding, i.e. the higher its resultant series capacitance compared to its resultant shunt capacitance. The series capacitance of normal disc windings can be increased by several orders of magnitude by interleaving the turns, i.e. by removing geometrically adjacent turns farther away from each other electrically, thereby increasing the voltage between adjacent turns. In spite of that it is not necessary to reinforce the interturn insulation as compared to the former practice, and savings can even be achieved in some cases. In earlier days it was a generally accepted practice to increase dimension on interturn insulation in normal disc windings to cope with the expected overvoltage stresses, for lack of a reliable method for calculating the voltage distribution within windings. Despite this increasing, the problem of providing adequate impulse voltage strength of insulation remained unsolved. By interleaving the turns, however, the voltage distribution can be made more uniform, so that the windings can be designed to have sufficient strength to withstand both the power-frequency and the impulse voltage stresses without the need for any reinforcement of interturn insulation. Dielectric strength considerations have shown that, up to about 36 kV rated voltage, normal disc windings can in every case, be designed for an adequate

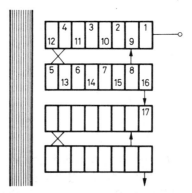

Fig. 5.10. Winding with increased series capacitance ($E = 2$). (English Electric Co.)

impulse voltage strength. In the range of 52 to 170 kV satisfactory impulse voltage strength levels can be obtained both with normal disc windings and with disc windings of increased series capacitance. For rated voltages of 245 kV and above, however, the required impulse voltage strength can generally be achieved only by employing windings of increased series capacitance [65].

The first winding with a series capacitance increased by means of interleaving the turns was introduced by the English Electric Co. in 1945 (Fig. 5.10). This disc-coil consisting of two sections was further developed by Ganz Electric Works by extending the interleaving of turns to winding elements containing 4, 6, 8, ..., i.e., any even number of sections.

When calculating the series capacitance of windings having interleaved turns, it is sufficient to consider the interturn capacitances only, the inter-section capacitances being negligible as compared to the former. A similar train of thought can be followed as for the case of normal disc-type windings when calculating the series capacitance of interleaved windings. Based on the case of a disc coil consisting of four sections ($E=4$) shown in Fig. 5.11, a general relation for disc coils consisting of

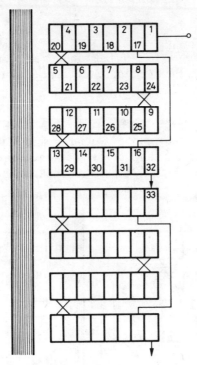

Fig. 5.11. Winding with increased series capacitance ($E=4$)

any number of sections will now be deduced. Let the number of turns be n, in each section, which means that in a disc coil consisting of E sections, the energy stored in $E(n-1)$ capacitances (interturn capacitances) each of magnitude C_t has to be taken into account. Assuming a uniform distribution within each section, the voltage between adjacent turns across $\dfrac{E}{2}(n-1)$ capacitances is $\dfrac{U_E}{E \cdot n} \dfrac{E \cdot n}{2} = \dfrac{U_E}{2}$, and across the remaining $\dfrac{E}{2}(n-1)$ capacitances is $\dfrac{U_E}{En}\left(\dfrac{E \cdot n}{2}-1\right)$, where U_E is the voltage appearing across the terminals of each winding element. Correspondingly the resultant series capacitance of each winding element is

$$C_r = \frac{C_t \left(\frac{U_E}{2}\right)^2 \frac{E}{2}(n-1) + C_t \left[\frac{U_E}{En}\left(\frac{En}{2}-1\right)\right]^2 \frac{E}{2}(n-1)}{U_E^2} =$$

$$= C_t \frac{E}{2}(n-1)\left[\frac{1}{4} + \frac{1}{4}\left(\frac{En-2}{En}\right)^2\right].$$

Considering the practical values of $n = 10$ to 30 and $E \geq 2$ with approximation this becomes

$$C_r = \frac{EC_t}{4}(n-1).$$

Summing up the energies of the N/E winding elements (N is the number of sections in the winding), the resultant series capacitance of the entire winding is

$$K = \frac{E^2 C_t}{4N}(n-1). \tag{5.29}$$

The series capacitance of windings wound of several wires in parallel can be influenced by modifying the arrangement of the parallel wires. If these parallel wires lie one next to the other (Fig. 5.12(a)), the series capacitance remains unchanged.

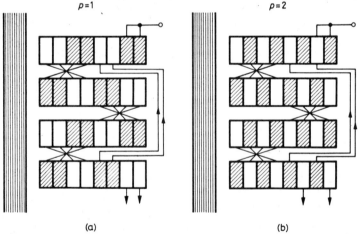

Fig. 5.12. Arrangement of parallel branches in windings of increased series capacitance. (a) $p = 1$; (b) $p = 2$

If, however, the wires are separated from one another (Fig. 5. 12(b)), the series capacitance of the winding increases considerably. Thus, in the latter case ($p = 2$), the number of interturn insulation stressed by voltage is twice that in the former case ($p = 1$).

The oil duct between turns, on the other hand, reduces the series capacitance. For the capacity between two turns separated from each other by an oil duct can be

210

neglected in comparison with the capacitance between turns lying on each other and separated by interturn insulation only. Denoting the number of non-adjacent parallel wires by p, and the number of oil ducts within a winding by o, and substituting the detailed expression of interturn capacitance according to Eqn. (5.26), the series capacitance of interleaved windings the following relation is obtained as

$$K = \frac{27.8D}{N} \left[\frac{\varepsilon_t (h + 2\delta_t)}{2\delta_t} \right] (pn - o - 1) E^2 \times 10^{-12} \ \mathrm{F}. \tag{5.30}$$

The dimensions are to be substituted in metres here.

A great advantage of interleaved windings is that their series capacitance can easily be modified within wide limits, according to dielectric strength and economic considerations, without changing their number of turns and geometrical dimensions. If the geometrical dimensions of the winding, N and n are all maintained constant, and no oil duct is assumed ($o = 0$) and $pn \gg 1$, then relation (5.29) takes the form of

$$K = kpE^2,$$

where k is a factor depending on the constants mentioned. For a winding consisting of two wires in parallel the value of p may be either 1 or 2. In practical designs the number of sections within a winding element is $E = 2, 4, 6$ or 8. Hence, the ratio of maximum to minimum series capacitances that can be realized is

$$\frac{K_{max}}{K_{min}} = \frac{2 \times 8^2}{1 \times 2^2} = 32.$$

Thus, the series capacitance of a winding can be varied in the ratio of $1:32$ without modifying either its geometrical dimensions and operating characteristics, or its rated voltage and current. An excessive increase of the number of sections in a winding element, however, would bring about a harmful increase of the interturn voltage. This would require reinforcement of interturn insulation, resulting in a decreased series capacitance, and would be undesirable both technically and economically. Therefore, it is expedient to increase the value of E only to a limit depending on the dimensions and characteristic data of the winding.

In interleaved windings the highest inter-section stress arises between the first and second winding element, i.e. between the Eth and $(E + 1)$th sections. More precisely, it arises between the $\left[\frac{n}{2}(E - 1) + 1 \right]$th turn of the first winding element and the $\left[nE + \frac{n}{2}(E + 1) \right]$th turn of the second winding element counted from the first turn of the first winding element, where the difference between the number of turns is $[n(E + 1) - 1]$. The highest stress between sections is $[n(E + 1) - 1]/nE$ times higher than the voltage imposed on one element. Regarding cases occurring in practice ($n \geq 10$), the maximum voltage between sections exceeds the voltage of one element by about 50% with $E = 2$, by about 25% with $E = 4$, by about 16% with $E = 6$ and by about 13% with $E = 8$. The maximum stress arising between turns, assuming uniform voltage distribution within each element, is half of the voltage imposed on one element.

As shown by measurements performed on experimental windings, however, this condition is not always fulfilled. Thus in the elements of interleaved windings high-frequency (several hundred kHz) oscillations occur, which tend to cause considerable increase of stresses between turns and sections by adding to the uniformly distributed fundamental voltage. In a winding of 18 turns per section ($n = 18$) the voltage rise due to internal oscillations remained below 10% with $E = 2$, 4 and 6, but for $E = 8$, however, the voltage between the two adjacent sections was found to be as high as 1.5 times that what would have corresponding to a uniform distribution! A similar phenomenon was also found to occur in windings containing two sections per element ($E = 2$), if the number of turns per section was excessively high ($n > 36$). Such oscillations of 5 to 10 µs duration may become specially hazardous in the case of chopped waves. This is another reason why the selection of too high a number of turns per section should be avoided. If compelled to adopt

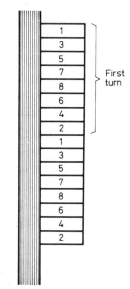

Fig. 5.13. Helical winding

such a solution the appearance of such stresses should be taken into account in the design of insulation.

Windings of interleaved turns have several variants in addition to the type already discussed here. Their series capacitance can be calculated in a similar way to that described, viz. by taking the energy of all part capacitances equal to the overall energy.

Series capacitance of a spiral winding. The spiral (helical) type of winding is used as the L.V. winding and the tapped winding of transformers. In the latter type of winding the successive sections constituting the tapping steps are wound beside one another as parallel branches, according to Fig. 5.13. In the diagram the first two turns of a helical winding containing 8 tapping steps are represented. When the turns are interleaved as shown, the voltage difference between adjacent turns is, in the majority of cases, equal to twice the step voltage. Of course, there are several other possibilities of interleaving the turns. The series capacitance of spiral windings, according to a deduction not detailed here, is given by the relation

$$K = C_t \frac{k_i n_t p}{\left(\dfrac{n_a}{n_t}\right)}, \tag{5.31}$$

where C_t is the turn capacity calculated from Eqn. (5.26), k_i is a factor for taking into account the measure of interleaving, n_t is the number of turns of a step and n_a is the number of all turns in the winding. For the winding of Fig. 5.13, the numerical values of these quantities are as follows: $k = 4$, $p = 1$, $n_t = 10$ and $n_a = 80$. Substituting these values:

$$K = 5 C_t,$$

In the case of a simple, untapped spiral winding $n_t = 1$ should be substituted into Eqn. (5.31).

In interleaved helical windings the series capacitance may increase due to the stray capacitances between non-adjacent turns (Fig. 5.14). As shown by experience

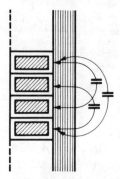

Fig. 5.14. Stray capacitances in helical windings

this effect can be taken into consideration by multiplying capacitance K by a correction factor $k_i > 1$. Substituting into Eqn. (5.31) the value of turn capacity C_t calculated from (5.26), and taking into account this correction factor, the following relation is obtained for calculating the series capacitance of spiral windings:

$$K = k_c \frac{27.8D}{\left(\dfrac{n_a}{n_t}\right)} \frac{\varepsilon_t(h + 2\delta_t)}{2\delta_t} k_i n_t p \times 10^{-12} \text{ F}. \tag{5.32}$$

(Quantities in metres.)

Parallel capacitances. For calculating the capacitances between adjacent concentric windings and between the innermost winding and core, the windings and

213

the core are considered as concentric cylinders with smooth surfaces and as the diameters being sufficiently large as compared to the spacing between windings, the capacitances are determined by means of the relation valid for plane condensers. The capacitance between two concentric cylindrical windings of heights H_1 and H_2 is given by the relation

$$C = \frac{27.8D \dfrac{H_1 + H_2}{2}}{\dfrac{\delta_o}{\varepsilon_o} + \dfrac{\delta_s}{\varepsilon_s}} \times 10^{-12} \text{ F}, \tag{5.33}$$

where D is the mean diameter of the oil duct between the two windings, δ_o is the thickness of oil insulation between the surfaces of the two cylinders, δ_s is the thickness of solid insulation between the two surfaces ($\delta_o + \delta_s$ = radial distance between the cylinder), ε_o and ε_s is the permittivity of the oil and solid insulation, respectively. For calculating the capacitance between innermost winding and core, $\dfrac{H_1 + H_2}{2}$ in relation (5.33) should be substituted by $1.1\,H$, where H is the height of the winding.

The capacitance between the outermost windings is calculated from the capacitance formula for parallel cylinders. Thus,

$$C = \frac{27.8\,H}{\ln \dfrac{\sqrt{a^2 - D^2} + s}{D}} \frac{(\delta_o + \delta_s)}{\dfrac{\delta_o}{\varepsilon_o} + \dfrac{\delta_s}{\varepsilon_s}} \times 10^{-12} \text{ F}, \tag{5.34}$$

where H is the height of (outermost) windings and D is the diameter, s is the spacing between axes of the two windings, ($\delta_o + \delta_s$) is the minimum distance between the two windings, where δ_o is the thickness of the oil and δ_s the thickness of solid insulation (transformer board). The quantities are to be substituted in metres. For the capacitance between the middle phase winding and the two adjacent outer windings, the value of C from formula (5.34) should be multiplied by 2. The capacitance between an outer winding and adjacent tank wall is also given by double the value obtained from formula (5.34) where, however, quantity s indicates double the value of the distance between winding axis and tank wall.

The capacitance between windings and further earthed parts (first of all the yoke and metal parts connected to it) can generally only be estimated. These capacitances never exceed 10 to 20% of the calculated earth capacitances.

5.2.4.2 Self and mutual inductances

The self- and mutual inductances can be determined less accurately than the capacitances. Inductivity of iron-core coils is a function of frequency: with increasing frequency the inductance decreases. The oscillations taking place in a winding are not of a single frequency; in addition to the fundamental harmonic mode, upper harmonics also appear. Depending on the type and arrangement of the winding these oscillation frequencies fall in the range of 1000 Hz to 1 000 000 Hz. Even if the inductance-to-frequency characteristics of the transformer were known, the calculation based on a mathematical model containing a frequency-dependent inductance would be unduly complicated. Experience has shown that in the calculation of impulse voltage phenomena of transformers, the frequency dependence of inductances need not be taken into account. Based on comparison of measurements and calculations performed, the inductance of transformer windings to be considered with regard to impulse voltage phenomena, is 1/15 to 1/40 part of the no-load inductance of the transformer. When calculating for larger trans-formers, the no-load inductance should be divided by 15 to 20, and for smaller transformers by 20 to 40.

The no-load inductance of large transformers can be calculated, with good approximation, from relation

$$L_{nl} = \mu_r \mu_o \frac{A}{l_c} N^2 H, \tag{5.35}$$

where μ_r is the relative permeability of the core, $\mu_o = 1.256 \times 10^{-6}$ is the permeability of vacuum in H/m^{-1}, A is the cross-sectional area of the core in m^2, l_c is the length of core in m, and N is the number of turns of the winding.

Thus, the inductance to be taken into account in calculating the impulse voltage distribution is

$$L = \frac{1}{15 \div 40} L_{nl}. \tag{5.36}$$

In the calculations, mutual inductive couplings to be considered are those existing within each winding and between concentric windings. According to experiments, the mutual inductance between two turns, defined by coordinates x and y, where x and y give the distance of the turns to be considered from one end of the homogeneous winding having a length of l

$$M(x, y) = L^{-v \left| \frac{x-y}{l} \right|}, \tag{5.37}$$

where L is the self-inductance of a section of the winding of unit length. Exponent v depends on the geometrical dimensions and magnetic conditions of the winding. For a finite model, the mutual inductance between the ith and jth equal elements, according to formula (5.37), is

$$M_{ij} = L_1 q^{|i-j|}, \tag{5.38}$$

where L_1 is the self inductance of one element, and q is a coefficient characterizing the closeness of mutual inductive coupling. For iron-core winding $q = 0.97$ to 0.99.

In practice, it is the self inductance of the entire winding that can be directly calculated on the basis of Eqn. (5.36). If, for the purpose of calculating the voltage distribution, the winding is divided into n equal elementary coil sections, then the self inductance of one such element is determined by taking the sum of the self inductances of all elements and of the mutual inductances as equal to the resultant inductance of the entire winding [138]. Thus, if the inductance of one element is L_1, the number of elements is n, the coefficient characterizing the mutual inductive coupling between individual winding elements is q, and the resultant inductance of the entire winding is L, then according to relation (5.38)

$$L = nL_1 \left\{ 1 + \sum_{k=1}^{n-1} \frac{n-k}{\frac{n}{2}} q^k \right\} \qquad (5.39)$$

must be true. From expression (5.39), with L, n and q being known, L_1 can be determined.

The mutual inductance between two concentric windings having inductances L_r and L_s, respectively, can be calculated from the relation

$$M_{rs} = q_{rs} \sqrt{L_r L_s}, \qquad (5.40)$$

where coefficient q_{rs} depends on the closeness of coupling between the windings. This inductive coupling is very close between two windings located on a common core, in such a case $q_{rs} = 0.97$ to 0.99. If one or both of the windings are divided in several elements, then the mutual inductive couplings between winding elements of the different windings have also to be determined. In the course of this procedure the following principle should be kept in mind: the division of windings must not modify the coupling between the two entire windings defined by Eqn. (5.40). The calculation is shown by an example.

Let a winding of inductance L_r be divided in two parts, the other winding of inductance L_s, concentrically arranged with respect to the former, being left undivided. The inductance of one element of the winding of L_r will be, according to formula (5.39), with $n=2$,

$$L_{r1} = \frac{L_r}{2(1+q)}. \qquad (5.41)$$

If the mutual inductive coupling between one element of the winding divided in two parts and the undivided winding is characterized by q^*, then the mutual inductance between the two windings will now be given by the relation

$$M_{rs} = 2q^* \sqrt{L_{r1} L_s}. \qquad (5.42)$$

From the identity of Eqns (5.40) and (5.42) substituting L_r from (5.41) gives

$$q^* = q_{rs} \sqrt{\frac{1}{2}(1+q)}. \qquad (5.43)$$

In the case of dividing one winding, or both windings, into several parts, the coefficient of mutual inductive coupling is thus determined.

The coefficient appearing in the denominator of formula (5.36) and the coefficients q characterizing mutual inductive couplings depend on the constructional design of the transformer and on the core material used. No rules of general validity can be given for their determination. When performing concrete calculations, it is appropriate to set out from data measured on completed units.

216

5.2.4.3 Damping factor

The voltage oscillations arising in a winding are damped by the winding resistance and the core losses. If the damping is also to be considered in the calculation, the model of Fig. 5.2 has to be completed with resistances. The resistance of the winding is taken into account by a resistance put in series with the inductances, whereas the core loss is represented by a resistance connected in parallel with them. A section of such a complemented infinitesimal model is shown in Fig. 5.15, where r is the resistance of a section of unit length of the winding, and R is the resistance corresponding to the iron loss in a unit-length section of the core.

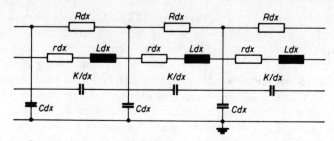

Fig. 5.15. Mathematical model of a transformer winding with damping resistances

The set of differential equations applicable to this model is too complicated for the purpose of determining the voltage distribution itself, but is, however, suitable for calculating the damping factor [136]. The damping factor of the harmonic oscillation of frequency $\beta_m = m \dfrac{\pi}{l}$ in space is

$$\delta = \frac{r_l}{2L_l} + \frac{m^2\pi^2}{2R_l(C_l + K_l m^2\pi^2)}, \tag{5.44}$$

where $r_l = rl$, $L_l = Ll$, $R_l = Rl$, $C_l = Cl$ and $K_l = K/l$ are resultant values relating to the entire winding, l is the length of winding and $m = 1, 2, 3, \ldots$ The first term of the damping factor is the damping caused by the winding resistance, while its second term is the damping due to the iron loss. For large transformers the second term is considerably larger than the first one.

In practice, the damping factor is usually determined on the basis of empirical data. Characteristic of the measure of damping is the quotient of amplitudes A_1 and A_2 of two subsequent half waves of identical polarity. According to measurements made on large transformers, this quotient remains approximately constant for all harmonics appearing in the oscillation pattern of a given transformer, i.e. its value is independent of the frequency. The damping factor relating to the kth harmonic can be calculated, with an accuracy satisfactory for the purpose, from the relation

$$\delta_k = \frac{\omega_k}{2\pi} \ln \frac{A_1}{A_2}.$$

217

The value of quotient A_1/A_2 lies in the range 1.2 to 1.4 depending on the design of the transformer, whereas the usual figures of the damping factor, depending on the frequency, are between a few 1000 and 20 000 s^{-1}. The effect of damping is taken into account by multiplying the oscillation components by the function $f_k(t)=e^{-\delta_k t}$.

At the beginning of voltage oscillations the effect of damping is hardly noticeable, and therefore it generally has no influence on the maximum voltages affecting the design of insulation.

5.2.5 Determination of number of elements in a transformer model

In establishing the model for the purpose of calculating the voltage distribution several, partly conflicting aspects have to be considered. As a requirement of primary importance, the voltage oscillations calculated on the basis of the model should as closely as possible approach the oscillations taking place in the real winding or winding system. This requirement can be formulated by stating that in the calculated voltages all harmonic oscillations should be present which play an important part in the voltage oscillations of the real winding. In the model voltages of a model consisting of a finite number of elements harmonic oscillations are present in a number corresponding to the number of independent nodes. (A node is considered as being independent, when its voltage is not fixed by the boundary conditions. In the model of a winding consisting of n elements and hit by an impulse wave at one end and earthed at the other end, the number of independent nodes is $n-1$.) The more susceptible is a winding or winding system to oscillation (viz. the higher the factor α of a homogeneous winding), the more elementary windings must be present in the model. In practice, it is appropriate to select the number of elementary windings so as to have an approximately uniform voltage distribution within individual elementary windings, viz. the value of factor α of individual elementary windings should possibly not be higher than 2 to 3. When dividing a homogeneous winding into n elementary sections, factor α of one such element, from formula (5.11), is obtained from the following relation

$$\alpha_1 = \sqrt{\frac{C_r/n}{K_r n}} = \frac{1}{n}\sqrt{\frac{C_r}{K_r}} = \frac{1}{n}\alpha_r,$$

where α_r relates to the entire winding.

When dividing the winding or winding system into elementary sections, all division points (all nodes in the model), should be selected to coincide with those points where knowledge of the voltage is required. In winding systems and in inhomogeneous windings or winding parts having characteristics differing from each other, the terminations of homogeneous sections should be selected as division points. In tapped regulating windings it is preferable to have the tapping points as division points, at least those whose voltage is expected to play an important part in the voltage distribution. The number of elements in the model should not be increased beyond a certain limit. The uncertainties in the calculation of coefficients of self- and mutual induction have been pointed out above. It has also been shown that the induction coefficient of the entire winding is determined from the respective coefficients of individual winding sections. So, the higher the number of winding sections, the higher the probability of making an error in determining the characteristics of individual elements. The number of self- and mutual induction coefficients increase proportionally with the square of the number of elements, and also the inaccuracy of calculation may increase correspondingly. Experience has shown that for the purpose of voltage distribution calculations, when not only the initial distribution but also the voltage oscillations are to be determined, it is not expedient to use models consisting of more than 10 to 12 elements. This number of elements is sufficient for the calculation of large transformers generally comprising windings of high series capacitances. On the basis of such a model a voltage distribution satisfying practical requirements can be calculated within an accuracy of $\pm 20\%$.

5.2.6 Voltage distribution in inhomogeneous windings

The discs next to the line terminals of disc type transformers often differ from the discs inside the winding. Generally, the outermost discs contain a reduced number of turns for constructional reasons, e.g., to provide space for end insulation and to reinforce the insulation to cope with the higher mechanical stresses arising here. It also occurs, however, mainly in older designs of disc windings having small series capacitances, that the end discs are manufactured with reinforced interturn insulation to withstand the increased overvoltage stresses expected here or, in order to reduce the stresses, these discs are made to have a higher series capacitance than other parts of the windings. In windings consisting of a high number of elements (i.e. having a high value of E) the required number of turns can often be achieved by using, at one of the winding ends, a winding section containing a number of elements

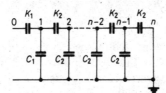

Fig. 5.16. Mathematical model of a winding with inhomogeneities in its initial section

differing from that of the others. This kind of inhomogeneous design of windings may considerably influence the voltage distribution. Generally, it is sufficient to examine the effect of inhomogeneity by checking the initial voltage distribution, i.e. on the basis of a network containing capacitances only. In unfavourable cases the method of modifying the capacitance conditions has to be resorted to, in order to obtain a more uniform voltage distribution.

The effect on the initial voltage distribution of an inhomogeneity arising at the initial section of the winding is calculated on the basis of the n-element capacitance network shown in Fig. 5.16, by assuming a step function of magnitude u_0 [64]. Except for the first element having a parallel capacitance C_1 and a series capacitance of K_1, all other elements are identical, having parallel capacitances of

$$C_2 = C_3 \ldots = C_n$$

and series capacitances of

$$K_2 = K_3 = \ldots K_n.$$

Introducing notations $C_1 = a_1 K_1$ and $C_2 = a_2 K_2$, and considering that for $t = 0$ the electrostatic charges must be zero, the following equations may be written:

$$-K_1(u_0 - u_1) + K_1 a_1 u_1 + K_2(u_1 - u_2) = 0,$$
$$-K_2(u_1 - u_2) + K_2 a_2 u_2 + K_2(u_2 - u_3) = 0,$$
$$\vdots$$
$$-K_2(u_{k-1} - u_k) + K_2 a_2 u_k + K_2(u_k - u_{k+1}) = 0,$$
$$\vdots$$
$$-K_2(u_{n-3} - u_{n-2}) + K_2 a_2 u_{n-2} + K_2(u_{n-2} - u_{n-1}) = 0,$$
$$-K_2(u_{n-2} - u_{n-1}) + K_2 a_2 u_{n-1} + K_2 u_{n-1} = 0.$$

By solving this set of equations the voltage of the kth node is obtained in the form

$$u_k = u_0 \xi r_k,$$

where

$$\xi = \frac{K_1}{K_2}\left(1 - \frac{xr_1}{1 + xr_1}\right),$$

$$x = \frac{K_1}{K_2}(1 + a_1) - (1 + a_2),$$

$$r_k = \frac{2}{n}\sum_{i=1}^{n-1} \frac{1}{a_2 + 4\sin^2\dfrac{i\pi}{2n}} \sin\frac{i\pi}{n} \sin k\frac{i\pi}{n}. \tag{5.45}$$

The inhomogeneity is characterized by the factor ξ whose value for a homogeneous winding is $\xi = 1$. The initial voltage distribution of a homogeneous winding can be approximated by Eqn. (5.10) instead of (5.45), according to which, with the notations used here

$$r_k = \frac{\sinh(n-k)\sqrt{a_2}}{\sinh n\sqrt{a_2}}. \tag{5.46}$$

The voltage distribution along a winding which is inhomogeneous in its last section can be determined in a similar way [60]. In that case it is the last element of the n-element model serving as basis of the calculation, that differs from the others. Writing in a way similar to that before, and solving a set of equations expressing the equilibrium of electrostatic charges, the voltage of the kth node is obtained from the following relation

$$u_k = u_0\left[r_{k1} - \frac{xr_{n-1}}{1 + xr_{n-1,n-1}} r_{k,n-1}\right],$$

220

where

$$r_{\nu\mu} = \frac{2}{n}\sum_{i=1}^{n-1}\frac{1}{a_1 + 4\sin^2\dfrac{i\pi}{2n}}\sin\nu\frac{i\pi}{n}\sin\frac{\mu i\pi}{n}, \tag{5.47}$$

$$x = \frac{K_n}{K_1}(1+a_2)-(1+a_1), \quad a_1 = \frac{C_1}{K_1}, \quad a_2 = \frac{C_n}{K_n}.$$

Example 5.1. Suppose the initial voltage distribution along a transformer winding consisting of 68 discs and earthed at one of its ends is to be determined. Because of reinforcement of interturn insulations, the first four discs differ from the others. The winding is represented by a 17-element model, each element consisting of 4 discs. The capacitance data of the winding is as follows:

$$K_1 = 3705 \times 10^{-12}\,F, \qquad K_2 = 5900 \times 10^{-12}\,F,$$
$$C_1 = 76.7 \times 10^{-12}\,F, \qquad C_2 = 87.2 \times 10^{-12}\,F.$$

With these data

$$a_1 = \frac{C_1}{K_1} = 0.0206, \quad \text{and} \quad a_2 = \frac{C_2}{K_2} = 0.0147.$$

Using the above values and relation (5.46)

$$r_1 = \frac{\sinh(n-1)\sqrt{a_2}}{\sinh n\sqrt{a_2}} = \frac{\sinh 16\sqrt{0.0147}}{\sinh 17\sqrt{0.0147}} = 0.885,$$

further

$$\frac{K_1}{K_2} = 0.629, \quad \text{and} \quad x = \frac{K_1}{K_2}(1+a_1)-(1-a_1) = -0.372.$$

From the above:

$$\zeta = \frac{K_1}{K_2}\left(1-\frac{xr_1}{1+xr_1}\right) = 0.629\left(1-\frac{-0.372 \times 0.885}{1-0.372 \times 0.885}\right) = 0.983.$$

The calculated values of $r_1, r_2, \ldots, r_{16}$ and $\zeta r_1, \zeta r_2, \ldots, \zeta r_{16}$ are represented in Fig. 5.17 as a function of division points of the winding. The voltage distribution would follow curve 1 (values of r_2, etc.) with no

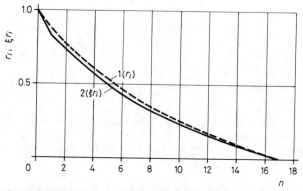

Fig. 5.17. Initial voltage distribution along a winding provided with reinforced insulation over its initial section: 1—homogeneous winding; 2—inhomogeneous winding

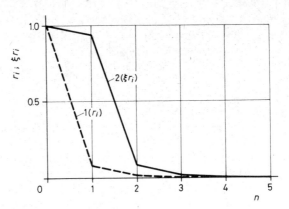

Fig. 5.18. Initial voltage distribution along a winding containing discs of increased series capacitance in its initial section: *1*—homogeneous winding; *2*—inhomogeneous winding

inhomogeneities present, whereas curve *2* (values of ξr_1, etc.) shows the voltage distribution along a winding inhomogeneous in its initial section. Under the effect of inhomogeneities the initial voltages will be lower, departing more from the straight line corresponding to uniform distribution. In addition, in the voltage distribution curve a break point appears after the initial section. Despite all this, the increase of initial voltage imposed on the first four discs is not hazardous, and the reinforced insulation will easily be capable of withstanding this stress.

Example 5.2. Let a winding consisting of 40 discs and earthed at one end be examined, the first 8 discs of which are of increased series capacitance. In this case, the winding is divided into five sections ($n = 5$). The capacitances are as follows:

$$K_1 = 4975 \times 10^{-12}\ \text{F}, \qquad C_1 = 276 \times 10^{-12}\ \text{F},$$
$$K_2 = 45 \times 10^{-12}\ \text{F}, \qquad C_2 = 276 \times 10^{-12}\ \text{F}.$$

From these values

$$a_1 = \frac{C_1}{K_1} = 0.0554 \quad \text{and} \quad a_2 = \frac{C_2}{K_2} = 6.12,$$

$$r_1 = \frac{\sinh(n-1)\sqrt{a_2}}{\sinh n\sqrt{a_2}} = \frac{\sinh 4\sqrt{6.12}}{\sinh 5\sqrt{6.12}} = 0.0845,$$

$$\frac{K_1}{K_2} = 110.4 \quad \text{and} \quad x = \frac{K_1}{K_2}(1 + a_1) - (1 + a_2) = 109.4.$$

With the above:

$$\xi = \frac{K_1}{K_2}\left(1 - \frac{xr_1}{1 + xr_1}\right) = 110.4\left(1 - \frac{109.4 \times 0.0845}{1 + 109.4 \times 0.0845}\right) = 11.04.$$

The initial voltage distributions of the homogeneous winding ($r_1, r_2, r_3, \ldots r_5$) and inhomogeneous winding ($\xi r_1, \xi r_2, \ldots, \xi r_5$) are shown in Fig. 5.18. It is apparent from that as a consequence of having increased the series capacitance of the initial section, the voltage imposed on this section has decreased, that of the adjoining section, however, has increased to the same extent, and so to a lesser degree, have the voltages imposed on the remaining sections. The position of maximum stress has only been displaced along the winding. Thus, an increase of the series capacitance of the initial section to such an extent is harmful.

From the curves of Figs 5.17 and 5.18 it can be stated that an inhomogeneity in the initial section of the winding primarily affects the voltage distribution prevailing

222

at the initial section of the winding, and has a minor influence on its remaining sections. Similarly, a winding-end inhomogeneity will mainly modify the stresses arising there, and will but slightly affect the voltage distribution of the initial section of the winding.

5.2.7 Calculation of transferred overvoltages

Generally, two or more windings are mounted on a limb of a transformer. An overvoltage directly affecting one of the windings will, due to existing capacitive and inductive couplings, also give rise to overvoltages in the other windings, thereby also indirectly jeopardising the latter. In some cases, transferred overvoltages tend to appear in windings never exposed to the direct effect of overvoltages, as in the case of the L.V. windings of generator-transformers. In other cases a transformer winding may be exposed to the effect of direct overvoltages, yet the overvoltages, transferred from another winding are expected to be even higher than the former. This may be the case, e.g., in network transformers of high transformation ratios and connected to overhead lines on both their sides. The magnitude of these transferred overvoltages should be limited by lightning arresters connected to the transformer terminals, or the insulation of the transformer windings concerned should be designed. to withstand foreseeable stresses.

The investigations described here, mainly serving the purpose of demonstrating the nature of the phenomenon, relate to a single-phase transformer containing two homogeneous concentric windings. In calculating the voltage transfer, the determination of capacitively transferred voltages is more involved, the inductively transferred voltages being proportional to the transformation ratio. Therefore, only the calculation of voltages transferred through capacitive couplings will be discussed, drawing a few general conclusions concerning these voltages. From the knowledge of capacitive voltage distribution, the upper limit of transferred voltages, including also those associated with voltage oscillations can be determined.

One winding of the transformer being investigated is that directly hit by the overvoltage (referred to as "impulsed" in the following), while the other is the winding not directly affected (referred to as "non-impulsed"). The upper end of the impulsed winding will be hit by the overvoltage, its other end being considered as earthed. Depending on whether one end of the non-impulsed winding is earthed or not, the five different connections shown in Fig. 5.19 will be distinguished:

(a) Neither end of the non-impulsed winding is earthed.

(b) Lower end of the non-impulsed winding is earthed, its upper end, i.e. that closer to the impulsed end of the impulsed winding is left open.

(c) Upper end of the non-impulsed winding is earthed, its lower end, i.e. that closer to the earthed end of the impulsed winding is left open.

(d) The non-impulsed winding is short-circuited, but unearthed.

(e) Both ends of the non-impulsed winding are earthed.

The sense in which the non-impulsed winding is wound may be either identical with, or opposed to, that of the impulsed winding.

The case of an open winding end does not differ much from that of terminating the winding with a very high resistance or a small capacitance, e.g., with a voltage divider. Short-circuiting of the winding may be employed in the course of an impulse voltage test to eliminate the appearance of transferred voltage across the terminals of non-impulsed windings. As regards transferred voltages, the open and earthed terminals represent the two extreme cases, since when terminating the winding ends

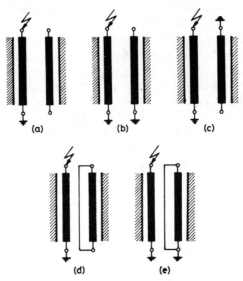

Fig. 5.19. Basic connections

by any finite impedance the stresses occurring within the winding always lie between the stresses arising in the two extreme cases.

The initial voltage distributions within the impulsed and non-impulsed windings are determined on the basis of the capacitance network shown in Fig. 5.20. Both windings are divided into n elementary sections. The resultant series capacitance of the impulsed winding is denoted by K_{1r}, that of the non-impulsed winding is denoted by K_{2r}, the resultant capacitance of the impulsed winding to earth by C_{1r}, that of the non-impulsed winding by C_{2r}, and the resultant capacitance between the two windings by C_{12r}. With these notations, the values of elementary capacitances appearing in Fig. 5.20 are as follows:

$$K_1 = K_{1r} n, \qquad K_2 = K_{2r} n,$$
$$C_1 = C_{1r}/n, \qquad C_2 = C_{2r}/n, \ C_{12r}/n.$$

The numbering of nodes of both the impulsed and the non-impulsed windings is started at 0, and the last nodes in both windings will thus be marked with serial number n. In the zero node of the impulsed winding a unit step function ($u_0 = 1$) is assumed, and its nth node is earthed ($u_n = 0$). The boundary conditions for the non-impulsed winding are as follows in the five basic cases:

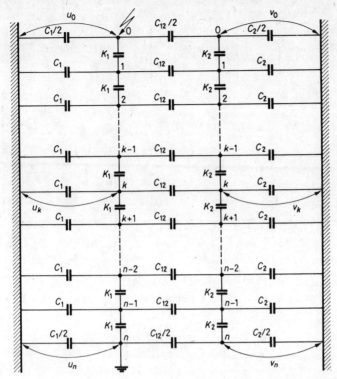

Fig. 5.20. Capacitance network of an impulsed and a non-impulsed winding

for connection (a): $v_0 \neq 0$ and $v_n \neq 0$,
for connection (b): $v_0 \neq 0$ and $v_n = 0$,
for connection (c): $v_0 = 0$ and $v_n \neq 0$,
for connection (d): $v_0 = v_n \neq 0$,
for connection (e): $v_0 = v_n = 0$.

As regards initial voltage distribution, Case (a) is practically identical with Case (b), and Case (c) with Case (e). Therefore, only the connections of (b), (d) and (e) will be examined in the following. For these three basic cases three different sets of equations can be written, considering the respective boundary conditions. It is convenient to treat the three cases in common by means of matrix calculus, and to express the result also in matrix form.

The sets of equations that can be written for the nodes of the capacitance network, on the basis of equilibrium of electrostatic charges, and applying to all the three cases, may be reduced to the following two matrix equations:

$$\mathbf{K}_u \mathbf{u} - \mathbf{T}^* \mathbf{v} = \beta u_0 \mathbf{e}_u \tag{5.48}$$

and

$$\mathbf{K}_v \mathbf{v} - \mathbf{T} \mathbf{u} = u_0 \mathbf{e}_v, \tag{5.49}$$

15

225

where

$$
\mathbf{u}=\begin{bmatrix} u_1 \\ u_2 \\ \vdots \\ u_{n-1} \end{bmatrix}; \qquad \mathbf{e}_u=\left.\begin{bmatrix} 1 \\ 0 \\ \vdots \\ 0 \end{bmatrix}\right\}(n-1)\text{ elements}
$$

$$
\mathbf{K}_u=\left.\begin{bmatrix} 2\beta+\gamma+1 & -\beta & 0 & \cdots & 0 & 0 & 0 \\ -\beta & 2\beta+\gamma+1 & -\beta & \cdots & 0 & 0 & 0 \\ \cdot & \cdot & \cdot & \cdots & \cdot & \cdot & \cdot \\ 0 & 0 & 0 & \cdots & -\beta & 2\beta+\gamma+1 & -\beta \\ 0 & 0 & 0 & \cdots & 0 & -\beta & 2\beta+\gamma+1 \end{bmatrix}\right\}(n-1)\text{ rows}
$$

$$\underbrace{}_{(n-1)\text{ columns}}$$

$$
\beta=\frac{K_1}{C_{12}}, \quad \gamma=\frac{C_1}{C_{12}}, \quad \delta=\frac{K_2}{C_{12}}, \quad \varepsilon=\frac{C_2}{C_{12}}
$$

in all cases, whereas $\mathbf{v}$, $\mathbf{e}_v$, $\mathbf{K}_v$ and $\mathbf{T}$ depend on the connection.
In respect of connection (b):

$$
\mathbf{v}=\begin{bmatrix} v_0 \\ v_1 \\ \vdots \\ v_{n-1} \end{bmatrix}, \qquad \mathbf{e}_v=\left.\begin{bmatrix} 1/2 \\ 0 \\ \vdots \\ 0 \end{bmatrix}\right\}n\text{ elements}
$$

$$
\mathbf{T}=\left.\begin{bmatrix} 0 & 0 & \cdot & \cdot & \cdot & 0 \\ 1 & 0 & \cdot & \cdot & \cdot & 0 \\ 0 & 1 & \cdot & \cdot & \cdot & 0 \\ \cdot & \cdot & \cdot & \cdot & \cdot & 0 \\ 0 & \cdot & \cdot & \cdot & 1 & 0 \\ 0 & \cdot & \cdot & \cdot & 0 & 1 \end{bmatrix}\right\}n\text{ rows}
$$

$$\underbrace{}_{(n-1)\text{ columns}}$$

$$
\mathbf{K}_v=\left.\begin{bmatrix} \delta+\dfrac{\varepsilon}{2}+\dfrac{1}{2} & -\delta & \cdots & 0 & 0 \\ -\delta & 2\delta+\varepsilon+1 & \cdots & 0 & 0 \\ \cdot & \cdot & \cdots & \cdot & \cdot \\ 0 & 0 & \cdots & 2\delta+\varepsilon+1 & -\delta \\ 0 & 0 & \cdots & -\delta & 2\delta+\varepsilon+1 \end{bmatrix}\right\}n\text{ rows}
$$

$$\underbrace{\phantom{\delta+\dfrac{\varepsilon}{2}+\dfrac{1}{2} \qquad -\delta \qquad \cdots \qquad 0 \qquad 0}}_{n\text{ columns}}$$

226

In respect of connection (d):

$$\mathbf{v} = \begin{bmatrix} v_0 \\ v_1 \\ \vdots \\ v_{n-1} \end{bmatrix}, \quad \mathbf{e}_v = \left.\begin{bmatrix} 1/2 \\ 0 \\ \vdots \\ 0 \end{bmatrix}\right\} n \text{ elements}$$

$$\mathbf{T} = \left.\begin{bmatrix} 0 & 0 & \ldots & 0 \\ 1 & 0 & \ldots & 0 \\ 0 & 1 & \ldots & 0 \\ \cdot & \cdot & \ldots & \cdot \\ 0 & 0 & \ldots & 1 \end{bmatrix}\right\} n \text{ rows}$$

$$\underbrace{}_{(n-1) \text{ columns}}$$

$$\mathbf{K}_v = \left.\begin{bmatrix} 2\delta+\varepsilon+1 & -\delta & \ldots & 0 & -\delta \\ -\delta & 2\delta+\varepsilon+1 & \ldots & 0 & 0 \\ \cdot & \cdot & \ldots & \cdot & \cdot \\ 0 & 0 & \ldots & 2\delta+\varepsilon+1 & -\delta \\ -\delta & 0 & \ldots & -\delta & 2\delta+\varepsilon+1 \end{bmatrix}\right\} n \text{ rows}$$

$$\underbrace{}_{n \text{ columns}}$$

In respect of connection (e):

$$\mathbf{v} = \begin{bmatrix} v_1 \\ v_2 \\ \ldots \\ v_{n-1} \end{bmatrix}, \quad \mathbf{e}_v = 0; \quad \mathbf{T} = \left.\begin{bmatrix} 1 & 0 & \ldots & 0 \\ 0 & 1 & \ldots & 0 \\ \cdot & \cdot & \ldots & \cdot \\ 0 & 0 & \ldots & 1 \end{bmatrix}\right\} (n-1) \text{ rows}$$

$$\underbrace{}_{(n-1) \text{ columns}}$$

$$\mathbf{K}_v = \left.\begin{bmatrix} 2\delta+\varepsilon+1 & - & \ldots & 0 & 0 \\ -\delta & 2\delta+\varepsilon+1 & \ldots & 0 & 0 \\ \cdot & \cdot & \ldots & \cdot & \cdot \\ 0 & 0 & \ldots & 2\delta+\varepsilon+1 & -\delta \\ 0 & 0 & \ldots & -\delta & 2\delta+\varepsilon+1 \end{bmatrix}\right\} (n-1) \text{ rows.}$$

$$\underbrace{}_{(n-1) \text{ columns}}$$

T* is the transpose of matrix **T**.

For the column matrices containing the nodal voltages, the following relations are obtained from relations (5.48) and (5.49) by simple calculation:

$$\mathbf{u} = u_0 [\mathbf{K}_u^{-1} \mathbf{T}^* (\mathbf{K}_v - \mathbf{T} \mathbf{K}_u^{-1} \mathbf{T}^*)^{-1} (\beta \mathbf{T} \mathbf{K}_u^{-1} \mathbf{e}_u + \mathbf{e}_v) + \beta \mathbf{K}_u^{-1} \mathbf{e}_u], \qquad (5.50)$$

$$\mathbf{v} = u_0 (\mathbf{K}_v - \mathbf{T} \mathbf{K}_u^{-1} \mathbf{T}^*)^{-1} (\beta \mathbf{T} \mathbf{K}_u^{-1} \mathbf{e}_u + \mathbf{e}_v). \qquad (5.51)$$

From relations (5.50) and (5.51) and with knowledge of the capacitances, the voltage distribution developing along the impulsed and non-impulsed windings can be determined by computer for all concrete cases.

In the following, a few general statements are made relating to the variation of transferred voltages along the windings. In a winding standing alone the initial voltage distribution is determined by coefficient α of the winding. In the present discussion the values of α differ in the cases of impulsed and non-impulsed windings, so they have to be given separately. By analogy with the definition contained in Eq. (5.11) let coefficient α for the impulsed winding be

$$\alpha_1 \equiv \sqrt{\frac{C_{1r} + C_{12r}}{K_{1r}}} = n \sqrt{\frac{\gamma + 1}{\beta}}$$

and for the non-impulsed winding

$$\alpha_2 \equiv \sqrt{\frac{C_{2r} + C_{12r}}{K_{2r}}} = n \sqrt{\frac{\varepsilon + 1}{\delta}}.$$

In the present two-winding system being investigated, the voltage distribution will depend on the coefficients α_1 and α_2.

The initial voltages developing in the basic connections of windings are illustrated by the diagrams of Fig. 5.21. The initial voltages indicated there have been calculated by means of Eqns (5.50) and (5.51) from concrete data ($n = 10$, $\alpha_1 = 5$, $\alpha_2 = 10$, $\gamma = 0.5$ and $\varepsilon = 1$). General conclusions as regards the character of voltage distribution curves can be drawn from the results shown. In connection (b), the initial voltage distribution along both the impulsed and non-impulsed winding is similar to the voltage distribution measured in a winding standing alone. The highest transferred voltage arises at the terminal of the non-impulsed winding. According to calculation, the voltage distribution will be similar in connection (a) as well, as already mentioned. The voltage distribution will be of entirely different character in connection (e). Here both ends of the non-impulsed winding are earthed, hence the value of initial voltage is obviously zero in these points. A high transferred voltage appears, however, inside the winding. The rate of rise is especially high in the first section of the voltage distribution curve, indicating the severity of the stress imposed on the first element of the winding. A similar curve also results from the calculation for case (c). The conformity of the voltage distribution curve calculated for case (a) with that of case (b), and that obtained for case (c) with that of case (d), indicates that no voltage is transferred capacitively to the non-impulsed winding end adjacent to the earthed end of the impulsed winding even if this winding end is not directly earthed. In arrangement (d) the voltage distribution differs from that of the others. Here both winding ends assume a finite potential.

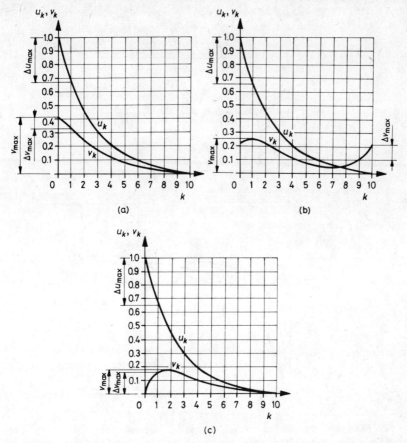

Fig. 5.21. Initial voltage distribution in basic connections of Fig. 5.19 (b), (d) and (e). ($\alpha_1 = 5$, $\alpha_2 = 10$)

From a comparison of the voltage distribution curves it will be seen that the transferred voltage with respect to earth (v_{max}) is highest in arrangement (b), and the transferred voltage imposed on one winding element (Δv_{max}) is maximum in arrangement (e). The dependence of these maximum voltages and the voltage imposed on the first element of the impulsed winding (Δu_{max}) depend on coefficients α_1 and α_2, as indicated in Figs 5.22 and 5.23. From these diagrams it can be seen that by selecting coefficients α, the transferred voltages can be varied within wide limits, and the presence of the non-impulsed winding has but a slight influence on the voltage distribution within the impulsed winding.

A similar method is used for calculating the initial voltages transferred from windings with their H.V. points at their centre (Fig. 5.24), as often employed in large transformers, to adjacent windings [133]. In that case also, the basis of calculation is a capacitance network as shown in Fig. 5.20, but now the unit step function is applied to the centre of the winding. For this reason it is appropriate to select $2n$ as the number of elements. Also here, matrix equations similar to (5.48) and (5.49) are

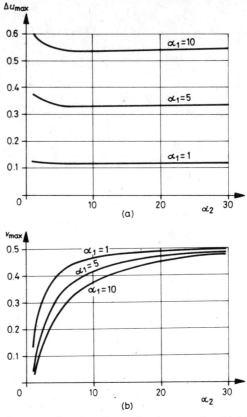

Fig. 5.22. Maximum voltages as a function of α_1 and α_2 in connection (b): (a) voltage appearing across the first element of the impulsed winding; (b) voltage to earth appearing on the terminal of the non-impulsed winding

obtained, but the matrices appearing in them differ from those valid for windings impulsed at their end. Ommitting any detailed description, voltage distribution curves well illustrating the nature of the phenomenon are presented here (Fig. 5.25) for a concrete case ($n = 5$, $\alpha_1 = 10$, $\alpha_2 = 10$, $\gamma = 0.5$ and $\varepsilon = 1$). Both the voltage against earth (v_{max}) transferred to the non-impulsed winding and the voltage (Δv_{max}) appearing across one winding element, are highest at the centre of the winding, opposite to the impulsed point of the centre-tapped winding. The dependence of maximum voltages on coefficients α_1 and α_2 for the case of $\gamma = 0.5$ and $\varepsilon = 1$ is demonstrated in Fig. 5.26. It can be seen that coefficient α of the impulsed winding (α_1) has an influence on v_{max} differing from that on Δv_{max}, with growing values of α_1 (for $\alpha_2 > 10$) Δv_{max} increases, whereas v_{max} decreases.

Because of voltage transfer, voltage oscillations also develop along the non-impulsed winding. The voltage of all points of the non-impulsed winding will oscillate around the voltage corresponding to the steady-state value defined by the turns ratio, starting from an initial value obtainable according to the foregoing. The

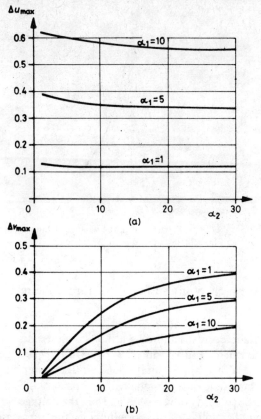

Fig. 5.23. Maximum voltages as a function of α_1 and α_2 in connection (e): (a) voltage appearing across the first element of the impulsed winding; (b) voltage appearing across the first element of the non-impulsed winding

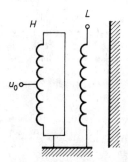

Fig. 5.24. Connection serving as basis for calculating the voltages transferred from the winding with its H.V. point at the centre

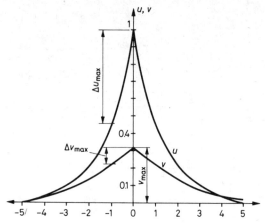

Fig. 5.25. Initial voltage distribution in the winding arrangement shown
in Fig. 5.24. ($\alpha_1 = 10$, $\alpha_2 = 10$, $\gamma = 0.5$ and $\varepsilon = 1$)

maximum of voltage oscillation can be assessed from the relation

$$|v_{max}| = |v_l| + |v_l - v_c|, \qquad (5.52)$$

where v_l is the voltage corresponding to a linear voltage distribution, i.e. to that resulting from the turns ratio, and v_c is the initial voltage defined by the capacitances. (In the above approximate relation a unit step function is assumed, and only the first harmonic of the oscillation is considered. Experience has shown that the value thus calculated is always higher than the voltage actually transferred.) When determining the voltage corresponding to the turns ratio, it should be considered whether the winding sense of the non-impulsed winding is the same as, or opposed to, that of the impulsed winding.

An approximate calculation is presented in Example 5.3.

Example 5.3. Suppose the voltage transferred from the H.V. winding to the L.V. winding is to be determined for a transformer having a turns ratio of 3 to 1, $\alpha_1 = 5$, $\alpha_2 = 20_F$ with a connection corresponding to diagram (b) of Fig. 5.19. In one case the non-impulsed winding is wound in the same sense as the impulsed winding, whereas in the other case the senses are opposed to each other. The initial voltage distributions (u_c and v_c) are given in Fig. 5.27. In the diagram the linear voltage distributions defining the axis of oscillation are also shown. The full line refers to the case of windings wound in an identical sense, the broken line to those wound in opposed senses. Maximum voltage will arise at the end of the winding. For windings wound in an identical sense, at this point $v_c = 0.47$ and $v_l = 0.33$. Hence, on the basis of relation (5.52)

$$|v_{max}| = |v_l| + |v_l - v_c| = |0.33| + |0.33 - 0.47| = 0.47,$$

i.e. the maximum voltage coincides with the initial voltage. On the other hand, for windings wound in opposite senses, substituting $v_c = 0.47$ and $v = -0.33$,

$$|v_{max}| = |-0.33| + |-0.33 - 0.47| = 1.13.$$

In this latter case, the maximum of transferred voltage oscillation will exceed even the peak value of voltage applied to the terminal of the impulsed winding. The polarity of this maximum is opposed to that of the voltage applied to the terminal.

232

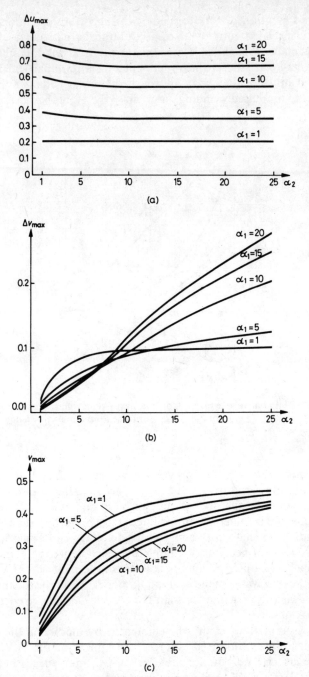

Fig. 5.26. Maximum voltages as a function of α_1 and α_2, in windings with their H.V. point in the centre: (a) voltage appearing across the first element of the impulsed winding; (b) voltage appearing across the middle element of the non-impulsed winding; (c) voltage with respect to earth appearing in the centre of the non-impulsed winding

As has been mentioned, transformers with high transformation ratios (turns ratios) deserve special attention with respect to transferred voltages. When voltage transfer takes place from the H.V. winding to the L.V. winding, the capacitively transferred voltages are those which represent a hazard. In particular, the value of the transferred initial voltage may approach the initial voltage appearing at the terminal of the impulsed winding, and may therefore be several times that

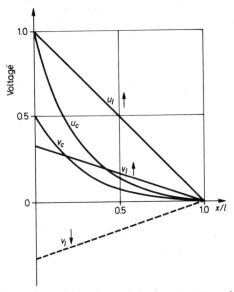

Fig. 5.27. Initial voltage distribution and steady-state voltages in connection (b) of Fig. 5.19. ($\alpha_1 = 5$, $\alpha_2 = 20$, turn ratio 3/1): x—distance from impulsed winding; l—total winding length

corresponding to the turn ratio. This voltage, however, can be reduced by selecting the winding capacitances properly. It can be adjusted to the value corresponding to the turn ratio, whereby the oscillations can be totally avoided (in relation (5.52) $v_l = v_c$). The case will be different with voltage transferred from the L.V. winding to the H.V. winding. In that case, the transferred voltage can attain a considerable value, because there is necessarily a large difference between the initial and final voltages. Namely, the initial maximum voltage transferred to the H.V. winding can be as high as the peak value of the overvoltage appearing at the terminal of the L.V. winding, i.e., generally much lower than the transferred voltage corresponding to the turn ratio.

In this case, oscillations cannot be eliminated by modifying the capacitances. Especially hazardous transferred voltages can appear at the H.V. terminals of autotransformers, not only under the effect of lightning surges but also due to switching impulses, as will be shown in further detail below.

The above method is not only suitable for describing the phenomenon of voltage transfer as it is shown above, but by its means the transferred voltages of given winding systems divided in not too many elementary sections and consisting of two

234

homogeneous windings can be calculated. In systems consisting of more than two windings or when the number of elementary windings is too high, instead of solving the matrix equations directly as presented above, it is more practical to employ some method utilizing computer capacity more economically, e.g. some iteration method or Gaussian elimination.

5.2.8 Influence of the wave form on voltage distribution

5.2.8.1 The front of the wave

So far, the voltage distribution developing under the effect of applying a unit step function has been investigated. This case is characterized by zero "front time" of the voltage applied to the winding terminal i.e. by the time elapsing from the appearance of voltage up to its maximum being zero ($T_0 = 0$). In the following the voltage conditions within a winding will be investigated for the case when it is hit by a wave of finite front time T_0. The investigation will be carried out by assuming the simplified wave form shown in Fig. 5.28 [45]. Omitting the mathematical deduc-

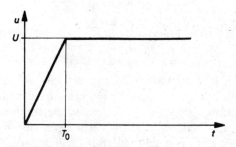

Fig. 5.28. Simplified waveform of finite front time

tion, only the phenomenon taking place will be described. The voltage at the winding terminal is

$$u(t) = \frac{U}{T_0} t, \quad \text{if} \quad t < T_0$$

and

$$u(t) = U, \quad \text{if} \quad T \geq T_0.$$

It can be shown that in a homogeneous winding like that of Fig. 5.2, when $T_0 \ll \dfrac{1}{\omega_{cr}}$, the initial voltage distribution along the winding develops as described in the foregoing. But now the voltage attains its initial value at all points of the winding in a finite time corresponding to the steepness of the front of the applied wave, and thereafter increases further at a lower rate. On oscillograms of transformer windings this initial voltage step is clearly observable (Fig. 5.29). With increasing values of T_0, this voltage step is decreasingly conspicuous, finally disappearing totally. At high values of T_0, there is no break in the initial section of the voltage curve and it rises steadily from its zero starting point.

Oscillations may occur within a winding, even if the steepness of the wave front is insufficient to permit their development in the initial voltage distribution. A unit step function, as is known, contains the entire frequency spectrum, so its application will cause all possible harmonic oscillations to appear in the winding. The longer the wave front, i.e. the longer the front time T_0, the more the higher harmonics will be

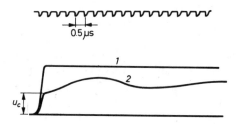

Fig. 5.29. Determination of initial voltage on the oscillogram: *1*—voltage at the transformer terminal; *2*—voltage at a point inside the winding; u_c=initial voltage

absent from the frequency spectrum. Simultaneously, in the voltage oscillations inside the winding the number of upper harmonics will decrease, and the oscillations will increasingly become smoother. Eventually, the fundamental harmonic will remain, then, the amplitude of that also will decrease. When very flat waves $(T_0 \gg 1/\omega_1)$ are applied, no voltage oscillations will occur, and a uniform voltage distribution will develop within the winding. As proved by experiments, with $T_{0\,cr}=(10\ldots12)\dfrac{1}{\omega_1}$, a practically uniform voltage distribution can be expected to develop in transformer windings. Thus, for an earthed winding, from Eqns (5.19) and (5.7), and with substitution of $n=1$,

$$T_{ocr}=(10\ldots12)\frac{1}{\pi}\sqrt{(\alpha^2+\pi^2)LK} \tag{5.53}$$

is obtained as a critical front time. Waves of front times longer than critical will no longer cause non-uniform voltage distribution and voltage oscillations in the winding. All this is clearly demonstrated by the oscillograms of Fig. 5.30. In that record voltage oscillations with respect to earth are shown appearing at the terminal of a winding characterized by $\alpha=6.2$, $L=16\times10^{-3}$ H and $K=10\times10^{-12}$ F and at 75, 50 and 25% of its full length, for waves of three different front times. The critical front time is

$$T_{cr}=\frac{12}{\pi}\sqrt{(6.2^2+3.14^2)16\times10^{-3}10\times10^{-12}}=10.7\times10^{-6}\ \text{s}.$$

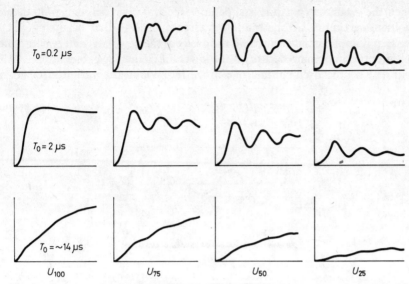

Fig. 5.30. Oscillograms of voltage oscillations taking place in a transformer winding following application of waves of different front times (time base 16 μs)

As can be seen from the oscillograms, when a wave of front time $T_0 = 0.2\,\mu$s is applied, many upper harmonics are present in the spectra.

Under the effect of the wave with front time $T_0 = 2\,\mu$s only the fundamental frequency can develop to any extent. Finally, the voltage of front time $T_0 \sim 14\,\mu$s, exceeding the critical front time defined above, is distributed along the winding uniformly and practically free of any oscillation.

5.2.8.2 The wave tail

If, instead of a voltage step, a voltage varying in time after reaching its peak value is applied to the terminal of a winding, then the voltage corresponding to the final or to the quasi-stationary state, respectively, in the case of an earthed-end, homogeneous winding can be described by the following function similar to (5.16)

$$u_f(t) = U(t)\left(1 - \frac{x}{l}\right). \tag{5.54}$$

Thus, if the applied voltage decreases as a function of time, the quasi-stationary voltage will also decrease in the same way throughout the entire winding.

Starting from its initial value, the voltage at each point of the winding will oscillate around the quasi-stationary voltage given in Eqn. (5.54) and, after decay of oscillations, will follow relation (5.54). The decay of overvoltage waves occurring in practice results in a reduction of the resultant voltage maximum, since by the time the voltage oscillation attains its maximum value, the quasi-stationary component

237

on which the oscillating voltage is superimposed will drop below its initial value. The voltage maximum will be the lower as compared to that occurring when a voltage step is applied, the more rapid the decay of the quasi-stationary component and the later this maximum occurs, i.e. the lower the frequency of oscillation. All this is illustrated by the oscillograms of Fig. 5.31. showing the variation of voltage

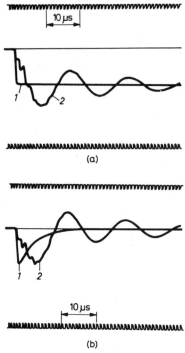

Fig. 5.31. Effect of wave tail on voltage oscillations: (a) voltage oscillogram of a wave form approaching that of a unit step function; (b) voltage oscillogram of a waveform with rapidly decaying tail: *1*—voltage at the winding terminal; *2*—voltage at an internal point of the winding

arising at an internal point of a transformer winding following the application of an overvoltage wave similar in its shape to a voltage step, varying but slightly along its tail and decaying rapidly after reaching its peak.

5.2.8.3 Chopped wave

A chopped impulse voltage wave can be regarded as the resultant of two superimposed full waves (Fig. 5.32). The initial section of a chopped wave is identical to a usual full wave $[U_1(t)]$ up to the instant of chopping, at which time another full wave $[U_2(t)]$ of opposite polarity and considerably higher rate of rise is superimposed on the first wave. A chopped wave, in itself, is generally less dangerous to any insulation than a full wave of equal amplitude, the duration of voltage stress imposed by a chopped wave being much shorter. The situation inside the winding is entirely different with respect to the voltages arising between internal points of the winding e.g. those imposed on a disc or a disc coil of two sections. The higher the factor α of a winding, i.e. the higher the amplitudes of internal voltage oscillations, the greater the difference between the shape of these voltages and that of the applied impulse voltage. Also the internal voltage differences arising under the effect of a

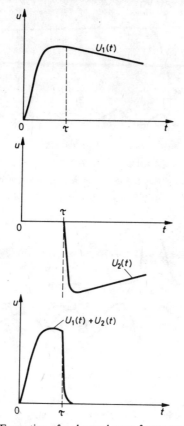

Fig. 5.32. Formation of a chopped wave from two full waves

chopped wave can be considered as produced by superposition of internal voltage differences brought about by two full waves [59].

Figure 5.33 shows the oscillogram of the voltage arising between two internal points (across a pair of disc) of a transformer winding. In (a) the oscillation brought about by a full wave is shown, and in (b) that produced by a wave chopped at instant τ. From the oscillograms it can be stated that the voltage imposed on a disc-coil is increased by a chopped wave for two reasons.

(a) The steepness of chopping exceeds that of the original wave, and an increase in the rate of rise of the wave front, as explained further above, opens the possibility for higher frequencies to appear in the oscillation pattern, and consequently to cause a less uniform voltage distribution and higher internal stresses. So, a higher stress will be caused by voltage $U_2(t)$ than by $U_1(t)$, although the amplitude of the latter is somewhat higher than that of the former.

(b) Whenever, at the instant of chopping, the oscillation caused by a superimposed wave (commencing at a later instant) is of polarity identical to that of

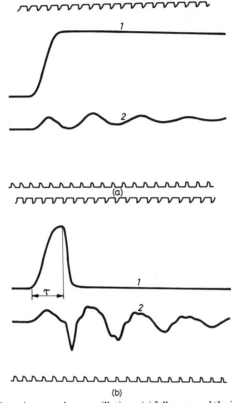

Fig. 5.33. Effect of wave chopping on voltage oscillations: (a) full wave and the internal voltage produced by it; (b) chopped wave and the internal voltage produced by it: *1*—voltage at the winding terminal; *2*—voltage appearing across a disc coil; τ=chopping time

the oscillation produced by the original wave, the maximum of the resultant oscillation will be higher than the voltage caused individually by either the original or by the superimposed wave. Thus if, at instant τ, the voltage oscillation caused by a positive $U_1(t)$ is negative, i.e. of the same polarity as $U_2(t)$, then the stress imposed by the chopped wave resulting from the supperposition of waves $U_1(t)$ and $U_2(t)$ will be higher than that which would be caused by either $U_1(t)$ or by $U_2(t)$ separately.

The voltage stresses arising between internal points of a homogeneous transformer winding under the effect of a chopped wave are higher than the stresses imposed by a full wave of equal amplitude only in windings of high values of α, i.e. in those very much inclined to oscillations.

5.2.8.4 Oscillating overvoltages

For the following investigation a single-frequency, undamped oscillating overvoltage wave of shape

$$u(t) = U \sin \sigma t \tag{5.55}$$

has been chosen. An oscillating overvoltage wave imposed on a transformer terminal can only cause hazardous stresses in the transformer winding if the oscillation frequency $\sigma/2\pi$ coincides with or lies close to one of the natural frequencies $\omega_n/2\pi$ of the transformer winding (see relation (5.19)).

Applying a unit step voltage of amplitude U to an earthed-end homogeneous transformer winding as shown in Fig. 5.2, and considering the damping effect of the winding and core, the voltage arising at a point characterized by coordinate x in the transformer winding at instant t will be given from Eqn. (5.18) by the relation

$$u(x, t) = U\left[\left(1 - \frac{x}{l}\right) - 2\sum_{n=1}^{\infty} \frac{A_n}{n\pi} \sin n\pi \frac{x}{l} \cos \omega_n t e^{-\delta_n t}\right], \tag{5.56}$$

where δ_n is the internal damping factor, all other notations being identical to the respective symbols used in (5.18). If instead of a unit step, the oscillating wave characterized by relation (5.55) is applied whose frequency coincides with one of the natural frequencies of the winding ($\sigma = \omega_n$) then the voltage of a point of the winding is obtained from the following relation partly similar to (5.56):

$$u_s(x, t) = u(x, 0) \sin \sigma t - U\left[2\frac{A_1}{\pi} \sin \pi \frac{x}{l}\left(\frac{1}{4} \sin \sigma t + \right.\right.$$

$$\left.\left. + \frac{\sigma}{2\delta_1} \cos \sigma t\right) + F(x, t)\right], \tag{5.57}$$

in which $F(x, t)$ is a function including the effects of higher harmonics and which is negligible when the damping is small [46]. From a comparison between Eqns (5.56) and (5.57) the following conclusions can be drawn:

(a) The stresses imposed by an oscillating wave of natural frequency, in the case of $\delta_1 \ll \sigma$, may be considerably higher than those caused by a voltage step.

(b) In both cases the amplitude of oscillations depends on the coefficient α of the winding, i.e. on the square root of the ratio of resultant shunt capacitance to the series capacitance (cf. relation (5.19) giving the value of A_n). Thus, the stresses imposed on windings of high series capacitance values (small coefficients α) both by impulse voltage waves and by resonant oscillating waves are always lower than those imposed on windings of small series capacitance values (of large coefficients α). Under the effect of an undamped oscillating wave an oscillation of several times higher amplitude than the applied voltage may arise (see Fig. 5.34). Fortunately, however, oscillating waves encountered in networks are always damped. This damping, causing a decrease of successive amplitudes of overvoltage waves, should not be confused with the internal damping dealt with so far which is due to the resistance of transformer winding and to core losses. The extent of damping of overvoltage waves depend on the characteristics of the network according to experience, the ratio of the first peak of an oscillating wave to the subsequent peak of equal polarity is always higher than 1.25. Such damped oscillating waves, do not generally give rise to excessive stresses in transformer windings (see Fig 5.35).

The above statements apply also to the case of winding systems composed of several windings. Thus, if a steep-front voltage wave applied to a winding system shows a uniform distribution within the system, then the stress arising in the system under the effect of oscillating waves of usual damping arriving at a transformer terminal surely remains below the hazardous stress.

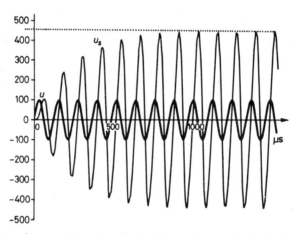

Fig. 5.34. Curves showing an undamped oscillation (u) according to relation (5.55) and oscillation (u_s) produced by it in the centre of the winding. (Calculated curves) [100]

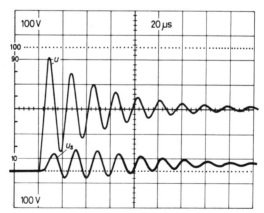

Fig. 5.35. Oscillograms of a damped oscillating wave (u) applied to a regulating transformer, and the voltage oscillation (u_s) produced by it in the tapped winding. The frequency of the oscillating wave is equal to the natural frequency of the transformer winding

5.2.8.5 Shape of voltage waves applied in dielectric tests

The shape of impulse voltage waves to be applied in the course of voltage tests is specified by the various Standards [47, 109]. The wave front is characterized by the virtual front time, which has to be determined on the basis of Fig. 5.36. Another characteristic of the wave is the virtual time to half value, which extends from point D of Fig. 5.36 (from the virtual origin of the impulse) up to the instant at which the voltage drops to 50% of its peak value. The front time of the standard impulse

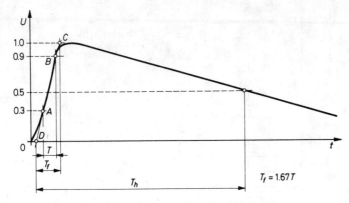

Fig. 5.36. Standard full wave

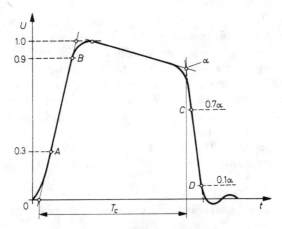

Fig. 5.37. Standard chopped wave

voltage wave is 1.2 μs, its time to half value is 50 μs. Unless otherwise agreed, ±30% deviation from the virtual front time and ±20% deviation from the virtual time to half value is acceptable.

The specified front time and time to half value of a standard chopped wave are equal to the respective figures for a full wave of the same peak value. The voltage is to be chopped at the wave tail after its peak and between 2 and 6 μs following the

16*

instant of virtual starting of the wave (Fig. 5.37). The points marked C and D in Fig. 5.37 serve for determining the virtual steepness of chopping.

The switching impulse to be used in transformer tests is characterized in a different way from the definition of the lightning impulse. The virtual front time (T_1) is again given but instead of the time to half value, here the time interval T_2 between the actual origin and the instant of the first zero is specified, as well as the "time above 90%" i.e. the time for which the voltage is required to remain above 90% of

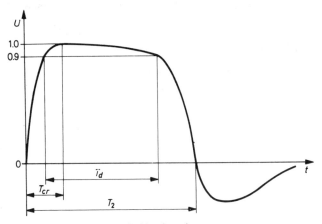

Fig. 5.38. Switching impulse wave

the peak value (T_d). The shape of the switching impulse wave is substantially influenced by saturation of the transformer core. The wave cannot develop as in the case of lightning impulse waves, but at the instant of core saturation, usually 1000 to 2000 µs after the origin of the wave, the voltage drops abruptly to zero. According to the Standard [109], the virtual front time should be at least 20 µs, the time elapsing from virtual origin to first zero at least 500 µs and the time above 90% at least 200 µs. It is common practice to give, instead of virtual front time, the time elapsing from the actual origin of the wave up to the instant of reaching the peak (T_{cr}), which should generally be longer than 100 µs. A normal switching impulse wave and its characteristic values as defined above are represented in Fig. 5.38.

5.2.8.6 Mathematical form of overvoltage waves

Aperiodic overvoltage waves, i.e. lightning and switching impulses can be described mathematically by a function composed of exponential function $U(t) = U_0 e^{-pt}$. So, the full lightning impulse wave can be given by the function

$$U(t) = U_0(e^{-p_1 t} - e^{-p_2 t}),\qquad(5.58)$$

where p_1 is a factor characterizing the wave tail and p_2 that relating to the wave front. The numerical values of U_0, p_1 and p_2 can be determined for the various wave shapes by means of nomograms [108]. For a standard wave of shape 1.2/50 µs and of unit amplitude the respective values are: $U_0 = 1.038$, $p_1 = 0.015 \times 10^6 \, \text{s}^{-1}$, $p_2 = 2.47 \times 10^6 \, \text{s}^{-1}$. The mathematical form of a wave chopped at an instant after the origin of the wave is as follows:

$$U(t) = U_{01}(e^{-p_1 t} - e^{-p_2 t}), \quad \text{if} \quad 0 \leq t < \tau, \quad \text{and}$$

$$U(t) = U_{01}(e^{-p_1 t} - e^{-p_2 t}) - U_{02}[e^{-p_3(t-\tau)} -$$

$$-e^{-p_4(t-\tau)}], \quad \text{if} \quad t \geq \tau. \tag{5.59}$$

For a wave of shape 1.2/50 µs and of unit amplitude, chopped at $\tau = 3$ µs:

$$\begin{aligned}
U_{01} &= 1.038, & U_{02} &= 0.986, \\
p_1 &= 0.015 \times 10^6, & p_3 &= 0.015 \times 10^6, \\
p_2 &= 2.47 \times 10^6, & p_4 &= 16 \times 10^6.
\end{aligned}$$

With the help of relation (5.59) wave shapes approximating the switching impulse shown in Fig. 5.38 can also be described. Thus if $U_{01} = 1.06$, $U_{02} = 0.8$, $p_1 = p_3 = 500 \, \text{s}^{-1}$, $p_2 = 50 \times 10^3 \, \text{s}^{-1}$, $p_4 = 10^4 \, \text{s}^{-1}$ and $\tau = 800 \times 10^{-6}$ s, then the characteristic values appearing in Fig. 5.38 are:

$$T_1 = 60 \, \text{µs}, \quad T_{cr} = 100 \, \text{µs}, \quad T_2 = 1050 \, \text{µs} \quad \text{and} \quad T_d = 300 \, \text{µs}.$$

For describing damped oscillating switching overvoltages the following relation is used:

$$u = U_0\left[e^{-pt} + e^{-Dft}\left(\frac{p}{2\pi f}\sin 2\pi f t - \cos 2\pi f t\right)\right]. \tag{5.60}$$

This function defines a damped periodic wave of frequency f and damping factor D, with its origin at point 0, superimposed on an exponential curve e^{-pt}. (A voltage wave of such shape appears on the H.V. terminal of a test transformer when a capacitor charged from a d.c. source is discharged into its L.V. winding.) The exponential curve is influenced by the value of factor p. With $p = 0$, the curve will represent a damped oscillation superimposed on a constant voltage. By selecting the values of D, p and f, the overwhelming majority of single frequency oscillating switching overvoltages occurring in practice can be simulated.

5.2.9 Calculation of voltage distribution within a homogeneous winding by considering mutual inductances

The Wagner model neglects the mutual inductive couplings between winding elements described by relation (5.37). By considering the mutual inductance within the winding a closer approach to the real voltage distribution is obtained. The mutual inductances have a considerable effect on the amplitude and frequency of harmonic oscillations and on their final distribution. In the development of the final voltage distribution discussed above a decisive part is played by the resistances so far disregarded in the model. If the resistances are neglected, the voltage distribution can only be termed pseudofinal. This pseudofinal distribution, around which the

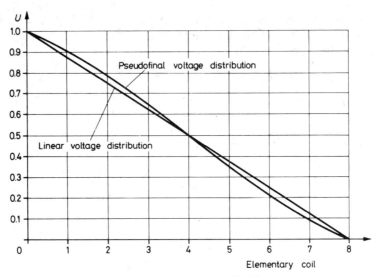

Fig. 5.39. Pseudofinal voltage distribution

oscillations occur, is identical with the linear, final voltage distribution determined by the resistances, if the mutual inductances are ignored. If however, corresponding to the real conditions, the inductances are considered then the pseudofinal distribution will deviate slightly from the linear voltage distribution [2, 61]. A pseudofinal voltage distribution is represented in Fig. 5.39 for an earthed-end winding divided in 6 elementary sections, with $q = 0.63$ (cf. formula (5.38)). As apparent from the diagram, the pseudofinal voltage in the first half of the winding is higher, and in its second half lower, than the linear voltage. The deviation caused by neglect of the mutual inductance is less than 10% in this case. In the calculation of amplitudes and frequencies of harmonic oscillations, however, this neglect may give rise to much greater errors reaching several hundred % in unfavourable cases.

It is advisable to perform the calculation of voltage distribution on the basis of a model consisting of a finite number of elements [88]. The mathematical model of an earthed-end homogeneous transformer winding consisting of n elements, where the mutual inductances between elementary coils are also considered, is shown in Fig. 5.40. Values L, C and K refer to one element of the winding. The current entering the kth node of the model through inductance L is denoted by i_k, and the voltage of the kth node with respect to the earth by u. The coefficient of mutual inductance between the ith and jth element of the model can be determined from relation (5.38).

Using Kirchhoff's laws, the following differential equations can be written for the nodal currents and voltages of each element, for $t \geqq 0$,

$$i_k - i_{k+1} = K \frac{d}{dt}(-u_{k-1} + 2u_k - u_{k+1}) + C \frac{d}{dt} u_k, \tag{5.61}$$

where $k = 1, 2, \ldots, n-1$

246

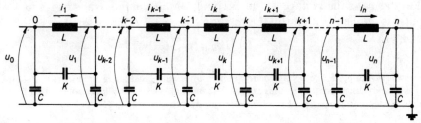

Fig. 5.40. Mathematical model of an n-element transformer winding

and

$$u_{k-1} - u_k = L \sum_{r=1}^{n} q^{|r-k|} \frac{d}{dt} i_r,$$ (5.62)

where $k = 1, 2, \ldots, n-1, n$.

The boundary conditions are:

$$u_n \equiv 0,$$
$$u_0 \equiv 0, \quad \text{if} \quad t < 0,$$
$$u_0 = U(t), \quad \text{if} \quad t \geq 0,$$

where $U(t)$ is a voltage varying as a function of time, applied to the beginning of the winding (to the 0th node) at instant $t=0$. Solving the differential equation with voltage function $U(t) = e^{-pt}$, the following relation is obtained for the voltage to earth appearing in the vth node:

$$u_v(t) = \sum_{k=1}^{n-1} y_{1k} w_{vk},$$ (5.63)

where

$$y_{1k} = w_{1k} \left[\frac{\dfrac{\omega_k^2}{\lambda_k} + \Phi_k p^2}{\omega_k^2 + p^2} e^{-pt} + \left(\frac{\dfrac{\omega_k^2}{\lambda_k} + \Phi_k p^2}{\omega_k^2 + p^2} - \Phi_k \right) \times \right.$$

$$\left. \times \left(\frac{p}{\omega_k} \sin \omega_k t - \cos \omega_k t \right) \right],$$

$$w_{vk} = \sqrt{\frac{2}{n}} \sin \frac{v k \pi}{n}; \quad \lambda_k = 4 \sin^2 \frac{k\pi}{2n},$$

$$\omega_k = \sqrt{\frac{\Phi_k \gamma_k}{LK}}; \quad \Phi_k = \frac{1}{\dfrac{C}{K} + \lambda_k},$$

$$\gamma_k = \frac{1}{1 - q^2} \left[(1 - 2q + q^2)\lambda_k + q\lambda_k^2 \right].$$

Considering also the factor p determining the wave shape as a variable, relation (5.63) can be written in the form

$$u_v(t) = F(p, t)$$ (5.64)

247

and in this case, the voltage at the vth point of the winding when e.g., a full-wave impulse is applied is given by the relation

$$u_{vl}(t) = U_0 [F(p_1, t) - F(p_2, t)]. \qquad (5.65)$$

Example 5.4. Suppose the voltage distribution is to be calculated along the experimental air-core winding already discussed in a previous example in conjunction with the influence of the wave front, the winding having an inductance of $L = 16 \times 10^{-3}$ H, series capacitance of $K = 10 \times 10^{-12}$ F and shunt

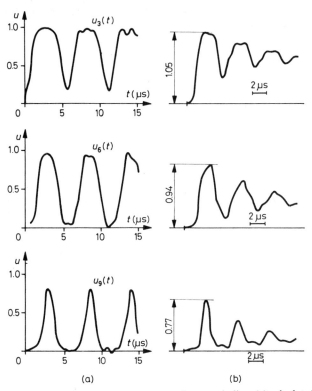

Fig. 5.41. Voltages inside a homogeneous transformer winding: (a) calculated voltages; (b) measured voltages

capacitance of $C = 400 \times 10^{-12}$ F. (C indicates the capacitance between the winding and an earthed metallic cylinder placed in it.) The mutual inductive coupling within the winding is rather loose: $q = 0.6$. Factor α of the coil, from formula (5.11), is

$$\alpha = \sqrt{\frac{400 \times 10^{-12}}{10 \times 10^{-12}}} = 6.3.$$

The winding is represented by a 12-element model ($n = 12$) to make higher harmonics also appear in the oscillation pattern. The winding consists of 48 discs, factor α of each element being

$$\alpha_1 = \frac{6.3}{12} \cong 0.5.$$

248

The series capacitance of each element is

$$K_1 = 12 \times 10 \times 10^{-12} = 120 \times 10^{-12}\,\text{F}$$

and its shunt capacitance is

$$C_1 = \frac{400}{12} \times 10^{-12}.$$

The self-inductance of each element from Eqn. (5.39) with the substitution of $L = 16 \times 10^{-3}$, $n = 12$ and $q = 0.6$, is

$$L_1 = 0.4 \times 10^{-3}\,\text{H}.$$

The calculations have been performed for the case of a very steep-front wave (of front time $T_f = 0.2\ \mu\text{s}$) applied in the course of measurements. Characteristic data of this wave are: $U_0 = 1$, $p_1 = 0.0064 \times 10^6$ and $p_2 = 8.1 \times 10^6$. The oscillations at 75, 50 and 25% of full length, i.e. in the 3rd, 6th and 9th nodes of the winding, as obtained by calculation, and those measured at the same points, are shown in Fig. 5.41. It can be seen that, as regards shape and amplitude, the first cycle of oscillation obtained by calculation (a) and that measured (b) are in good agreement. The maximum voltages on which the design of insulation is based are calculated sufficiently accurately for practical purposes. In the subsequent sections of the calculated curve considerable errors are caused by disregarding the effect of damping in the calculation.

5.2.10 Simplified method for calculating the voltage distribution between windings of transformers having a tapped winding (regulating transformers)

Regulating transformers of conventional design have at least three windings per phase; the H.V. winding (H), the tapped winding (T) and the L.V. winding (L). The windings may be connected in several ways, the most common practice being the connection of the tapped winding to the neutral terminal of the H.V. winding. The tapped winding can be connected to the main (H.V.) winding either in the same or in the inverse winding sense. In the first case, the voltage of the tapped winding adds to that of the main winding, whereas in the second case, it is subtracted from the same. The desired transformation ratio is adjusted by earthing the required tapping of tapped winding. The number of tap positions is equal to twice the number of taps provided on the tapped winding. The following four "characteristic" tap positions deserve special attention (Fig. 5.42).

Max.: connection giving the highest output voltage. The main and tapped windings are connected in series in the same winding sense, and the "lower" end of the tapped winding is earthed.

+0: connection giving rated output voltage. The "upper" end of the tapped winding is connected to the main winding and is earthed, its "lower" end is left open.

−0: connection giving rated output voltage. The "lower" end of the tapped winding is connected to the main winding and this end is earthed, whereas its "upper" end is open.

Min.: connection giving the lowest output voltage. The main and tapped windings are connected in series in opposite winding sense, and the "upper" end of the tapped winding is earthed.

249

For such a winding system consisting of several windings, the voltage distribution within individual windings can only be determined, if the voltage distribution between windings is known. The stresses imposed by the lightning impulse waves are different in the various tapping positions. Strictly speaking, the voltage distribution ought to be determined for each position separately. In the case of windings of small series capacitances ($\alpha > 3$) the windings are divided into several elementary sections, so the calculation ought to be based, in every tapping position, on a model composed of a large number of elements. In the majority of practical

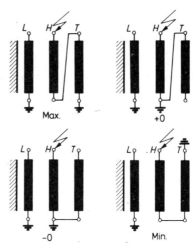

Fig. 5.42. Characteristic tap positions of regulating transformers

cases, however, the calculation may be simplified, and the calculation of lightning impulse stresses required for designing the insulation can be based on a simple two element model.

Conditions for the application of such a model are as follows:

(a) Both the main winding and the tapped winding have large series capacitances ($\alpha < 3$), so that each of them can be represented by a single inductance and a single series capacitance.

(b) Determination of the voltage distribution within the L.V. winding is not required, and this winding may be considered as being at earth potential. (It has been seen already in the course of investigating the transferred voltages that the voltage distribution within the non-impulsed winding, i.e. within the L.V. winding in the present case, has practically no influence on the voltage distribution within the impulsed winding. Therefore, this simplification does not materially impair the accuracy of calculating the voltage distribution between main and tapped windings or that within each of them.)

(c) The highest stresses arise in the four "characteristic" tapping positions, or at least the stresses occurring in these positions differ only slightly from the highest ones. According to experience this is true for all tapped windings in which the

250

capacitance between two adjacent tappings is higher than the resultant capacitance of the entire tapped winding.

When these conditions are fulfilled, the lightning impulse voltage distribution between windings can be calculated on the basis of the model of Fig. 5.43. In the model, the shunt capacitances have been taken into consideration by assuming

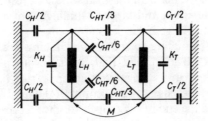

Fig. 5.43. Simplified mathematical model of the regulating transformer

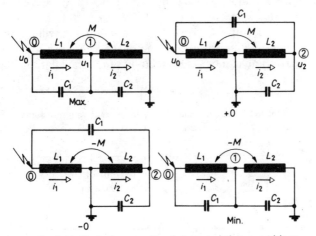

Fig. 5.44. Equivalent circuits of characteristic tap positions

them to being connected to the ends of the winding elements. The shunt capacitances to earth have been divided into two equal halves, and between the windings, "cross capacitances" have also been included in the model. As proved by experience, the division of inter winding capacitance into two halves alone does not yield acceptable results for the initial voltage distribution. This deficiency could be eliminated by representing each winding by several series elements, i.e. by increasing the number of nodes in the model. Thereby, however, the advantage gained by the simplicity of calculating with a model consisting of a few elements would be lost. Satisfactory initial voltage distribution is obtained, without increasing the number of nodes, by dividing the inter winding capacitances into four parts as shown in Fig. 5.43. The division of inter winding capacitance according to the ratios $1/3$–$1/6$–$1/6$–$1/3$ is not a result of theoretical calculation, but is based on experience gained in the course of measurements, and in the case of windings having high series capacitances it represents a good approach to real initial voltage distributions.

The notations used in Fig. 5.43 are as follows: L_H and L_T are the self-inductances of the main winding and the tapped winding, respectively; M is the mutual inductance between main and tapped windings; K_H and K_T are the resultant series capacitances of the main and tapped windings, respectively; C_H and C_T are the capacitances to earth of the main and tapped windings, respectively, and C_{HT} is the capacitance between the main and tapped windings. From Fig. 5.43, for the four basic switching variants of Fig. 5.42, the simple circuits shown in Fig. 5.44 can be deduced. The scheme belonging to the Min switching position differs from that belonging to the Max position, and that belonging to position -0 from that belonging to position $+0$ in the negative sign appearing before the coefficient of mutual induction. This "$-$" sign takes account of the modification of the schemes by having interchanged the end of the tapped winding as compared to the first two positions. The self-inductances and mutual inductances are the same for all four positions ($L_1 = L_H$, $L_2 = L_T$); the capacitances, however, are different. The relations between capacitances C_1 and C_2 appearing in the model and the capacitances of the coils are listed in Table 5.5.

The voltage of the non-impulsed winding of the circuits shown in Fig. 5.44, under the effect of an impulse voltage applied to point 0 varying arbitrarily with time, is marked with an arrow and can be determined by simple calculation. Voltage $U(t)$ is

Table 5.5

Capacitances in the mathematical model of regulating transformers

Connection	C_1	C_2
Max	$K_H + C_{HT}/3$	$K_T + (C_H + C_T)/2 + C_{HT}/3$
$+0$	$C_{HT}/6$	$K_T + C_T/2 + C_{HT}/3$
-0	$C_{HT}/3$	$K_T + C_T/2 + C_{HT}/6$
Min	$K_H + C_{HT}/6$	$K_T + (C_H + C_T)/2 + C_{HT}/6$

connected to the network at instant $t = 0$. Correspondingly,

$$u_0 \equiv 0, \quad \text{if} \quad t < 0$$

and

$$u_0 = U(t), \quad \text{if} \quad t \geq 0.$$

With these conditions, the following differential equations can be written for the four cases.

For connection Max:

$$u_0 - u_1 = L_1 \frac{di_1}{dt} + M \frac{di_2}{dt},$$

$$u_1 = M \frac{di_1}{dt} + L_2 \frac{di_2}{dt},$$

$$i_1 - i_2 = -C_1 \frac{du_0}{dt} + (C_1 + C_2) \frac{du_1}{dt}.$$

252

For connection $+0$:

$$u_0 = L_1 \frac{di_1}{dt} + M \frac{di_2}{dt},$$

$$-u_2 = M \frac{di_1}{dt} + L_2 \frac{di_2}{dt},$$

$$i_2 = -C_1 \frac{du_0}{dt} + (C_1 + C_2) \frac{du_2}{dt}.$$

For connection -0:

$$u_0 = L_1 \frac{di_1}{dt} - M \frac{di_2}{dt},$$

$$-u_2 = -M \frac{di_1}{dt} + L_2 \frac{di_2}{dt},$$

$$i_2 = -C_1 \frac{du_0}{dt} + (C_1 + C_2) \frac{du_2}{dt}.$$

For connection Min:

$$u_0 - u_1 = L_1 \frac{di_1}{dt} - M \frac{di_2}{dt},$$

$$u_1 = L_2 \frac{di_2}{dt} - M \frac{di_1}{dt},$$

$$i_1 - i_2 = -C_1 \frac{du_0}{dt} + (C_1 + C_2) \frac{du_1}{dt}.$$

From a practical point of view, the voltage function of shape $U(t) = e^{-pt}$ is mainly of importance. When this voltage function is applied, the voltage against earth appearing at the non-earthed end of the tapped winding, i.e. the voltage across the tapped winding, is obtained for all four fundamental connections by the relation

$$u_n(t) = u_f e^{-pt} - (u_f - u_c) \left(\frac{p}{\omega} \sin \omega t - \cos \omega t \right). \quad n = 1,2. \tag{5.66}$$

In the formula, u_f is the value of the pseudofinal voltage referred to instant $t = 0$; u_c is the initial voltage as defined by the capacitances; and ω is the angular frequency of the voltage oscillation ($\omega = 2\pi f$). The values of u_f and u_c can be calculated, in all four switching positions alike, from the relations

$$u_f = \frac{u_f^* + u_c p^2}{\omega^2 + p^2} \tag{5.67}$$

and

$$u_c = \frac{C_1}{C_1 + C_2}. \tag{5.68}$$

u_f^* and ω, for the four switching positions, are given in Table 5.6, whereas C_1 and C_2 can be calculated with the aid of Table 5.5. In Table 5.6 $N = (C_1 + C_2) \times (L_1 L_2 - M^2)$, and the voltage across the main winding is $u_m(t) = u_0(t) - u_n(t)$.

Table 5.6

Data required for simplified calculating of lightning impulse
voltage distribution of regulating transformers

Connection	u_f^*	ω
Max	$\dfrac{L_2 + M}{N}$	$\sqrt{\dfrac{L_1 + L_2 + 2M}{N}}$
+0	$-\dfrac{M}{N}$	$\sqrt{\dfrac{L_1}{N}}$
−0	$\dfrac{M}{N}$	$\sqrt{\dfrac{L_1}{N}}$
Min	$\dfrac{L_2 - M}{N}$	$\sqrt{\dfrac{L_1 + L_2 - 2M}{N}}$

On the basis of the above relations, the voltage oscillations (variations in time of voltages appearing across the main and tapped windings) of regulation transformers which can be represented by the model of Fig. 5.42, can be determined for various wave shapes given by equations (5.58) and (5.59) by using relations (5.64) and (5.65).

Example 5.5. Suppose the voltage oscillations which occur in the main and tapped windings of a 120 kV 55 MVA transformer, with a tapping range of $\pm 15\%$ following the application of a 1.2/50 lightning impulse voltage wave are to be determined. The calculation is performed by utilizing the following data:

$$L_1 = 1.46 \text{ H}, \quad K_H = 1425 \text{ pF}, \quad C_H = 2080 \text{ pF},$$
$$L_2 = 0.032 \text{ H}, \quad K_T = 7640 \text{ pF}, \quad C_{HT} = 1570 \text{ pF},$$
$$M = 0.21 \text{ H}, \quad C_T = 1040 \text{ pF},$$
$$U_0 = 1.038, \quad p_1 = 0.015 \times 10^6 \text{ s}^{-1}, \quad p_2 = 2.47 \times 10^6 \text{ s}^{-1}.$$

The voltage oscillations obtained by calculation are shown in Fig. 5.45 and the oscillogram of oscillations taking place in the transformer is represented in Fig. 5.46. A comparison shows that the calculated and the measured values of maximum voltages serving as the basis for designing do not differ by more than 20% in all the four cases. Calculations performed in the cases of several other transformers have shown that this degree of accuracy is characteristic of the simplified calculation method shown, and this accuracy is adequate for the designing of insulation. The damping effect of losses shows only after the first cycle of oscillation, as in the case of a single winding.

Voltage distribution within the windings. The voltages imposed on the main and tapped windings are different in the four basic schemes. These voltages are the highest in the schemes denoted by Min. The situation is well demonstrated in Fig. 5.47. As shown there, the highest voltage ($U_{H \max}$) appearing across the main winding is considerably above the peak value of the lightning impulse voltage (U_0) applied to

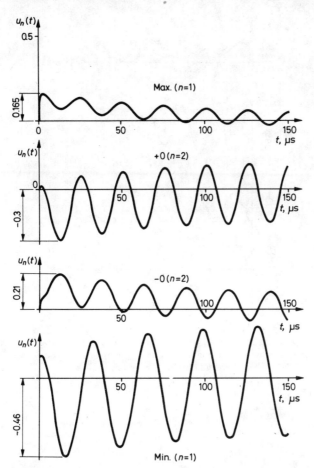

Fig. 5.45. Voltages appearing across the tapped winding of a regulating transformer in the four characteristic tap positions, when applying a 1.2/50 μs lightning impulse voltage wave. (Calculated curves), n: serial number of non-impulsed node

the terminal, whereas the highest voltage of the tapped winding ($U_{T\max}$) exceeds twice the voltage U_{Tf} proportional to the number of turns. The voltages for which the insulation within the winding have to be designed are $U_{H\max}$ and $U_{T\max}$.

Our simplified method is suitable only for the calculation of regulating transformers consisting of main and tapped windings having high series capacitances, and provided that factor α does not exceed the value of 3 for either winding. Within the windings, however, a certain degree of non-uniformity in the voltage distribution has to be reckoned with even in this case, since a fully uniform voltage distribution would only develop in the case of $\alpha = 0$, i.e. if no shunt capacitances were present. In the designing of insulation within the winding, this nonuniformity has to be taken into consideration. Experience shows that the calculation will include a sufficient safety margin if this unevenness of voltage distribution within the winding is taken into account by introducing a factor of non-

255

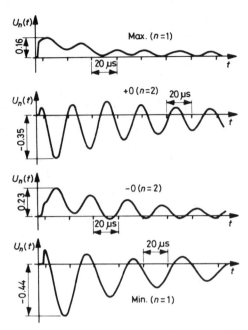

Fig. 5.46. Voltages appearing across the tapped winding of the same regulating transformer in the four characteristic tap positions, when applying a 1.2/50 μs lightning impulse voltage wave (oscillograms). n: serial number of non-impulsed node

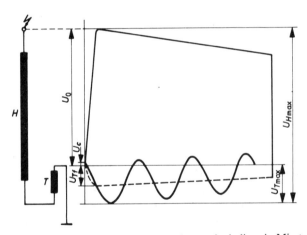

Fig. 5.47. Voltage appearing across the main and tapped windings in Min tap position

uniformity given by Eqn. (5.13), related to the beginning of the winding and valid for the case of a unit pulse. Thus, for $\alpha \leq 3$, the design voltage of insulation to be provided on each winding section, e.g. on each disc is determined by multiplying the equally distributed $U_{H\,max}$ and $U_{T\,max}$ each section by the factor $\delta_H = \alpha_H \coth \alpha_H$ and $\delta_T = \alpha_T \coth \alpha_T$, respectively.

256

5.2.11 Non-oscillating winding systems.
Settings of series capacitances

Winding systems consisting of several windings (such as are used in regulating transformers, and autotransformers) cannot be freed from oscillations merely by providing each winding with a high series capacitance, i.e. by making it oscillation free in itself. A high series capacitance only ensures a closely uniform distribution of voltage within the winding. The entire winding system will only be free from oscillation, if the voltage is uniformly distributed also among the windings, i.e. the overvoltage is shared by the individual windings within the system in the same proportion as the service voltage. This condition is fulfilled, as in a single winding, if the initial voltage appearing across the windings agrees with the pseudofinal voltage defined by the self and mutual inductances, the latter voltage being, to a good approximation, equal to that corresponding to the turns ratio in the case of windings linked by the same flux. The ratio being fixed by the specification data, the only way the designer can ensure freedom from oscillations is by the proper selection of capacitances. The insulation with respect to earth and between windings, and thereby also the shunt capacitances, are all definitely fixed by the specified insulation levels. The series capacitance, however, as shown in the course of calculating the capacitances, may be varied within wide limits without modifying the geometrical dimensions, number of turns, etc. of the winding. Thus, the oscillation-free condition of winding systems consisting of several windings can be achieved by properly selecting and matching to each other the series capacitances of the windings. The way of ensuring freedom from oscillations of a system is not by employing the highest possible values, but by using properly selected high values of series capacitances.

By setting the series capacitances in relation to each other, however, freedom from oscillations cannot be achieved in every case. As has already been pointed out in the discussion of transferred overvoltages, the initial voltage transferred from the H.V. winding to the L.V. winding can, in principle, be adjusted to the value required by the turn ratio by selecting the values of α, i.e. the series capacitances of the windings correspondingly, thereby avoiding the occurrence of oscillations. The oscillation of the voltage transferred from the L.V. to the H.V. winding, however, could only be eliminated if the initial voltage transferred to the H.V. winding were higher then the peak value of voltage applied to the L.V. winding, which is obviously impossible. The situation is the same in all such tapping positions of regulating transformers, in which the phase of the voltage of the tapped winding is opposed to that of the voltage of the main winding. Thus in such tapping positions the pseudofinal voltage of the tapped winding is of polarity opposed to the applied voltage, as illustrated in Figs 5.45 and 5.46 for the +0 and Min schemes. Since the initial voltage may necessarily be of the same polarity as the applied voltage, the oscillations cannot be avoided by any setting of capacitances. Nevertheless, efforts should be made to reduce the amplitude of these oscillations. Therefore, the initial voltage in the tapped winding should be kept as low as possible, as is apparent from the oscillation patterns. Again, as it follows from Eqn. (5.68) and is shown in Table 5.4, the initial voltage in the tapped winding is the lower the higher the series capacitance of this winding.

It should be noted additionally that, in the course of design work, the optimum solution is not always a fully oscillation-free winding system, even if it were realizable in principle. There are transformers in which the practical realization of high-capacitance windings is not expedient for some other, e.g. technological reason, and where it would cause more trouble in the design or manufacture than the expected voltage oscillations. In addition, an excessive increase of series capacitance brings about harmful growth of the interturn voltage, as explained earlier, so an increase of the series capacitance beyond a certain limit is to be avoided also for this reason. In such a case, realization of the oscillation-free condition should be abandoned and reinforcement of insulation should be adopted as the protective means against overvoltages.

Example 5.6. In the scale model of a regulating transformer there are 384 turns in the main winding and 64 turns in the tapped winding. In the model, the L.V. winding has been replaced by an earthed cylindrical sheet to represent the capacitance between main winding and L.V. winding. In the position marked Max (Fig. 5.42) corresponding to maximum voltage, the number of turns of the tapped winding is $[64/(384+64)] \times 100 = 14.3\%$ of the full number of turns, and that of the main winding is 85.7% of the same. According to the explanation given above, in this scheme the freedom from oscillation is ensured if in the tapped winding the initial voltage (u_c) defined by the capacitances is equal to the voltage resulting

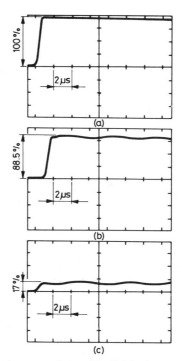

Fig. 5.48. Voltages with correctly set capacitances ($E = 2$): (a) voltage at the terminal of the main winding; (b) voltage across the main winding; (c) voltage across the tapped winding

from the number of turns, i.e.

$$u_c = 0.143 u_0, \tag{5.69}$$

where u_0 is the peak value of applied voltage. With the shunt capacitances and with the series capacitance of the tapped winding given, let the series capacitance of the main winding required to ensure its freedom from oscillations be determined. With the notation used earlier, let be $C_H = 400$ pF, $C_{HT} = 330$ pF, $C_T = 0$ and $K_T = 2250$ pF. (The interleaved tapped winding containing elements consisting of 8 sections ($E = 8$) and of one parallel conductor ($p = 1$). Using relation (5.68) and the data of Table 5.4:

$$u_c = u_0 \frac{K_H + C_{HT}/3}{K_H + K_T + (C_H + C_T)/2 + 2C_{HT}/3}. \tag{5.70}$$

From this relation, after substituting the values of known capacitances and that of u_c/u_0 calculated from (5.69), for the capacitance of the main winding, the following value is obtained:

$$K_{H\,\mathrm{opt}} = 318 \text{ pF.}$$

This value cannot be realized exactly by the winding of dimensions and number of turns given, but it can be approximated, if preparing the main winding as an interleaved winding of discs consisting of two sections ($E = 2$). Then, so, the series capacitance of the main winding will be

$$K_{H(E=2)} = 420 \text{ pF.}$$

Now, ratio u_c/u_0 deviates but slightly from the value corresponding to the turn ratio. Thus, from formula (5.70), with $K_H = 420$ pF:

$$\frac{u_c}{u_0} = 0.172.$$

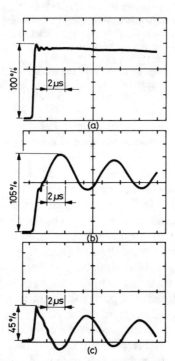

Fig. 5.49. Voltages with incorrectly matched capacitances ($E = 4$): (a) voltage at the terminal of the main winding; (b) voltage across the main winding; (c) voltage across the tapped winding

Considering the oscillations, the maximum value of the relative voltage appearing across the tapped winding is given by relation (5.52) as

$$\frac{u_{T\max}}{u_0} = |0.143| + |0.143 - 0.172| = 0.172$$

and the relative voltage across the main winding can exceed

$$\frac{u_{H\max}}{u_0} = 0.857 + 0.857 - 0.828 = 0.886.$$

In this case, the value of factor α of the main winding will be $\alpha_H = \sqrt{400/420} = 0.97$, and that of the inequality factor, according to relation (5.13), will be $\delta_H = \alpha_H \coth \alpha_H = 0.97 \times 1.34 \approx 1.3$. If the main winding is designed as an interleaved winding to comprise elements consisting of four sections ($E = 4$) then, according to relation (5.29), the series capacitance of the winding will increase fourfold, thus

$$K_{H(E=4)} = 1680 \text{ pF}.$$

In this case, $\alpha_H = 400/1680 = 0.485$, and the value of the inequality factor $\delta_H = 0.485 \times 2.22 = 1.08$. Thus, the voltage distribution within the main winding is more uniform than in the previous case. However, the initial voltage appearing across the tapped winding, which is at the same time also the maximum voltage is obtained from equation (5.70) as $u_{T\max} = 0.412u_0$, i.e. almost 2.5 times that of the previous value. The maximum voltage appearing across the main winding, calculated as before, is

$$u_{H\max} = 1.126u_0.$$

It can be seen that by selecting too large a value for the series capacitance of the main winding, the voltage imposed on the tapped winding has increased. The stress within the main winding has not changed much, since the voltage across the full winding has increased but the inequality factor has decreased.

The correctness of the calculations has been proved also by the measurement performed on the model. It can be read from the oscillograms of Fig. 5.48 that in the case of nearly approaching the optimum setting ($E = 2$), the voltage appearing across the main winding is 88.5%, and that across the tapped winding is 17%. (The respective calculated values are 88.6 and 17.2%.) On the other hand, with improper setting ($E = 4$), as shown by the oscillograms of Fig. 5.49, the voltage is 105% across the main winding and 45% across the tapped winding. The calculated values deviate by less than 10% from those measured.

5.2.12 General method for calculating voltage distribution within windings and winding systems [30]

So far, the lightning impulse voltage distribution has been determined for certain special cases (for a single transformer winding or a winding system consisting of not more than two windings). In most practical calculations, winding systems comprising several windings are encountered. The transformer itself may consist of several windings, such as an autotransformer having a tapped winding, or a regulating transformer having more than one tapped winding. In the case of small series capacitance windings each winding has to be represented by a multi-element mathematical model. Also, the constructional design of the transformer may require the division of a winding into several parts for the purpose of calculation, such as in the course of investigating windings with the H.V. point at the centre or tapped regulating windings. In the cases mentioned, the determination of lightning impulse voltage distribution can be reduced to the calculation of a general network of n elements. It is appropriate to reduce the system of differential equations of the network into the form of matrix differential equations. The solution of the inhomogeneous matrix equation obtained by reducing the latter equations leads to the generalized eigenvalue problem of matrices.

The model is a network containing arbitrary inductances and capacitances. Its characteristic features are as follows:

(a) The network consists of m elements arbitrarily connected or only inductively coupled with one another. M_{ij} is the coefficient of mutual inductance between the ith and jth element of the network, when $i = 1, 2, \ldots n$, and $j = 1, 2, \ldots, m$. If $i = j$, so the self-inductance of the element concerned is obtained: M_{ii} is the self-inductance of the ith element. Obviously, for mutual inductances $M_{ij} = M_{ji}$. Correspondingly, there are altogether

$$N_M = \frac{m}{2}(m+1)$$

inductances present in the network.

(b) The number of independent nodes is n. With these go two nodes of definite potential: the impulsed node of the network and the node directly earthed. A function of known variation in time is applied to the impulsed node denoted by serial number 0. The independent nodes are marked with serial numbers ranging from 1 to n, with the earthed node obtaining the serial number $(n+1)$. Thus, the overall number of all, independent and definite-potential nodes is $(n+2)$. C_{ij} indicates the capacitance between ith and jth node of the network, where $j = 1, 2, \ldots, n$ and $i = 1, 2, \ldots, j-1, j+1, \ldots, n, n+1$. Obviously, $C_{ij} = C_{ji}$. The capacitance between the 0th and $(n+1)$th node is connected parallel with the entire network, so it does not have any role in the phenomenon to be investigated. Correspondingly, there are altogether

$$N_c = \frac{n}{2}(n+3)$$

capacitances characterizing the entire network. Let at instant $t = 0$ to the 0th node of the general network described voltage $U(t)$ varying arbitrarily in time be applied.

$$u_0 \equiv 0, \quad \text{if} \quad t < 0$$

and

$$u_0 = U(t), \quad \text{if} \quad t \geq 0.$$

Based on Kirchhoff's laws, m differential equations can be written to obtain the voltage of each element (i.e. the voltage difference between adjacent nodes) and n differential equations to find the current of each node. (The establishment of differential equations and of matrices based on them appearing in the calculation are demonstrated on concrete examples, whereas the calculation procedure is only described in general). The two systems of differential equations can be reduced to the following matrix equations:

$$\mathbf{Du} = \mathbf{Qi'} - \mathbf{h}u_0 \tag{5.71}$$

and

$$\mathbf{D^*i} = \mathbf{Ku'} - \mathbf{l}u_0'. \tag{5.72}$$

261

The above matrices have the following meaning:

$$\mathbf{u} = \begin{bmatrix} u_1 \\ u_2 \\ \vdots \\ u_n \end{bmatrix}$$: column matrix of n elements, containing the node voltages,

$$\mathbf{i} = \begin{bmatrix} i_1 \\ i_2 \\ \vdots \\ i_m \end{bmatrix}$$: column matrix of m elements, containing the current of each element,

$$\mathbf{Q} = \begin{bmatrix} M_{11} & M_{12} & \cdots & M_{1m} \\ M_{21} & M_{22} & \cdots & M_{2m} \\ \cdot & \cdot & \cdots & \cdot \\ M_{m1} & M_{m2} & \cdots & M_{mm} \end{bmatrix}$$

quadratic, symmetrical, non-singular matrix of mth order, built up to the inductances of the network ($M_{ij} = M_{ji}$),

$$\mathbf{K} = \mathbf{K}^{(n,n,)} = [c_{ij}]$$

a quadratic symmetrical matrix of n-th order, built up of the capacitances of the network, where

$$c_{ij} = -C_{ji} = -C_{ij} \qquad \text{for} \qquad i \neq j$$

and $\qquad c_{ii} = C_{0i} + C_{1i} + \ldots C_{i-1,i} + C_{i+1,i} + \ldots + C_{ni} + C_{n+1,i}.$

$\mathbf{D}$ is a matrix dependent on network arrangement, and connections of elements, consisting of m rows and n columns. Its elements may be 1, -1 or 0. Matrix $\mathbf{D}$ can be set up on the basis of the system of differential equations relating to the voltage of each element in the way shown in the examples

$\mathbf{D^*}$ = transpose of matrix $\mathbf{D}$,

$\mathbf{h}$ = column matrix of m elements, also depending on the network arrangement and having elements that may be 1, -1 and 0, as before. From the system of differential equations set up for the voltages of elements, $\mathbf{h}$ may be established, simultaneously with $\mathbf{D}$.

$$\mathbf{l} = \begin{bmatrix} C_{01} \\ C_{02} \\ \vdots \\ C_{0n} \end{bmatrix}$$: column matrix of n elements. In this matrix the capacitances between the 0th node and all other nodes appear.

By eliminating current matrix **i** from matrix differential equations (5.71) and (5.72), after simple conversion, the following inhomogeneous matrix differential equation of second order is obtained:

$$\mathbf{Ku''} + \mathbf{D^*Q^{-1}Du} = \mathbf{l}U''(t) + \mathbf{D^*Q^{-1}h}U(t). \tag{5.73}$$

Before dealing with the solution of differential equation (5.73), let us see how the initial and pseudofinal voltages arising in the nodes of the network can be determined separately. As is known, the initial voltage is defined by the capacitances. Omitting from Eqn. (5.73) the terms containing inductances:

$$\mathbf{Ku''_c} = \mathbf{l}U''(t)$$

from which, for instant $t=0$, the initial voltage is calculated from the matrix equation

$$\mathbf{u}_c = \mathbf{K^{-1}l}U(0). \tag{5.74}$$

Equation (5.74) for a unit step voltage can be obtained directly from the equilibrium of charges. In fact, a concrete form of this general matrix equation is represented by matrix equations (5.50) and (5.51) deduced for calculating the transferred initial voltages.

The pseudofinal voltages, on the other hand, are defined by the self and mutual inductances, based on Eqn. (5.73), these are obtained from

$$\mathbf{D^*Q^{-1}Du}_f = \mathbf{D^*Q^{-1}h}U(t). \tag{5.75}$$

Introducing notations

$$\mathbf{D^*Q^{-1}D} = \mathbf{A}_0$$

and

$$\mathbf{D^*Q^{-1}h} = \mathbf{q},$$

equation (5.75) can be rewritten in the form

$$\mathbf{A}_0\mathbf{u}_f = \mathbf{q}U(t),$$

from which the following matrix equation is obtained for the pseudofinal voltages appearing in the various nodes:

$$\mathbf{u}_f = \mathbf{A}_0^{-1}\mathbf{q}U(t). \tag{5.76}$$

From the chain of thought explained above, the conditions of freedom from oscillations are expressed by the matrix equation

$$\mathbf{K^{-1}l} = \mathbf{A}_0^{-1}\mathbf{q}.$$

Assuming a unit step voltage, and considering the first harmonic only, the maximum voltages arising in the various nodes can be calculated from the relation

$$|u_{max}| = |u_f| + |u_f - u_c|,$$

where u_f and u_c indicate related elements of column matrices (5.76) and (5.74), respectively, belonging to the node concerned.

Now, let us deal with solving of matrix differential equation (5.73). This differential equation may be rewritten in the form

$$A_2 u'' + A_0 u = g(t), \tag{5.77}$$

where

$$A_2 = K,$$
$$A_0 = D^* Q^{-1} D$$

and

$$g(t) = lU''(t) + D^* Q^{-1} hU(t).$$

Here, A_2 and A_0 are symmetrical and in practical cases definite matrices. The generalized eigenvalue problem [143] associated with matrix differential equation (5.77) may be written in the form:

$$|A_2 \lambda - A_0| v = 0.$$

This matrix equation has a non-trivial solution $v \neq 0$ only if the so-called characteristic determinant disappears, i.e.:

$$|A_2 \lambda - A_0| = 0.$$

After developing the determinant, a polynomial of nth order in λ is obtained which is termed the generalized characteristic equation of the problem. If single eigenvalues are assumed, than a single $v_k \neq 0$ solution belongs to each eigenvalue λ_k. λ_k is called the kth generalized eigenvalue and v_k the kth generalized eigenvector. Since A_0 and A_2 are positive definite matrices all λ_k eigenvalues will be positive and the case of resonance may be excluded. Let the diagonal matrix of nth order

$$\Lambda = \langle \lambda_1, \lambda_2, \ldots \lambda_n \rangle$$

built up of the eigenvalues and the quadratic matrix of nth order

$$V = [v_1, v_2, \ldots v_n]$$

containing the eigenvectors be introduced. The generalized eigenvectors can be determined from normalizing conditions

$$V^* A_2 V = E$$

and

$$V^* A_0 V = \Lambda.$$

Here, V^* is the transpose of matrix V and E is a unit matrix of order n. Introducing now transformation $u = Vy$ and multiplying from the left Eqn. (5.77) by V^*, then differential equation

$$V^* A_2 V y'' + V^* A_0 V y = V^* g(t)$$

is obtained. By considering the normalizing conditions we get

$$y'' + \Lambda y = V^* g(t).$$

Thus, n independent scalar differential equations of second order are generated which have the form

$$y_k'' + \lambda_k y = \gamma_k(t) \quad (k = 1, 2, \ldots, n),$$

264

where $\gamma_k(t)$ represents the kth element of column matrix $\mathbf{V}^*\mathbf{g}(t)$. y_k can be calculated, in the known way, from the relation

$$y_k = y_{0k} \cos \sqrt{\lambda_k} t + y'_{0k} \frac{\sin \sqrt{\lambda_k} t}{\sqrt{\lambda_k}} +$$

$$+ \int_{\tau=0}^{\tau=t} \frac{\sin \sqrt{\lambda_k}(t-\tau)}{\sqrt{\lambda_k}} \gamma_k(\tau) d\tau,$$

in which

$$y_{0k} = y_k(0) \quad \text{and} \quad y'_{0k} = y'_k(0).$$

Considering initial condition

$$\mathbf{K}\mathbf{u}(0) = \mathbf{l}U(0)$$

resulting from the equilibrium of charges for instant $t=0$, and substituting $\mathbf{u}(0) = = \mathbf{V}\mathbf{y}(0)$ and $\mathbf{K} = \mathbf{A}_2$, then multiplying from the left by $\mathbf{V}^*$

$$\mathbf{V}^*\mathbf{A}_2\mathbf{V}\mathbf{y}(0) = \mathbf{V}^*\mathbf{l}U(0)$$

is obtained.

Taking into consideration the normalizing conditions, the transformed initial conditions are obtained in the following form:

$$\mathbf{y}(0) = \mathbf{V}^*\mathbf{l}U(0)$$

and

$$\mathbf{y}'(0) = \mathbf{V}^*\mathbf{l}U'(0).$$

If, now, notations

$$\mathbf{r} = \begin{bmatrix} r_1 \\ r_2 \\ \vdots \\ r_n \end{bmatrix} = \mathbf{V}^*\mathbf{l} \quad \text{and} \quad \mathbf{s} = \begin{bmatrix} s_1 \\ s_2 \\ \vdots \\ s_n \end{bmatrix} = \mathbf{V}^*\mathbf{D}^*\mathbf{Q}^{-1}\mathbf{h}$$

are also introduced, then the following relations result:

$$y_{0k} = r_k U(0),$$
$$y_{0k}' = r_k U'(0)$$

and

$$\gamma_k(t) = r_k U''(t) + s_k U(t).$$

Now, for the various components of y_k the following relations are obtained:

$$y_k = U(0)r_k \cos \sqrt{\lambda_k} + U'(0)r_k \frac{\sin \sqrt{\lambda_k} t}{\sqrt{l_k}} +$$

$$+ \int_{\tau=0}^{\tau=t} \frac{\sin \sqrt{\lambda_k}(t-\tau)}{\sqrt{\lambda_k}} \left[r_k U''(\tau) + s_k U(\tau) \right] d\tau. \qquad (5.78)$$

265

Considering the transformation $\mathbf{u}=\mathbf{V}\mathbf{y}$, the voltage arising in the vth node of the network will be obtained in the form

$$u_v(t)= \sum_{k=1}^{n} v_{vk}y_k,$$ (5.79)

where v_{vk} is the kth element in the vth row of matrix $\mathbf{V}$, i.e. the vth element of generalized and normalized eigenvector $\mathbf{v}_k$; with voltage function $U(t)=U_0 e^{-pt}$, after performing the integration, the following relation for y_k is obtained:

$$y_k=U_0\left\{ \frac{s_k+p^2 r_k}{\lambda_k+p^2} e^{-pt} + \left(\frac{s_k+p^2 r_k}{\lambda_k+p^2} r_k \right) \times \right.$$

$$\left. \times \left(\frac{p}{\sqrt{\lambda_k}} \sin \sqrt{\lambda_k}t - \cos \sqrt{\lambda_k}t \right) \right\}$$ (5.80)

in which r_k and s_k is the kth element of vector $\mathbf{r}$ and $\mathbf{s}$, respectively. For waves occurring in practice composed of several exponential curves, the nodal voltages can be calculated, by analogy with relations (5.64) and (5.65), by summing up the voltages arising under the effect of the various components.

By means of the calculation procedure described in principle, voltage distributions within a winding or winding system which can be simulated by a however complicated a mathematical model consisting of an arbitrary number of elements, can be determined for voltage waves showing any arbitrary variation with time. In

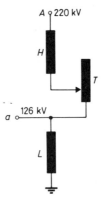

Fig. 5.50. Connection diagram of a regulating auto-transformer

practice, it is not advisable to increase the number of elements beyond a certain limit, for reasons already explained in Section 5.2.5. Further simplification lies in performing the calculation, in the case of three-phase, star-connected, or earthed-neutral transformers, for one phase only, i.e. for the windings located on one limb of the transformer. As proved in practice in the case of a single-phase impulse, the mutual effects of windings belonging to different phases can be left out of consideration. Another way of reducing the number of elements, when investigating winding systems consisting of several windings, is to assume in the calculation all

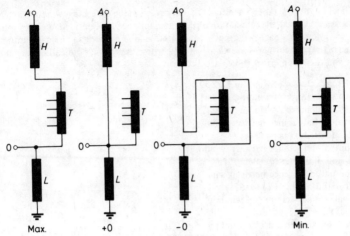

Fig. 5.51. Characteristic tap positions of regulating auto-transformers

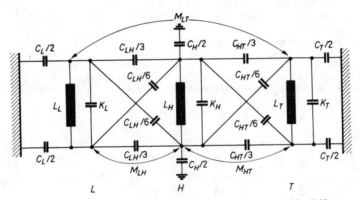

Fig. 5.52. Mathematical model of the auto-transformer of Fig. 5.50

L.V. windings not connected directly to the investigated windings are at earth potential, as it has already been done for the regulating transformers.

Example 5.7. Determine the voltages to earth at the winding ends of a three-phase autotransformer of ratio 220/126 kV, in one of its tap positions corresponding to the rated voltage ratio, under the effect of an assumed 1.2/50 μs lightning impulse wave hitting its 126 kV terminal.

The windings of one phase of the auto-transformer is shown schematically in Fig. 5.50. The voltage is regulated on the 220 kV side of the transformer. The lower end of the 220 kV winding (marked H) is connected to the required tap point of the tapped winding (marked T) by means of which the output voltage can be regulated in the range of $\pm 15\%$. The ends of winding T can be interchanged, and the resulting characteristic schemes are shown in Fig. 5.51. The symbols are the same as those used in the description of regulating transformers.

The tertiary winding located nearest to the core of the transformer is treated in the calculation as an earthed cylinder. The 126 kV winding (L) is beside the tertiary winding directly followed by the 220 kV

267

winding (H), and finally, the tapped winding (T) is arranged as the outermost winding. Since the neutral of the transformer is earthed, the calculation is performed for one phase only. All the three windings are of high series capacitance, so each winding is considered as one element in the calculation. The three-element model of the transformer is shown in Fig. 5.52, where the symbols indicating the inductances and capacitances and their numerical values are as follows:

L_L: self inductance of winding L: 2.52 H,
L_H: self inductance of winding H: 1.41 H,
L_T: self inductance of winding T: 0.129 H,
M_{LH}:   mutual inductance between windings
             L and H: 1.83 ($q = 0.97$),
M_{LT}:   mutual inductance between windings
             L and T: 0.53 H ($q = 0.93$),
M_{HT}:   mutual inductance between windings
             H and T: 0.412 H ($q = 0.97$),
K_L: resultant capacitance of winding
             L: 530 pF,
K_H: resultant capacitance of winding H: 940 pF,
K_T: resultant capacitance of winding T: 9600 pF,
C_L: capacitance of winding L to earth (mainly to the tertiary winding regarded earthed): 2700 pF,
C_{LH}:   capacitance between windings L and H: 1900 pF,
C_{HT}:   capacitance between windings H and T: 2500 pF,
C_H: capacitance of winding H to earth: 770 pF,
C_T: capacitance of winding T to earth: 1540 pF.

The voltage oscillations are to be determined in tap position 0, assuming an impulse applied to the 126 kV terminal with the 220 kV terminal kept open. The relevant mathematical model is given in Fig. 5.53, considering also Figs 5.51 and 5.52. From Fig. 5.53, by the reduction of capacitances, the network shown

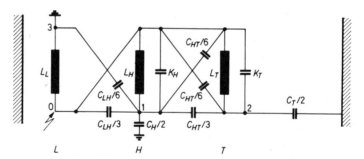

Fig. 5.53. Mathematical model for tap position 0

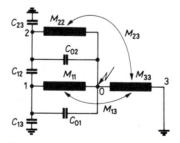

Fig. 5.54. Compact mathematical model

268

in Fig. 5.54 is derived. By comparing Figs 5.53 and 5.54 the following relation prevail between the elements appearing in the network and the characteristic data of the transformer:

$$M_{11}=L_H, \quad M_{12}=-M_{HT}, \quad M_{13}=M_{LH},$$
$$M_{22}=L_T, \quad M_{23}=-M_{LT}, \quad M_{33}=L_L,$$

$$C_{01}=\frac{C_{LH}}{3}+K_H+\frac{C_{HT}}{6}, \quad C_{02}=\frac{C_{HT}}{6}+K_T,$$

$$C_{12}=\frac{C_{HT}}{3}, \quad C_{13}=\frac{C_{LH}}{6}+\frac{C_H}{2}, \quad C_{23}=\frac{C_T}{2}.$$

Based on Kirchhoff's laws, the following equations may be written for the various network elements:

$$u_1-u_0=M_{11}i_1'+M_{12}i_2'+M_{13}i_3',$$
$$u_2-u_0=M_{12}i_1'+M_{22}i_2'+M_{23}i_3',$$
$$u_0=M_{13}i_1'+M_{23}i_2'+M_{33}i_3'.$$

The same may be rewritten in matrix form:

$$-\begin{bmatrix} -1 & 0 \\ 0 & -1 \\ 0 & 0 \end{bmatrix}\begin{bmatrix} u_1 \\ u_2 \end{bmatrix}=\begin{bmatrix} M_{11} & M_{12} & M_{13} \\ M_{12} & M_{22} & M_{23} \\ M_{13} & M_{23} & M_{33} \end{bmatrix}\begin{bmatrix} i_1 \\ i_2 \\ i_3 \end{bmatrix}-u_0\begin{bmatrix} -1 \\ -1 \\ 1 \end{bmatrix}. \qquad (5.81)$$

There is no need to write the nodal equations for the currents, since matrices **D** and **h** characterizing the network can already be determined on the basis of Eqn. (5.81). Namely, by comparing Eqns (5.71) and (5.81):

$$\mathbf{D}=\begin{bmatrix} -1 & 0 \\ 0 & -1 \\ 0 & 0 \end{bmatrix} \quad \text{and} \quad \mathbf{h}=\begin{bmatrix} -1 \\ -1 \\ 1 \end{bmatrix}.$$

In this case, based on the relations given in the description of the general calculation procedure, the remaining matrices appearing in the calculation take the following form

$$\mathbf{K}=\begin{bmatrix} C_{01}+C_{12}+C_{13} & -C_{12} \\ -C_{12} & C_{02}+C_{12}+C_{23} \end{bmatrix},$$

$$\mathbf{Q}=\begin{bmatrix} M_{11} & M_{12} & M_{13} \\ M_{12} & M_{22} & M_{23} \\ M_{13} & M_{23} & M_{33} \end{bmatrix},$$

$$\mathbf{D^*}=\begin{bmatrix} -1 & 0 & 0 \\ 0 & -1 & 0 \end{bmatrix}, \quad \mathbf{l}=\begin{bmatrix} C_{01} \\ C_{02} \end{bmatrix}.$$

In the knowledge of these matrix elements, the data required for computing can be given, hence the voltage in nodes 1 and 2 of the network, i.e. the voltages arising with respect to the earth at the free ends of the 220 kV winding and tapped winding can be determined. Characteristic data of the wave shape obtained for a 1.2/50 µs standard wave are: $U_{01}=1.038$, $p_1=0.015\times 10^6$, $p_2=2.47\times 10^6$.

The voltage oscillation appearing at the 220 kV terminal as obtained by calculation is represented in Fig. 5.56(b), and that actually measured is given in Fig. 5.56(b). The 1.2/50 µs voltage wave applied to the 126 kV terminal is also shown (Fig. 5.55(a) and Fig. 5.56(a)). The calculated curve closely approximates the measured voltage wave, both in shape and amplitude. For a switching impulse wave of 220 µs front time and 2000 µs time to half value, also applied to the 126 kV terminal in the case of connection scheme

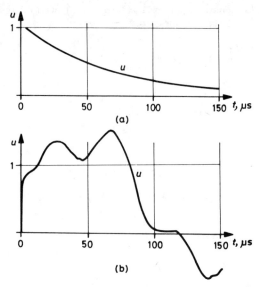

Fig. 5.55. Voltage (u_1) appearing at the terminal of winding H on application of an impulse (u_0) of shape 1.2/50 to winding L. (Calculated curve)

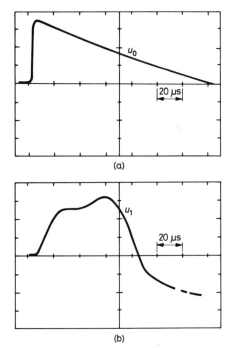

Fig. 5.56. Oscillograms of an impulse (u_0) of shape 1.2/50 applied to winding L and of the voltage (u_1) appearing at the terminal of winding H

$+0$, the calculated shape of oscillation appearing at the 220 kV terminal is plotted in Fig. 5.57, and the oscillogram taken during a measurement performed under similar conditions is represented in Fig. 5.58. In that case also, a good approximation is given by the calculated curve (disregarding the effect of damping).

It should be noted that the determination of voltage transferred from the 126 kV to the 220 kV terminal with the 220 kV terminal left open is of no practical importance. Because in reality, an earthing resistance is connected to this terminal, during the performance of impulse voltage tests, whereas in service a lightning arrester and, generally, also an overhead line of finite wave impedance is attached, so that the voltage is prevented from attaining the value assumed in the example. Apart from illustrating the course of calculation, the example has been presented mainly to characterize the nature of oscillations taking place in winding systems.

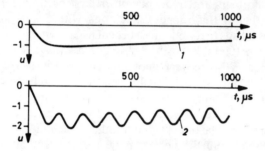

Fig. 5.57. Voltage appearing at the terminal of winding H (Curve 2) when a switching impulse wave is applied to winding L (Curve 1). (Calculated curve)

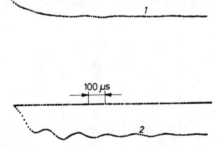

Fig. 5.58. Oscillogram of the voltage appearing at the terminal of winding H (Curve 2) when a switching impulse wave is applied to winding L (Curve 1)

As made apparent by Fig. 5.58, oscillations may be produced, in certain cases, even by flat-front switching impulse waves. Therefore, the voltage distribution caused by a switching impulse wave cannot be expected to be uniform, in every case. Uniformity is dependent on the critical front time further defined above. For winding systems consisting of windings of high series capacitance, the critical front time is considerably longer than that of a single winding. The frequency of oscillations being lower, the front time is, according to Eqn. (5.53):

$$T_{0\,cr} = (10 \text{ to } 12)\frac{1}{\omega_1}.$$

According to the general calculation procedure described in this section, the first harmonic can be computed from the first generalized eigenvalue, as $\omega_1 = \sqrt{\lambda_1}$. From this relation, the frequency is obtained as $f_1 = \dfrac{\omega_1}{2\pi}$, and the critical front time is

$$T_{0\,cr} = (10 \text{ to } 12)\frac{1}{\sqrt{\lambda_1}}.$$

For large transformers, the frequency of the first harmonic lies between 5 and 20 kHz, or even below this range. Calculating with 5 kHz, $\omega_1 = 31.5 \text{ s}^{-1}$ and $T_{0\,cr} = 380 \text{ μs}$. This means that under the effect of usual switching impulse waves with front times between 100 to 300 μs, the developing voltage distribution will be far from being uniform. Moreover, another remarkable factor should also be considered. In the case of such oscillations of relatively low frequency when applying a normal lightning impulse wave, the voltage corresponding to the pseudofinal distribution (i.e. the voltage around which the oscillations develop) will already be very small by the time the amplitude of oscillation attains its maximum. Thus, the voltage maximum developing in the course of oscillation cannot reach an excessively high value. On the contrary, in the case of a switching impulse wave, the pseudofinal voltage component reaches its maximum close to the peak value of the applied voltage wave, so that the resultant voltage in test conditions may be higher than that brought about by a steep-fronted but rapidly decaying overvoltage wave of atmospheric origin.

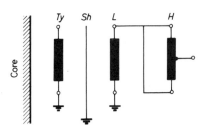

Fig. 5.59. Connection scheme of an auto-transformer

Example 5.8. Applying a 1.2/50 μs lightning impulse voltage to the 400 kV and then to the 126 kV terminals of a $\dfrac{400}{\sqrt{3}} \Big/ \dfrac{126}{\sqrt{3}} \Big/$ 18 kV single-phase auto-transformer, find what voltage will appear on its non-impulsed 126 kV and 400 kV winding, respectively, and across its 18 kV tertiary.

The transformer circuit is given in Fig. 5.59. Closest to the core lies the tertiary winding marked *Ty*. This is followed by an earthed shielding cylinder having the purpose of preventing the initial voltage transfer from the autotransformer windings to the tertiary, lest—due to the high turn ratio—dangerously high transferred voltages should appear on the tertiary winding. Then, the $126/\sqrt{3}$ kV winding marked L follows, and the $400/\sqrt{3}$ kV winding with its H.V. point at the centre lies outside. Windings H and L of the transformer are of high series capacitance, and each of them could be represented by one element in the model. Since, however, winding H is centre-tapped, its division into two parts is quite obvious: the upper and lower parts of the winding are represented by separate elements. For dielectric strength considerations, knowledge of the maximum stress imposed on the insulation between windings H and L is of importance. The highest voltage is expected to arise between the two windings at the middle of the limb, between the centre (impulsed) point of the centre-tapped winding and the centre point of winding L lying opposite. Therefore, the voltage in the centre of winding L should also be known, i.e. this point in the model should also be selected to be a node. Winding Ty is of low series capacitances ($\alpha_{Ty} = 18$). Should the voltage distribution within that winding be required, then its division into several elementary parts would be required in the model. Now, however, the voltage appearing across the entire winding, i.e. that arising at the non-earthed end of it, is wanted, therefore the winding is represented by a single element in the model. Thus, altogether a five-element mathematical model is obtained (Fig. 5.60).

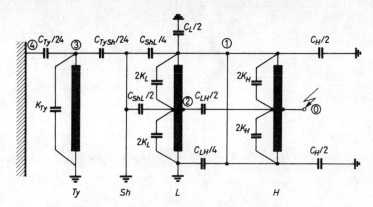

Fig. 5.60. Mathematical model of the transformer of Fig. 5.59

When a winding is to be represented by several elements, the series capacitances have to be divided among individual elements proportionally. Thus if the resultant series capacitance of the winding is K_r, and the winding is divided into n elementary parts, then the series capacitance of each elementary part is $K_e = nK_r$. Parallel capacitances also have to be divided proportionally among the nodes. Since the number of nodes, including the beginning and end of the winding, is one more than the number of elements, only half of the capacitance resulting from the proportional division falls to the beginning and end point of the winding.

For winding Ty, having a small series capacitance, no knowledge of voltage distribution within the winding is required, as already mentioned. Only the effect of the voltage appearing across the entire winding on the system is wanted. The shunt capacitances of the winding, i.e. the capacitances toward the core and earthed shielding cylinder, can now be considered in one point only, viz. in the non-earthed end terminal of the winding. Therefore, a capacitance of such magnitude should be put at that point to make the charge transferred from the winding towards the earth correspond to the charge valid in the case of the real distributed capacitances. The charge transferred in reality can be determined from the voltage distribution curve. Denoting the shunt capacitance per unit length of the winding by C, the charge transferred along the winding toward the earth, in the case of an earthed-end winding, will be on the basis of relation (5.10):

$$Q = UC \int\limits_0^1 \frac{\sinh \alpha \left(1 - \dfrac{x}{l}\right)}{\sinh \alpha} \, d\left(\frac{x}{l}\right).$$

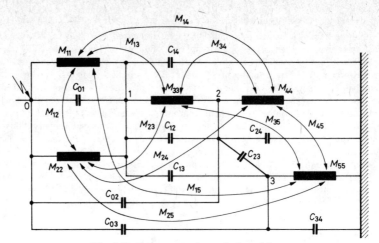

Fig. 5.61. Compact mathematical model

After performing the integration:

$$Q = UCl \frac{\cosh \alpha - 1}{\alpha \sinh \alpha}$$

and, for $\alpha > 10$, the following relation is obtained with good approximation

$$Q = \frac{1}{\alpha} UCl.$$

The magnitude of reduced capacitance C_{red} to be placed at the end point of the winding should be such as required for transferring a charge equal to the that obtained above. Correspondingly, substituting Cl by C_r:

$$C_{red} U = \frac{1}{\alpha_{Ty}} C_r U,$$

i.e.

$$C_{red} = \frac{1}{\alpha} C_r.$$

Thus, at the upper end point of winding Ty, with good approximation, only $1/\alpha_{Ty} = 1/18$ part of the capacitance falling between winding Ty and the adjacent windings is assumed in the calculation.

From the above, after reducing the capacitances, the network of Fig. 5.61 is obtained for the case of impulsing the 400 kV terminal. Using this network, the differential equation can already be written and the matrices determined. The interrelations between the capacitances appearing in the model of Fig. 5.60 and in the capacitance network of Fig. 5.61, and the values of capacitances, are given below:

$$C_{01} = 2K_H + 2K_H = 692 + 692 = 1384 \text{ pF},$$

$$C_{02} = \frac{C_{LH}}{2} = 1000 \text{ pF},$$

$$C_{03} = 0,$$

$$C_{12} = 2'K_H = 6300 \text{ pF},$$

$$C_{13} = 0,$$

274

$$C_{14} = \frac{C_H}{2} + \frac{C_H}{2} + \frac{C_L}{2} + \frac{C_{ShH}}{4} + \frac{C_{LH}}{4} =$$

$$= 850 + 850 + 1200 + 700 + 500 = 4100 \text{ pF},$$

$$C_{23} = 0,$$

$$C_{24} = 2K_L + \frac{C_{ShH}}{2} = 6300 + 1400 = 7700 \text{ pF},$$

$$C_{34} = K_{Ty} + \frac{C_{Ty}}{18} + \frac{C_{TySh}}{18} =$$

$$= 50 + 400 + 470 = 920 \text{ pF}.$$

The values of self and mutual inductances appearing in the calculation (see Section 5.2.4) are as follows:

$$M_{11} = M_{22} = 4.18 \text{ H},$$

$$M_{12} = 4.05 \text{ H},$$

$$M_{13} = M_{14} = M_{23} = M_{24} = 0.9 \text{ H},$$

$$M_{15} = M_{25} = 0.442 \text{ H},$$

$$M_{33} = M_{44} = 0.213 \text{ H},$$

$$M_{34} = 0.207 \text{ H},$$

$$M_{35} = M_{45} = 0.103 \text{ H},$$

$$M_{55} = 0.054 \text{ H}.$$

As already done in Example 5.7, for the various elements of the network the following differential equations can be written:

$$u_0 - u_1 = M_{11}i_1' + M_{12}i_2' + M_{13}i_3' + M_{14}i_4' + M_{15}i_5',$$

$$u_0 - u_1 = M_{12}i_1' + M_{22}i_2' + M_{23}i_3' + M_{24}i_4' + M_{25}i_5',$$

$$u_1 - u_2 = M_{13}i_1' + M_{23}i_2' + M_{33}i_3' + M_{34}i_4' + M_{35}i_5',$$

$$u_2 = M_{14}i_1' + M_{24}i_2' + M_{34}i_3' + M_{44}i_4' + M_{45}i_5',$$

$$u_3 = M_{15}i_1' + M_{25}i_2' + M_{35}i_3' + M_{45}i_4' + M_{55}i_5'.$$

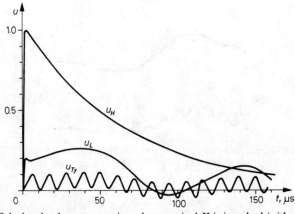

Fig. 5.62. Calculated voltages appearing when terminal H is impulsed (with a 1.2/50 wave)

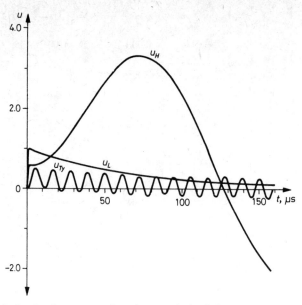

Fig. 5.63. Calculated voltages appearing when terminal L is impulsed (with a 1.2/50 wave)

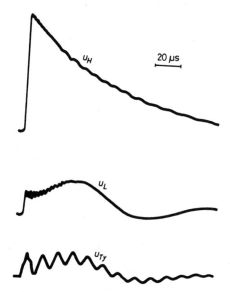

Fig. 5.64. Oscillographic records of voltages when terminal H is impulsed (with a 1.2/50 wave). Voltage scale of the third oscillogram is double of that of the first two

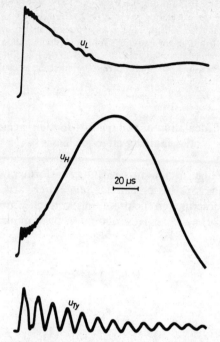

Fig. 5.65. Oscillographic records of voltages when terminal L is impulsed (with a 1.2/50 wave). Voltage scale of the second oscillogram is half that of the first and third

From the above, as in the previous example, matrices $\mathbf{D}$ and $\mathbf{h}$ can be written:

$$\mathbf{D} = \begin{bmatrix} 1 & 0 & 0 \\ 1 & 0 & 0 \\ -1 & 1 & 0 \\ 0 & -1 & 0 \\ 0 & 0 & -1 \end{bmatrix}, \quad \mathbf{h} = \begin{bmatrix} 1 \\ 1 \\ 0 \\ 0 \\ 0 \end{bmatrix}.$$

As it can be seen, for determining matrices $\mathbf{D}$ and $\mathbf{h}$, it would in fact, be sufficient to write the left-hand side of the system of differential equations referring to the voltages of elements. The other matrices are arrived at in a similarly simple way.

$$\mathbf{K} = \begin{bmatrix} C_{01}+C_{12}+C_{13}+C_{14} & -C_{12} & -C_{13} \\ -C_{12} & C_{02}+C_{12}+C_{23}+C_{24} & -C_{23} \\ -C_{13} & -C_{23} & C_{03}+C_{13}+C_{23}+C_{34} \end{bmatrix},$$

$$\mathbf{M} = \begin{bmatrix} M_{11} & M_{12} & M_{13} & M_{14} & M_{15} \\ M_{12} & M_{22} & M_{23} & M_{24} & M_{25} \\ M_{13} & M_{23} & M_{33} & M_{34} & M_{35} \\ M_{14} & M_{24} & M_{34} & M_{44} & M_{45} \\ M_{15} & M_{25} & M_{35} & M_{45} & M_{55} \end{bmatrix},$$

277

The matrices appearing in the calculation can also be determined in a similar way for the case of impulsing the 126 kV terminal. After substituting the corresponding matrix elements the data required for computing the voltage distribution are ready. The voltage oscillations obtained by computation are shown in Figs 5.62 and 5.63, while the oscillograms measured on the transformer are given in Figs 5.64 and 5.65. Only slight differences can be observed between computed and measured voltages, and also the accuracy of calculation is satisfactory.

5.2.13 Calculation of voltage distribution including the damping effect of losses

The damping of voltage oscillations within transformer windings, caused by winding resistance and core losses can be taken into account in the general calculation procedure described in the preceding section by multiplying expression (5.80) by the function $f_k(t) = e^{-\delta_k t}$ corresponding to the damping effect. Since, now,

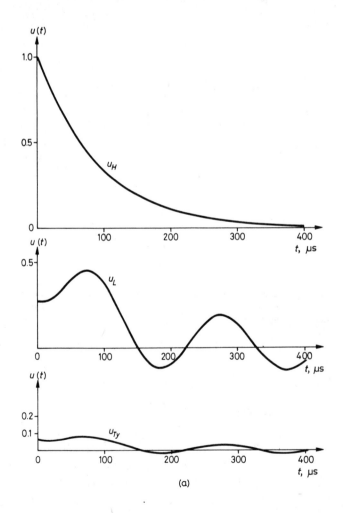

(a)

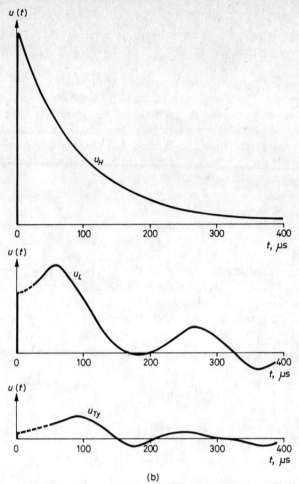

(b)

Fig. 5.66. (a) Calculated damped voltage oscillations, taking place in a 750 kV transformer on application of a 1.2/50 lightning impulse voltage wave; (b) oscillographic record of voltage oscillations in the same 750 kV transformer on application of a 1.2/50 wave

$\omega_k = \sqrt{\lambda_k}$, introducing the notation $\ln A_1/A_2 = D$, the damping factor can be written in the form

$$\delta_k = \frac{\sqrt{\lambda_k}}{2\pi} D$$

and for y_k the following relation is obtained:

$$y_k = U_0 \left\{ \frac{s_k + p^2 r_k}{\lambda_k + p^2} e^{-pt} + \left(\frac{s_k + p^2 r_k}{\lambda_k + p^2} - r_k \right) \left(\frac{p}{\sqrt{\lambda_k}} \sin \sqrt{\lambda_k} t - \cos \sqrt{\lambda_k} t \right) e^{-\delta_k t} \right\}. \tag{5.82}$$

For large transformers the value of D, found by experience, lies in the range 0.18 to 0.34.

The effect of damping on voltage oscillations is well illustrated by the calculated curves of Fig. 5.66 (a), which shows a 1.2/50 lightning impulse voltage wave (U_H) applied to the 750 kV terminal of a $\dfrac{750}{\sqrt{3}}\left(\dfrac{410}{\sqrt{3}}\right)$ 15 kV single-phase autotransformer and the resulting voltage oscillation appearing at the 410 kV terminal (U_L) and at the 15 kV terminal (U_{Ty}). The decay of oscillations can be easily observed. The oscillographic records of voltage oscillations are given in Fig. 5.66 (b). The calculated curves show good agreement with the measured results. It can also be stated, on the evidence of Figs 5.66 (a) and 5.66 (b), that the damping does not significantly modify the first peak, whose height is indicative of the expected voltage stresses and influences the design of insulation.

5.2.14 Calculation of stresses imposed by oscillating and damped switching waves

As demonstrated in the description of the general calculation procedure, the method given is suitable for determining the voltage distribution developing within a winding under the effect of a voltage wave varying arbitrarily in time. Substituting into Eqn. (5.78) the function under (5.60) describing oscillating switching waves

$$U(t)=U_0\left[e^{-pt}+e^{-Dft}\left(\frac{p}{2\pi f}\sin 2\pi f t-\cos 2\pi f t\right)\right]=$$

$$=U_0\left[e^{-pt}+e^{-\kappa t}A\sin(\sigma t-\varphi)\right] \tag{5.83}$$

(with $\kappa=Df$; $\sigma=2\pi f$; $\varphi=\arctan \sigma/p$ and $A=1/\sin\varphi$) an expression consisting of two terms is obtained for y_k:

$$y_k=(y_k)_1+(y_k)_2 . \tag{5.84}$$

Here, taking into account also the damping effect of transformer winding and core,

$$(y_k)_1=U_0\left[\left(\frac{s_k+p^2 r_k}{\lambda_k+p^2}\right)e^{-pt}+\left(\frac{s_k+p^2 r_k}{\lambda_k+p^2}-r_k\right)\left(\frac{p}{\sqrt{\lambda_k}}\sin\sqrt{\lambda_k}t-\cos\sqrt{\lambda_k}t\right)e^{-\delta\kappa t}\right]$$

and

$$(y_k)_2=U_0\left\{\frac{A}{2\sqrt{\lambda_k}}\left[(P_k+Q_k)\sin(\sigma t-\varphi)-\right.\right.$$

$$-(R_k-S_k)\cos(\sigma t-\varphi)]e^{-\kappa t}+$$

$$+\left\{\frac{1}{2\sqrt{\lambda_k}}\left[P_k\sin(\sqrt{\lambda_k}t+\varphi)-Q_k\sin(\sqrt{\lambda_k}t-\varphi)+\right.\right.$$

$$+R_k\cos(\sqrt{\lambda_k}t+\varphi)]-S_k\cos(\sqrt{\lambda_k}t-\varphi)+$$

$$+\left[(p+\kappa)\frac{\sin\sqrt{\lambda_k}t}{\sqrt{\lambda_k}}-\cos\sqrt{\lambda_k}t\right]r_k\right\}e^{-\delta\kappa t}\right\},$$

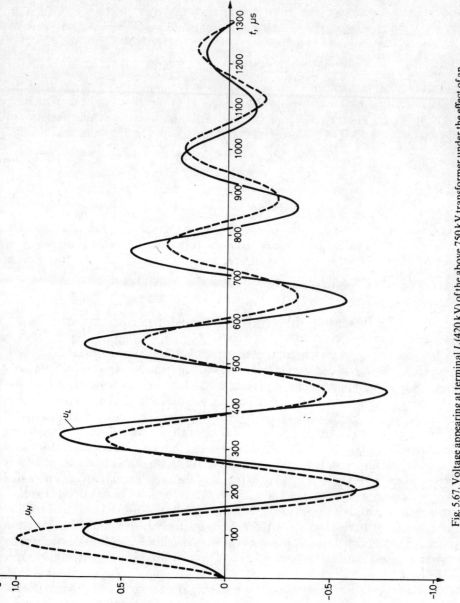

Fig. 5.67. Voltage appearing at terminal L (420 kV) of the above 750 kV transformer under the effect of an oscillating switching wave. (Calculated curve)

where

$$P_k = \frac{B_k(\sqrt{\lambda_k} + \sigma) - E_k \kappa}{\kappa^2 + (\sqrt{\lambda_k} + \sigma)^2}, \quad Q_k = \frac{B_k(\sqrt{\lambda_k} - \sigma) + E_k \kappa}{\kappa^2 + (\sqrt{\lambda_k} - \sigma)^2},$$

$$R_k = \frac{B_k \kappa + E_k(\sqrt{\lambda_k} + \sigma)}{\kappa^2 + (\sqrt{\lambda_k} + \sigma)^2}, \quad S_k = \frac{B_k \kappa - E_k(\sqrt{\lambda_k} - \sigma)}{\kappa^2 + (\sqrt{\lambda_k} - \sigma)^2},$$

$$B_k = (\kappa^2 - \sigma^2) r_k + s_k, \quad E_k = 2\kappa\sigma r_k.$$

The oscillating, damped switching impulse wave applied to terminal H of the 750 kV transformer mentioned in the discussion of damping, and the oscillating wave appearing at terminal L under the effect of the applied switching impulse are shown in Fig. 5.67. The calculation is performed with the following data:

$$\begin{aligned} U_0 &= 1, & (\kappa &= 1496, \\ p &= 1.5 \times 10^4, & \sigma &= 27\,646, \\ D &= 0.34, & \varphi &= 1.0737, \\ f &= 4400 \text{ Hz.} & A &= 1.1377). \end{aligned}$$

As apparent from the diagram, the amplitude of the wave appearing at terminal L is higher than would be expected from the turn ratio. In fact, the frequency of the oscillating wave approaches the resonant frequency of the transformer ($f_r = 5221$ Hz).

5.3 Insulating materials and transformer insulation

The most important insulating materials, of optimum dielectric strength properties, that have been used in large high-voltage transformers through almost a century are the oil-impregnated, cellulose-based products: primarily paper and pressboard. In the bushings used for leading out the ends of transformer windings from the tank oil-impregnated paper, or resin-bonded or resin-impregnated paper, are also employed as insulation in the inside of the ceramic shell. In the internal insulation, mainly where high mechanical strength is also required, phenol-based paper is still employed. However, the dielectric strength of this material, mainly its creepage strength, is much poorer than that of the pressboard. Parts of laminated wood are practically only mechanically stressed. The low electric stress that may arise in these parts is insignificant, if they are carefully processed (i.e. with all voids of air eliminated). In recent designs, oil resistant enamel is also employed as interturn insulation, although it is less capable of withstanding mechanical stresses (caused by thermal displacements and short-circuit forces) than oil-impregnated paper insulation.

In recent years the use of high-pressure (5 bar) SF_6 gas has been proposed instead of oil as an insulating medium in transformers. According to experiments performed, such transformers can be manufactured with foil windings, with plastic foil insulation placed between turns, and with cooling provided by vertical cooling ducts running between the foil windings. The most significant advantage of SF_6

insulation for large transformers is the much higher interturn dielectric stress that may be imposed on the interturn insulation than on a conventional oil-impregnated paper interturn insulation. Thereby, a considerable decrease in the dimensions of windings can be achieved. Also the thermal stress of insulation may be much higher, permitting the application of significantly increased current densities with unchanged conditions of cooling. Thus, the quantity of materials built into the transformer also decreases, resulting in a reduction of weight for transport. A further advantage of SF_6 gas insulated transformers is that hardly any aging of pressboard insulation occurs in SF_6 gas [94].

A problem is imposed, however, in designing the tank that has to withstand an internal pressure exceeding 5 bar. In addition, certain operating experiences have indicated that, from an economical point of view, a transformer of such design is not necessarily the optimum solution. Nevertheless, it has been concluded from model tests that even 500 MVA, 500 kV transformers could be realized by this insulation system.

In smaller power transformers, silicone oil is also used as an insulating medium instead of mineral oil; and Kraft paper can be replaced by "nylon-paper" (a synthetic sheet material). Such nylon-papers are made by a process similar to that employed in the manufacture of Kraft paper, but use aromatic polyamide fibres and flakes (e.g. NOMEX Aramide paper). Since aromatic cyclical elements are introduced into the polymeric ring, several properties of the materials thus obtained will be superior to those of linear polycondensates, such as polyamides. The most important property of these materials is their high thermal stability. Their dielectric strength is excellent, being above that of Kraft paper, and they retain this property for a sustained period even at 180 to 200 °C. Their application should be considered in large power transformers expected to carry frequent, very large overloads of short duration, causing temperature rises in the windings even exceeding 200 °C for short periods of time. In such cases it may prove more economical to install, instead of a transformer of higher rating, a smaller transformer provided with "nylon-paper" interturn insulation. Because of the higher costs involved, as in the case of silicone oil, this type of insulation is seldom adopted, and even then is used mainly in transformers of smaller ratings.

In any case, in the manufacture of large power transformers, no kind of plastic insulation has so far succeeded in superseding the oil-impregnated paper and oil-impregnated pressboard insulations. Due to their high dielectric strength, their excellent cooling properties and the possibility of their easy regeneration, these insulation systems are being used even in transformers of the presently highest voltage ratings, and it is most improbable that any other kind of insulation system could take over their place in the near future. Therefore, only the oil-paper and oil-pressboard insulation systems will be dealt with in the following sections.

5.3.1 The oil

The physical, chemical and electrical properties of insulating oils applicable in transformers are specified by the relevant standards. In the IEC standard dealing with the insulating oils of transformers and apparatus [122], three classes of oil are distinguished according to the flash point, pour point and kinematic viscosity figures. Oils of Class I have flash points $\geq 140\,°C$, pour points $\leq -30\,°C$, and kinematic viscosities $\leq 16.5 \, mm^2 \, s^{-1}$ (cSt) measured at $40\,°C$, the respective figures for oils belonging to Class II are: $\geq 130\,°C$, $\leq -45\,°C$ and $\leq 11 \, mm^2 \, s^{-1}$, and for those of Class III $\geq 95\,°C$, $\leq -60\,°C$ and $\leq 3.5 \, mm^2 \, s^{-1}$. The density at $20\,°C$ of insulating oil of satisfactory quality should be below $0.895 \, g \, cm^{-3}$. The oil must not contain any inorganic acid or alkali soluble in water, corrosive sulphur or mechanical impurities and should be transparent. These properties depend on the crude oil used and on the refining process, and most of them do not change either during the treatment following filling of the transformer with oil, or under influences occurring in service.

On the other hand, the breakdown voltage, interfacial tension, the dissipation factor and permittivity of the oil are all subject to changes. These properties in transformer oils of good quality are as follows: breakdown voltage at 50 or 60 Hz in the condition when the transformer is filled at least $200 \, kV \, cm^{-1}$, interfacial tension at $25\,°C$ at least $40 \times 10^{-3} \, N \, m^{-1}$, loss factor (tan δ) at $90\,°C$ lower than 50×10^{-4}, and permittivity (ε) 2.2. These factors are highly influenced by the environmental conditions (moisture and gas absorption, mechanical contamination, etc.), so they are liable to change after the transformer has been filled, during treatment of transformer insulation and later, under the conditions of service.

For the design of transformer insulation the most important property of the oil is its dielectric strength, which may reach $4000 \, kV \, cm^{-1}$ in the case of oils purified to an extremely high level under laboratory conditions, whereas in engineering practice, the highest value expected is about $330 \, kV \, cm^{-1}$ in the case of maximum attainable purity. In the standards the figure specified for new oils is even lower (about $200 \, kV \, cm^{-1}$, as already stated above).

When determining the dielectric strength of the oil by laboratory measurement, the volume effect should also be taken into consideration. The "volume effect", a phenomenon verified by experiments, is that the dielectric strength of a smaller volume of oil is found to be considerably higher than that measured on a large quantity of oil. For example, a breakdown voltage of about $600 \, kV \, cm^{-1}$ measured in an oil sample having a volume of $10^{-6} \, cm^3$ corresponds to a breakdown voltage of about $60 \, kV \, cm^{-1}$ only, when measured in an oil sample of the same quality but having a volume of $1 \, m^3$ (see Fig. 5.68). Special attention should be paid to this phenomenon, which is explained by the statistical distribution of minute-size impurities always present in the oil, if it is required to draw a conclusion from scale model experiments on the dielectric strength to be expected in the tank of large transformers containing a huge quantity of oil [129].

The influence of oil velocity on its breakdown strength can also be explained by the volume effect. The dielectric strength of oil streaming with the usual velocity of about $10 \, cm \, s^{-1}$ is slightly higher than that of stationary oil, since flowing oil sweeps away the particles which could bridge the electrodes. At higher flow velocities, however, the dielectric strength of the oil decreases, because the number of particles passing between the electrodes increases, and this effect becomes predominant over the counteracting one resulting from the shortening of time of stressing due to the oil flowing between the electrodes. Increasing the flow velocity from $10 \, \text{cm s}^{-1}$ to $200 \, \text{cm s}^{-1}$ may cause a drop of 20 to 25% in the dielectric strength of the oil [129].

During the service life of the transformer, further deterioration of dielectric strength has to be reckoned with. The breakdown voltage of the transformer oils in operation, however, should not be less than $140 \, \text{kV cm}^{-1}$. If the breakdown voltage drops below this value, the oil should either be regenerated, or changed. A fall of

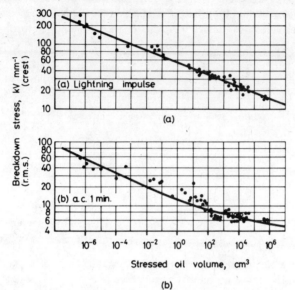

Fig. 5.68. Breakdown voltage of oil as a function of oil volume examined

dielectric strength of insulating oils to a fraction of the ideal value is caused by moisture, gases or solid impurities present in the oil.

Water may be present in the oil in three different ways: in dissolved states, finely distributed, (forming an emulsion), and roughly distributed (forming droplets). More moisture can be kept dissolved by the oil, the higher its temperature and the higher the partial pressure of water vapour present in its environment. In the temperature range 0 to 80 °C, oil in contact with air at atmospheric pressure is capable of retaining 30 to 600 ppm dissolved moisture (1 ppm = 1 mg water/1 kg oil). The dielectric strength of the oil is influenced considerably by the moisture dissolved in it. As apparent from Fig. 5.69, above a moisture content of 40 ppm the dielectric strength of the oil deteriorates rapidly [15].

Water is expected to be present in the oil in the non-dissolved state, i.e. finely or roughly distributed, when further moisture enters the oil already saturated with moisture, or the state of the oil changes in such a way that less water is needed for saturation, e.g. because the oil cools down. The oil is rendered so turbid and its dielectric strength so poor by moisture present in the finely distributed state that the oil becomes unsuitable for further use. Moisture present in rough distribution, because its specific density is higher than that of the oil, tends to accumulate at the deepest point of the tank, where it causes then, a drastic drop in local dielectric strength. Since, however, the regions of elevated field strengths within the transformer are generally situated near the upper parts of the tank, the water accumulating at the bottom seldom causes any trouble.

At temperatures below the water freezes, and if the density of the oil is then above that of water, the ice will float in the uppermost layers of the oil. Generally, the pieces of ice in themselves represent no hazard. Trouble may be caused when the ice melts and the resulting water, invading the regions of high field strength, causes a considerable reduction of dielectric strength, thereby leading to direct injury and

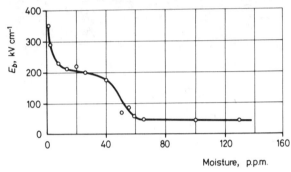

Fig. 5.69. Variation of breakdown strength of oil as a function of moisture content [15]

total breakdown of the dielectric strength of insulation. An especially dangerous situation may arise if the ambient temperature is cycling a few degrees above and below the freezing point. This hazard can be avoided by using, in transformers critical in this respect, an oil of lower specific density than that of ice. The specific density of the ice at $0\,^\circ\mathrm{C}$ can usually vary in a rather wide range. According to measurements, it may lie between 0.88 and 0.92 g cm^{-3}, so in this case, the density of the transformer oil should be in the range 0.86 to 0.87 at $0\,^\circ\mathrm{C}$ [99].

A high moisture content may be especially hazardous if the oil contains at the same time foreign particles, such as longer fibres (e.g. pieces of paper or yarn). Such fibres assume the direction of the field and, in the presence of moisture, constitute a conductive bridge between parts of different potentials. Resulting flashovers may severely damage the transformer.

Moisture in the oil has no effect on its useful life. Even oils which have been allowed to become damp over a prolonged period can be regenerated by vacuum drying and filtering. Moisture in the oil, however, may penetrate the paper, thereby causing its aging. This topic will be discussed in further detail below.

The dielectric strength of the oil may be reduced also by gases present in the oil. As with moisture, these can be present in three physical states: dissolved, finely distributed, and roughly distributed. At $25\,^\circ\mathrm{C}$ and at normal atmospheric pressure, the oil is capable of keeping the following substances dissolved, expressed as a percentage of the oil volume: 10% air, or 8.5% nitrogen, or 16% oxygen, or 100% carbon dioxide. The percentage of gas soluble increases with rising temperature and pressure [103].

286

The dielectric strength of an oil saturated with gas is not much lower than that of an oil free of dissolved gas. The situation becomes hazardous if a change in the state of the oil (decrease of temperature or pressure, or a chemical process taking place in the oil) causes a reduction of the quantity of gas the oil can dissolve, so that some of the dissolved gas is released. Generally, this released gas appears in the oil in a microscopic distribution, in the so-called finely distributed form. This state can be recognized from the turbid colour of the oil. The gas present in the finely distributed state, by itself reduces the dielectric strength of the oil. This state, however, is of a transitory character, because the air bubbles of microscopic size tend to unite to form larger voids of air. Their size keeps growing until the flow resistance is overcome by the lifting force and a bubble starts ascending. The gas being already in the roughly distributed state, tends to approach the surface of the oil, and if reaching in the vicinity of parts subject to high electrical stresses, will cause electrical defects in the transformer. Roughly distributed gas bubbles may enter the oil due to some defect caused by improper mounting, e.g. a defective seal in the piping system of oil pumps.

The presence of gases causes not only a dangerous local deterioration of dielectric strength but, in contrast to the effect of moisture, it is also harmful with respect to the life of the oil. Air or oxygen present in the oil give rise to chemical changes which will bring about permanent deterioration of electrical properties of the oil and its aging. In addition, these gases and carbon dioxide also promote the aging of cellulose-based materials surrounded by the oil.

Gases may enter the oil either from outside due to its direct contact with the air, or due to the decomposition of the oil itself or of the cellulose-based substances immersed in it. Such decomposition may take place under the effect of higher temperatures or electric discharges. Oil and cellulose are organic substances with little resistance to heat, their molecular decomposition—to whatever small extent—being already detectable at room temperature. The decomposition of oil starts to accelerate considerably at 130 °C and that of cellulose at even lower temperatures. One part of the decomposition products consists of gases which remain dissolved in the oil until their quantity reaches the saturation limit. Then the gases released from the oil get into the Buchholz relay. In the case of rapid gas development (e.g. caused by an interturn fault, flashover or a similar defect) the oil in the vicinity of the fault becomes supersaturated, the quantity of gas in the Buchholz relay increases rapidly and the protection operates.

The solubilities in oil of the gases developing differ so much that the composition of the gases released (due to saturation) and gathering in the Buchholz relay differs entirely from that of the gases present in the oil in the dissolved state. Thus, by analysing the composition of the gases dissolved in the oil, conclusions can be drawn about the presence of possible defects in the transformer (hot spots or electric discharges), whereby prevention of serious disturbances is made possible [33].

On the basis of the composition of gases dissolved in the oil, three main types of defects may be distinguished. Discharges taking place release gas bubbles consisting mainly of hydrogen. The development of acethylene is characteristic of high-intensity discharges (sparking and arcing). An increase in the quantity of ethylene gas indicates to overheated spots within the transformer. The method of analysing the dissolved gases is widely practised, both for the supervision of transformers in service and, as a test-room procedure, for the detection of possible hot spots.

The aging of oil is indicated, in addition to the decreasing dielectric strength, by the reduction of the interfacial tension. As proved by experiments, at the interface between a new oil of good quality and water a much higher interfacial tension can be measured than between an aged, highly acidified oil and water. According to relevant statistical data, oils in service showing an interfacial tension exceeding 24×10^{-3} N m^{-1} are sludge-free. The interfacial tension of strongly sludged oils lies below 16×10^{-3} N m^{-1}. As shown by experience, the lowest value with which a transformer may be kept in service is 18×10^{-3} N m^{-1}.

Less information concerning the aging of the oil is furnished by the value of the discipation factor (loss tangent). The tan δ values of artificially aged oils or those in use for a prolonged time may be a multiple of the loss factors measured on new oils. In the course of supervisory checkings performed on satisfactorily operating transformers values as high as 7000×10^{-4} have been encountered. However, there is no unequivocal relation between the magnitude of the dissipation factor and the moisture content, dielectric strength and acidity.

The forced oil circulation generally used in large transformers has practically no influence on the aging of oil. On the contrary, the better temperature conditions obtainable by forced oil cooling, as compared to natural circulation, exposes the oil to a slightly, but detectably lower rate of aging. In the case of forced oil circulation, the air sucked in by the oil pump through a defective seal in the piping may represent a risk, the oxidizing effect of this air causing additional aging of the oil.

The susceptibility of the oil to degradation caused by oxidation can be reduced by antioxidant additives. The degradation of oil by oxidation is decreased or delayed by such inhibitors, whereby the good insulation properties of the oil can be preserved for a longer time [121].

5.3.2 The paper

Papers used for insulation in transformers have thicknesses of 30 to 120 µm and densities of 0.7 to 0.8 g cm^{-3}. It is common practice to specify the mass per square m (g m^{-2}), this being the product of thickness and density; e.g. the mass per unit surface area of a 30 µm thick paper of 0.7 g cm^{-3} density is $30 \times 10^{-6} \times 0.7 \times 10^6 = = 21$ g m^{-2}. (Conductor insulation on wires of circular section is made of 20 to 30 g m^{-2} paper, on those of rectangular section 40 to 60 g m^{-2} paper, and the main insulation of 100 g m^{-2} paper.) An important characteristic of the paper is its tensile strength. This is not expressed directly but in units of the so-called tensile length. The tensile length is the length of a suspended paper tape of uniform thickness and width, the weight of which just causes rupture of the tape under the tensile force produced by its dead weight. For papers used for insulation purposes its value is 6 to 8 km. From among the electrical specification data of a paper the most important property is its dielectric strength at power frequency (50 to 60 Hz), dielectric strength, having a value of 100 to 150 kV cm^{-1} in the dry condition and when applied in a thin layer. Its dissipation factor (loss tangent: tan δ) is 10 to 30×10^{-4}, its resistivity is 100 to

$600 \times 10^6 \, M\Omega$ cm and its permittivity is about 2. (The permittivity of the base material of paper, i.e. that of the cellulose is higher, lying between 5 and 6. The difference is due to non-uniform and incomplete filling of the volume of the paper by the cellulose fibres, leaving innumerable gaps whose permittivity is equal to 1 between the fibres. The same explanation applies to the dielectric strength of the paper, which is inferior to that of the cellulose fibres.)

The properties of paper deteriorate under the effects of heat, water and oxygen: the paper is subject to aging. The role of other factors, even that of the electric field, is of secondary importance compared to these three factors.

The aging of paper first causes a deterioration of mechanical properties, and a reduction of dielectric strength only afterwards. As long as a paper retains its mechanical strength and does not get charred and break up into pieces, it remains electrically satisfactory.

The progress of aging can be judged, apart from the reduction of tensile strength, by a chemical property, namely, the degree of molecular polymerization. The degree of polymerization of paper (the number of coupled glycose rings of cellulose) is directly related to the breaking strength of the paper: the lower the degree of polymerization, the lower the breaking strength. A reduction of the degree of polymerization from the 1000 to 1300 characteristic of a new Kraft paper to 150 practically means full insulation damage. Another characteristic used for indicating the aging of a paper is the so-called factor of degradation [29], which is proportional to the number of rings split in the cellulose chain structure under the effect of aging. Dependence of the degradation factor η on the duration of exposure to a temperature below 130 °C is expressed by the relation

$$\eta - \eta_0 = k_\eta t,$$

where η_0 is an experimentally determined constant and k_η is a factor characteristic of the rate of degradation. The temperature dependence of factor k_η is shown in Fig. 5.70 where it can be seen that below 130 °C the factor k_η approximately doubles per 10 °C increase of temperature. From the above formula it follows that increasing the temperature by 10°C, cuts to half the time during which a certain degree of aging is reached or, in other words, the useful life of insulation is halved when its temperature is increased by about 10 °C. According to an earlier formula proposed by Montsinger a temperature increase of 8 °C is already sufficient to reduce the service life to half. In practice a still smaller temperature rise, of 6 °C, is reckoned with when calculating the life of the entire insulation, and the same value is also taken as the basis for the international loading guides dealing with the load carrying capacity of transformers.

Such relations for the expected service life of insulation, however, are mainly of theoretical importance and assessments based on them are rather uncertain. It is sufficient to note that, according to Fig. 5.70, decreasing the temperature by 3 °C increases the service life of insulation by about 30%. Not even the expected average temperature over a future period of several years can be predicted, with an accuracy of 3 °C, let alone the local temperature maxima by which the life of insulation is primarily determined.

19

One aging process is considerably accelerated by the presence of water. Paper, when brought into contact with air of room temperature and 100% relative humidity, is capable of absorbing about 15% water; at 60% relative humidity the water absorption is about 8%, and at 20%, about 4%. (The percentage values refer to weights.) At 4% moisture content, however, the aging rate of paper at 130 °C is twenty times higher than at 0.5% moisture content. (The quoted percentage values of moisture indicate the quantity of water present at the beginning of the aging process, because in the course of aging decomposition of paper water is also formed. The ratio of water, carbon dioxide and carbon monoxide formed is 70:12:18.)

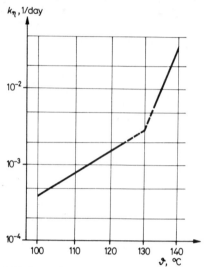

Fig. 5.70. Dependence of factor k_η on temperature.
k_η is a factor characterizing the rate of degradation of paper

Moisture present in the paper also has other effects, beside accelerating the aging process. Thus the loss factor of the paper increases under the effect of moisture. With high water content, this increase of tan δ may be so high that thermal instability is caused and breakdown will occur. Reducing the moisture content to a minimum and protecting the transformer from absorbing water during service is very important for increasing the service life. The harmful effects of moisture can be avoided by reducing the average water content of the paper below 0.5%.

The third factor causing aging of paper is the presence of oxygen. This factor is less important than the other two. According to laboratory experiments, the increase of aging rate of paper is at most doubled by the presence of oxygen [29].

5.3.3 Oil–paper insulation

The dielectric strength of laminated paper impregnated with oil is considerably above that of either of the two components. In fact, the role of the paper in the combination is to split the oil into minute gaps. The paper is a tangled mass of fibres

290

not fully compressed, permitting the oil to assume—between fibres and sheets—the form of extremely thin oil spaces whose dielectric strength is much higher than that measured between thick-layers of the same oil. The dielectric strength of oil-impregnated paper insulation is defined by the strength of these small oil spaces. For, on the one hand the dielectric strength of the oil is lower than that of the paper fibres, and on the other hand, also the permittivity of the oil is lower, so that the oil takes the greater share of the electric stress. The dielectric strength of the oil-paper combination depends mainly on how perfect was the impregnation of the paper with oil, and also how successful was the elimination of the effects (of moisture and gases) which reduce the strength of the insulation. The dielectric strength of oil-

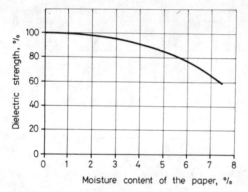

Fig. 5.71. Effect of moisture content on the dielectric strength of oil-impregnated paper at room temperature

impregnated paper insulation of new transformers is 200 to 400 kV/cm, and its permittivity about 3.5.

The dielectric strength of oil-impregnated paper insulation is affected by the moisture content mainly in the hot condition. At room temperature the breakdown strength, up to 3% moisture content, is practically constant, and even at 6% water content, it is still around 80% of the original value (see Fig. 5.71) [103]. At higher temperature, the situation is much worse. In Fig. 5.72 the breakdown voltage of oil-impregnated paper samples is shown as a function of moisture content, at 80 °C. It can be seen that, when only the mechanism of dielectric breakdown acts (e.g. under the effect of a lightning impulse voltage) the breakdown strength drops to about 50% of its original value at 6% moisture content, whereas in the case of combined thermal and electrical breakdown (power frequency stresses) the breakdown strength is only 15% of the original at the same moisture content [120].

In addition to the dielectric strength, the conductivity of the system is also influenced by the moisture content. The conductivity of the oil–paper system increases slightly between 0 and 1% moisture content and considerably above 1% moisture content; so the measured value of insulation resistance indicates the moisture content. The accumulation of aging products has a lesser effect than the moisture content on resistivity.

Both the moisture content and aging have a considerable effect on the dielectric polarization of oil–paper insulation. By observing the dielectric polarization spectrum of oil–paper insulation systems, the effect of moisture and of aging can be segregated, i.e. it can be determined whether the deterioration of insulation is due to an excessive moisture content or to aging of the insulation.

The moisture content also influences the resistance of oil–paper insulation against partial discharges. As proved by experiments, no considerable changes in

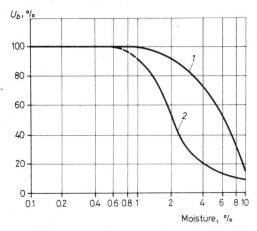

Fig. 5.72. Breakdown voltage of oil-impregnated paper samples as a function of moisture content at 80 °C [24]: *1* — in the case of purely dielectric breakdown mechanism; *2* — in the case of combined thermal and dielectric breakdown mechanism

the behaviour of oil–paper insulation under the effect of partial discharges can be observed up to 3% moisture content of the paper.

A further consequence of moisture is the accelerated aging of insulation. In the two-component oil–paper insulation system the effects of thermal stresses are determined by the aging of the paper, and aging, as shown in the discussion of the properties of the paper, is highly promoted by the presence of moisture.

Both in order to maintain the dielectric strength at a given level and to reduce the liability of the transformer to aging, the moisture content in the oil–paper insulation system should be maintained below a given level, preferably below 0.5% in the paper and below 20 ppm in the oil. However, in the three-component oil–paper–water system a definite water content may be present at a given temperature both in the paper and in the oil; if the state of equilibrium is upset, water starts to migrate from the oil into the paper or from the paper into the oil. The behaviours of the two substances are different as regards water absorption. With increasing temperature, the moisture absorbing capability of the oil increases and that of paper decreases, therefore the distribution of the overall quantity of water in a transformer at a lower temperature will differ from that developing at a higher temperature. The conditions of humidity equilibrium are illustrated by the isothermal curves of Fig. 5.73. According to the curves, e.g. at 90°C and with an oil of 20 ppm water content, a state of equilibrium prevails (i.e. there is no migration of moisture between oil and paper) if the moisture content of the paper is 0.8%. It is difficult to judge the conditions purely on the basis of Fig. 5.73 as the water content of the oil is usually expressed in ppm and that of the paper as a percentage. Water migration will, therefore, be illustrated by an example.

Suppose the quantity of oil in a transformer is 50 000 kg and that of the paper (and pressboard) is altogether 5000 kg. At the time the oil was introduced into the

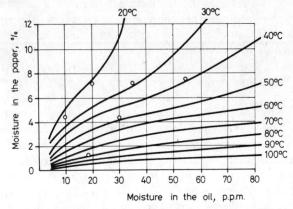

Fig. 5.73. States of equilibrium of moisture contents of oil and paper at different temperatures

transformer at 20 °C, the moisture content of the paper was 0.5% (25 kg) and that of the oil was 10 ppm (0.5 kg). Thus, by themselves both of the two components contain a sufficiently small amount of water. According to Fig. 5.73, however, no equilibrium is possible at such moisture content at 20 °C. If it is assumed that the overall quantity of water in the transformer does not change, then with the given weight conditions the relation between moisture contents (in % and ppm) in the paper and in the oil are indicated by line *a* of Fig. 5.74. The state of equilibrium

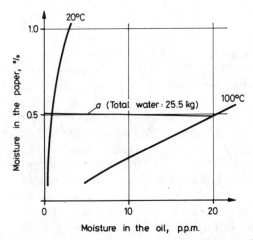

Fig. 5.74. States of equilibrium of moisture contents of oil and paper at a given moisture content

293

occurs where the isotherm pertaining to the given temperature is intersected by line *a*. Hence in our case, at 20 °C, the moisture content of the paper hardly changes, increasing from 0.5% to 0.509%, whereas the water content of the oil drops from 10 ppm to 1 ppm. Quantitatively, this means that 450 g of the water present in the oil will migrate into the paper, so that 50 g of water will remain in the oil, and the water content of the paper increases to 25.45 kg. When the transformer is loaded, the temperature of the insulation system may rise to 100 °C. The state of equilibrium will then be given by the intersection point of line *a* with the 100 °C isotherm of Fig. 5.74: the moisture content of the oil will be 20 ppm and that of the paper 0.49%. Such a distribution of water between the oil and the paper will cause no problem, since a water content of 20 ppm in the oil is of no concern.

The situation will be much worse if the paper can absorb moisture through the oil. For example, if the paper of the transformer just dealt with in the previous example absorbs a further 225 kg water, whereby its water content increases to 250 kg or 5%, then the 10 ppm moisture of the oil can just maintain equilibrium at 20 °C with this water content. The 5% moisture content in the paper is rather hazardous even in itself reducing the dielectric strength of the paper to an inadmissible degree. Suppose now that a load is imposed on the transformer and its temperature rises to 100 °C. At this temperature, in the state of equilibrium of the overall 250.5 kg of water 13 kg (260 ppm) will be contained in the oil and 237.5 kg (4.75%) will remain in the paper, i.e. 12.5 kg water will migrate from the paper into the oil. Thus, in addition to the reduction of dielectric strength of the paper, the dielectric strength of the oil will also drop, due to the high moisture content, well below 100 kV/cm, which can by no means be tolerated in high-voltage transformers.

Moreover, a sudden migration of moisture may be associated with a further hazardous consequence, viz. when water is separated from the oil in the form of droplets. This is why a sudden load increase may lead to a breakdown (diffusion catastrophe).

Similar harmful effects are to be attributed to the air or other gases present in the oil–paper insulation system. If perfect degasification of the paper is neglected, gas bubbles and voids remain in the insulation. Bakelite requires special attention in this respect, because the presence of such residual gas cavities is almost unavoidable. The dielectric strength of the paper decreases due to these gas bubbles and, since the permittivity of air is lower than that of paper, a considerably higher electric stress will arise in the voids than in the paper. The flashovers and discharges occurring in the voids do not cause breakdown of the dielectric strength of the entire insulation, and are therefore termed partial discharges. Under the effect of partial discharges, however, the local deterioration of insulation may propagate and, after a shorter or longer period of time, may lead to the total breakdown of insulation.

Another source of partial discharges is represented by the gas bubbles developing under the effect of high field strengths, from the moisture present in the paper. Such gas bubbles may appear, depending on the moisture content, at a field strength as low as 10 kV cm^{-1}, but in the case of paper very carefully dried out the critical field strength rises to extraordinarily high levels. Under the effect of moisture, partial discharge may occur in insulations of geometry and quality which would otherwise exclude the possibility of their occurrence. The process, when once started, makes itself independent of the initiating conditions, and starts to spread owing to its physical and chemical effects [25].

It follows from the above that, for the sake of safe operation and long service life of transformers, moisture and gas contents should be kept below admissible limits as early as possible in the course of manufacture, and the degradation of this state during operation should be prevented or, at least, the deterioration taking place through the long years of service should be confined within reasonable limits.

5.3.4 The pressboard

Among the cellulose-based insulating materials, beside the soft paper dealt with so far, pressboard has acquired the most important role. Insulating cylinders and moulded pieces (e.g. angle rings, insulating discs with long creepage paths to bring out the leads through the tank wall, stress rings, etc.) are made of it. The kind of special quality pressboard, the so-called tranformerboard [93] generally used in large power transformers is made of high-grade sulphate cellulose (from scrap paper and rag). The most suitable raw material is pine wood. This kind of pressboard combines the excellent electrical properties of soft paper with the outstanding mechanical properties of hard-paper insulation. It can easily be dried out and impregnated.

Pressboard is produced by compressing several thin paper layers in a wet condition, without using any cementing or bonding material. The thinner the individual layers the better will be the quality of the pressboard. (For example, in "transformerboard" there are about 35 layers of 30 μm thick paper sheets per mm.)

According to production technology, two kinds of pressboard are distinguished. Calendered pressboard is obtained by pressing, followed by drying without pressure, then by rolling under high pressure. Density of the end product is 1.15 to 1.3 $kg\,dm^{-3}$. The precompressed transformerboard is dehydrated, solidified and dried under a special hot press. Its density is about 1.25 $kg\,dm^{-3}$.

The dielectric strength of pressboard impregnated with oil under a pressure of 2.7×10^{-2} mbar lies in the range 200 to 250 $kV\,cm^{-1}$, and even higher in thicknesses of 1 to 2 mm. The field strength for inception of a partial discharge is about 70 to 80% of this value. The loss factor tan δ of pressboard at 20 °C is about 40×10^{-4}, and at 130 °C about 70×10^{-4}. Its permittivity is 4.4. to 4.5. The surface field strength permissible for pressboard, depending on the creepage distance, falls in the range of 5 to 30 $kV\,cm^{-1}$ (the higher values apply to shorter distances). The permissible surface field strength for a section 15 cm long is about 10 $kV\,cm^{-1}$. At this field strength applying 50 Hz voltage, no partial discharge yet appears.

Regarding the hazards associated with partial discharges, a distinction should be made between those appearing on the surface of the pressboard and those taking place inside the insulation. The former appear in the measurements with a much higher intensity, but are less dangerous, since their time to breakdown is considerably longer than that of discharges of other kinds. The latter are of lower intensity, but are detrimental to the insulation and may, sooner or later, cause a breakdown. This difference in the behaviour of the two kinds of discharge fully explains the experience that transformers exhibiting a lower partial discharge level during tests soon become defective in service, whereas those of much higher partial discharge levels operate satisfactory through many years of service.

The mechanical properties of pressboard are also excellent. The tensile strength of a precompressed pressboard of 1.25 kg dm^{-3} density, depending on the sheet thickness and direction of rolling, lies in the range of 95 to 125 N mm^{-2} and its elongation is 2.3 to 4.6%. More important, however, is the behaviour of the pressboard under compression. The permanent deformation of pressboard is small both in the cases of static and dynamic loads (such as occurring in the course of assembly, and as imposed by short-circuit stresses, respectively), and its elasticity is high. In this respect, however, there is a considerable difference between calendered and precompressed pressboards. During an experiment the permanent deformation of a calendered pressboard after 10 consecutive applications of a dynamic load was found to have increased to a value higher than 6%, and its modulus of elasticity was about 500 N mm^{-2}. For a precompressed pressboard values of 1% and 1750 N mm^{-2}, respectively, were measured after applying the same load as in the former case. These figures can be improved further by means of treatment and calibration of the surface.

The aging properties of the pressboard are also favourable. The degree of polymerization of a precompressed transformerboard, after curing the sample at 130 °C for 200 hours, was found to drop from 1700 to 700, which is still an acceptable value.

Water present in the oil has the same effect on pressboard as on soft paper. All that has been described in relation to soft paper applies to pressboard insulation as well.

5.3.5 Vacuum treatment of oil transformers

To remove the harmful water content from the paper insulation of new transformers a drying process termed vacuum treatment is applied [10]. The windings mounted on the core are first subjected to pre-drying, consisting of heating the insulation of the transformer to 97 to 100 °C by means of forced circulating air at atmospheric pressure. During this process the core temperature rises to about 85 °C. This is essential because, as proved by experiments [24], the temperature distribution within the mass of paper has an important role in the quality of drying; the situation is optimal when the insulation is so heated as to attain uniform temperature outside and inside alike. By the pre-drying process, however, moisture can be removed only from the surface and from the layers near to the surface. The moisture can escape from inner layers through the outer layers that have been dried out already, but these layers, being hygroscopic, will tend to bind the moisture passing through and at least part of it will be absorbed again by the paper layers that had already been dried out. Thereby the drying process will be retarded. Thus, by this method, using hot air alone, the moisture could not be removed from the winding in a finite length of time.

The drying process can be accelerated by reducing the pressure of the external medium considerably—and together with it the partial pressure of water vapour present in the air. Therefore, for the period of coarse drying—which follows predrying, the pressure is decreased to as low as 1.3 to 2.6×10^{-3} bar, and then in the course of fine drying, further down to 1.3×10^{-5} bar. Practice has shown that a vacuum of this magnitude is sufficient even for drying of the thick paper insulation of large transformers. During the 10 to 16 days period of coarse and fine drying, 300 to 450 dm^3 of water escapes from the paper mass weighing 10 to 12 metric tons in a large power transformer. Thereafter, the mechanical structures of the transformer are fixed in their final position and locked, the transformer is put into its tank, and then subjected to a process termed final drying in a vacuum of about 1.3×10^{-5} bar. Finally, the transformer is filled with previously degassed and dried transformer oil.

The time of drying can be considerably shortened further, and the process made more effective by applying a method which has become popular in the past decade termed the vapour phase process [104]. This consists of using, instead of air, the vapour of some kind of light oil for heating the transformer. According to experiences gained so far, kerosene (a hydrocarbon compound) has proved to be the best for this purpose. The advantages of the vapour phase treatment may be listed as follows. The drying process takes place in an atmosphere practically free of oxygen, so the insulation may be heated to a much higher temperature (up to 130 °C) than usual, which will considerably accelerate drying. The use of vapour as the heat transfer medium has a further advantage in condensing on the surface of the insulation, whereby its condensation heat is transferred to the latter. Further, this condensation takes place also on the inner winding surfaces of poor accessibility in the case of heating by hot air, and the maximum intensity of condensation occurs on

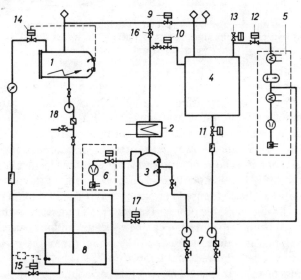

Fig. 5.75. Schematic drawing of vapour phase drying equipment: *1* — evaporator; *2* — condenser; *3* — collecting container; *4* — vacuum oven; *5* — main vacuum pump; *6* — vent pump; *7* — condensate pump; *8* — kerosene tank; *9* — steam inlet valve; *10* — steam return valve; *11* — condensate valve; *12* — vacuum valve; *13* — ventilation valve; *14* — filling valve; *15* — outlet valve; *16* — steam bypass valve; *17* — changeover valve; *18* — oil and water drain pump

297

the coldest parts of surfaces. Correspondingly, the heat transfer surface is larger than in the conventional process. Finally, the vapour pressure of the heat transfer medium being lower than that of water is also of importance permitting continuous extraction of moisture from the insulating material, i.e. from the beginning of heating until the end of fine vacuum drying.

A schematic arrangement for vapour phase drying equipment is shown in Fig. 5.75. The drying process can be divided into four phases (Fig. 5.76).

In the preparatory phase (I) the air is removed from the system, kerosene vapour is sucked in and heated to the required temperature. In the oven (in the space around the transformer) the pressure is gradually decreased to 6.7×10^{-3} bar. (There is no kerosene vapour in the oven yet.) The preparatory phase lasts 2 to 4 hours, depending on the size of the oven and the transformer.

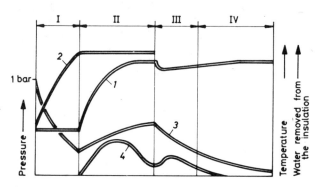

Fig. 5.76. Process of drying: I — preparatory period; II — heating up period; III — pressure reduction period; IV — fine drying; 1 — temperature of trans-former; 2 — temperature of kerosene; 3 — pressure; 4 — water discharged per unit of time

Kerosene vapour is allowed to flow into the oven in the heating phase (II). The vapour condenses on the transformer and its latent heat is transferred to the insulation, thereby continuously warming it. With increasing temperature, the pressure of the kerosene vapour in the oven and that of the water in the insulation become steadily higher. Since, however, the pressure of water vapour is considerably higher than that of kerosene vapour, moisture present in the insulation evaporates already at a relatively low temperature, at the beginning of the phase, and escapes from the insulation. The condensed kerosene vapour residual air and water vapour leave the oven through suitable piping. The water is condensed so that the quantity of water removed from the insulation can be collected and measured. The quantity of water leaving per unit time increases at the beginning of the process. When the transformer attains the required final temperature, it is held at this temperature to permit remnants of water to escape. During this period the quantity of water leaving per unit time starts to decrease. Depending on the size of the transformer, the duration of the heating phase is 20 to 30 hours.

In the pressure reduction phase (III) the major part of the kerosene taken up by the insulation is again evaporated, precipitated in the condenser and led back into the vaporizer. The duration of the pressure reduction phase is 3 to 5 hours, and lasts until the pressure in the oven drops to 20 to 27×10^{-3} bar. In the course of this phase, first, the temperature of the transformer drops slightly, then it slowly increases again, while the rate of water extraction from the insulation initially increases, and then decreases.

The last phase of drying corresponds to the fine drying of the traditional process, during which the pressure is reduced to 1.3×10^{-5} bar. In this phase, lasting 24 to 48 hours, the residual moisture is removed from the insulation.

The time saving obtainable by vapour phase drying is illustrated by Fig. 5.77. As can be seen, the drying of a 400 kV 200 MVA transformer is requiring about 15 days

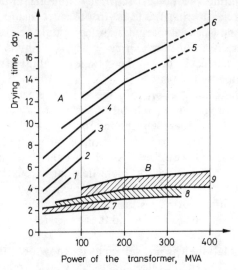

Fig. 5.77. Drying times: A — vacuum drying with hot air; $1 = 90$ kV; $2 = 110$ kV; $3 = 130$ kV; $4 = 150$ kV; $5 = 220$ kV; $6 = 400$ kV; B — drying with kerosene; $7 = 150$ kV; $8 = 220$ kV; $9 = 400$ kV

with the traditional method reduced to about 4 days when the vapour phase process is employed. Transformers of very high power ratings (300 to 400 MVA) are today dried exclusively by the vapour phase method.

In addition to the advantages pointed out above, a further favourable feature of the vapour phase process lies in the cleaning effect of vapour condensing in the heating phase, by which any dust and other contamination that may have been deposited on the windings are rinsed away.

The transformer may take up moisture from the ambient atmosphere during service, so the oil of the transformer should be prevented from coming into contact with the external air. The original method for achieving this aim was to connect the air space of the conservator with the external atmosphere through a dehydrating breather and an oil trap. A newer and more effective method consists of circulating the air contained in the conservator in a closed-circuit piping system and removing the moisture from this air by means of cooling or even by freezing. (Due to the varying oil volume caused by temperature variations, there is a connection between the air in the condenser and the outer atmosphere through the oil trap and breather.) Contact with external air can be fully eliminated if a nitrogen atmosphere is maintained in the conservator above the oil. A resilient membrane laid on the oil surface can fully separate the latter from the air. The characteristic properties of these methods and designs applied for sealing off the oil are covered in detail in the Section 7.

5.3.6 Design of internal insulation

The primary requirement to be satisfied in designing insulation is to make it capable of withstanding the power frequency, lighting impulse and switching impulse stresses applied during the tests. In order to ensure good utilization of insulation, approximately equal margins of safety should be applied in all the three kinds of imposed stresses. This aim can be achieved mainly by properly designed construction, and by correct setting of the voltage distribution within each winding and between individual windings.

In recent years the general effort to achieve a more economical utilization of insulation and to improve service reliability has raised a new requirement to be satisfied by transformer designers. The reduction introduced in the power frequency test voltages and the simultaneous introduction of partial discharge measurements have made it necessary to consider also the partial discharges in the design. Correspondingly, the occurrence of high field at strength spots within the transformer should be avoided by constructional measures, as they could give rise to internal discharges, either during testing or in subsequent service, damaging the insulation and decreasing its useful life and the service reliability of the transformer [117].

For designing insulation, it is not sufficient to know only the properties of the substances to be used. For on the one hand, the dielectric strength data of materials are based on assuming a surrounding homogeneous electric field, one scarcely occurring within a transformer. On the other hand, the stress arising in the insulating materials of a transformer is more complex and involved than that imposed during the standard testing of materials: e.g., an insulating plate may be exposed to a stress arising across the material and to a surface (creepage) stress. It is even less possible to design an insulating structure to withstand partial discharges knowing only the properties of the material. The dielectric strength and partial discharge inception and extinction voltages of insulation are considerably influenced by the manufacturing technology of drying-out and impregnation, and the precision and reliability of production. Pointed tips and insufficiently rounded-off edges may become sources of partial discharges under the effect of high field strengths present in their vicinity.

From the above it follows that the designing of insulation should be based, primarily, on the results of experiments carried out—under laboratory conditions closely simulating real conditions—on scale models, winding elements, and typical structural parts. If the measurements are performed on a sufficient number of specimens, then expected threshold voltages of discharge and dielectric strength can be determined by evaluating the results and applying the methods of mathematical

300

statistics. If the scatter of the scale model measurements is sufficiently small, the magnitude of voltage allowed to be imposed on the insulation is that at which the probability of breakdown or (in the case of power frequency stress) of the inception of partial discharge is lower than 1%. For larger scatters, or when the test is performed on a limited number of samples, or on just one 1:1 scale model, the applied test voltage should not be higher than at most 65 to 70% of the minimum breakdown voltage i.e. the design should be carried out with a safety factor of about 1.4 to 1.5.

However, it is not always necessary to determine the inception and extinction voltages of partial discharge by using a scale model to simulate real conditions. In a transformer, there are spots even among those exposed to such voltage stresses where due to the relatively simple geometry of the structure, the electric field can be determined in a simpler way with an accuracy satisfying practical requirements. Such simpler methods are:

(a) field calculation based on some simple geometrical configuration (cylinder-plane, cylinder-cylinder, ring-cylinder, etc. electrodes) closely approaching the arrangements of the original structure;

(b) digital computation of the field by a numerical method;

(c) measurement on a plane model consisting of a conducting (silver-content) paint applied to a graphite paper base (on a simplified figure obtained by simulating the real configuration over a plane);

(d) measurement in an electrolytic tank.

After determining the field strength at critical points by one of the methods listed above and from knowledge of the permissible value of local field strength, the structural part concerned can be checked with respect to partial discharges.

When designing a new construction in the course of transformer development work, it is generally necessary to assemble a model of certain especially critical parts (usually to the scale of 1:1) in order to determine by measurement performed on it the inception and extinction voltages of partial discharges and the dielectric strength of the part concerned. It may be necessary as well to determine the field configuration by calculation (or by plane-model measurement or, perhaps, by electrolytic tank experiment) for checking the construction on such a basis. Since, however, the majority of insulations and insulating structures are identical in all transformers, it is possible to give generally applicable design data for such parts. These insulations, according to their location and purpose, may be classified in four categories.

The interturn insulation in a transformer winding serves for separating from each other the turns energized to different potentials. Generally, a higher voltage than that has to be withstood by the winding insulations or inter-section insulations (those between winding sections and layers). The highest electrical stress is imposed on the main insulation separating the windings of different voltage ratings and the windings from earthed parts. Further insulations that require designing are those provided on the inner connections of windings and winding ends. In the following sections dielectric strength data of these insulations are given, together with diagrams directly utilizable in design work and general rules for insulation design.

5.3.6.1 Interturn insulation

In the overwhelming majority of cases, copper or aluminium conductors of transformers are insulated with Kraft paper, though the use of wires with enamel insulation is gaining ground. The paper strip is applied overlapped over the conductor. The layer thickness of interturn insulation is between 0.2 and 1.5 mm. In small series-capacitance windings, i.e. in the case of uneven lightning impulse voltage distribution, the designing is generally governed by the expected lightning

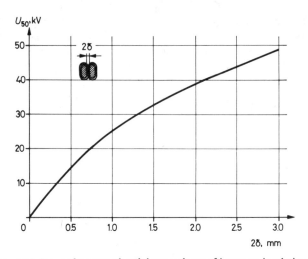

Fig. 5.78. Power frequency breakdown voltage of interturn insulation

impulse stresses. In windings with high series capacitance the distribution of lightning impulse voltage is uniform but, due to interleaving, a multiple of the power frequency interturn voltage (which would result from uniform voltage distribution) may arise between geometrically adjacent turns, so that in many cases the power frequency stress will be decisive in the designing of insulation.

The power frequency breakdown voltage of oil-impregnated Kraft paper interturn insulation is shown in Fig. 5.78 as a function of the thickness of the insulation layer between adjacent windings, i.e., of double the thickness of insulation on a conductor. The values of breakdown lightning impulse voltage are given in Fig. 5.79. The voltage of Figs 5.78 and 5.79 are minimum values resulting from a series of experiments. It should be noted that these breakdown voltages refer to turn insulations manufactured and treated (dried, oil-impregnated) under usual conditions and intended for use in large power transformers. Under laboratory conditions, treated and tested in small tanks, considerably higher breakdown voltages even reaching double those shown can be obtained. Experience has shown that it is not advisable to base the design on higher voltage values than those indicated in the diagram. It is appropriate to select the thickness of interturn insulation so that, in the course of the voltage test, the voltage imposed on interturn insulation should be less than 70% of the value given in the diagrams. The

302

permissible magnitude of switching impulse waves is 1.35 to 1.5 times the peak value of power frequency stress.

Up to an insulation layer of about 1.2 mm thickness (i.e. to 0.6 mm thick insulation on one conductor), the inception voltage of partial discharges is 85 to 90% of the breakdown voltage, and the extinction voltage is 75 to 80% of the same. For insulation thicknesses exceeding 1.2 mm, a wider scatter of inception and extinction voltages can be observed and the extinction voltage decreases. According to the above, in the case of insulations of thicknesses below 1.2 mm, 70% of the

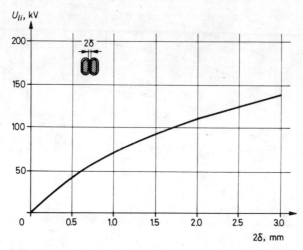

Fig. 5.79. Lightning impulse breakdown voltage of interturn insulation

breakdown voltages are lower than the extinction voltage of partial discharges (up to this thickness the permissible electric stress is 16 kV mm^{-1}). Also with larger thicknesses, 70% of the breakdown voltages are below the inception voltage. Thus, the value of 70% satisfies the conditions of proper design for partial discharges.

5.3.6.2 Inter-section insulation

A 4 to 12 mm oil duct is generally provided between sections of the disc-type windings of transformers to ensure the required dissipation of heat. Due to the creepage distance formed by the pressboard spacers placed between the sections, the voltage stress permitted to arise in the duct is much lower than that tolerable for an oil layer of equal thickness. From a knowledge of voltage imposed on the duct, the required width of the duct can be determined from Fig. 5.80 for the power frequency stress and from Fig. 5.81 for the lightning impulse stress. In the course of the voltage test, the stress imposed on the insulation should not be higher than 70% of the value taken from the diagrams. As in the case of interturn insulation, the permissible switching impulse stresses in the inter-section insulation are 1.35 to 1.5 times the peak value of the power frequency stress.

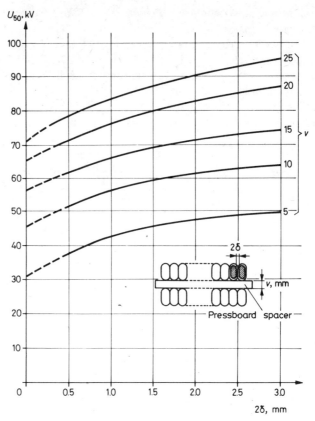

Fig. 5.80. Power frequency breakdown voltage of inter-section insulation

Assuming average conditions of manufacturing, the inception and extinction voltages of partial discharges show a much wider scatter for inter-section insulation than for interturn insulation. As proved by scale model measurements, the inception voltage of partial discharges—depending on the interturn insulation and thickness of the pressboard spacer—is 50 to 80% and the extinction voltage 35 to 65% of the breakdown voltage. An interesting phenomenon is that above a certain width of insulation the inception and extinction voltages do not increase with growing width but decrease. Nevertheless, for reasons of economy, it is not advisable to reduce the voltage stress permissible in the voltage test below 70% of the breakdown voltage. The observed discharges—generally occurring during application of the test voltage—are mainly surface discharges and are no danger to the insulation, as has already been pointed out in Section 5.3.4

The scatter in the value of inception and extinction voltages of partial discharges can be reduced considerably and the voltages themselves increased by better matching of the spacers to the electric field, by rounding off their sharp edges. In the case of transformers of voltage ratings above 300 kV this requirement is especially important.

In disc-type windings composed of wires having smaller cross-sectional areas, an alternative arrangement is used, for the sake of better utilization of space, in which the oil duct and pressboard spacers are replaced by a pressboard disc in every other inter-section spacing. When designing the insulation between adjacent sections, the overall thickness of interturn insulation and pressboard disc should be taken into

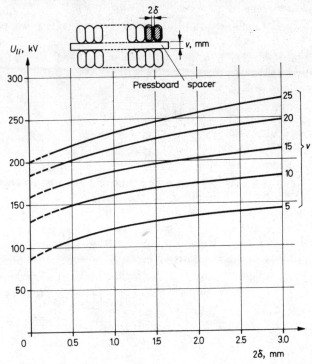

Fig. 5.81. Lightning impulse breakdown voltage of inter-section insulation

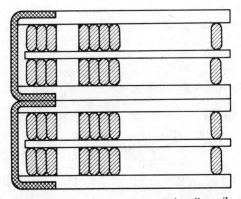

Fig. 5.82. Reinforcement of two-section disc-coils

account. The pressboard disc extends beyond the winding edges by 4 to 5 mm, providing for the required creepage distance between the sections.

Protection against lightning impulse stresses arising between adjacent sections at the beginning of windings can be afforded, as shown in Fig. 5.82, by reinforcing the insulation of disc-coils (two sections) situated at the beginning third part of the winding: the dielectric strength is increased by U-shaped segments of pressboard pulled over the disc-coils. In this case, the permissible value of ligthning impulse voltage that may be applied per mm thickness of overall solid insulation between copper conductors is 30 kV.

Along the disc and spacers, a field strength of 50 to 60 kV cm^{-1} (at power frequency) may be permitted to arise at the surface of the oil.

5.3.6.3 Main insulation

Between cylindrical windings mounted on a common limb but having different voltage ratings, the layer insulation generally consists of oil and some kind of solid insulation. Up to about 35 kV rating, the oil layer is divided into ducts of 3 to 15 mm widths by paper-based phenol tubes, whereas above 35 kV the tubes are made of pressboard. Except for the winding ends, the electric field between cylindrical windings may be regarded, with good approximation, as homogeneous, so that the voltages appearing across the layers can easily be determined by calculation. Namely, in such a case, the electric fields imposed on the various layers will be inversely proportional to their permittivities. If the permittivities of the layers of thicknesses $d_1, d_2, \ldots, d_n$ are $\varepsilon_1, \varepsilon_2, \ldots \varepsilon_n$, respectively, then on applying voltage U to the entire width of insulation, the field strength in the part of thickness d_1 will be

$$E_1 = \frac{U}{\varepsilon_1 \left(\dfrac{d_1}{\varepsilon_1} + \dfrac{d_2}{\varepsilon_2} + \ldots \dfrac{d_n}{\varepsilon_n} \right)}. \tag{5.85}$$

In the other layers:

$$E_2 = E_1 \frac{\varepsilon_1}{\varepsilon_2}, \qquad E_3 = E_1 \frac{\varepsilon_1}{\varepsilon_3}, \qquad \ldots \qquad E_n = E_1 \frac{\varepsilon_1}{\varepsilon_n}. \tag{5.86}$$

Since the permittivity of the oil ($\varepsilon_0 = 2.2$) is lower than that of paper-based phenol ($\varepsilon_{pph} = 5$ to 6) or pressboard ($\varepsilon_{pb} = 4$ to 5), the field strength in the oil will always be higher than that in the solid insulation. In addition, the dielectric strength of the oil is lower than that of the solid insulation, so that the stress imposed on the oil will by all means be the higher. Under the effect of an excessively high electric stress, first, sporadic low-intensity discharges of short duration occur in the oil, the frequency of which gradually increases, then the process goes over into intense partial discharges, leading to total breakdown of one of the oil ducts. Thereby, the stress imposed on the remaining parts of the insulation grows, causing further breakdowns in the oil, whereby an increasingly higher stress is imposed on the solid insulation. All oil ducts having suffered a breakdown, the solid insulation will have to withstand the full voltage. Consequently, the two basic principles of correct designing of the main insulation are as follows:

306

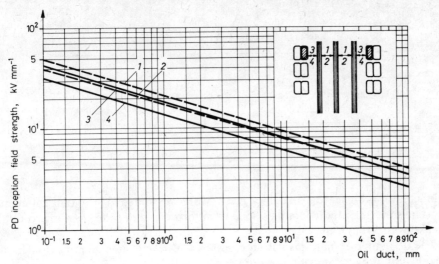

Fig. 5.83. Inception voltages of partial discharges in oil ducts: *1* — degassed oil, insulated electrodes; *2* — gas-saturated oil, insulated electrodes; *3* — degassed oil, non-insulated electrodes; *4* — gas-saturated oil, non-insulated electrodes

(a) The field strength permitted to occur in the oil ducts, assuming a new oil filling free from moisture and gas, should be sufficiently low to prevent the phenomenon described above from starting even at the test voltage level.

(b) After a shorter or longer time of service, a gradual drop in the dielectric strength of oil ducts, due to the contamination of the oil, should be reckoned with. Therefore, the solid insulation should be designed so that it remains capable of withstanding the full voltage in the case of breakdown of all oil ducts.

Based on the results of experiments, the lines of Fig. 5.83 give, for four different cases, the field strength values at which the partial discharge initiating the described breakdown process of oil ducts starts in an oil duct of given width [93]. The four cases are as follows:

1. Oil duct confined by solid insulation, e.g. by pressboard cylinders; degassed oil.
2. Oil duct confined by solid insulation; gas-saturated oil.
3. Oil duct adjacent to windings; degassed oil. (Interturn insulation is left out of consideration, constituting an additional factor of safety.)
4. Oil duct adjacent to windings; gas-saturated oil.

In addition, there are radial spacing sticks present in the oil ducts, where a creepage stress thus also arises. The spacing sticks may be stressed even by 60 to 80 kV cm^{-1} during the power frequency tests, the sections concerned being short.

Up to 35 kV rating, generally, a paper-based phenol cylinder is accomodated between the H.V. and L.V. windings. Insulation spacings established in practice are given in Table 5.7.

At higher voltage several pressboard cylinders are applied. Dividing the oil into several ducts of small widths is favourable, since the dielectric strength of the oil in a

Table 5.7

Interwinding insulation at 35 kV rated voltage

Rated voltage, kV	6	10	20	30	35
Distance between windings (oil + solid insulation), mm	6	8	13	18	20.5
Thickness of paper based phenol cylinder, mm	2.5	3.0	4.0	6.0	7

thinner layer is higher than in a thick layer. Moreover, the solid insulating cylinders prevent impurities and contaminations from forming long chains leading to total breakdown through the oil.

The designing of main insulation consisting of several pressboard cylinders is demonstrated by means of an example.

Example 5.9. Let the test voltage of a 220 kV transformer be $U_t = 395$ kV. As explained above, the solid insulation between the H.V. and L.V. windings is required to withstand in itself the entire test voltage. Dielectric strength of the pressboard cylinder used is $E_p = 250$ kV cm^{-1}, of which 70% or 175 kV cm^{-1} is permitted to arise in the transformer, so the overall thickness of the pressboard cylinders is

$$d_p = \frac{U_t}{E_p} = \frac{395}{175} \approx 2.3 \text{ cm} = 23 \text{ mm}.$$

The maximum field strength permissible for the oil depends on the number of oil ducts selected. The narrower each duct, the higher the field strength which may be allowed, according to Fig. 5.83. Correspondingly, the smaller will be the overall width of the oil ducts, the distance between windings (i.e. the leakage field gap). From the point of view of cooling, the role of the oil duct adjacent to the winding is decisive, therefore its width should be selected to be 6 to 12 mm in the case of larger transformers. Let the width of the oil duct adjacent to the winding be 6 mm. According to line 3 of Fig. 5.83, in a 6 mm wide oil duct situated adjacent to the winding, a partial discharge starts at about 93 kV cm^{-1}. For determining the overall width of the oil ducts 70% of this value, i.e. 65 kV cm^{-1} is reckoned with. From relations (5.85) and (5.86), the thickness of the entire oil layer is

$$d_0 = \frac{U_t}{E_0} - d_p \frac{\varepsilon_0}{\varepsilon_o},$$

where E_0 is the field strength permissible for the oil, ε_0 is the permittivity of the oil and ε_p that of the pressboard. In the present case $U_t = 395$ kV, $E_0 = $ kV cm^{-1}, $d_p = 2$ cm, $\varepsilon_0 = 2.2$, and $\varepsilon_p = 4.4$. With these values

$$d_0 = \frac{395}{65} - 2.3 \frac{2.2}{4.4} \approx 5 \text{ cm} = 50 \text{ mm}.$$

Thus, the overall width of the leakage field gap will be

$$d = d_p + d_0 = 7.3 \text{ cm} = 73 \text{ mm}.$$

One possible arrangement of this leakage field gap is shown in Fig. 5.84. The thickness of the cylinder supporting the H.V. winding is selected to be 7 mm for mechanical reasons, the remaining pressboard cylinders are 4 mm thick (the dielectric strength of a thinner pressboard cylinder is higher than that of one thicker). The width of oil ducts adjacent to the windings is 6 mm as mentioned and the remaining oil ducts are 9 mm and 10 mm wide, respectively.

The main insulation designed for power frequency voltage has to be checked for lightning impulse voltage and switching impulse wave as well. A lightning impulse voltage 1.7 times the peak value of the 50 Hz stress, and a switching impulse wave 1.4 times the same, are permissible.

In designing the insulation between the winding closest to the core and the core itself the role of mechanical strength is decisive. This winding is usually the one of lower voltage rating, therefore the electric stress imposed on it is not excessively high, whereas the short-circuit forces which tend to compress the inner winding may

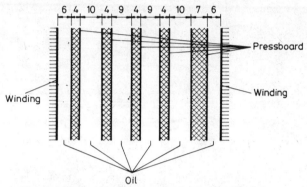

Fig. 5.84. Inter-winding insulation of a 220 kV transformer

reach substantial levels. For this reason, the innermost insulating cylinder (the core-to-winding cylinder) is usually a 6 to 10 mm thick bakelite tube for up to 35 kV voltage rating. At voltages of 60 kV and above, a shielding cylinder surrounds the limb, and the insulation has to be designed as in the case of inter-winding insulation.

5.3.6.4 End insulation

Insulation of winding ends from each other and from the yoke is the task of the so-called end-insulation. The stress imposed on the end insulation is partly like that of the main insulation, but at the ends of the windings the field is not homogeneous, and therefore the permissible specific stress has to be reduced, i.e. the insulation distances have to be increased. Along the other edges of insulation the creepage distances are of importance, and the geometrical shape of insulation is determined mainly by the required creepage path. Its usual length in fuction of power frequency test voltage is given—up to 400 kV test voltage—in Fig. 5.85. When determining the creepage distance, it is common practice to multiply the paper or pressboard insulations covering the electrodes (copper conductors, iron core) by 10 to 20 times their thickness (dimension perpendicular to the field).

In transformers of lower voltage ratings, up to about 35 kV, the paper-based phenol cylinder between the windings extends beyond the latter, and an insulating disc is placed between the windings and the yoke (see Fig. 5.86). The extension (d_1) of

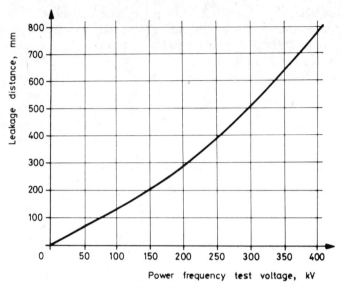

Fig. 5.85. Creepage distance required in end insulations as a function of power frequency test voltage

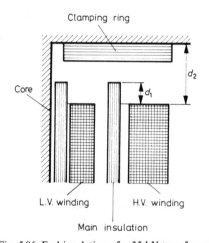

Fig. 5.86. End insulation of a 35 kV transformer

the paper-based phenol cylinder should be as long in mm, as the rated voltage of the transformer in kV, i.e. in the case of 20 kV transformers it should be about 20 mm. It is appropriate to choose the distance d_2 of the winding edge from the yoke three times longer than d_1.

The end insulation in transformers of 60 kV and higher ratings is more complicated. Here the arrangement used at lower voltages would result in excessively large spacings, therefore the creepage distance is increased by means of pressboard collars, termed angle rings interleaved with the cylinders. The angle

310

rings are fitted to the tubes placed between the windings, and have to be arranged so as to form extensions of the tubes. The higher the voltage rating, the more angle rings are used. The parts of angle rings bent over the windings (the "skirts") must not be arranged too close to each other, because experiments have shown that only about three times the distance of angle rings can be regarded as an effective creepage path between two angle rings. If the skirts are too close to each other, then flashovers will occur in their vicinity in spite of increasing their number.

In properly designed end insulation the surface of angle rings is, as far as possible, parallel with the equipotential lines, whereby a higher dielectric strength can be

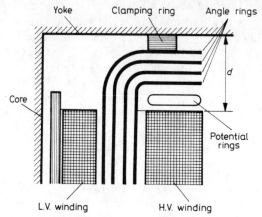

Fig. 5.87. End insulation of transformers above 60 kV

achieved, since a lower stress is imposed on the solid insulation as regards creepage distance. In correctly constructed end insulation the surface stress along the creepage path, caused by the power frequency test voltage can be kept below 10 kV cm^{-1} throughout. It is usual practice, in critical cases, to determine the field pattern expected to develop in the vicinity of end insulation by calculation, plane model measurement or an electrolytic tank experiment.

For angle ring applications, the values of distance d (of Fig. 5.87) are listed as a function of power frequency test voltage in Table 5.8. An end insulation designed according to these data will withstand the standard lightning impulse voltage test as well.

The field developing at the winding ends is favourably influenced by a potential ring placed on the uppermost section and connected to the potential of the first turn (see Fig. 5.87). In large transformers, in order to reduce the losses caused by the stray flux, an insulating ring provided with a conducting band or graphite coating is used for this purpose.

In uniformly insulated transformers both winding ends should have identical dielectric strength, whereas in non-uniformly insulated transformers only the line end of the winding has to be designed to withstand the test voltage of line terminals, while the neutral end need only withstand the considerably lower voltage specified

311

for the neutral point. Consequently, the distance between the line end of the winding and the yoke will be longer than that between the neutral end of the winding and the yoke. The insulation of the windings is designed to diminish in two or more steps from the line end toward the neutral point end.

The distance between winding ends and yoke and with it the limb height can be reduced by arranging the H.V. point at the centre of the winding. In effect, such a winding consists of two halves connected in parallel and wound in opposite direction around the limb, one half being situated on the upper half of the limb and the other half on its lower half. The inner ends of the two halves connected together at the centre of the limb constitute the line end of the winding, and the outer ends, i.e. those at the top and botton of the full winding, also connected together form the neutral points end or in autotransformers, the winding end connected to the other winding. Thus, only the entry at the centre of the limb has to be designed to the test voltage corresponding to the line voltage, whereas the insulation of the winding ends are determined by the test voltage of the neutral point or that of the L.V.

Table 5.8

Distance between winding and yoke as a function of power frequency test voltage

Power frequency test voltage U, kV	Distance of end insulation d, mm
140	90
185	100
230	115
275	130
360	170
395	190
630	380

Table 5.9

Distance between windings of different phases as a function of power frequency test voltage

Power frequency test voltage kV	Spacing between windings in the case of windings with the H.V. point at the end, mm	Spacing between windings in the case of windings with H.V. point at the centre, mm
140	40	40
185	55	55
230	65	65
275	80	80
360	125	110
395	145	120
630	300	200

winding of the autotransformer. In the case of windings with their H.V. point at the centre, the leakage field gap can also be reduced, since fewer angle rings are required at the winding ends. Thus, for very high voltage ratings, the designing of insulation between windings depends on the shape of end insulation. It should be noted that whether a winding has its H.V. point at the end or at the centre, the upper end insulation is larger than the lower end insulation, because of the presence of the clamping ring.

To increase the dielectric strength, pressboard barriers are placed into the oil duct located between windings of different phases. The structure of interphase insulation is the same as that of the main insulation, and the spacings to be maintained between phases have to be selected according to Table 5.9. The barriers, as in the case of angle rings, are preferably also arranged along the equipotential lines.

5.3.6.5 Examples for main and end insulation

The proper design of main insulation and winding ends are illustrated by a few examples. In Fig. 5.88. the upper end of windings mounted on one limb of a 750/420 kV single-phase auto-transformer is shown, with the schematic section of windings and end insulation. The 750 kV winding is centre tapped, so that the winding should be insulated to 420 kV from the yoke. In the diagram, the core is marked by letter C, the 15 kV tertiary winding by Ty, L is the 420 kV common winding, and H the 750 kV series winding of the autotransformer. The main insulation between core and winding Ty is a bakelite tube with a Kraft paper cylinder wound on it. The main insulation between windings Ty and L and that between windings L and H, are composed of a large number of thin-walled pressboard tubes, with oil ducts between them braced by axial sticks.

The electric stress imposed on the insulation between winding Ty and the core is low, whereas the short-circuit forces tending to compress the winding give rise to serious mechanical stresses in that part of the insulation. That is why a bakelite tube of high mechanical strength is employed here. On the other hand, between windings L and H the electrical stress is high, and the mechanical stress is relatively low, because the short-circuit forces tend to expand the high-voltage winding H radially. Therefore, the use of many thinner pressboard as insulation is more advantageous there. The insulation between windings Ty and L is exposed to rather high stresses both electrically and mechanically. The rigidly fixed pressboard tubes employed here are capable of fulfilling both duties.

The ends of windings having their H.V. point at the centre are brought out as in the two examples shown in Figs 5.89 and 5.90. Figure 5.89 illustrated a 400 kV winding end [93]. In this case the tapped winding is arranged outside the 400 kV winding, the winding end being isolated from the tapped winding by pressboard collars. The conductor coming from the winding end leads into a pressboard funnel clamped to the tank over and sealed against ingress of oil. Also the 400 kV bushing protrudes into this funnel. To replace the bushing the oil in the tank need not be drained off, since the oil surrounding the protruding part of the bushing is separated from the oil in the tank by the pressboard funnel. The arrangement illustrated is suitable for the application of a re-entrant type condenser bushing.

313

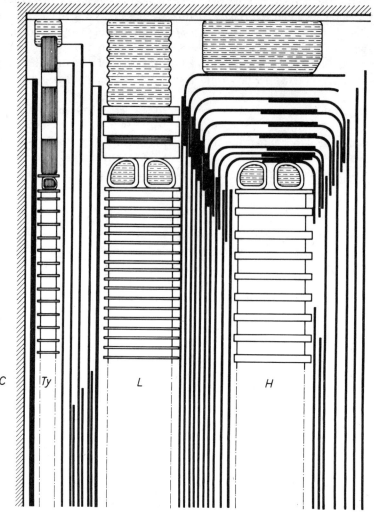

Fig. 5.88. Schematic cross section of upper end of windings mounted on one limb of a 750/420 kV transformer

Figure 5.90 shows how the end of a 750 kV winding having its H.V. point at the centre is brought out through the tank wall. The 750 kV bushing extends into a bushing seat (base) that can easily be dismantled from the tank wall. The oil space of this bushing seat is separated from the inner oil space of the tank by a pressboard diaphragm remaining in place throughout the entire service life. This pressboard insulating structure is fitted into a circular opening of the tank wall and consists of concentric conical and cylindrical pressboard elements. The fitting for connection of the 750 kV bushing is passed through the centre of the pressboard structure. The oil-tight wall has to be designed, on the one hand, to withstand the test voltage specified

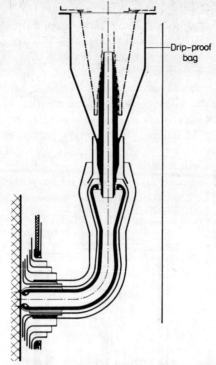

Fig. 5.89. 400 kV lead brought out from the centre of the main winding through the tapped winding to re-entrant type bushing

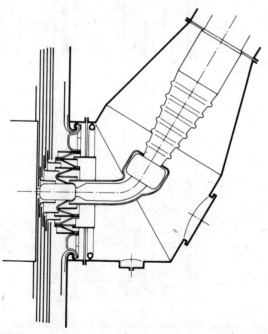

Fig. 5.90. 750 kV winding end led through the oil sealing wall and connected to the condenser bushing

for the H.V. terminal and, on the other hand, for the prevailing oil pressure. It is clearly shown in Fig. 5.90 how the angle rings attached to the pressboard insulation sheets enveloping the winding column fit in between the pressboard elements of the oil sealing wall. The pressboard sheets and the angle rings are arranged in conformity with the electric field, i.e. parallel with the equipotential lines. During transportation, the bushing seat can easily be dismantled without need for draining the oil from the transformer tank.

As the example, Fig. 5.91 shows the schematic section of windings mounted on one of the limbs of the 120 kV ± 15%/11 kV, 25 MVA regulating transformer dealt with already in conjunction with the calculation of short-circuit forces. In the diagram C designates the core, L the 11 kV winding, H the 120 kV winding and T the tapped winding. The 120 kV winding is star connected, the tapped windings are

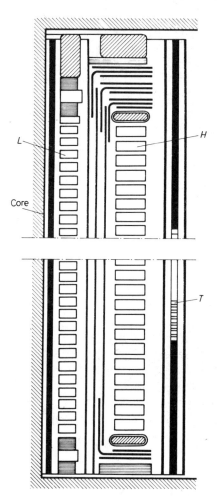

Fig. 5.91. Schematic cross section of windings mounted on one limb of a 120 kV ± 15%/11 kV regulating transformer

316

arranged at the neutral point end, and the neutral point is earthed. The 11 kV winding is connected in delta.

The material of the cylinder arranged between winding L and the core is also bakelite in this case, providing the required mechanical and electrical strength. Insulation of both ends of winding L is identical, except for the clamping ring located at the upper end of the winding as already mentioned above, and the winding is connected in delta. Insulation between windings L and H consists of pressboard tubes possessing a dielectric strength superior to that of bakelite. The radial spacing between winding and the closest pressboard tube is ensured by bracing sticks. In view of the higher service and test voltages the upper end insulation, i.e. the insulation of the line end of winding H, is of much higher rating than the insulation of the lower end attached to the tapped winding.

The tapped winding marked T is arranged outside the winding H, separated from it by a pressboard tube and by a Kraft paper cylinder wound over the latter. This helical winding of edgewise bent turns is confined between bakelite tubes of thicknesses equal to that of the winding, the tubes fulfilling here the role of the clamping ring. The bakelite tube outside the tapped winding ensures the mechanical strength of the winding and isolates the windings of different phases from each other.

5.3.6.6 Inner connections

For connecting the winding end to the bushing and for linking the inner winding ends with each other paper insulated copper straps, rods or tubes are used. In transformers with tapped windings the tap points are connected in the same way with the tap changing mechanism. If rods or tubes are used, their ends are fitted with a short length of stranded wire to provide flexibility. For designing the insulation of inner connections the expected electric field and the maximum field strength developing at the surface of connections have to be determined. The spacings between connections and toward earth parts (core, tank and frame), and the required thickness of insulations can only be determined with knowledge of the field strength data. The insulation of connections will safely withstand the stresses occurring during the voltage test, if no partial discharges take place either in the paper insulation or in the oil. In moisture-free and gas-free oil paper insulation, this requirement is fulfilled if the field strength remains below the breakdown strength of the oil both at the electrode surfaces and at the contact surfaces between paper and oil. A sufficient safety margin in designing is obtained if only 50 to 70 per cent of this value is permitted to occur in the insulation.

In a homogeneous field, with bare electrodes, the field strength is given as the quotient of voltage between electrodes (U) to the spacing between them (a):

$$E_h = \frac{U}{a}.$$

In an inhomogeneous field the field strength increases at the electrode surfaces. This increase can be taken into account by the inhomogeneity factor, depending on the electrode arrangement and on local conditions. The values of the inhomogeneity

Table 5.10

Factors of inhomogeneity

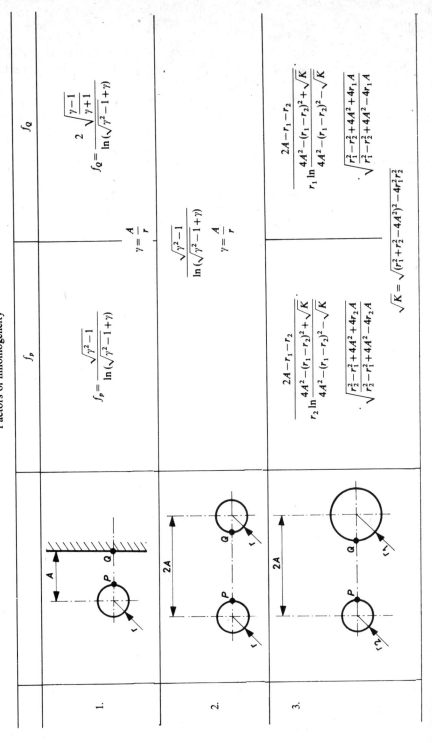

		f_P		f_Q
1.	(sphere below a hatched plane, distance A, load P and point Q)	$f_P = \dfrac{\sqrt{\gamma^2-1}}{\ln\left(\sqrt{\gamma^2-1}+\gamma\right)}$	$\gamma=\dfrac{A}{r}$	$f_Q = \dfrac{2\sqrt{\dfrac{\gamma-1}{\gamma+1}}}{\ln\left(\sqrt{\gamma^2-1}+\gamma\right)}$
2.	(two equal spheres, centre distance $2A$, loads P and Q)	$\dfrac{\sqrt{\gamma^2-1}}{\ln\left(\sqrt{\gamma^2-1}+\gamma\right)}$	$\gamma=\dfrac{A}{r}$	
3.	(two unequal spheres, centre distance $2A$, loads P and Q)	$\dfrac{2A-r_1-r_2}{r_2\ln\dfrac{4A^2-(r_1-r_2)^2+\sqrt{K}}{4A^2-(r_1-r_2)^2-\sqrt{K}}}\cdot\sqrt{\dfrac{r_2^2-r_1^2+4A^2+4r_2A}{r_2^2-r_1^2+4A^2-4r_2A}}$	$\sqrt{K}=\sqrt{(r_1^2+r_2^2-4A^2)^2-4r_1^2 r_2^2}$	$\dfrac{2A-r_1-r_2}{r_1\ln\dfrac{4A^2-(r_1-r_2)^2+\sqrt{K}}{4A^2-(r_1-r_2)^2-\sqrt{K}}}\cdot\sqrt{\dfrac{r_1^2-r_2^2+4A^2+4r_1A}{r_1^2-r_2^2+4A^2-4r_1A}}$

№			
4.		$\dfrac{r_1-r_2}{r_2\ln\dfrac{r_1}{r_2}}$	$\dfrac{r_1-r_2}{r_1\ln\dfrac{r_1}{r_2}}$
5.		$\dfrac{\sqrt{\gamma^2-1}}{\ln(\sqrt{\gamma^2-1}+\gamma)}\cdot\dfrac{R_{1a}-R_{1i}}{R_{1a}\ln\dfrac{R_{1a}}{R_{1i}}}$	$2\sqrt{\dfrac{\gamma-1}{\gamma+1}}\cdot\dfrac{1}{\ln(\sqrt{\gamma^2-1}+\gamma)}\cdot\dfrac{R_{1a}-R_{1i}}{R_{1i}\ln\dfrac{R_{1a}}{R_{1i}}}$
6.		$\dfrac{\sqrt{\gamma^2-1}}{\ln(\sqrt{\gamma^2-1}+\gamma)}\cdot\dfrac{R_1-R_2-D}{R_1\ln\dfrac{(R_1+R_2)^2-D^2+\sqrt{m}}{(R_1+R_2)^2-D^2-\sqrt{m}}}\cdot\sqrt{\dfrac{R_1^2-R_2^2+D^2+2R_1D}{R_1^2-R_2^2+D^2-2R_1D}}$	$2\sqrt{\dfrac{\gamma-1}{\gamma+1}}\cdot\dfrac{1}{\ln(\sqrt{\gamma^2-1}+\gamma)}\cdot\dfrac{R_1-R_2-D}{R_2\ln\dfrac{(R_1+R_2)^2-D^2+\sqrt{m}}{(R_1+R_2)^2-D^2-\sqrt{m}}}\cdot\sqrt{\dfrac{R_1^2-R_2^2-D^2+2R_2D}{R_1^2-R_2^2-D^2-2R_2D}}$

$$\sqrt{m}=\sqrt{(R_1^2+R_2^2-D^2)^2-4R_1^2R_2^2}$$

$$\gamma=\frac{A}{r}$$

factor are compiled in Table 5.10 for a few basic cases often encountered in transformers [92]. Arrangement 1 gives the factors required for calculating the field developing between a cylindrical conductor and a plane parallel with the conductor. Such a field appears in the case of an inner connection and a tank wall. The field between two conductors of equal diameters can be calculated with the factors of inhomogeneity of arrangement 2, and that between two conductors of different diameters with the factors of arrangement 3. For calculating the field between two concentric cylinders, e.g. between an outer winding and a tank wall surrounding it, the factors of inhomogeneity are given in row 4 of the table. Arrangement 5, the field between a ring and a cylinder passing through its centre, approaches the practical case when a cylindrical conductor of given diameter is passed through a circular hole of rounded-off edges, cut into a metal sheet. In arrangement 6 the conductor intersects the plane of the ring not at its geometrical centre. This arrangement, selecting a sufficiently large diameter for the ring, provides the case of the field developing between two non-intersecting perpendicular conductors. The factors of inhomogeneity relating to cases 1, 2 and 5 can be read directly from the curves of Fig. 5.92 for the ratios of diameters occurring in practice.

When the expected field strengths and the breakdown voltages of oil and paper are known, the dielectric strength of inner connections can be determined and a conductor arrangement of the required strength can be designed. If the insulation is not thick, then the field strength figures tabulated may be taken as valid for the boundary layer between paper and oil, where a power frequency voltage of 40 to

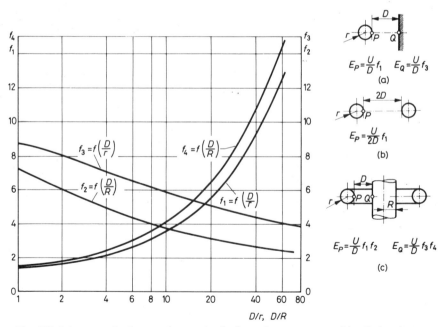

Fig. 5.92. Inhomogeneity factors of some simple electrode arrangements: (a) cylinder-plane; (b) two cylinders of equal radii; (c) concentric ring and cylinder

50 kV cm^{-1} is permissible. In the case of thick paper insulation the field strengths developing at the surface of the electrode and at the boundary layer between paper and oil should be checked separately. At the electrode surface 60 to 80 kV cm^{-1}, and at the paper/oil boundary layer field strength values of 40 to 50 kV cm^{-1} are permissible in the course of the power frequency test.

As an example, let us examine the field strength developing, in the course of the power frequency test, in the vicinity of a conductor of diameter $d = 60$ mm and insulated with paper $\delta = 15$ mm thick, led out from the winding end of a 220 kV transformer, the edge of the conductor being 200 mm away from the tank wall. The test voltage is 395 kV.

For the sake of simplicity, the calculation is based on the relations given for concentric cylindrical electrodes. The field strength calculated this way will be 10 to 20% higher than the true value, so the error committed will be in the direction of a higher degree of safety.

Between two concentric cylindrical electrodes of radii r_0 and r_2, where the permittivity of material between radii r_0 and r_1 is ε_1, and that between r_1 and r_2 is ε_2

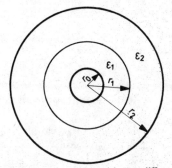

Fig. 5.93. Concentric cylindrical electrodes with two different insulating materials

(see Fig. 5.93), the field strength in the region of permittivity ε_1 varies according to the relation

$$E_{1r} = \frac{U}{r} \frac{\varepsilon_2}{\varepsilon_2 \ln \dfrac{r_1}{r_0} + \varepsilon_1 \ln \dfrac{r_2}{r_1}}$$

and in the region of permittivity ε_2 according to the relation

$$E_{2r} = \frac{U}{r} \frac{\varepsilon_1}{\varepsilon_2 \ln \dfrac{r_1}{r_0} + \varepsilon_1 \ln \dfrac{r_2}{r_1}}$$

as a function of radius r, where U is the voltage connected across the two electrodes.

In the present case let $U = 395 \text{ kV}$, $r_0 = 3$ cm, $r_1 = 4.5$ cm, $r_2 = 24.5$ cm, $\varepsilon_1 = 4.5$ and $\varepsilon_2 = 2.2$. With these values, the maximum field strength at the conductor surface $(r = r_0)$ will be

$$E_{1r_0} \approx 34 \text{ kV cm}^{-1}$$

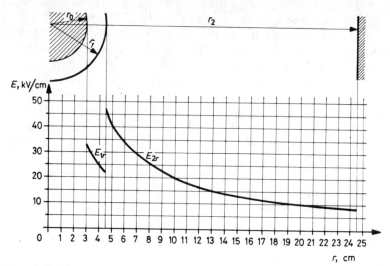

Fig. 5.94. Electric field in the vicinity of conductor connected to the bushing of a 220 kV transformer

and at the paper/oil boundary surface $(r = r_1)$

$$E_{2r_1} \approx 46 \text{ kV cm}^{-1}.$$

Both values are acceptable. The variation of field strength along the radius is shown in Fig. 5.94.

For some cases experimental data are also available. Thus for the arrangement represented by row 2 of Table 5.10, the value of the 50 Hz 1 min test voltage permissible between bare copper conductors of 10 mm diameter in the case of new transformer oil is given in Fig. 5.95 as a function of spacing between the conductors. The dielectric strength of the arrangement increases considerably under the effect of paper insulation wound on the conductor. The permissible values of the 50 Hz 1 min test voltage for the cases of 5 and 10 mm thick paper wraps are also shown in Fig. 5.95.

In the course of designing the inner connections, it is an essential requirement to check the creepage paths developing along the clamping elements, insulating plates and wooden parts in contact with the connections. Along longer creepage distances, extending over a length of a few times 10 cm, a creepage stress of 10 kV cm^{-1} may be safely permitted when applying the specified power frequency test. Along shorter distances of a few centimetres, as high as 3 to 5 times this value may be considered as permissible.

5.3.6.7 External insulation of transformers

The winding ends of transformers up to a rated voltage of 35 kV are usually brought out from the inside of the tank through simple porcelain bushings. At higher voltage ratings condenser bushings are used for that purpose. Metal cylinders

322

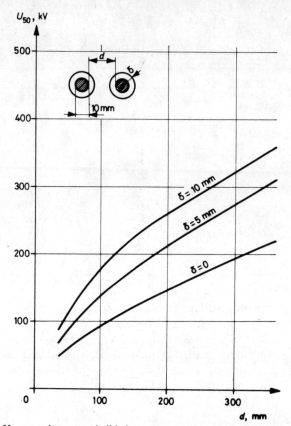

Fig. 5.95. 1 min 50 Hz test voltage permissible between paper-insulated copper conductors of 10 mm diameter

placed between oil-impregnated, resin-impregnated or resin-bonded paper insulation within the porcelain insulation of the condenser bushing constitute a chain of series capacitors by which a uniform voltage distribution between the energized current carrying rod or tube and earthed parts (tank) is realized. The condenser bushing is designed so that its partial discharge intensity at the test voltage specified for the transformer should not exceed half the intensity permitted for the transformer. The bushing itself is dealt with in detail in Section 7.

The pattern of the electric field around the part of the condenser bushing located inside the tank should be given special attention. In critical cases it is advisable to determine the potential distribution developing here by calculation or scale model experiment performed in advance.

The clearances in air to be maintained between the outer end of the condenser bushing and the earthed parts are compiled in Table 5.11. The clearances given in the list are based on experience.

The porcelain insulators of transformers are provided with protective spark gaps or some other arc protection devices. Their task is to divert the electric arc from the

Table 5.11

Minimum clearances for external insulation

Highest voltage for equipment kV	Minimum clearance	
	line to earth, mm	line to line, mm
≦ 12	200	250
24	280	350
36	400	460
72.5	700	780
123	1000	1200
145	1170	1380
170	1350	1600
245	1850	2250
420	3100	3500
765	5800	6700

surface of the porcelain insulator protecting it against the detrimental effects of external flashovers. Up to 245 kV rated voltage protective rods are adopted in most cases, and above this voltage level shielding rings are used.

In the design of bushing insulators their shape, number of skirts and creepage distances must be determined so that the external insulation should safely resist the effects of rain, fog and pollution to which the insulator may be exposed in service.

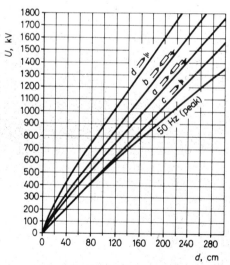

Fig. 5.96. 50% lightning impulse breakdown voltage and 50 Hz breakdown voltage of rod–rod and rod–plane electrodes as a function of electrode spacing (barometric pressure 98×10^3 Pa, temperature 20 °C, absolute moisture content of air 11 g m^{-3}). Curves *(a)* and *(c)*: positive impulse; curves *(b)* and *(d)*: negative impulse

324

The external insulation of a transformer designed with the clearances given in Table 5.11 need not be tested separately, since the distances given have been tested by experiments and they satisfy the standard specifications. When employing distances deviating from those specified by the Standards, it is advisable to check the clearances by means of the electric strength curves of rod–plane and rod–rod type spark gaps. The values of 50% lightning impulse breakdown voltages valid for these basic arrangements (in the case of 1.2/50 µs waves), as well as those of the power frequency breakdown voltages, are illustrated by Fig. 5.96.

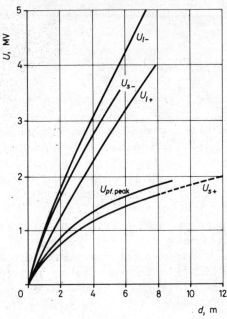

Fig. 5.97. 50% breakdown voltage of a rod–plane arrangement as a function of electrode spacing [22]. U_{l-}: negative lightning impulse voltage; U_{l+}: positive lightning impulse voltage; U_{s-}: negative switching impulse voltage; U_{s+}: positive switching impulse voltage; U_{pf}: peak value of power frequency voltage

For higher voltage ratings (above 300 kV) and for correspondingly longer clearances, the stresses imposed by the switching impulse waves are to be considered in the design of external insulation (see Fig. 5.97). Values to breakdown voltage in the case of lightning impulse waves and, especially, in that of the switching impulse waves, depend markedly on the polarity of the voltage.

In critical cases it is advisable to prepare a 1:1 scale model of the external insulation and to check the correctness of clearances on this model.

5.4 Testing of transformer insulation

The purpose of dielectric tests is to check transformer insulation. Dielectric tests are used to determine whether or not the transformer will be capable of withstanding the voltage stresses to which it may be exposed in service. In the course of power-frequency, lightning-impulse and switching-impulse tests, the transformer is subjected to voltage stresses higher than those expected to occur in service. The power-frequency test of high-voltage transformers ($U_m = 300$ kV) is generally associated with partial discharge measurements. From these results conclusions can be drawn about the state of insulation, the quality of manufacture and any concealed defects of insulation.

5.4.1 Power–frequency voltage tests

The traditional method of checking transformer insulation is the power-frequency voltage test. Its purpose is to verify the dielectric strength of major insulation between individual windings, between windings and the earthed parts, and of the insulation within individual windings (interturn insulation and that between discs, coils, etc.). On windings with uniform insulation the test is performed in two parts: the dielectric strength of the main insulation is checked by applying a separate-source, power-frequency voltage withstand test, and that of the insulation within the windings by an induced-overvoltage withstand test.

For the separate-source voltage withstand test (using a single-phase testing transformer) the tested winding is short-circuited so that all parts are at the same potential. All non-tested windings are also short-circuited, connected to each other and earthed together with the transformer tank. In compliance with the practice followed in most European countries, the standard test voltage U_p (shown in Table 5.1), applied between the tested winding and earthed parts, must be withstood by the insulation for 60 s.

In the induced-overvoltage withstand test one winding of the transformer is fed by an a.c. voltage, and the test voltage is produced in the other windings by the tested transformer itself. For this test one point of each winding and of the transformer tank are earthed, and between the line terminals of the H.V. winding double the rated voltage is generally produced as the test voltage. Should, however, this test voltage cause a voltage exceeding the test voltage of the separate-source voltage test at any point of the winding within the transformer, the voltage of the induced-overvoltage test must be reduced correspondingly.

It is expedient to feed three-phase windings with a symmetrical three-phase voltage. Where there is a neutral terminal, it may be earthed.

The frequency of the test voltage is higher than the rated one, to prevent the development of inadmissibly large flux densities in the core and to avoid an excessive excitation current causing a harmful temperature rise in the winding supplied during the test. If the frequency of the test voltage does not exceed double the rated value, the duration of the test is 1 min. For higher frequencies, duration t_t of the test should be determined from the equation

$$t_t = 120 \frac{f_n}{f_t},$$

where f_n is the rated frequency and f_t the test frequency. This duration, however, should not be shorter than 15 s.

Windings with non-uniform insulation can be tested only with the relatively low test voltage specified for the testing of the neutral terminal when using the separate source test method. This voltage is insufficient for testing the major insulation. Therefore, the insulation between the windings and between windings and earthed parts is checked simultaneously with the insulation within the windings, by an induced-voltage test. During the test, the specified test voltage has to appear between the line terminals as well as between line terminals and earthed parts. The test voltages specified by IEC corresponding to the practice followed by most European countries (U_p) are given in Table 5.1 for transformers with rated voltages below 300 kV ($U_m < 300$ kV), and in Table 5.2 for transformers with rated voltages of 300 kV and above ($U_m \geq 300$ kV) and tested by Method 1. The winding is earthed at a suitable point, so that the stress imposed on the main insulation uniformly decreases from the tested terminal to the earthed point of the winding. As regards interturn insulation, it would be desirable to make the double value of interturn voltage apply also in this case, as that for the induced tests specified for transformer windings of uniform insulation. Generally, this voltage cannot be obtained accurately, but by deliberately selecting the test scheme, the closest possible approach to the double interturn voltage should be aimed for (e.g., in the case of transformers having tapped windings, by suitable selection of the tapping position). The test duration should be determined depending on the frequency, as mentioned in relation with transformers of uniform insulation.

When testing single-phase transformers, the voltage of the neutral terminal may be raised to the required level by connecting it to an auxiliary transformer or the tertiary (low voltage) winding of the main transformer. An example is given in Figure 5.98 which shows the arrangement for the overvoltage testing of a single-phase autotransformer having a ratio of $\frac{400}{\sqrt{3}}$ kV $\Big/ \frac{126}{\sqrt{3}}$ kV $\Big/ 18$ kV. The common winding of the autotransformer is connected in series with the tertiary winding. Applying a supply voltage of 45.5 kV at 150 Hz to the tertiary winding, the required test voltages (230 kV and 630 kV, respectively), will appear both at the 126 and the 400 kV terminals. During the induced-voltage test, 2.53 times the rated interturn voltage appears in each winding. Although this voltage considerably exceeds the normally 2 times interturn voltage, it may generally be applied without risk to any conventionally insulated high-voltage transformer.

Three-phase transformers are tested in single-phase connection. A scheme applicable for testing star/delta-connected transformers is shown in Fig. 5.99. When excitation is applied to one of the outer phase windings, the flux Φ in the limb belonging to that winding returns through the middle and the other outer limb, each carrying a flux of $\Phi/2$. Hence, the voltage of the winding on the excited limb will be

327

twice that of the non-excited limbs. (In order to ensure the desired flux distribution, the line terminals of the phases not tested must also be connected together in case these terminals are left unearthed, as against the case shown in Fig. 5.99.) In the connection shown, one third of the test voltage appears on the neutral terminal during the test if the full test voltage is present between the tested terminal and earthed parts. Therefore, the test may only be performed in this connection if the insulation of the neutral terminal is designed to withstand this voltage. The voltage imposed on the tested phase winding is 2/3 of the test voltage. For instance if the rated voltage of the transformer is 123 kV ($U_m = 123$ kV) then, according to Table 5.1 (considering the higher insulation level given), the respective test voltage will be 230 kV. Thus, the lower-voltage winding has to be excited with a voltage to obtain 230 kV at the tested terminal of the high-voltage winding. The voltage of the neutral terminal will be 77 kV, so its insulation level will have to be at least equal to this value. (This corresponds to 52 kV rated voltage, the insulation level pertaining to

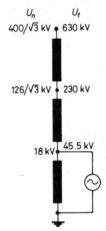

Fig. 5.98. Connection of a $\dfrac{400}{\sqrt{3}}$ kV $\Big/$ $\dfrac{126}{\sqrt{3}}$ kV $\Big/$ 18 kV auto-transformer for power-frequency test

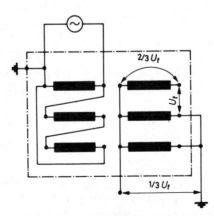

Fig. 5.99. Connection of a transformer winding with non-uniform insulation for power frequency test

328

the latter being 95 kV.) The voltage imposed on the tested winding is 153 kV, i.e. about 2.2 times the rated voltage ($123/\sqrt{3} = 71$ kV). Thus, by this test scheme, a good approximation of double the rated voltage is obtained, which is considered as desirable for checking the strength of interturn insulation.

The test should be performed with a voltage of sinusoidal waveform as nearly as possible, and its peak value divided by $\sqrt{2}$ should be considered as the test voltage. The test should start at a voltage not more than one third of the test value, and from then on the voltage should be raised at a uniform rate. At the end of the test, the voltage should be lowered continuously to at least 1/3 of the test value, before switching off. These specifications serve for avoiding harmful inrush currents on making the circuit and over-voltages on breaking it.

The test is successful if no breakdown or flashover occurs either within or outside the transformer during it. Inductions of failure may be the collapse of the test voltage, or the occurrence of some audible or visible phenomenon. When testing large transformers it is advisable to take oscillographic records of the shape of voltage prevailing on the H.V. transformer terminal (or terminals, in the case of autotransformers) and that of the current flowing into the earth. In the case of a failure, the oscillograms may yield information as to the location and nature of the defect.

Another test procedure (Method 2) is proposed by the IEC Recommendations for transformers having voltages rated above 300 kV. Instead of the short-duration

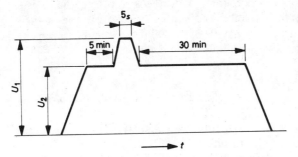

Fig. 5.100. Voltage versus time according to test method No. 2. of IEC

power–frequency withstand voltage test, this second method requires a lower-voltage, long-duration withstand voltage test associated with a partial discharge measurement to be performed on such transformers. According to the time sequence for the application of the test voltage shown in Fig. 5.100, the voltage should be raised to a certain level U_2 and held there for a period of 5 minutes. Then the voltage should be further raised to level U_1, held there for 5 s, thereafter reduced again to U_2, and held there for a duration of 30 minutes. Level U_1 is equal to the highest service voltage (U_m) of the equipment, whereas level U_2 may be either $1.3\, U_m/\sqrt{3}$ or $1.5\, U_m/\sqrt{3}$, as agreed upon by the manufacturer and user. Through the entire duration of the measurement, at voltage U_2, the intensity of partial discharges shall be monitored. The limiting value of the "apparent charge" is 300 pC at $U_2 = 1.3\ U_m/\sqrt{3}$ and 500 pC at $U_2 = 1.5\ U_m/\sqrt{3}$.

In the course of testing the neutral of star-connected windings and one terminal of delta-connected windings must be earthed. Three-phase transformers may be tested phase by phase in a single-phase connection, or in symmetrical three-phase connection. For the former case, the voltages appearing at the terminals are shown in Fig. 5.101.

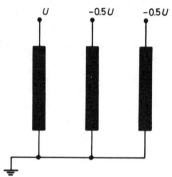

Fig. 5.101. Voltages appearing at the terminals during single-phase testing of a three-phase transformer

The transformer is deemed to satisfy the test requirements if no collapse of the test voltage occurs, the apparent charge stays below the specified limit, and no significant, steadily rising trend can be observed in the intensity of discharge.

The most important details of partial discharge measurement and evaluation of the results are dealt with in Section 5.4.2.

5.4.2 Measurement of partial discharges

A partial discharge occurring in the insulation of a transformer, i.e. a breakdown of a small part of the insulation (e.g. through an inclusion or a cavity), may be regarded as an instantaneous shunting or shorting of a small partial capacitance of the entire insulation. Owing to the discharge of this partial capacitance, the distribution of charges within the insulation changes suddenly, causing a short current pulse. The partial capacitance becomes charged again and, when the voltage across it attains the level required to bring about a repeated breakdown it is again discharged. The lost charge is made up by the supply voltage source. The current pulses may take place in rapid succession and the height, shape and frequency of pulses depend on the character, size and spatial arrangement of the inclusion. As regards their frequency spectrum, the current pulses contain oscillations with frequencies ranging from a few tens of thousand Hz up to several MHz. If there are several discharge spots in the insulation, each of them may be regarded as a current source generating high-frequency pulses.

The purpose of partial discharge measurements is to measure the discharge current pulses. From the nature of the phenomenon, the charge or current of each

330

individual partial discharge cannot be measured directly, and only the changes of charges and voltages appearing in the measuring circuit outside the transformer which occur under the effect of the partial discharges can be detected. The measuring circuit may be set up in several different ways, the three schemes most often used being shown in Fig. 5.102. For the discharge current pulses a closed circuit has to be provided in every case. In Fig. 5.102(a), the transformer winding being tested is earthed through a measuring impedance (Z_m). The current pulses generated in the winding flow through the measuring impedance and are led back into the winding through the capacitance to earth (C_t) of the winding, the fittings connected thereto and the supply transformer. The high-frequency impedance may be reduced and the sensitivity of the method improved thereby, if the terminal of the transformer is earthed through a coupling capacitor C_c. This capacitance should not be lower than 1000 pF.

When testing large high-voltage transformers, it is convenient to earth the neutral of the winding directly for safety reasons. In such cases, the measuring impedance is connected in series with the coupling capacitor (Fig. 5.102(b)). This method is

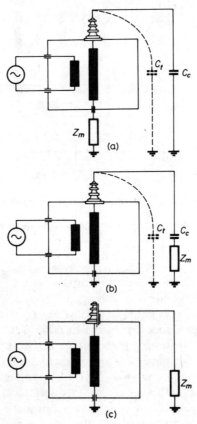

Fig. 5.102. Connections used during partial discharge measurements of transformers: (a) Measuring impedance in neutral of the transformer winding; (b) measuring impedance in series with the coupling capacitor; (c) measuring impedance connected to the bushing

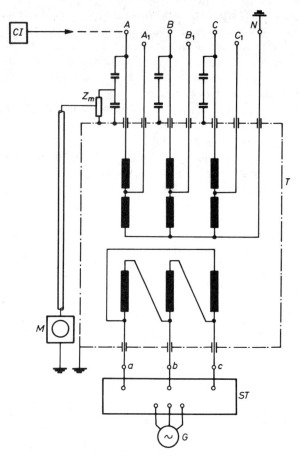

Fig. 5.103. Connection used during the partial discharge measurement of a 400/132/18 kV auto-transformer: G — generator; ST — supply transformer; T — transformer to be tested; M — measuring instrument; CI — calibrating instrument

equivalent to that of Fig. 5.102(a) if the role of capacitance C_t is left out of consideration. The use of a coupling capacitor may be cumbersome, so for transformers fitted with condenser bushings an obvious solution is provided by the use of the bushing itself as coupling capacitor and by connection of the measuring impedance to the capacitance tap of the bushing (Fig. 5.102(c)). This scheme is generally adopted for the testing high-voltage transformers.

Figure 5.103 shows the connection used during partial discharge measurement of an autotransformer having a 400/132/18 kV voltage ratio. The measuring instrument is connected to the capacitance tap of the condenser bushing of terminal A. If possible, it is of course convenient to perform the measurement on several terminals simultaneously; primarily on the terminals of the common and series windings of the same phase of an autotransformer (on A and A_1, on B and B_1, and on C and C_1, respectively).

332

By means of an instrument connected to the measuring impedance the intensity of partial discharge is measured, i.e. quantities characterizing the discharge current pulses are measured: the magnitude of individual discharges (C), the repetition rate of pulses (n), charge quantities of all impulse observed within a given period of time, i.e. the current (C/s), the discharge power (W), and the quadratic rate (C^2/s). In the transformer specification of IEC, measurement of the charge of individual pulses is recommended. The charge measured as flowing through the measuring impedance, however, is not a real charge but an apparent charge that has a rather loose relation with the real charge. According to the IEC standard dealing with the measurement of partial discharges, the apparent charge of a partial discharge is that charge which, if injected instantaneously between the terminals of the test object would momentarily change the voltage between its terminals by the same amount as the partial discharge itself.

The measuring equipment should be calibrated with this definition in mind. A charge of known magnitude is applied to the tested winding either between the terminals of the winding ends, or between the terminal concerned and earth, and the value (usually a voltage value) shown by the measuring instrument connected to the measuring impedance is determined. Thus the conversion factor in terms of pC mV^{-1}, pC μV^{-1} or some other unit can be obtained. Then if, during a partial discharge measurement a charge q is measured, this means that a discharge of unknown magnitude taking place within the transformer exerts the same effect on the measuring instrument as charge q entering the terminals of the tested winding.

Since apparent charges only can be measured, the relation between the real and apparent charge being unknown, it is virtually impossible to assess on the basis of the measured discharge intensity the damage caused by partial discharges. Also the limit of 300 pC and 500 pC, respectively, specified by the IEC Recommendation for these charges are rather arbitrary. Their selection is supported by the experience that, in service, transformers satisfying this requirement undergo no deterioration or damage of insulation caused by partial discharge.

The instruments used for measurement of partial discharges may be classified in two groups as narrow-band and wide-band instruments.

The bandwidth of narrow-band instruments is 10 kHz (or less) around some adjusted frequency. The radio noise meters belong to this group. They offer the advantage of eliminating possible interference effects from local broadcasting stations by suitable selection of the measuring frequency. The sensitivity of the measurement can be considerably improved by tuning the natural frequency of the measuring circuit to the measuring frequency. The measuring circuit is tuned to the desired frequency by means of a variable inductance connected in parallel with the lower capacitance of the condenser bushing (Fig. 5.104). With narrow-band measurements the measuring frequency should not be higher than 500 kHz, and below 300 kHz if feasible. The main reason for this is that the higher-frequency components of the discharge current are considerably attenuated by the transformer winding.

The frequency-band of wide-band instruments is relatively broad, their upper frequency limit being usually in the 150 to 400 kHz range and their lower limit between about 40 and 50 kHz. Some versions [91] of wide-band instruments, which are extremely suitable for the testing of transformers, measure individual discharge pulses. The bandwidth of the type best suited to transformer measurements ranges from 10 to 300 kHz. It is suitable for measurement of the peak value of individual pulses. The measurement consists in comparing a reference pulse with the measured pulse displayed on the screen of an oscilloscope integral with the instrument. The pulses appear along an ellipse corresponding to one cycle of the supply voltage, so their distribution within the cycle can easily be determined. From the distribution of

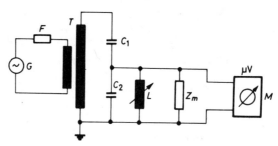

Fig. 5.104. Partial discharge measurement by means of radio noise meter: G — generator (supply voltage); F — filter to eliminate high frequency noise produced by the generator; T— transformer to be tested; C_1 — capacitance of H.V. part of condenser bushing; C_2 — capacitance of L.V. part of condenser bushing; L — tuning inductance; Z_m — measuring impedance; M — measuring instrument

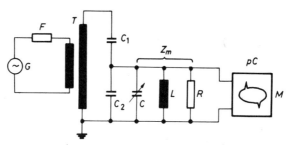

Fig. 5.105. ·Partial discharge measurement by means of apparent charge meter (ERA). C — tuning capacitance matched to the capacitance of the object to be tested; L — inductance of measuring impedance; R — resistance of measuring impedance. For explanation of further symbols see Fig. 5.104

pulses, conclusions can be drawn as to the type of defect. Connection of the instruments into the measuring circuit is shown in Figure 5.105.

No generally valid relation can be detected between the pC value measured by the charge measuring instrument and the μV value measured by the radio noise meter. According to measurements performed on various large transformers the ratio $pC\mu V^{-1}$ may attain values between 1 and 20.

Evaluation of the results of partial discharge measurements is a complicated problem. The value of the measured apparent charge in itself is not suitable for judging partial discharge hazards. Such a conclusion may be drawn from oscilloscopic analysis of discharge pulses, and from location of the defect. On the basis of characteristic properties (amplitude, distribution in time, variation as a

334

function of voltage and time) of pulses generated by the discharge pulses and displayed on the oscilloscope screen, four main types of internal partial discharges may be distinguished as follows:

(a) Discharges in solid insulation (paper, pressboard, wood, etc.). Such discharges may occur within the insulation (due to cavities, inhomogeneities etc.), and also on the boundary surface of insulating materials or between metallic and insulating surfaces.

(b) Discharges caused by bubbles, moisture or contamination. Discharges of this kind generally occur between the oil and some solid insulating material or between oil and metal.

(c) Discharges in the oil between pointed or inadequately rounded-off live metal parts and some other nearby metallic objects of different potential.

(d) Discharges between metallic objects of undefined potential ("floating objects") to parts of fixed potential.

The dielectric strength of a transformer is jeopardized by discharges affecting solid insulation (paper, pressboard). Such discharges exert a destructive effect over a prolonged period, thereby causing deterioration of the dielectric strength of the insulation. They may directly cause a severe damage, e.g. turn-to-turn fault, but it more often happens that the damage is brought about by some abnormal effect, e.g. short-circuit stress at a spot within the transformer where the dielectric strength has been weakened by previous partial discharges. Therefore, partial discharges taking place within the solid insulation, or on its surface, must be avoided by all possible means, since these may become sources of severe disturbances. Hence, partial discharges belonging to category (a) are undoubtedly hazardous.

The situation is relatively simple for partial discharges occurring in the oil under heading (b). During the period of rest following the drying and vacuum treatment of a transformer, the gas bubbles that may have remained in the oil escape upward from the inside of the oil, or are dissolved. The discharge intensity measured after a longer period of rest is generally lower than that obtained before this period. It is another interesting observation that the partial discharge intensity measured immediately after the power-frequency testing of transformers is higher than that measured later. This is attributed to gassing taking place in the transformer during the voltage test. Dissolution of gas cavities, especially in insulations of larger thickness, may take several weeks.

Only the oil is affected by the partial discharges (c) and (d) and as long as such discharges do not fall into the more critical sections of the electric field, they will not jeopardize the dielectric strength of the transformer.

The damage associated with a defect is more easily assessed if its location is known. For instance, a fault affecting the transformer winding is generally more critical than a discharge at the connecting leads of windings, although the oscillogram of the latter shows a pattern almost identical with that of the former. For locating the fault, both electrical and acoustical methods are available [5]. In favourable cases a defect may be located to an accuracy of a few centimetres.

As shown by practical experiences, the 300 pC and 500 pC apparent charge values proposed by the IEC Recommendations as permissible upper limits of partial discharge intensity for power transformers are realistic requirements, and they can generally be met. These values, however, should not be considered as rigid limits. The recommended figures may be exceeded by 200 or 300 pC or even more in all cases without introducing undue hazards. Transformers which exhibit higher partial discharge intensities then the upper limits specified by the standard, but have successfully withstood all other tests, need not be rejected, not even according to the relevant IEC Recommendations. In such cases, however, further investigations have to be performed to check the severity of the defect.

First, it should be ascertained that the discharges originate from the tested transformer and are not caused by the voltage source, objects of floating potential (unearthed objects), welding performed in the vicinity, etc., and whether the interference is not due to external discharges (occurring at the transformer terminals or connections).

If it is ascertained that the source of discharge is within the transformer, then determination of the hazardous nature of the defect should be attempted on the basis of the oscillogram and by the "geographical" location of the source. The observed phenomenon may be due to insufficient drying or oil impregnation of the transformer. In such a case, the situation may be improved by re-processing the transformer.

If the discharges do not seem to be hazardous, then a repeated test, or a test of longer duration than the original, or performed at an elevated voltage, can be resorted to. If no substantial change in the intensity of partial discharges can be brought about by increasing the voltage or by extending the time of application of the test, that may be taken as a justification for putting the transformer into service without risk. In such a case, however, it is expedient to make checks at regular intervals, e.g. analyses of dissolved gases in the oil, to make sure that no excessive deterioration has taken place during service.

When partial discharges are detected that are only slightly higher than the specified limits, but are nevertheless considered to be hazardous, or have a very high intensity (of several thousand pC), the defect should be eliminated by all possible means.

5.4.3 Lightning impulse test

The purpose of the lightning impulse test is to check the dielectric strength of transformer insulation against overvoltages of atmospheric origin. This test is to be performed as a type test on transformers below 300 kV, and as a routine test on transformers of higher voltage ratings, by applying the standard negative-polarity 1.2/50 μs impulse voltage wave described earlier.

Before the voltage test, the correctness of the calculated distribution of impulse voltage may be ascertained by means of voltage distribution measurements carried out by a low-voltage recurrent surge generator. With this equipment adjustable rise-time and time to half value impulse voltage waves of a few hundred volts amplitude can be produced, generally at a frequency equal to half the system frequency, i.e. in the case of a 50 Hz network, 25 impulse waves per second. Applying this voltage to the terminals to be tested, before putting the transformer into its tank, oscillographic records can be taken of the shape of impulse voltage oscillations. Since the impulse voltage phenomena are of linear character (the phenomena hardly being influenced by core saturation), the impulse voltage distribution measured at low voltage will be practically identical with that to be expected in the high-voltage test. By this measurement the voltages of different points of the transformer with respect to earth, those appearing across the sections of windings and those between the different windings, can be determined.

When performing impulse voltage tests on single-phase transformers, the test voltage is applied to the terminal being live in service, while the terminal being earthed in service is earthed also during testing. If both terminals of a single-phase transformer are normally live, the impulse voltage test should be performed from the direction of both terminals. The terminal not-tested should either be earthed directly, or through a low impedance such as a current measuring shunt (1–10 Ω). Three-phase transformers are tested by phases, i.e. the test voltage is applied to one phase of the transformer, whereas the other two and the neutral (if any) are earthed, either directly or through a low impedance.

The terminals of the separate (non-impulsed) windings of the transformer should be earthed through resistors to prevent any voltage above those to be expected in service from being transferred from the impulsed winding to the non-impulsed windings. (During service, the terminals are protected against direct or transferred overvoltage of inadmissible magnitudes by lightning arresters or protective spark gaps.) According to the IEC Recommendation for power transformers, the non-impulsed terminals should be connected to earth through resistors to restrict the peak value of voltage appearing there to 75% of the voltage specified for these terminals. The magnitude of transferred voltages and the required value of the earthing resistor can be determined by a low-voltage recurrent surge generator before performing the test. It should be borne in mind, however, that by reducing the resistance only the voltage appearing at the terminal can be lowered, and the internal voltages are not thereby prevented from increasing. (It has been shown in the discussion of transferred voltages that higher voltages may arise within the short-circuited windings than in the open-circuited ones.) Hence, a reduction of the earthing resistance to an unjustifiably low value can be harmful. Too low an earthing resistance is unfavourable in another respect as well, in that it reduces the fault detection sensitivity of the oscillographic method.

There are cases in which due to the earthing of the non-impulsed terminals or neutral either directly or through a low-impedance resistor, the load imposed on the surge generator becomes so high that it will not be able to produce the standard wave shape (e.g., in the case of autotransformers and very low-impedance windings). In such cases, the value of the earthing resistor may be increased to 400 to 500 Ω. Obviously, care must also be taken in that case to maintain the voltage of the terminals concerned below 75% of their impulse withstand voltage.

The three typical connection schemes used for the lightning impulse testing of three-phase transformers are represented in Figs 5.106, 5.107 and 5.108. In Fig. 5.106 the scheme for a transformer is shown, where the neutral of the high-voltage winding is earthed, and the low-voltage winding is connected in delta. The two non-impulsed high-voltage terminals are earthed through resistances R_1, the neutral terminal is

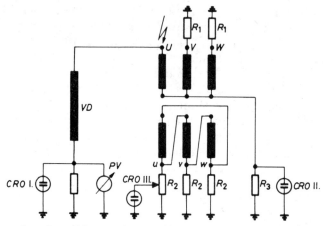

Fig. 5.106. Connection of star/delta transformer with brought out neutral for the lightning impulse voltage test

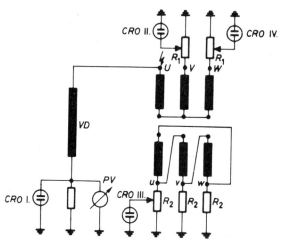

Fig. 5.107. Connection of star/delta transformer with isolated neutral for the lightning impulse voltage test

connected to earth through a low-value (1 to 100 ohms) measuring resistor (R_3) suitable for fault indication. The earthing resistors connected to the three low-voltage terminals are R_2. Figure 5.107 differs from the connection shown in Fig. 5.106 only so far that here the neutral point of the high-voltage winding is not earthed, and in this case the two non-impulsed high-voltage terminals are earthed through low-impedance resistors. Finally, in Fig. 5.108, the high-voltage winding is connected in delta, and the neutral of the low-voltage winding is directly earthed. In the diagrams, the voltage divider *(VD)* serving for impulse voltage measurement, the associated cathode-ray oscilloscope *(CRO I)* and peak voltmeter *(PV)*, and the oscilloscopes serving for fault detection *(CRO II, CRO III, CRO IV, CRO V)* are also shown.

338

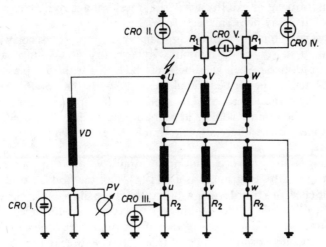

Fig. 5.108. Connection of delta/star transformer with brought out neutral for the lightning impulse voltage test

Since all the three phases have to be tested, the position of impulsed and non-impulsed windings are changed cyclically during the test.

When testing transformers having tapped windings, the choice of tapping must be decided. As a general principle it may be stated that the test should be performed in the tap position which is the most critical from the insulation point of view. It is, however, not always easy to declare which of the tap positions is to be taken as the most critical. In the case of transformers with one tapped winding, the highest voltage across the tapped winding generally appears in the connection marked Min. At the same time, however, the upper end of the tapped winding is at earth potential, so its dielectric strength to earth cannot be verified by a test performed in the Min connection. Within the tapped winding, the most critical voltage will not necessarily arise in one of the characteristic connections. From a knowledge of the impulse voltage distribution, the manufacturer and purchaser should agree upon the connection to be applied. The lightning impulse test may also be performed for each phase in a different tapping. Also several tap positions may be tested in one phase, but the number of impulses applied should not exceed the number specified by the Standard for the testing of one phase.

Lightning impulse testing of the neutral terminal of transformers may be performed in two ways, if necessary. The first method consists of earthing the neutral through a suitable impedance, and applying to one of the line terminals of the transformer an impulse which makes a test voltage of the desired value appear at the neutral terminal. The amplitude of the impulse applied to the line terminal must not exceed 75% of the test voltage specified for that terminal. According to the second method, the specified test voltage may be applied directly to the neutral terminal. In such a case, all line terminals should be earthed. Since in this connection the surge generator is excessively loaded by the transformer, the standard front time of 1.2 µs generally cannot be attained. In such cases, a test voltage of a much longer front time, up to 13 µs, may be permitted.

It would not be justified to perform direct lightning impulse tests on windings not directly exposed in service to lightning voltages (e.g. low-voltage windings of generator-transformers). Such windings may be tested with the transferred surge method. This test is performed simultaneously with that of the high-voltage winding. The low-voltage terminals are connected to earth through resistances of such value that the desired voltage appears on them when the specified impulse voltage is applied to the high-voltage terminal. The earthing resistor should not exceed 5000 Ω.

For detecting failures that occur during an impulse voltage test, an oscillographic procedure proposed by Hagenguth is used [44]. This method is based on the fact that under the effect of a transformer winding, the curve shape of the current flowing to earth at the other end of the winding remains constant with time as long as no failure (e.g. turn-to-turn fault, breakdown between discs, breakdown or flashover to an earthed part, etc.) occurs. Under the effect of a breakdown between two conductive parts normally isolated from each other, the winding characteristics (capacitances and inductances) influencing the lightning impulse phenomenon will be subject to a sudden change, and consequently the current will also change. Thus, from the change observable in the oscillogram showing the shape of current flowing to earth at the end of the winding, conclusion can be drawn about the fault occurring within the winding. In the case of momentary faults lasting for a time short compared to the duration of an impulse voltage wave, the shape of the oscillograms is dominated by the change of capacitances. If, however, the fault persists for a longer time, then the effect of the change of inductance also shows up. Correspondingly, a fault of short duration will cause a minor distortion in the shape of the curve, whereas a longer fault will result in an increased current and a displacement of the current curve with respect to the zero line.

Thus, preceding the test, and using the same connection as during the test, a voltage of shape identical to that of the test voltage but of lower amplitude (between 50 to 75% of it) is applied to the winding, and the wave shape of the neutral current is recorded (in the connection of Fig. 5.106, on resistor R_1). It is assumed that the winding is perfect at that voltage. If the oscillogram recorded at the test voltage differs from that recorded at the reduced voltage in amplitude only, then the impulse-voltage characteristics of the winding have suffered no change under the effect of the test voltage, and the winding may therefore be regarded as having remained perfect.

If, however, the oscillogram recorded with full test voltage and that recorded for some reduced voltage differ also in their shape, then some change must have taken place in the winding, and the winding has thus suffered some kind of damage. Defects may be revealed not only by the neutral currents proposed by the original Hagenguth method, but other currents or voltages may also be used, as demonstrated by the oscilloscopes CRO II, CRO III, CRO IV, and CRO V represented in Figs 5.106, 5.107 and 5.108. For fault detection it is sufficient to take, in addition to the impulse applied to the terminal an oscillographic record of some fault detecting signal (a supplementary voltage or current). The signal to be recorded

340

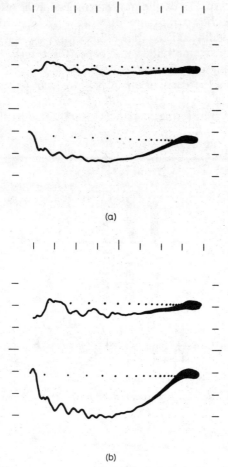

(a)

(b)

Fig. 5.109. Fault detecting oscillograms recorded during the lightning impulse voltage test of a star/star transformer. Connection is shown in Fig. 5.110: (a) record taken at 60% of test voltage; (b) record taken at test voltage

by the oscilloscopes should be selected from the available possibilities so as to obtain maximum sensitivity of fault detection.

The methods outlined above are demonstrated on a few oscillograms. Figure 5.109 shows two oscillograms recorded during the lightning impulse testing of a star/star connected transformer with neutral terminals that can be earthed during the test. The transformer was tested in the connection of Fig. 5.110. One beam (*CRO I*) of the double-beam oscilloscope has recorded the neutral current of the winding tested (lower trace), whereas the other beam *(CRO II)* has been used for recording the voltage transferred to the winding not-tested (upper trace). The sweep time of both beams was 40 µs. The logarithmic time base of the oscillograms has made it possible to observe and evaluate with equal convenience both the high-frequency oscillations appearing at the beginning of the wave, and the subsequent smooth section of the same wave. The oscillograms recorded for the tests carried out at 60% of the test voltage are shown in Figure 5.109 (a), while those of Figure 5.109 (b) are recorded at full test voltage. The latter oscillograms differ from those taken at a reduced voltage in their amplitude only, hence the transformer has successfully withstood the lightning impulse voltage test.

Figure 5.111 shows the oscillogram of voltage transferred from an impulsed lower-voltage winding to the higher-voltage winding of the same phase of a delta/delta connected transformer. No defect is indicated by the two lower curves,

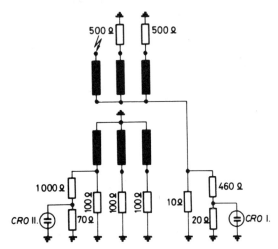

Fig. 5.110. Connection of a star/star transformer for the lightning impulse voltage testing

Fig. 5.111. Distortion in the oscillogram caused by a turn-to-turn fault in a delta/delta connected transformer

which differ from the reference curves (not shown) taken at lower voltage only in their amplitudes. The distortion, of short duration only, observable on the upper curve relates to a turn-to-turn fault.

The oscillogram of Fig. 5.112 has been taken during the tesing of a star/star connected transformer, showing the transferred voltage (upper curve) and the neutral current (lower curve) plotted at reduced voltage and in a sound state, as well as that taken at the standard test voltage showing a major—disc to disc—internal flashover. The distortion is much more marked than in the previous case, and the course of oscillations shows a complete change after the instant of breakdown.

342

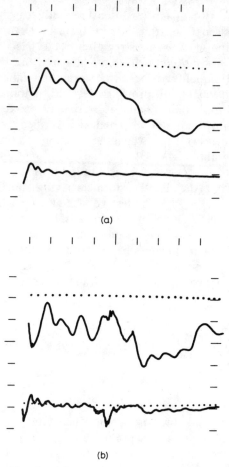

Fig. 5.112. Distortion caused by a breakdown of a larger winding section: (a) sound state; (b) faulty state

The existence of a fault can be established without any difficulty on the basis of oscillograms plotted during the tests, but not so the location of the defect. Accumulated experience can however indicate the kind of defect, e.g. whether it is a fault involving a few turns of a winding, or a breakdown between discs or layers, or a fault within the tap changer, or an earth fault. As regards a more accurate location of the fault, e.g. in the case of a turn-to-turn fault, as to which part of a winding (its beginning, middle, or end) is affected, it is extremely seldom possible to form a reliable prediction. Such information is usually obtainable only after dismantling the transformer, and it remains difficult to pinpoint the defect even then, since a minor puncture occurring under the effect of the impulse voltage hardly leaves any trace in the winding. By applying further pulses the damage can be made more extensive and the defect can be located more easily.

The magnitude of the test voltage and the number of impulses to be applied during a test are specified in the various standards. It is convenient to find a combination of these two quantities to ensure, with reasonable probability, an adequate dielectric strength of transformers against overvoltages of atmospheric origin [142]. The impulse withstand strength is not a "fixed" value, but a random variable, since breakdowns occur at different voltages with different probabilities. It is desirable to make the quality of transformer insulation sufficiently high that, when a single voltage pulse equal to the protective level V_{PL} is applied to its terminal, the impulse breakdown probability shall not exceed 0.2 per cent. Since such a low breakdown probability cannot be practically verified by measurement, the Tolerable Quality Level (TQL) on which the rejection of a transformer is based has been chosen considerably higher, at 2%. A conventional figure for the rejection probability is 90%. Consequently, for the impulse testing of transformers the aim may be to reject in 90% of the cases transformers having a breakdown probability of 2% at the protection level V_{PL}. The test voltage (V_T) and the number (n) of test impulses applied should then be selected to satisfy this requirement.

The selection of related values of V_T and n may be determined on the following basis. Assume that the breakdown probability of a transformer subjected to a single impulse at a given voltage level V_T is $P_1(V_T)$. Applying n impulses at the same voltage, the breakdown voltage according to the laws of mathematical statistics is

$$P_n = 1 - [1 - P_1(V_T)]^n. \tag{5.87}$$

Let P_n be 90%, i.e. let $P_n = 0.9$, then the number n of impulses at which breakdown will occur with 90% probability can be calculated from the relation

$$n = - \frac{1}{\lg [1 - P_1(V_T)]} \tag{5.88}$$

(see Fig. 5.113).

The actual impulse breakdown probability distribution of a given transformer is never known, in order to simplify the problem, it may be assumed for the purpose of the present calculation that the insulation of the transformer may be characterised by the properties of one of its weak points. On the basis of measurements on test pieces simulating such critical parts of the insulation, the breakdown probability is best described by a Weibull distribution having a shape parameter of 3.5 ($K = 3.5$) and a ratio of the standard mean deviation of the experimental results to the mean value being equal to 12% ($S/V = 0.12$). This distribution curve is then regarded as characteristic of the dielectric strength of the entire transformer.

With such conditions, the relation between breakdown voltage (V) and breakdown probability (P_1) can be given. It is reasonable to express the breakdown voltage as a relative value (v) (Fig. 5.114). The protection level $(V_{PL}$ is taken as the reference voltage level and $v = V/V_{PL}$, with a breakdown probability of 2% attributed to V_{PL}. Hence, curve $v(P_1)$ will show with what probability a transformer breakdown will be caused by an impulse voltage v times higher than the protective level V_{PL} for which the breakdown probability is 2%.

By means of relation (5.88) linking n with P_1, and Weibull's distribution curve $v(P_1)$, interconnected pairs of values v and n can be determined at which the breakdown probability is 90% (see Fig. 5.115). There it can be seen that e.g. for

344

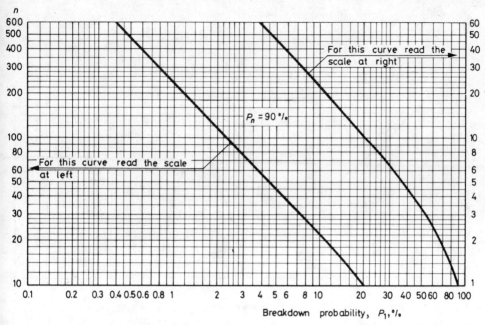

Fig. 5.113. Number of applied voltage pulses as a function of breakdown probability

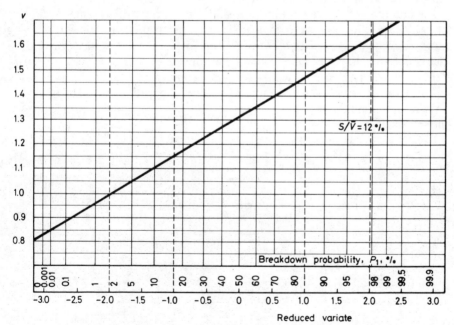

Fig. 5.114. Relation between breakdown voltage and probability of breakdown
(Weilbull distribution, $K = 3.5$)

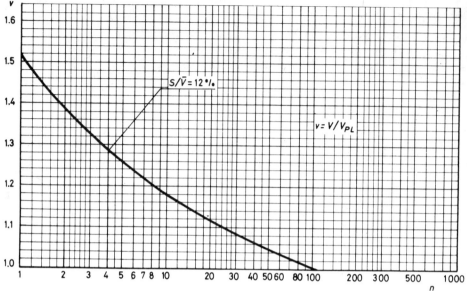

Fig. 5.115. Relation between breakdown voltage and number of applied pulses

$n=3: v=1.33$, for $n=5: v=1.26$, etc. For the rated voltage levels of 420 kV and 765 kV, some characteristic data are shown in Table 5.12.

The impulse voltage values may be read from the respective columns of Tables 5.1, 5.2 and 5.3. If a full wave impulse test is performed on the transformer, after a few reduced voltage pulses between 50 and 75% of the test voltage, three of the specified test voltage impulses should be applied to each line terminal of the transformer. If, as a special test, a test with chopped lightning impulses is also specified, then in addition to the reduced, full and chopped impulses, one full wave, two chopped-wave and again two full-wave test impulses should be applied to each line terminal.

The lightning impulse test voltage of 420 kV transformers lies in the range of 1050 to 1425 kV, and that of 765 kV transformers is between 1550 and 1950 kV (see

Table 5.12

Voltage levels for transformers with rated voltages $U_m = 420$ kV and $U_m = 765$ kV, respectively

Highest voltage for equipment, kV	Protective level, U_{PL}, kV	Lightning impulse voltage U_T, kV			
		$n=3$ $(v=1.33)$	$n=5$ $(v=1.26)$	$n=7$ $(v=1.22)$	$n=9$ $(v=1.19)$
420	878	1168	1106	1071	1045
765	1440	1915	1814	1757	1714

346

Table 5.3). When comparing these IEC test voltages with the statistical values pertaining to the 3 impulses given in Table 5.12 (i.e. with 1168 and 1915 kV, respectively), it can be seen that the values obtained by the statistical method are in very good agreement with the standard values.

Conclusions concerning the test results should be drawn on the basis of the oscillograms. The oscillograms taken during a chopped-wave test may be utilized for fault detection purposes, providing the accuracy of the chopping time is at least $\pm 0.2 \, \mu s$, otherwise discrepancies will appear in the fault signals due to the differences between the times of chopping. If the required accuracy cannot be attained, the result of the chopped-wave test should be assessed by the oscillogram of the subsequent full-wave test.

5.4.4 Switching impulse test

The aim of the switching impulse test is to verify the dielectric strength of transformers against switching surges. The test should be performed according to Method 2 as a routine test, with a wave of negative polarity and of a shape defined further above, on transformers having voltage ratings of 300 kV and above. The test voltage should be selected from the values of Table 5.3.

The usual aperiodic switching wave is produced by the same impulse generator as that used for lightning impulse testing, only the resistors serving for the adjustment of front time and time to half value have to be replaced by others of higher resistance. Damped oscillating switching waves can be generated by discharging a capacitor bank, charged to the required potential, through the primary winding of the testing transformer used for power frequency testing.

The impulse voltage phenomena are of such short duration that no saturation of the transformer core can take place. In the case of switching surges, the role of saturation becomes substantial. Thus, both the lightning impulse and the switching impulse may be considered as voltage surges bringing about unidirectional changes of flux. The duration of a switching impulse is sufficiently long to permit the flux to reach saturation level in the iron core.

The variation with time of a switching impulse and that of the current flowing through the winding under the effect of the same surge are shown by the oscillograms of Fig. 5.116. The flux waveform has been included with the current and voltage curves in the figure assuming that no remanence was present in the core, the initial flux then being zero. As the flux in the core approaches saturation, the current shows a sudden increase and the voltage rapidly decays to zero. If a higher voltage is applied to the winding the wave tail will be shorter since saturation will set in earlier. No considerable extension of the wave tail is possible by increasing the energy of the surge generator, the limiting factor being the saturation of the core. When a surge generator of higher energy is applied, only the decay in the section of the wave beyond its peak value can be retarded, the occurrence of the sudden change of voltage will not take place until a considerably later instant, and then the voltage will decay even more rapidly.

From the above it follows that during a switching impulse test, the shape of the test voltage is influenced much more by the transformer than in the case with a lightning impulse test. The wave shape depends on the voltage: the higher the test voltage, the shorter the wave tail. The development of the wave is also affected by the previous magnetic state of the core: the remanence of the same polarity as that of the test voltage will shorten the wave tail, whereas by de-magnetizing the core, or by producing a remanence of polarity opposite to that of the test voltage, the wave tail can be extended. Consequently, there are fewer constraints concerning the waves

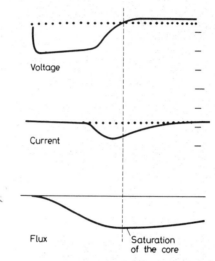

Fig. 5.116. Switching impulse wave applied to a transformer winding and consequent variation of current and flux as functions of time

shapes to be applied to the terminals than in the case of the standard lightning impulse voltage waves, as already discussed in the first part of this Chapter.

Another difference as compared to the lightning impulse test is that during the switching impulse test the terminals of the non-impulsed windings are not earthed through resistors, because the latter would affect the shape of voltage surges generated by surge generators and appearing at the transformer terminals, viz. by shortening the wave tail. This shortening effect on the wave tail is the greater the lower the resistance of earthing relative to the discharge (shunt) resistors of the surge generator. In the case of conventional surge generators, the distortion of standard lightning impulse voltage waves caused by the presence of earthing resistors is normally not so great that the wave shape fails to meet the specifications. The switching impulse waves are much more sensitive to the presence of earthing resistors, the series and parallel resistances of the surge generator then being higher than in the former case by 1 to 2 orders of magnitude: the usual earthing resistors would cause total distortion of the wave shape. This is why the terminals of non-impulsed windings are left unearthed during the switching impulse tests.

The transferred switching impulse voltage appearing at the open terminals, like the transferred lightning impulse voltage, consists of a pseudofinal voltage proportional to the turns ratio and of a superimposed oscillation. The oscillation frequency of winding systems composed of high series-capacitance windings superimposed on the transferred voltage component proportional to the turn ratio, may result in voltage peaks of considerable amplitudes. For instance, due to the superimposed oscillation, in the case of a 750/420 kV autotransformer under the effect of a switching impulse voltage wave applied to the 420 kV terminal, a transferred voltage having a crest value 30% higher than that calculated from the turn ratio has been found (Fig. 5.117). Therefore, the voltages appearing at the open winding

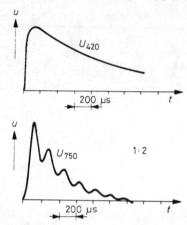

Fig. 5.117. Switching wave transferred from the L.V. terminal to the H.V. terminal of a 750/420 kV autotransformer. The voltage scale of the second oscillogram (U_{750}) is half that of the first

terminals not-tested should always be checked during measurements, since great errors may be expected in calculating the transferred voltages from the turn ratio of windings of the transformer tested.

For testing, the neutral point of the tested windings and windings not used in the test should be suitably earthed. The non-tested terminals must not be earthed either through a resistor or directly. Three-phase transformers are tested by phases in the test connection shown in Fig. 5.101, as applied also during power frequency tests. The test voltage is either directly applied to the terminal of the tested winding, or produced as a transferred voltage.

It should be borne in mind that in such a case, 1.5 times the test voltage will appear between the terminal of the tested winding and the terminals of non-tested windings.

The detection of a breakdown occurring in the course of a switching impulse test is much simpler than in the case of a lightning impulse test. Any minor defect of the winding, e.g. a turn-to-turn fault, causes a very conspicuous change in the switching surge impedance of the transformer, and gives rise to considerable distortion not only in the wave shape of the neutral current, but also in that of the terminal voltage. This is to be seen in Fig. 5.118, where oscillograms are shown for the current and voltage appearing in the lower-voltage winding of a 60 MVA three-phase transformer in sound condition, and for the case when 0.8% of the turns are artifically shorted [4]. Therefore when a switching impulse test is performed, it is sufficient to investigate the oscillogram of the voltage applied to the transformer, for this method will provide a reliable proof of any breakdown that may have occurred within the transformer.

As in the lightning impulse test, testing begins with a switching impulse of reduced voltage (50 to 70% of the full test voltage) followed by applying three impulses with

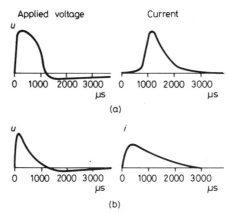

Fig. 5.118. Distortion in the shape of switching impulse voltage (u) and current (i) caused by a turn-to-turn fault: (a) sound winding; (b) fault extending over 0.8% of the turns

their peak values equal to the full test voltage. In order to observe possible voltage dependent changes in the wave shape, it is advisable to take oscillographic records at a few intermediate voltages in the range between the reduced and full values of the test voltage.

5.4.5 Testing of external insulation

The tests discussed so far have related to the checking of the internal insulation of transformers. No separate testing of any external insulation is required if the clearances are satisfactory, e.g. the requirements specified by the relevant IEC Recommendations (Pub. 71–2, 71–3) or by some other approved standards are satisfied (see also Table 5.5). If the actual clearances are less than those specified therein, a 1:1 scale model of the transformer should be built, but containing only those parts necessary to allow the dielectric strength of the external insulation to be checked. The test voltages for the external insulation are generally higher than those specified for the internal insulation.

6. TEMPERATURE RISE
AND COOLING OF TRANSFORMERS

Electrical, thermal and hydraulic problems are of equal importance in the design of transformers. Generally, the solution of thermal problems are those which cause difficulties for an electrical engineer. In this Chapter the most important items of information that should be known for a better understanding of physical phenomena of the warming and cooling processes taking place during the service of transformers are summarized.

As known, the role of the oil in oil-immersed transformers, in addition to serving as insulation, is to act as a heat-transmitting cooling medium. The heat, of which the major part is generated in the active parts of the transformer, is taken up by the oil and carried into some kind of heat-exchanger. In most cases this heat-exchanger is an oil–air cooler, or less frequently an oil–water cooler. Therefore, this Chapter deals with the cooling of an active part, the winding, and with radiator-type and compact oil–air coolers.

The subject outlined above is discussed in the following four sections: general concepts and phenomena, cooling of winding, radiator cooling, and compact heat exchangers.

6.1 General concepts and phenomena

6.1.1 Temperature rises and temperatures in the transformer

According to design standards, the temperature rise is the difference between the temperature of a given part of the transformer and that of the cooling medium. The temperature limits permissible in the active parts influence the constructional design, size, cost, load carrying capacity and operating conditions of the transformer.

The maximum continuous load carrying capacity of the transformer depends on its rating (to which the guaranteed data refer), on the temperature of the cooling medium, and on whether or not the user wants to slow down the aging of insulations which takes place under the effect of temperature and time. In addition, the magnitude and duration of periodic maximum load carrying capacity depends on the constructional features of the transformer, on the operating conditions preceding the given overload and on the magnitude of time constants.

According to IEC Publ. 76, 1976 [110], the average temperature rise of a transformer winding, determined by the resistance method, above that of the coolant should not exceed 65 °C. The average temperature rise of the winding is denoted by $\Delta\vartheta_w$ and is expressed in °C.

The temperature rise of the oil above the temperature of the cooling medium, measured below the tank cover, should not exceed 60 °C. The maximum oil temperature rise is denoted as $\Delta\vartheta_{om}$, °C.

The maximum ambient temperature is 40 °C for air cooling, and 25 °C for water cooling. The temperature of air as cooling medium is denoted by ϑ_a, °C. The permissible temperature rise levels are equal for both air cooling and water cooling. This means that the average temperature of a transformer winding is by 15 °C lower with water cooling than with air cooling. The increased safety margin applied in the case of water cooling is justified by the frequent contamination of the water side of oil–water coolers during service.

The temperature rise limits given above are valid for transformers installed at an altitude not exceeding 1000 m above sea level. If the transformer is intended for use at altitudes above 1000 m, and is tested at an altitude lower than 1000 m, then the temperature rise limits should be reduced by 2% in the case of natural air cooling and by 3% in the case of forced air cooling for every extra 500 metres (or fraction thereof) above the 1000 metres altitude. By this rule the Standard takes into account the reduced cooling effect of air due to its decreasing density with increasing altitude. With water cooling the temperature rise limits need not be reduced in consideration of the altitude of the site of installation.

With ambient air temperatures exceeding 40 °C or with cooling water temperature exceeding 25 °C, the temperature rise limits specified by the Standard should be reduced by the same extent as by which the temperature of the cooling medium exceeds 40 °C in the case of air and 25 °C in that of water.

The average temperature of the uppermost winding means the average temperature of the winding located at the top of the winding stack, when the transformer is loaded with its rated load. The average temperature of the uppermost winding is denoted by ϑ_{wm}, and its temperature rise by $\Delta\vartheta_{wm}$, both expressed in °C. The average temperature of the uppermost winding is

$$\vartheta_{wm} = \vartheta_a + \Delta\vartheta_w + \frac{\Delta\vartheta_{wo}}{2}, \tag{6.1}$$

where $\Delta\vartheta_{wo}$ is the difference between inlet and outlet temperature of the oil.

By the temperature of the hottest spot of the winding, the temperature of the hottest inner layer of the disc located at the top of the winding stack is to be understood. The temperature of the hottest spot is ϑ_c, °C. The hottest layer is called the hottest spot of the winding, and its temperature is always higher than the average temperature of the uppermost winding.

In the IEC Publ. 76, 1976 the rate of aging of interturn insulation of transformers under the effect of time and temperature is referred to the hottest spot temperature of 98 °C, i.e. to a value which normally occurs with 20 °C cooling air temperature and under the continuous rated load.

The temperature of the hottest spot or "hot spot" of the winding is not specified by the IEC Publ. 76, 1976. For insulating materials of Class A, 115 °C is permitted as the highest service temperature. The checking of this temperature limit, however, comes up against practical difficulties in the case of transformers, so not even a calculation method is recommended by the standards for that purpose.

In transformers which satisfy the standard temperature rise specifications even the average temperature of the innermost winding may exceed the 115 °C limit, if the manufacturer has failed to proceed with due care. For example, let the ambient temperature be $\vartheta_a = 40$ °C, the average temperature rise in the winding $\Delta\vartheta_w = 65$ °C, and the difference between inlet and outlet temperatures of the oil $\Delta\vartheta_{wo} = 40$ °C (a figure considerably greater then usual, but not forbidden by the Standards). With these values the average service temperature of the winding is

$$\vartheta_{wm} = \left(40 + 65 + \frac{40}{2}\right) = 125 \text{ °C}.$$

The hot-spot temperature lies a few centigrades above this value.

For overloads of short duration, in the IEC Publ. 354, 1972 [84] "Loading Guide for Oil-Immersed Transformers", 140 °C is permitted as the highest hot spot temperature, whereas in short-circuit conditions, 250 °C is an average winding temperature in copper windings, and 200 °C an average winding temperature in aluminium windings. In practical cases, due to the short fault clearing times and low service current densities, the short-circuit temperature rise limits are not reached.

In order to obtain the value of average temperature of the uppermost winding permitted for the case of short-circuits, half of the difference between temperatures of the oil leaving and entering the winding has to be added to the 250 °C or 200 °C, respectively. Correspondingly, during a short-circuit, the average temperature of the uppermost winding in the case of copper may rise to 260 °C, and to 210 °C in the case of aluminium.

Temperature rises and temperature drops not limited
by the Standards

By the mean oil temperature rise the mean temperature rise of the oil streaming along the surfaces of the winding is to be understood. The temperature of the oil entering the winding at the bottom of the winding stack may be assumed to increase to a first approximation, in proportion to the length of flow path within the winding. For natural oil flow the mean average oil temperature rise is obtained by subtracting from its maximum temperature rise one half of the temperature difference of the oil on leaving and entering the winding. The temperature of the oil leaving the winding, ϑ_{oow}, °C, is equal to the maximum oil temperature:

$$\vartheta_{oow} = \vartheta_{om}.\tag{6.2}$$

The mean oil temperature rise of the oil of transformers with forced oil cooling may be determined by means of calculation, from the results of temperature-rise tests. The average temperature rise of the oil is denoted by $\Delta\vartheta_{oaw}$, °C.

The logarithmic-mean temperature difference between cooled oil and cooling air is $\Delta\vartheta_{o-a}$, °C. In the case of radiator-type cooling this log–mean temperature difference consists of two parts, viz. the air-side and the oil-side temperature drops. (The temperature drop taking place in the radiator wall is neglected.) For a given cooling method the amount of heat to be dissipated and the logarithmic-mean temperature difference define the size of the cooler.

By adding to the ambient temperature the mean temperature rise of the air flowing through the cooler and the logarithmic-mean temperature difference between oil and cooling air, the average temperature of the oil in the cooler, ϑ_{oac}, °C, is obtained:

$$\vartheta_{oac} = \vartheta_a + \frac{\Delta\vartheta_a}{2} + \Delta\vartheta_{o-a}. \tag{6.3}$$

With natural oil circulation, this value differs but slightly from the average temperature of the oil flowing in the winding.

$$\vartheta_{oaw} \cong \vartheta_{oac}. \tag{6.4}$$

In the case of forced oil circulation, due to the mixing of the oil in the transformer tank, the average temperature of oil flowing in the winding is higher than the average temperature of the oil in the cooler:

$$\vartheta_{oaw} > \vartheta_{oac}. \tag{6.5}$$

The winding average–oil average temperature gradient is obtained by subtracting the oil average temperature rise from the winding average temperature rise. This

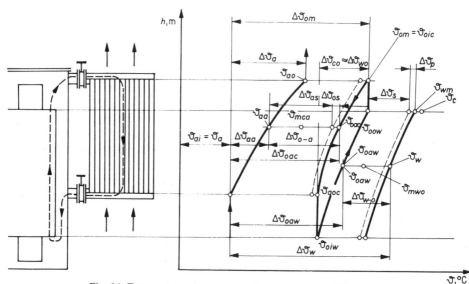

Fig. 6.1. Temperatures, temperature rises, temperature differences

354

Table 6.1

Explanation of symbols indicating temperatures, temperature differences
and temperature rise values appearing in Fig. 6.1 and in the text

Notation	Description
ϑ	temperature
$\varDelta\vartheta$	temperature difference or temperature rise over ambient temperature
ϑ_a	ambient temperature
ϑ_{ai}	inlet air temperature
ϑ_{aa}	average air temperature
ϑ_{ao}	outlet air temperature
ϑ_m	mean boundary layer temperature
$\varDelta\vartheta_a$	temperature rise of air in the cooler, difference of air outlet and inlet temperatures
ϑ_{oiw}	winding inlet oil temperature
ϑ_{oaw}	winding average oil temperature
ϑ_{oow}	winding outlet oil temperature
$\varDelta\vartheta_{oaw}$	average temperature rise of oil over ambient temperature in the windings
$\varDelta\vartheta_{wo}$	difference of outlet and inlet oil temperatures in the windings
$\varDelta\vartheta_{o-a}$	logarithmic-mean temperature difference between oil and air in the cooler
ϑ_{oic}	cooler inlet oil temperature
ϑ_{oac}	cooler average oil temperature
ϑ_{ooc}	cooler outlet oil temperature
$\varDelta\vartheta_{co}$	difference of cooler inlet and outlet oil temperatures
$\varDelta\vartheta_{oac}$	cooler average-oil temperature rise over ambient temperature
ϑ_{om}	maximum oil temperature
$\varDelta\vartheta_{om}$	maximum oil temperature rise over ambient temperature
$\varDelta\vartheta_w$	average temperature rise of winding over ambient temperature
ϑ_{wm}	average temperature of uppermost disc
ϑ_c	hot-spot temperature of winding
$\varDelta\vartheta_c$	hot-spot temperature rise of winding over ambient temperature
$\varDelta\vartheta_{w-o}$	difference between average winding and average oil temperatures
$\varDelta\vartheta_p$	temperature drop through insulation
$\varDelta\vartheta_s$	temperature drop from the surface of the insulation to the surrounding oil
$\varDelta\vartheta_{os}$	surface temperature drop from the oil to the cooler wall
$\varDelta\vartheta_{wm}$	average temperature rise of uppermost disc over ambient temperature
ϑ_{mwo}	mean boundary layer temperature of oil in the windings
ϑ_{mca}	mean boundary layer temperature of air in the cooler
$\varDelta\vartheta_{as}$	surface temperature drop from the cooler surface to the ϑ_{aa} average air temperature in the cooler

Table 6.2

Physical properties of transformer oil

Temper-ature ϑ, °C	Density ρ, kg m^{-3}	Specific heat c_p, W s kg^{-1} °C^{-1}	Thermal conductivity γ, W m^{-1} °C^{-1}	Coefficient of thermal cubic ex-pansion β, °C^{-1}	Kinematic viscosity ν, m^2 s^{-1}	Temperature distribution coefficient α m^2 s^{-1}	Prandtl number Pr
−15	902	1700	0.1341	6.20×10^{-4}	450×10^{-6}	8.758×10^{-8}	5130
−10	899	1720	0.1337	6.35×10^{-4}	290×10^{-6}	8.663×10^{-8}	3340
−5	896	1740	0.1333	6.40×10^{-4}	180×10^{-6}	8.568×10^{-8}	2100
0	893	1764	0.1330	6.55×10^{-4}	120×10^{-6}	8.453×10^{-8}	1419
5	890	1785	0.1326	6.70×10^{-4}	85×10^{-6}	8.359×10^{-8}	1016
10	887	1805	0.1322	6.80×10^{-4}	62×10^{-6}	8.270×10^{-8}	749
15	884	1825	0.1318	6.90×10^{-4}	45×10^{-6}	8.182×10^{-8}	549
20	882	1848	0.1314	7.00×10^{-4}	35×10^{-6}	8.074×10^{-8}	433
25	879	1870	0.1310	7.15×10^{-4}	27×10^{-6}	7.989×10^{-8}	337
30	876	1890	0.1306	7.25×10^{-4}	21×10^{-6}	7.906×10^{-8}	265
35	873	1910	0.1303	7.40×10^{-4}	17×10^{-6}	7.823×10^{-8}	217
40	870	1930	0.1299	7.50×10^{-4}	14×10^{-6}	7.742×10^{-8}	180
45	867	1950	0.1295	7.65×10^{-4}	11.5×10^{-6}	7.662×10^{-8}	150
50	864	1975	0.1291	7.75×10^{-4}	9.7×10^{-6}	7.583×10^{-8}	127
55	861	1995	0.1287	7.80×10^{-4}	8.2×10^{-6}	7.524×10^{-8}	108
60	858	2015	0.1283	7.90×10^{-4}	7.0×10^{-6}	7.447×10^{-8}	93.9
65	855	2040	0.1280	7.90×10^{-4}	6.1×10^{-6}	7.371×10^{-8}	82.7
70	852	2060	0.1276	7.95×10^{-4}	5.4×10^{-6}	7.297×10^{-8}	74
75	849	2080	0.1272	7.95×10^{-4}	4.7×10^{-6}	7.223×10^{-8}	65
80	847	2100	0.1268	7.95×10^{-4}	4.2×10^{-6}	7.151×10^{-8}	58.7
85	844	2120	0.1264	8.00×10^{-4}	3.8×10^{-6}	7.079×10^{-8}	53.6
90	841	2140	0.1260	8.00×10^{-4}	3.5×10^{-6}	7.025×10^{-8}	49.8
95	838	2160	0.1257	8.00×10^{-4}	3.2×10^{-6}	6.955×10^{-8}	46
100	835	2180	0.1253	8.00×10^{-4}	3.0×10^{-6}	6.887×10^{-8}	43.5

Table 6.3

Physical properties of dry air at pressure
$P = 101\,325\ \mathrm{Nm^{-2}}$

Temper-ature ϑ, °C	Density ρ, kg m^{-3}	Specific heat c_p, W s kg^{-1} °C^{-1}	Thermal conductivity γ, W m^{-1} °C^{-1}	Coefficient of thermal cubic ex-pansion β, °C^{-1}	Kinematic viscosity v, m^2 s^{-1}	Temperature distribution coefficient a, m^2 s^{-1}	Prandtl number Pr
−20	1.365	1008	2.26×10^{-2}	253^{-1}	11.93×10^{-6}	16.50×10^{-6}	0.724
0	1.252	1008	2.37×10^{-2}	273^{-1}	13.70×10^{-6}	18.95×10^{-6}	0.723
10	1.206	1008	2.455×10^{-2}	283^{-1}	14.70×10^{-6}	20.40×10^{-6}	0.722
20	1.164	1012	2.525×10^{-2}	293^{-1}	15.70×10^{-6}	21.80×10^{-6}	0.722
30	1.130	1012	2.580×10^{-2}	303^{-1}	16.61×10^{-6}	23.05×10^{-6}	0.722
40	1.092	1012	2.650×10^{-2}	313^{-1}	17.60×10^{-6}	24.40×10^{-6}	0.722
50	1.056	1016	2.720×10^{-2}	323^{-1}	18.60×10^{-6}	25.80×10^{-6}	0.722
60	1.025	1016	2.800×10^{-2}	333^{-1}	19.60×10^{-6}	27.20×10^{-6}	0.722
70	0.996	1016	2.865×10^{-2}	343^{-1}	20.45×10^{-6}	28.40×10^{-6}	0.722
80	0.968	1020	2.930×10^{-2}	353^{-1}	21.70×10^{-6}	30.10×10^{-6}	0.722
90	0.942	1020	3.000×10^{-2}	363^{-1}	22.90×10^{-6}	31.80×10^{-6}	0.722
100	0.916	1020	3.070×10^{-2}	373^{-1}	23.78×10^{-6}	32.95×10^{-6}	0.722

Table 6.4

Physical properties of water on the saturation line

Temperature ϑ, °C	Density ρ, kg m^{-3}	Specific heat c_p, W s kg^{-1} °C^{-1}	Thermal conductivity γ, W m^{-1} °C^{-1}	Coefficient of thermal cubic expansion β, °C^{-1}	Kinematic viscosity ν, m^2 s^{-1}	Temperature distribution coefficient α m^2 s^{-1}	Prandtl number Pr
0	999.8	4236	0.551	-0.63×10^{-4}	1.790×10^{-6}	13.0×10^{-8}	13.70
10	999.6	4211	0.574	$+0.88 \times 10^{-4}$	1.300×10^{-6}	13.6×10^{-8}	9.56
20	998.2	4202	0.599	2.07×10^{-4}	1.000×10^{-6}	14.2×10^{-8}	7.06
30	995.2	4198	0.617	3.04×10^{-4}	0.805×10^{-6}	14.7×10^{-8}	5.50
40	992.2	4198	0.634	3.90×10^{-4}	0.659×10^{-6}	15.3×10^{-8}	4.30
50	988.0	4198	0.648	4.60×10^{-4}	0.556×10^{-6}	15.5×10^{-8}	3.56
60	983.2	4202	0.659	5.30×10^{-4}	0.479×10^{-6}	16.1×10^{-8}	3.00
70	977.7	4211	0.667	5.80×10^{-4}	0.415×10^{-6}	16.1×10^{-8}	2.56
80	971.8	4215	0.674	6.30×10^{-4}	0.366×10^{-6}	16.4×10^{-8}	2.23
90	965.3	4223	0.680	7.00×10^{-4}	0.326×10^{-6}	16.7×10^{-8}	1.95
100	958.3	4228	0.682	7.50×10^{-4}	0.295×10^{-6}	16.9×10^{-8}	1.75

temperature gradient consists of two parts, viz. from the temperature drop taking place in the insulation of the winding material and from the surface temperature drop between the outer surface of the insulation and the oil surrounding the winding.

$$\Delta \vartheta_{w-o} = \Delta \vartheta_w - \Delta \vartheta_{oaw},\tag{6.6}$$

$$\Delta \vartheta_{w-o} = \Delta \vartheta_p + \Delta \vartheta_s.\tag{6.7}$$

Calculation of thermal phenomena of the transformer winding begins with determining the above temperature drop. Subtracting it from the average temperature rise of the winding permitted by the Standard, we get for natural oil flow a sum composed of the average temperature rise of cooling air in the cooler and of the logarithmic-mean temperature differences between cooled oil and cooling air. For forced oil flow, the effect of mixing within the transformer tank should also be considered. The winding average–oil average temperature gradient is

$$\Delta \vartheta_{w-o}, {}^{\circ}C.$$

In designing the transformer and transformer coolers, in addition to the above, several further temperature gradients and drops are used. Explanation of the notations are given in Fig. 6.1 and Table 6.1. In Fig. 6.1 the temperatures, temperature rises and drops of a natural air cooled, natural oil flow transformer are shown in the coordinate system of temperature ϑ and height h.

The physical characteristic data of transformer oil, air and water will be required later, and are listed in Table 6.2, 6.3 and 6.4.

6.1.2 Aging of insulating materials

The structure of insulating materials built into transformers, mainly those based on cellulose, is subject to aging, which is due to the effects of heat and oxygen and depends on the duration of their influence.

The original electrical, mechanical and chemical properties of insulating paper are modified by aging. Although the reduction of dielectric strength due to aging is not considerable, the modifications to mechanical properties (folding endurance and rip strength) render the transformer sensitive to the displacements of windings caused by the electrodynamic effects of short-circuits, as a result of which a transformer with aged insulation becomes to suffer turn-to-turn faults. The breaking strength, as characterized by the tensile length, of but slightly aged insulation drops to a fraction of its original value. Among the chemical properties, the change in the degree of polymerization, i.e. shortening of the length of the molecular chain is the most characteristic modification (degree of polymerization DP).

From laboratory experiments on insulating materials and several decades of experience, it has been found that no unequivocally determinable dangerous temperature limit can be found for an insulating material. But only an exponential relation between temperature and duration of the thermal effect and the extent of aging can be established for a given insulating material. Considering this empirical law, the time of amortization, and the technical obsolescence of the transformer, a temperature limit can be assigned below which the rate of aging can be expected to permit an economic service life. Aging takes place at a higher rate in the presence of moisture. There are three ways in which moisture may access the transformer. It can either be left at the end of the drying process as residual moisture, or it can enter the transformer through leaks or through the conservator, or it may appear as a by-product of the aging of insuling materials. Oxygen enters the transformer either through leaks where the pressure in the cooling circuit is lower than atmospheric, or through the conservator.

Service life expectancy and deterioration

The period of time passing until deterioration of an insulating material becomes critical is called the service life or life expectancy. According to the Arrhenius law, the life duration of an insulating material is

$$E = Ce^{A-\frac{B}{T}} \quad \text{years}, \tag{6.8}$$

where $C = 1$ year and A and B are experimentally determined constants valid for the insulating material in question, A is a dimensionless number, B is expressed in K, T is the absolute temperature in K.

The life expectancy formula of Montsinger applicable to the 80 to 140 °C range of temperatures occurring in transformers is

$$E = De^{-p\vartheta} \quad \text{years}, \tag{6.9}$$

where p is a constant in °C^{-1}, ϑ is the temperature in °C and D is a constant expressed in years.

The value of p required for the determination of the service life has not been specified in the IEC recommendation "Loading Guide for Oil-Immersed Transformers", because no agreement has been reached due to the different views prevailing in the formulation of physical and chemical properties of insulating material which has deteriorated.

Loss of life expectancy

Opinions agree in stating that in the temperature range 80 to 140 °C, the life expectancy is halved under the effect of an increase in temperature $\Delta\vartheta = 6$ °C.

This means that if a service life of E years applies for temperature ϑ in the temperature range of 80 to 140 °C, a temperature of $\vartheta + 6$ °C will result in a lifeduration of $E/2$ years. With these data the numerical value of constant p °C^{-1} appearing in Montsinger's formula can be determined:

$$E = De^{-p\vartheta} \quad \text{years},$$

$$\frac{E}{1} = De^{-p(\vartheta + 6^{\circ}C)} \quad \text{years}.$$

360

Dividing one equation by the other

$$2 = e^{p6\,°C}$$

is obtained, whence

$$p = 0.1155\,°C^{-1}.$$

(6.10)

Relative reduction of life expectancy

If the life expectancy at some temperature ϑ_b is chosen to be "normal", then that applicable to an arbitrary temperature ϑ can be related to this "normal" life expectancy. The ratio of these two life expectancies is called the relative reduction of life and is denoted by V (a dimensionless number). Using Montsinger's formula:

$$V = \frac{De^{-0.1155\vartheta_b}}{De^{-0.1155\vartheta}} = e^{0.1155(\vartheta - \vartheta_b)}.$$

(6.11)

The dimension of exponent 0.1155 appearing in the formula is $°C^{-1}$.

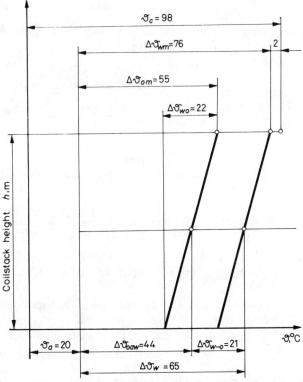

Fig. 6.2. Hot-spot temperature, average winding and average oil temperature rises of normally aging transformer

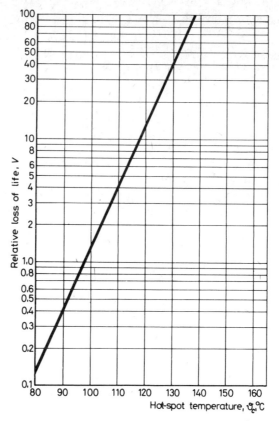

Fig. 6.3. Relative loss of life as a function of hot-spot temperature

Let ϑ_b be chosen to be 98 °C. This temperature corresponds to the hottest spot temperature of a transformer operated at an ambient temperature of $\vartheta_a = 20$ °C, the average winding temperature rise of which is $\Delta\vartheta_w = 65$ °C, and the difference between temperatures of the oil entering and leaving the winding is $\Delta\vartheta_{wo} = 22$ °C, to which a further 2 °C temperature differential is added to obtain the temperature of the hot-spot (see Fig. 6.2).

$$\vartheta_b = \vartheta_c = \vartheta_a + \Delta\vartheta_w + \frac{\Delta\vartheta_{wo}}{2} + 2\,°C = \left(20 + 65 + \frac{22}{2} + 2\right)°C = 98\,°C .$$

Hence, the insulation of a transformer with 98 °C hot-spot temperature will deteriorate at the normal rate. The service life of such a transformer, due to aging, is a few decades. Obviously, during this period the transformer may be damaged for some other reasons, or it may happen that it will be scrapped due to technical obsolescence.

Substituting $\vartheta_b = 98$ °C into the above formula of relative reduction of service life, a relation suitable for practical calculations is obtained:

362

Table 6.5

Relative loss of life as a function of hot-spot
temperature

Hot-spot temperature, °C	Relative loss of life V
80	0.125
86	0.25
92	0.5
98	1.0
104	2.0
110	4.0
116	8.0
122	16.0
128	32.0
134	64.0
140	128.0

$$V = e^{0.1155(\vartheta - 98°C)}. \qquad (6.12)$$

Plotting this relation obtained for V in a coordinate system with linear ϑ axis and logarithmic V axis, a straight line results, (see Fig. 6.3). Using the numerical data of Table 6.5 the relative loss of life pertaining to a given hot-spot temperature and to a given life-duration can easily be calculated.

Example 6.1 On a summer day, let the hot-spot temperature of a transformer be 98 °C through 12 h. 104 °C through 8 hours and 116 °C through 4 hours. What will be the loss of life relative to normal life loss?

$$V = \frac{12 \cdot 1 + 8 \cdot 2 + 4 \cdot 8}{24 \cdot 1} = 2.5.$$

Thus, operation of the transformer for one such day will cause a reduction of life equivalent to 2.5 days of normal operation.

Time of operation at constant hot-spot temperature causing 1 day reduction of life

Suppose a transformer operates for t_c hours during a 24-hour period with a hot-spot temperature of ϑ_c, which is considerably higher than 98 °C, whereas through $(24 - t_c)$ hours the hit-spot temperature is much lower than 98 °C. If we want to have the same reduction of life as in normal operation for one day, t_c can be calculated from the following formula:

$$t_c = \frac{24}{V} \text{ h.} \qquad (6.13)$$

Example 6.2 The load imposed on a transformer is such that the hot-spot temperature does not exceed 80 °C for the greater part of the day, but for a few hours during the peak load period we want to overload the transformer so that its hot-spot temperature rise to 122 °C. How long may the period of overload last, if the resulting life loss is to be kept within that associated with normal operation? The value of V for ϑ_c = 122 °C is read from Table 6.5 and is substituted in the formula (6.13):

$$t_c = \frac{24}{16} h = 1.5 \, h.$$

Naturally, when the time constant of the transformer is also considered, the duration of overload may well exceed that obtained above. In the case of hot-spot temperature not exceeding 80 °C the relative loss of life is taken as zero.

Weighted ambient temperature

If the ambient temperature ϑ_a during a period t of prolonged operation under load is subject to substantial changes, then the value of the hot-spot temperature ϑ_c will change though with a delay depending on the time constant of the transformer. The ambient temperature required for calculating ϑ_c in such cases is not the arithmetical average ambient temperature taken over the prolonged period concerned, but a weighted ambient temperature ϑ_{awt}. This weighted ambient temperature is obtained by applying the following simplifying assumption.

The useful life of a transformer depends on the hot-spot temperature ϑ_c, and changes occurring in the ambient temperature influence the hot-spot temperature just like variations of load. An increase $\Delta\vartheta = 6$ °C in the ambient temperature halves the life expectation, just as if the 6 °C increase in the hot-spot temperature were due not to change of ambient temperature but to an increased load.

The weighted ambient temperature ϑ_{awt} is that temperature which lasts for the same t period of time and causes the same loss of life expectancy as an ambient temperature $\vartheta_a(t)$ which is varying in magnitude with time. The following equation can be written for equal reduction of life expectancy:

$$t e^{0.1155 \vartheta_{awt}} = \int_0^t e^{0.1155 \vartheta_a(t)} dt .$$

Writing the logarithms of both sides,

$$\vartheta_{awt} = \frac{\left(\ln \int_0^t e^{0.1155 \vartheta_a(t)} dt \right) - \ln t}{0.1155} \tag{6.14}$$

is obtained.

Let period t be divided into n equal intervals and, instead of integration, apply the method of summation.

After rearrangement, the following expression is obtained for the weighted ambient temperature:

$$\vartheta_{awt} = 8.65 \cdot \ln \frac{1}{n} \sum_{i=1}^{n} e^{0.1155 \vartheta_{ai}} . \tag{6.15}$$

Example 5.3. During a year, suppose the monthly average ambient temperatures are as follows:

$$\vartheta_{a1} = 30 \text{ °C for 2 months,}$$
$$\vartheta_{a2} = 20 \text{ °C for 6 months,}$$
$$\vartheta_{a3} = 10 \text{ °C for 2 months,}$$
$$\vartheta_{a4} = 0 \text{ °C for 2 months.}$$

The yearly average ambient temperature:

$$\vartheta_{aa} = \frac{30 \times 2 + 20 \times 6 + 10 \times 2 + 0 \times 2}{12} = 16.67\,°C.$$

The yearly weighted ambient temperature:

$$\vartheta_{awt} = 8.65 \ln\left[\frac{1}{12}(2e^{0.1155 \times 30} + 6e^{0.1155 \times 20} + 2e^{0.1155 \times 10} + \right.$$

$$\left. + 2e^{0.1155 \cdot 0})\right] = 20.8\,°C.$$

Thus the yearly weighted ambient temperature is higher by 4.13 °C than the yearly average ambient temperature.

6.1.3 Mode of cooling

Natural oil flow

In the closed-circuit cooling system of the transformer the oil is kept in circulation by the gravitational buoyancy. The heat developing in the active parts passes, through surface transfer, into the oil surrounding the active parts. The oil warms up under the effect of heat, its specific gravity decreases and it flows upwards in the surroundings of active parts, while the ascending oil is replaced by cold oil streaming in below. The hot oil dissipates its heat along the colder walls of the coolers, its specific gravity increases and it flows downwards. In the coolers, a flow of downward direction, i.e. opposed to that prevailing along the active parts, is produced.

Air flow in the coolers

The air flow in the coolers may be brought about either naturally by the lower specific gravity of the hot air, or by means of fans. The system is called natural air cooling in the former case, and, forced air cooling in the latter.

Natural oil flow, natural air cooling

A schematic diagram of this cooling system is given in Fig. 6.4. The physical phenomenon is represented, according to established usage, in the coordinate system of temperature ϑ and height h. The oil entering the winding at point A is heated within the winding, streams upwards, and leaves the winding at point B. Along the section between B and C, i.e. up to the point where it enters the cooler, the oil is cooled slightly, due to heat dissipation occurring at the top of the tank side and toward the tank cover. The oil is cooled in the radiators, between points C and D, and it descends. The cooled oil flows from D to A, enters the winding, and the process repeats itself. In the diagram, $\Delta\vartheta_{o-a}$ is the logarithmic—mean temperature difference between cooling down oil and warming up air, $\Delta\vartheta_{wo}$ is the temperature difference between oil leaving and entering the winding, and $\Delta\vartheta_{co}$ equal to the former value, is the temperature difference between oil entering and leaving the radiator.

365

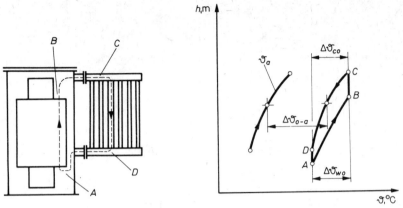

Fig. 6.4. Schematic diagram of natural oil flow, natural air cooling. Area $ABCDA$ is proportional to the buoyancy maintaining the oil flow

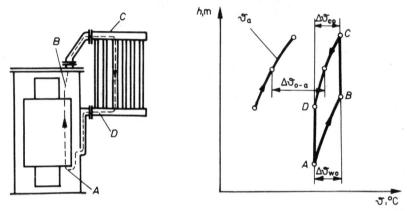

Fig. 6.5. Schematic diagram of natural oil flow, natural air cooling, with radiators placed high

By mounting the radiators at an elevated height, as shown in Fig. 6.5, as compared to a low arrangement, the buoyancy acting in the cooling circuit can be increased, assuming an equal amount of heat produced. Correspondingly, the value $\Delta\vartheta_{wo} = \Delta\vartheta_{co}$ diminishes, but that of $\Delta\vartheta_{o-a}$ remains virtually unchanged. The flow rate of the oil in the cooling circuit will be higher.

Natural oil flow, forced air cooling

An outline of this cooling system is shown in Fig. 6.6. The cooling air is blown over the radiators by a fan. Due to the high air velocity, the air-side heat transfer coefficient is increased, thus compared to natural air cooling a smaller temperature drop on the air side will suffice for transferring an equal quantity of heat. Again as

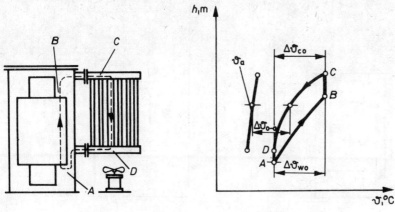

Fig. 6.6. Schematic diagram of natural oil flow, forced air cooling

compared to natural air cooling, assuming the same amount of loss to be dissipated, the cooling of the oil will be faster due to the better heat transfer coefficient on the air side, and branch $C-D$ will be somewhat more convex. As a first approximation $\Delta\vartheta_{wo} = \Delta\vartheta_{co}$ remains unchanged in magnitude, or perhaps rises slightly, due to the higher viscosity of the oil circulating in the complete heating system which is the result of the substantial reduction of $\Delta\vartheta_{o-a}$ achieved by introducing the system of forced air cooling. This latter phenomenon follows from the fact that for achieving a given heat dissipation, a lower temperature drop is sufficient, because of the better heat transfer coefficient on the air side. By changing over from natural air cooling to forced air cooling, the cooling can be improved about 2.6-fold, with an identical value of $\Delta\vartheta_{o-a}$.

Axial-flow fans are generally used for forced air cooling, mostly with vertical blow, and with horizontal blow in the remaining cases.

Forced oil flow

The transfer of substantial amounts of waste heat generated in the active parts of large transformers to the air, requires a voluminous and costly radiator battery. Constructional considerations prevent, in many cases, to mount radiator groups high enough to ensure the buoyancy by which the velocity of oil flow in the cooling circuit would be improved and, thereby, the values of $\Delta\vartheta_{wo}$ and $\Delta\vartheta_{co}$ reduced.

It is convenient in the case of large power transformers to change over from the expensive and bulky radiator batteries to the use of compact coolers. In such coolers, high oil flow velocities are employed in order to obtain good oil-side heat transfer coefficients. To achieve the required high flow rates of oil and to overcome the increased oil-side flow resistance, installation of an oil pump on the oil side of the cooling circuit is necessary. Within the windings the oil flows practically in the natural way, at a velocity defined by gravitational buoyancy and viscous resistance

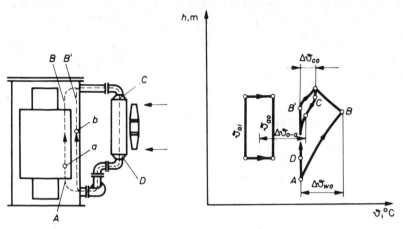

Fig. 6.7. Schematic diagram of forced oil flow, forced air cooling

of the streaming oil. The insertion of oil pumps influences mainly the coolers. For a better survey of the problem let Fig. 6.7 be examined.

In Fig. 6.7 the principle of cooling of a forced oil flow transformer is represented in the temperature–height coordinate system. As can be seen, the heat transfer process taking place inside the winding is but very slightly affected by the pumping of the oil. This method of cooling is characterized by the mixing of the oil heated up in the winding while flowing along the path denoted by a with the oil streaming along the tank wall through the shunt path denoted b. Because of this mixing, the temperature of the oil ascending from the winding cannot be measured directly, since the temperature of a mixture can only be measured by thermometers accommodated in the tank cover. The oil entering the coolers is also of lower temperature, thus the temperature of the oil cooled by the coolers is lower than the maximum oil temperature. Another effect of oil mixing is that, with identical heat loss, the pumps have to deliver more oil of lower temperature than if there were no oil mixing. Because the flow velocity of the oil in the winding changes but very little as compared to that of natural oil flow, heat flux per unit transfer area permitted for the windings must not be higher than in the case of natural oil flow. Advantages of the cooling system with compact coolers and forced oil flow are the limited space requirement low investment cost and the possibility of mounting the coolers at an arbitrary height.

Forced-directed oil flow

If the entire volume of oil kept in circulation is led through the active parts of the transformer, i.e. the shunt path shown in Fig. 6.7 is abolished, then the cooling system of forced-directed oil flow represented in Fig. 6.8 is obtained.

Only the oil heated by the losses generated in the active parts and that filtering through the leaks of the system is kept in circulation by the pump. Thus, the oil entering the cooler is always at maximum temperature. Consequently, due to the

368

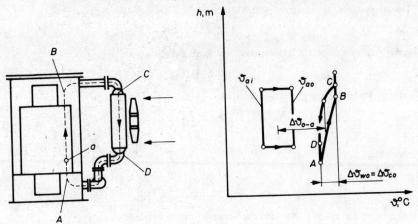

Fig. 6.8. Schematic diagram of forced-directed oil flow forced air cooling

greater temperature difference $\Delta\vartheta_{o-a}$, a cooler of smaller volume is required than in the case of the cooling system of Fig. 6.7, assuming the same amount of heat loss is to be dissipated. With the velocity of oil flow becoming higher in the active parts, first the magnitude of heat transfer coefficient increases inside the windings and thus the heat flux per unit transfer area can be increased without exceeding the average temperature rise permissible in the winding. Obviously, there are also economic and technical limits imposed on increasing the heat flux per unit transfer area.

6.1.4 Quasistationary warming-up and cooling phenomena

Temperature rise of a homogeneous body versus time
with surface heat transfer coefficients
being independent of the surface temperature drop

From the point of view of temperature rise, a body is considered homogeneous if it exhibits an infinite internal thermal conductivity, i.e. if it has a uniform temperature and constant specific heat throughout.

Consider a homogeneous body of mass m, surface A, specific heat c and heat transfer coefficient α, which is heated to a temperature exceeding the ambient temperature by $\Delta\vartheta_\infty$, at instant $t=0$.

The quantity of heat stored in the body at instant $t=0$ is

$$Q_0 = mc\Delta\vartheta_\infty . \tag{6.16}$$

Left alone, i.e. possessing no internal heat sources, the body starts to cool down, its temperature and therewith its stored heat decreases, and it transfers its heat to the surroundings.

During period dt the quantity of heat stored in the body decreases by dQ.

$$-dQ = mcd(\Delta\vartheta) .$$ (6.17)

The same amount of heat is transferred to the surroundings:

$$dQ = \alpha A\, \Delta\vartheta\, dt .$$ (6.18)

Combining and rearranging the two equations:

$$-\frac{mc}{\alpha A}\frac{d(\Delta\vartheta)}{\Delta\vartheta} = dt$$ (6.19)

is obtained.

Introducing the notation:

$$\tau = \frac{mc}{\alpha A},$$ (6.20)

the expression for τ, which is termed the time constant, is obtained.

Let us now integrate the right-hand and left-hand sides of the equation obtained for dt:

$$-\tau \ln \Delta\vartheta = t + C .$$ (6.21)

At instant $t = 0$ $\Delta\vartheta = \Delta\vartheta_\infty$, hence the integration constant will be

$$C = -\tau \cdot \ln \Delta\vartheta_\infty .$$ (6.22)

Substituting the value of C into Eqn. (6.21) and rearranging the latter:

$$-\tau \ln \Delta\vartheta + \tau \ln \Delta\vartheta_\infty = t .$$ (6.23)

Since $\ln\left(\dfrac{\Delta\vartheta_\infty}{\Delta\vartheta}\right)^\tau = t$, we may write $\left(\dfrac{\Delta\vartheta_\infty}{\Delta\vartheta}\right)^\tau = e^t$,

hence:

$$\Delta\vartheta = \Delta\vartheta_\infty e^{-\frac{t}{\tau}} .$$ (6.24)

By a similar deduction, the following relation valid for a body warming up is obtained:

$$\Delta\vartheta = \Delta\vartheta_\infty\left(1 - e^{-\frac{t}{\tau}}\right) .$$ (6.25)

$\Delta\vartheta_\infty$ denotes the temperature rise attained by the body on reaching the steady-state condition. If the body of heat transfer surface and heat transfer coefficient α dissipates power P with a surface temperature drop of $\Delta\vartheta_\infty$ to the surroundings, then on the basis of the relation:

$$\alpha A = \frac{P}{\Delta\vartheta_\infty}$$ (6.26)

the time constant can be calculated without knowing the surface area concerned or the magnitude of the surface heat-transfer coefficient:

$$\tau = \frac{mc}{\dfrac{P}{\Delta\vartheta_\infty}} .$$ (6.27)

In practical cases, the dissipated power of the body and the associated surface temperature drop, the mass and the specific heat of the body is known. The temperature rise of a warming-up body versus time follows the exponential law (see Fig. 6.9).

370

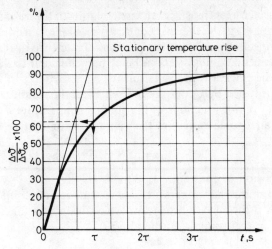

Fig. 6.9. A body warming up according to the exponential law

*Temperature rise of a homogeneous body
in the case of heat transfer coefficient dependent on the
surface temperature drop*

The exponential relation between the temperature rise of a warming-up or cooling body and time is valid only in the case of a constant heat transfer coefficient. If heat transfer coefficient depends on the surface temperature drop then this means—since the surface temperature drop is subject to a continuous change during the warming-up or cooling process—that also the surface heat transfer coefficient varies during warming-up or cooling. In such cases a non-exponential power function steps into the place of the relation between temperature rise and time. On the preceding pages in each formula α can be replaced by the overall conductance for heat transfer k. The overall conductance for heat transfer characterizes the resultant heat transfer capability of several thermal conductancies connected in series. Its dimension is identical to that of the heat transfer coefficient α. Suppose a winding is considered as a homogeneous body, which is cooled down in an oil tank of infinite thermal capacity. Then, by considering the surface temperature drop occurring across the interturn insulation between the convection surface of the winding and the cooling oil, the overall conductance for heat transfer referred to the outer convection surface A_c of the winding will be obtained, from the known relation:

$$\frac{1}{k_{A_c}} = \frac{1}{\alpha} + \frac{\delta_p}{\lambda_p \dfrac{A_\delta}{A_c}}, \tag{6.28}$$

as

$$k_{A_c} = \frac{\alpha \lambda_p A_\delta}{\lambda_p A_\delta + \delta_p A_c \alpha}, \tag{6.29}$$

24*

371

where α is the surface heat transfer coefficient relating to the boundary layer between insulation and oil, λ_p is the thermal conductivity of interturn insulation, A_δ is the mean area of projections of the surface sections of interturn insulation surrounding the disc toward the oil, said projections being perpendicular to the direction of heat flow and δ_p is the thickness of insulation.

The replacement of α by k does not in itself affect the validity of the relations, as long as α and k are not functions of the surface temperature drop.

For natural air cooled radiators the value of α (or that of heat transmission coefficient k) is

$$\alpha = a\Delta\vartheta^{1/3} , \tag{6.30}$$

where the dimension of constant a is $\mathrm{Wm}^{-2}\,{}^\circ\mathrm{C}^{-4/3}$.

Let us now examine the temperature rise vs. time relation obtained for a cooling body of mass m, if the surface heat-transfer coefficient is a function of the surface temperature drop according to formula (6.30).

The quantity of heat transferred during time dt is

$$dQ = a\Delta\vartheta^{1/3} \, A\Delta\vartheta \, dt , \tag{6.31}$$

$$dQ = aA \, \Delta\vartheta^{4/3} \, dt . \tag{6.32}$$

Reduction of heat content of the body is

$$-dQ = mcd(\Delta\vartheta) . \tag{6.33}$$

Combining Eqns. (6.32) and (6.33), after rearrangement

$$-\frac{mc}{aA} \Delta\vartheta^{-4/3} \, d(\Delta\vartheta) = dt \tag{6.34}$$

is obtained.

Integrating both sides:

$$\frac{3\,mc}{aA} \frac{1}{\sqrt[3]{\Delta\vartheta}} = t + K . \tag{6.35}$$

Let $\Delta\vartheta$ be equal to $\Delta\vartheta_\infty$ at instant $t=0$, thus

$$K = \frac{3\,mc}{aA} \frac{1}{\sqrt[3]{\Delta\vartheta_\infty}} . \tag{6.36}$$

Let us introduce the notation

$$\tau' = \frac{3\,mc}{aA} \tag{6.37}$$

and substitute the value of K into the equation obtained after integration:

$$\tau' \frac{1}{\sqrt[3]{\Delta\vartheta}} = t + \tau' \frac{1}{\sqrt[3]{\Delta\vartheta_\infty}} .$$

The temperature rise will be

$$\Delta\vartheta = \left(\frac{\tau'}{t + \dfrac{\tau'}{\sqrt[3]{\Delta\vartheta_\infty}}} \right)^3 . \tag{6.38}$$

372

As apparent from the above, a power function relating time and temperature rise is obtained for the cooling of a homogeneous body whose temperature rise is $\Delta\vartheta_\infty$, at $t = 0$.

6.1.5 Warming and cooling of a three-mass composite body

The warming and cooling of a transformer is analogous to the process taking place in a three-mass composite body. The three masses in this case are the winding of the transformer, the core and the oil. In the process of warming up, one part of the heat developing in the winding and the core increases their heat contents, and the other part is transferred to the oil through the surfaces of the winding and core in contact with the oil. Again, one part of the heat transferred to the oil increases its heat content, and the other part is transferred to the surroundings via the coolers. The increasing heat content of the three masses brings about their respective temperature rises. Direct heat transfer between winding and core is neglected, this being prevented by the insulating cylinders placed between them. In large transformers, one part of the additional load loss caused by the stray flux develops in the tank, another part in the frame structure and the third part occurs in the core. The part developing in the core causes a temperature rise of the core itself. Among the losses arising in the inactive parts that developing in the frame directly heats the oil, whereas that arising in the tank causes partly heating of the oil and partly of the air streaming along the outer surface of the tank. The additional load loss can be ascribed to its site of origin only be means of calculation which cannot be verified by measurements. No great error is introduced by assuming that the entire loss generated in the inactive parts causes heating of the oil.

Each of the temperature rises of the various parts, assuming constant voltage, load and ambient conditions, tends towards a limit value. These temperature rise limits are termed the steady-state temperatures of the various masses, which approach these temperatures at different rates of rise, but, according to the exponential law, asymptotically.

Let the following notations be introduced:

P, W	loss
A, m^2	convection surface to which
k, Wm^{-2}°C^{-1}	overall conductance for heat transfer refers to
c, Ws kg^{-1}°C^{-1}	specific heat
m, kg	mass
$Q = mc = a$, Ws °C^{-1}	heat content
$Ak = b$ W °C^{-1}	thermal conductivity
$\Delta\vartheta = x$ °C	temperature rise
$d\Delta\vartheta = dx$ °C	infinitesimal temperature rise
t, s	time
dt, s	infinitesimal time
$\dfrac{dx}{dt} = \dot{x}$, °C s^{-1}	derivative of temperature rise with respect to time
∞	index to indicate the steady-state condition
1, 2, 3	indices to denote the winding, the core and the oil.

Starting from the beginning of the warming period, when $t=0$, suppose the temperature rise of the winding is $\Delta\vartheta_1$, that of the core is $\Delta\vartheta_2$, and that of the oil is $\Delta\vartheta_3$ after time t.

Using the notations introduced, the following three differential equations can be written:

$$m_1 c_1 d\,\Delta\vartheta_1 + (\Delta\vartheta_1 - \Delta\vartheta_3)A_1 k_1\,dt = P_1\,dt\,, \tag{6.39}$$

$$m_2 c_2 d\,\Delta\vartheta_2 + (\Delta\vartheta_2 - \Delta\vartheta_3)A_2 k_2\,dt = P_2\,dt\,, \tag{6.40}$$

$$m_3 c_3 d\,\Delta\vartheta_3 + \Delta\vartheta_3\,A_3 k_3\,dt = (\Delta\vartheta_1 - \Delta\vartheta_3)A_1 k_1\,dt +$$

$$+ (\Delta\vartheta_2 - \Delta\vartheta_3)A_2 k_2\,dt + P_3\,dt\,. \tag{6.41}$$

Equation (6.39) describes the warming process of the winding. The right-hand side of the equation is the heat loss developing in the winding in a period of dt. The first term on the left-hand side is the increase of heat content of the winding, while the second term is the heat transferred to the oil.

By Eqn. (6.40) the warming process of the core is described, in a way analogous to the case of the winding.

Equation (6.41) describes the warming up of the oil. The first term on the right-hand side of the equation indicates the heat received from the winding, the second term is that received from the core, and the third term is the heat corresponding to the stray loss developing in the inactive materials. The first term on the left-hand side is the increment of heat content of the oil, and the second term is the heat transferred to the surroundings.

Introducing a mathematical system of notations and rearranging the following inhomogeneous linear differential equations of first order are obtained:

$$\dot{x}_1 - \left(-\frac{b_1}{a_1}x_1 + \frac{b_1}{a_1}x_3\right) = \frac{P_1}{a_1}\,, \tag{6.42}$$

$$\dot{x}_2 - \left(-\frac{b_2}{a_2}x_2 + \frac{b_2}{a_2}x_3\right) = \frac{P_2}{a_2}\,, \tag{6.43}$$

$$\dot{x}_3 - \left(\frac{b_1}{a_3}x_1 + \frac{b_2}{a_3}x_2 - \frac{b_1+b_2+b_3}{a_3}x_3\right) = \frac{P_3}{a_3}\,. \tag{6.44}$$

Rewriting the above set of equations in matrix form:

$$\dot{x} - Ax = g(t)\,, \tag{6.45}$$

where the dimension of x is $^\circ$C, that of $\dot{x}$ and $g(t)$ is $^\circ$C s^{-1}, that of matrix A is s^{-1} and t is s. Writing the above equation in expounded form:

$$\begin{bmatrix} \dot{x}_1 \\ \dot{x}_2 \\ \dot{x}_3 \end{bmatrix} - \begin{bmatrix} -\dfrac{b_1}{a_1} & 0 & \dfrac{b_1}{a_1} \\ 0 & -\dfrac{b_2}{a_2} & \dfrac{b_2}{a_2} \\ \dfrac{b_1}{a_3} & \dfrac{b_2}{a_3} & -\dfrac{b_1+b_2+b_3}{a_3} \end{bmatrix} \begin{bmatrix} x_1 \\ x_2 \\ x_3 \end{bmatrix} = \begin{bmatrix} \dfrac{P_1}{a_1} \\ \dfrac{P_2}{a_2} \\ \dfrac{P_3}{a_3} \end{bmatrix}. \tag{6.46}$$

The cooling process of the three-mass system is described by the homogeneous form of the above set of equations, as

$$\dot{x} - Ax = 0. \tag{6.47}$$

The column matrix of the initial conditions is

$$x(0) = \begin{bmatrix} x_{10} \\ x_{20} \\ x_{30} \end{bmatrix} = x_0. \tag{6.48}$$

The solution of the set of equations $\dot{x} - Ax = 0$ satisfying the initial condition $x(0) = x_0$ is

$$x(t; x_0) = e^{At}x_0 + \int_0^t e^{A(t-\tau)}g(\tau)\, d\tau, \tag{6.49}$$

where the dimension of $g(\tau)$ is $°C\, s^{-1}$ and that of τ and $d\tau$ is s.

First, let us solve the homogeneous set of equations:

$$x = e^{At}x(0). \tag{6.50}$$

The calculation of e^{At} is performed through the following steps:

(a) The roots (eigenvalues) λ_1, λ_2, λ_3 of the characteristic equation of A are searched for.

(b) The Lagrange polynomial belonging to each root is determined.

(c) Symbols λ are everywhere substituted by A.

(d) Having calculated matrices $L_1 A$, $L_2 A$, $L_3 A$, they are multiplied by $e^{\lambda_1 t}$, $e^{\lambda_2 t}$, $e^{\lambda_3 t}$, respectively, whereby e^{At} is obtained.

(e) After having determined e^{At}, the solution satisfying the initial conditions is arrived at by multiplying e^{At} by $x(0)$ from the right.

(f) The solution is rewritten in scalar form.

(g) Expression $(e^{At})^{-1} = (e^{-t})^A$ is calculated.

(h) Expression $(e^{At})^{-1} f(\tau)$ is calculated.

(i) Integration

$$\int_0^t e^{-A\tau} f(\tau)\, d\tau$$

is performed.

(j) The result thus obtained is multiplied by e^{At} from the left.

(k) To the result obtained the solution of the homogeneous equation satisfying the initial condition is added:

$$x = e^{At} x(0) + e^{At} \int_0^t e^{At} f(\tau)\, d\tau. \tag{6.51}$$

(1) Reductions are performed.

(m) The final result is written in scalar form.

The solution of the set of differential equations describing the cooling process has the following form:

$$x_1 = a_{11}e^{-\lambda_1 t} + a_{12}e^{-\lambda_2 t} + a_{13}e^{-\lambda_3 t},$$

$$x_2 = a_{21}e^{-\lambda_1 t} + a_{22}e^{-\lambda_2 t} + a_{23}e^{-\lambda_3 t}, \qquad (6.52)$$

$$x_3 = a_{31}e^{-\lambda_1 t} + a_{32}e^{-\lambda_2 t} + a_{33}e^{-\lambda_3 t},$$

where the temperature rise values pertaining to $t = 0$s indicating the starting of the cooling process:

$$x_{10} = a_{11} + a_{12} + a_{13},$$

$$x_{20} = a_{21} + a_{22} + a_{23}, \qquad (6.53)$$

$$x_{30} = a_{31} + a_{32} + a_{33}.$$

The solution of the set of differential equation describing the process of warming up takes the form:

$$x_1 = b_{11}e^{-\lambda_1 t} + b_{12}e^{-\lambda_2 t} + b_{13}e^{-\lambda_3 t} + x_{1\infty},$$

$$x_2 = b_{21}e^{-\lambda_1 t} + b_{22}e^{-\lambda_2 t} + b_{23}e^{-\lambda_3 t} + x_{2\infty}, \qquad (6.54)$$

$$x_3 = b_{31}e^{-\lambda_1 t} + b_{32}e^{-\lambda_2 t} + b_{33}e^{-\lambda_3 t} + x_{3\infty},$$

where the initial temperature values pertaining to the instant $t = 0$s are

$$x_{10} = b_{11} + b_{12} + b_{13} + x_{1\infty},$$
$$x_{20} = b_{21} + b_{22} + b_{23} + x_{2\infty}, \qquad (6.55)$$
$$x_{30} = b_{31} + b_{32} + b_{33} + x_{3\infty},$$

where $x_{1\infty}$, $x_{2\infty}$, $x_{3\infty}$ are the steady-state temperature rise values for the given constant load which are attained after the lapse of time $t = \infty$.

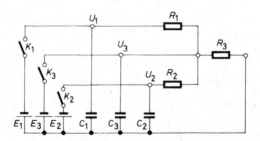

Fig. 6.10. Equivalent electric circuit for simulating the warming-up of a three-mass body, with the losses arising in the inactive parts considered

The physical process of warming up of a three-mass composite body can be demonstrated in the most suitable way by an electrical analogue model (see Fig. 6.10). Capacitors C_1, C_2, C_3 correspond to the heat content of winding, core and oil denoted by $= 1, 2$ and 3, respectively. Electric resistors are the respective thermal resistances, and voltages U_1, U_2, U_3 represent the temperature rise values of the three masses. Storage batteries E_1, E_2, E_3 supply electric powers P_1, P_2, P_3 through terminals 1, 2, 3 into the network on closing of the switches K_1, K_2, K_3, which initiates the warming-up process.

Example 6.4. Mass data of a 250 MVA, 242/15.75 kV generator-transformer are as follows: Mass of copper winding: $m_c = 28\,000$ kg, specific heat of copper $c_c = 390$ W s kg^{-1} °C^{-1}. Mass of interturn insulation (oil–paper): $m_p = 1800$ kg, specific heat of interturn insulation (oil–paper): $c_p = 1600$ W s kg^{-1} °C^{-1}.

Products of masses and specific heats:

$$a_1 = m_c c_c + m_p c_p = (28\,000 \times 390 + 1800 \times 1600) \text{ W s °C}^{-1} = 13.8 \times 10^6 \text{ W s °C}^{-1}.$$

The product of mass and specific heat of interturn insulation is added to the respective product for the winding, because their temperatures are nearly equal and both cool together.

Instead of giving the area of the convection surface and its heat transfer coefficient, the value of b_1 is calculated from the fact that, in the case of the rated load, the average winding temperature is by 20 °C above the average temperature of the oil, by which 520 kW is transferred from the winding to the oil. The amount of thermal power transferred per each °C of temperature is just the constant b_1 required.

$$b_1 = \frac{520\,000 \text{ W}}{20 \text{ °C}} = 26 \times 10^3 \text{ W °C}^{-1}.$$

Mass of core: $m_i = 103\,000$ kg, specific heat of core: $c_i = 470$ W s kg^{-1} °C.

$$a_2 = m_i c_i = 103\,000 \times 470 \ Ws \text{ °C}^{-1} = 48.4 \times 10^6 \text{ W s °C}^{-1}.$$

It is known from measurements that at the rated voltage and under the rated load the core attains a temperature 25 °C higher than that of the oil, while transferring $P'_0 = 300$ kW to the oil. The actual no-load loss is only $P_0 = 220$ kW, a part of stray loss which is mainly eddy-current loss caused by the stray flux, however, according to our estimation, half of it, i.e. 80 kW develops in the core. The other half of the stray or eddy-current loss, i.e. 80 kW, arises in the inactive steel parts such as frame, tank, etc., and this portion of the loss indirectly contributes to the heating of the oil. These are the reasons why the heat dissipated by the core has been increased to 300 kW.

The d.c. and eddy-current losses developing in the transformer winding are in total $P_{l1} = 520$ kW. The stray loss is $P_{l2} = 160$ kW, hence the entire load loss is $P_l = 680$ kW. The no-load loss is $P_0 = 220$ kW. Overall loss of the transformer under rated conditions of service is $P_t = P_l + P_0 = 900$ kW.

As regards the core:

$$b_2 = \frac{300\,000 \text{ W}}{25 \text{ °C}} = 12 \times 10^3 \text{ W °C}^{-1}.$$

There are 43 t of oil in the transformer of specific heat $c = 2014$ W s kg^{-1} °C^{-1}, the mass of the steel structure is 57 t of specific heat $c = 470$ W s kg^{-1} °C^{-1}, 4 t of pressboard with a specific heat of $c = 1600$ W s kg^{-1} °C^{-1}, 1.75 t of wood with a specific heat $c = 1400$ W s kg^{-1} °C^{-1}, 0.4 t of phenol-base paper with a specific heat $c = 1600$ W s kg^{-1} °C^{-1}.

Products of masses and specific heats:

Oil	$43\,000 \times 2014$ W s °C^{-1} = 86.6×10^6 W s °C^{-1}
Steel	$57\,000 \times 470$ W s °C^{-1} = 26.8×10^6 W s °C^{-1}
Pressboard and phenol-base paper	$4\,400 \times 1600$ W s °C^{-1} = 7.04×10^6 W s °C^{-1}
Wood	$1\,750 \times 1400$ W s °C^{-1} = 2.45×10^6 W s °C^{-1}
Total:	122.9×10^6 W s °C^{-1}

$$a_3 = 122.9 \times 10^6 \text{ W s °C}^{-1}.$$

The materials listed are considered as cooling together with the oil, their initial temperature being equal to the average temperature of the oil. In the case of an average temperature rise of coolers $\Delta \vartheta_{o-a} = 35$ °C, 900 kW is transferred to the surroundings by the transformer coolers.

$$b_3 = \frac{900\,000}{35} \text{ W s °C}^{-1} = 25.7 \times 10^3 \text{ W °C}^{-1}.$$

It is assumed that the transformer operates at an ambient temperature of 20 °C until disconnection, and it is loaded with its rated load.

Let us find by calculation how the winding, core and oil will cool down as a function of time, after disconnecting the transformer. For convenience, the data obtained by the above calculation are listed here for easier starting of the computation:

$$a_1 = 13.8 \times 10^6 \text{ W s °C}^{-1}, \quad b_1 = 26 \times 10^3 \text{ W °C}^{-1},$$
$$a_2 = 48.4 \times 10^6 \text{ W s °C}^{-1}, \quad b_2 = 12 \times 10^3 \text{ W °C}^{-1},$$
$$a_3 = 122.9 \times 10^6 \text{ W s °C}^{-1}, \quad b_3 = 25.7 \times 10^3 \text{ W °C}^{-1}.$$

At the instant $t=0$; $x_{10}=(20+35)$ °C $=55$ °C; $x_{20}=(25+35)$ °C $=60$ °C; $x_{30}=35$ °C. For the sake of simplifying the calculation, the coefficients of the set of differential equations are calculated first:

$$\frac{b_1}{a_1} = \frac{26 \times 10^3 \text{ s}^{-1}}{13.8 \times 10^6} = 1.884\,058 \times 10^{-3} \text{ s}^{-1},$$

$$\frac{b_2}{a_2} = \frac{12 \times 10^3 \text{ s}^{-1}}{48.4 \times 10^6} = 0.247\,934 \times 10^{-3} \text{ s}^{-1},$$

$$\frac{b_1}{a_3} = \frac{26 \times 10^3 \text{ s}^{-1}}{122.9 \times 10^6} = 0.211\,554 \times 10^{-3} \text{ s}^{-1},$$

$$\frac{b_2}{a_3} = \frac{12 \times 10^3 \text{ s}^{-1}}{122.9 \times 10^6} = 0.097\,640 \times 10^{-3} \text{ s}^{-1},$$

$$\frac{b_1+b_2+b_3}{a_3} = \frac{(26+12+25.7) \times 10^3 \text{ s}^{-1}}{122.9 \times 10^6} = 0.518\,308 \times 10^{-3} \text{ s}^{-1}.$$

With the given data, matrix A takes the following form:

$$A = 10^{-3} \times \begin{bmatrix} -1.884\,058 & 0 & 1.884\,058 \\ 0 & -0.247\,934 & 0.247\,934 \\ 0.211\,554 & 0.097\,640 & -0.518\,308 \end{bmatrix}$$

The form of matrix A^2 is:

$$A^2 = 10^{-6} \times \begin{bmatrix} 3.948\,255 & 0.183\,959 & -4.526\,197 \\ 0.052\,451 & 0.085\,679 & -0.189\,977 \\ -0.508\,230 & -0.074\,816 & 0.691\,431 \end{bmatrix}$$

(a) Let us find the roots of the characteristic equation of matrix A:

$$|\lambda E - A| = \begin{vmatrix} \lambda + 1.884\,054 \times 10^{-3} & 0 & -1.884\,058 \times 10^{-3} \\ 0 & \lambda + 0.247\,934 \times 10^{-3} & -0.247\,934 \times 10^{-3} \\ -0.211\,554 \times 10^{-3} & -0.097\,640 \times 10^{-3} & \lambda + 0.518\,308 \times 10^{-3} \end{vmatrix} =$$

$$= \lambda^3 + 2.650\,373 \times 10^{-3}\lambda^2 + 1.149\,395 \times 10^{-6}\lambda + 0.097\,684 \times 10^{-9} = 0.$$

The roots:

$$\lambda_1 = -0.113\,342 \times 10^{-3},$$
$$\lambda_2 = -0.404\,056 \times 10^{-3},$$
$$\lambda_3 = -2.132\,974 \times 10^{-3}.$$

(b) Now, let us determine the Lagrange polynomial pertaining to each root:

$$L_1(\lambda) = \frac{(\lambda - \lambda_2)(\lambda - \lambda_3)}{(\lambda_1 - \lambda_2)(\lambda_1 - \lambda_3)} = \frac{(\lambda + 0.404\,056 \times 10^{-3})(\lambda + 2.132\,974 \times 10^{-3}) \times 10^6}{(-0.113\,342 + 0.404\,056)(-0.113\,342 + 2.132\,974)}$$

$$L_1(\lambda) = \frac{10^6}{0.587\,135}(\lambda^2 + 2.537\,030 \times 10^{-3}\lambda + 0.861\,841 \times 10^{-6}).$$

Substituting λ^2 by A^2 and λ by A:

$$L_1(A) = \frac{10^6}{0.587\,135}\left(10^{-6}\begin{bmatrix} 3.948\,255 & 0.183\,959 & -4.526\,197 \\ 0.052\,451 & 0.085\,679 & -0.189\,077 \\ -0.508\,230 & -0.074\,916 & 0.691\,431 \end{bmatrix} + \right.$$

$$+ 2.537\,030 \times 10^{-3} \times 10^{-3}\begin{bmatrix} -1.884\,058 & 0 & 1.884\,058 \\ 0 & -0.247\,934 & 0.247\,934 \\ 0.211\,554 & 0.097\,640 & -0.518\,308 \end{bmatrix} + $$

$$\left. + 0.861\,841 \times 10^{-6}\begin{bmatrix} 1 & 0 & 0 \\ 0 & 1 & 0 \\ 0 & 0 & 1 \end{bmatrix}\right) = \frac{1}{0.587\,135}\begin{bmatrix} 0.030\,184 & 0.183\,959 & 0.253\,715 \\ 0.052\,451 & 0.318\,504 & 0.439\,039 \\ 0.028\,489 & 0.172\,900 & 0.238\,309 \end{bmatrix}.$$

$$L_1(A)x(0) = \frac{1}{0.587\,135}\begin{bmatrix} 0.030\,184 & 0.183\,959 & 0.253\,715 \\ 0.052\,451 & 0.318\,504 & 0.439\,039 \\ 0.028\,489 & 0.172\,900 & 0.238\,309 \end{bmatrix}\begin{bmatrix} 55 \\ 60 \\ 55 \end{bmatrix} = $$

$$= \frac{1}{0.587\,135}\begin{bmatrix} 21.577\,685 \\ 37.361\,410 \\ 20.281\,710 \end{bmatrix} = \begin{bmatrix} 36.750\,807 \\ 63.633\,423 \\ 34.543\,521 \end{bmatrix}.$$

$$L_1(A)x(0)e^{\lambda_1 t} = \begin{matrix} 36.750\,807e^{-0.113\,342 \times 10^{-3}t} \\ 63.633\,423e^{-0.113\,342 \times 10^{-3}t} \\ 34.543\,521e^{-0.113\,342 \times 10^{-3}t} \end{matrix}$$

$$L_2(\lambda) = \frac{(\lambda - \lambda_1)(\lambda - \lambda_3)}{(\lambda_2 - \lambda_3)(\lambda_2 - \lambda_1)} = \frac{(\lambda + 0.113\,342 \times 10^{-3})(\lambda + 2.132\,974 \times 10^{-3}) \times 10^6}{(-0.404\,056 + 2.132\,974)(-0.404\,056 + 0.113\,342)}.$$

$$L_2(\lambda) = \frac{10^6}{0.502\,621}(\lambda^2 + 2.246\,316 \times 10^{-3}\lambda + 0.241\,756 \times 10^{-6})$$

$$L_2(A) = -\frac{10^6}{0.502\,621}\left(10^{-6}\begin{bmatrix} 3.948\,255 & 0.183\,959 & -4.526\,197 \\ 0.052\,451 & 0.085\,679 & -0.189\,977 \\ -0.508\,230 & -0.074\,816 & 0.691\,431 \end{bmatrix} + \right.$$

$$+ 2.246\,316 \times 10^{-6} \times 10^{-3}\begin{bmatrix} -1.884\,058 & 0 & 1.884\,058 \\ 0 & -0.247\,934 & 0.247\,934 \\ 0.211\,554 & 0.097\,640 & -0.518.308 \end{bmatrix} + $$

$$\left. + 0.241\,756 \times 10^{-6}\begin{bmatrix} 1 & 0 & 0 \\ 0 & 1 & 0 \\ 0 & 0 & 1 \end{bmatrix}\right) = -\frac{1}{0.502\,621}\begin{bmatrix} -0.042\,179 & 0.183\,959 & -0.294\,007 \\ 0.052\,451 & -0.229\,503 & 0.366\,961 \\ -0.033\,013 & 0.144\,514 & -0.231\,097 \end{bmatrix}.$$

$$L_2(A)x(0) = -\frac{1}{0.502\,621}\begin{bmatrix} -0.042\,179 & 0.183\,959 & -0.294\,007 \\ 0.052\,451 & -0.229\,503 & 0.366\,961 \\ -0.033\,013 & 0.144\,514 & -0.231\,097 \end{bmatrix}\begin{bmatrix} 55 \\ 60 \\ 35 \end{bmatrix}$$

$$= -\frac{1}{0.502\,621}\begin{bmatrix} -1.572\,550 \\ 1.958\,260 \\ -1.233\,270 \end{bmatrix} = \begin{bmatrix} 3.128\,699 \\ -3.896\,097 \\ 2.453\,678 \end{bmatrix}.$$

$$L_2(A)x(0)e^{\lambda_2 t} = \begin{matrix} 3.128\,699e^{-0.404\,056 \times 10^{-3}t} \\ -3.896\,097e^{-0.404\,056 \times 10^{-3}t} \\ 2.453\,678e^{-0.404\,056 \times 10^{-3}t} \end{matrix}.$$

$$L_3(\lambda) = \frac{(\lambda - \lambda_1)(\lambda - \lambda_2)}{(\lambda_3 - \lambda_1)(\lambda_3 - \lambda_2)} = \frac{(\lambda + 0.113\,342 \times 10^{-3})(\lambda + 0.404\,056 \times 10^{-3}) \times 10^6}{(-2.132\,974 + 0.113\,342)(-2.132\,974 + 0.404\,056)}.$$

$$L_3(\lambda) = \frac{10^6}{3.491\,778}(\lambda^2 + 0.517\,398 \times 10^{-3}\lambda + 0.045\,797 \times 10^{-6}).$$

$$L_3(A) = \frac{10^6}{3.491\,778}\left(10^{-6}\begin{bmatrix} 3.948\,255 & 0.183\,959 & -4.526\,197 \\ 0.052\,451 & 0.085\,679 & -0.189\,977 \\ -0.508\,230 & -0.074\,816 & 0.691\,431 \end{bmatrix}\right. +$$

$$+ 0.517\,398 \times 10^{-3} \times 10^{-3}\begin{bmatrix} -1.884\,068 & 0 & 1.884\,058 \\ 0 & -0.247\,934 & 0.247\,934 \\ 0.211\,554 & 0.097\,640 & -0.518\,308 \end{bmatrix} +$$

$$\left.+ 0.045\,797 \times 10^{-6}\begin{bmatrix} 1 & 0 & 0 \\ 0 & 1 & 0 \\ 0 & 0 & 1 \end{bmatrix}\right) = \frac{1}{3.491\,778}\begin{bmatrix} 3.019\,244 & 0.183\,959 & -3.551\,389 \\ 0.052\,451 & 0.003\,195 & -0.061\,696 \\ -0.398\,772 & -0.024\,297 & 0.469\,056 \end{bmatrix}.$$

$$L_3(A)x(0) = \frac{1}{3.491\,778}\begin{bmatrix} 3.019\,244 & 0.183\,959 & -3.551\,389 \\ 0.052\,451 & 0.003\,195 & -0.061\,696 \\ -0.398\,772 & -0.024\,297 & 0.469\,056 \end{bmatrix}\begin{bmatrix} 55 \\ 60 \\ 35 \end{bmatrix} =$$

$$= \frac{1}{3.491\,778}\begin{bmatrix} 52.797\,345 \\ 0.917\,145 \\ -6.973\,320 \end{bmatrix} = \begin{bmatrix} 15.120\,476 \\ 0.262\,658 \\ -1.997\,069 \end{bmatrix}.$$

$$L_3(A)x(0)e^{\lambda_3 t} = \begin{matrix} 15.120\,476e^{-2.132\,974 \times 10^{-3}t}, \\ 0.262\,858e^{-2.132\,974 \times 10^{-3}t}, \\ -1.997\,069e^{-2.132\,974 \times 10^{-3}t}. \end{matrix}$$

As seen, matrices $L_1(A)$, $L_2(A)$, $L_3(A)$ found by the above calculation have been multiplied from the right by matrix $x(0)$ specifying the initial conditions, and the products thus obtained have been multiplied by $e^{\lambda_1 t}$, $e^{\lambda_2 t}$, $e^{\lambda_3 t}$.

The rounded-off solution in scalar form is as follows:

$$x_1 = 36.75e^{-0.113 \times 10^{-3}t} + 3.13e^{-0.404 \times 10^{-3}t} + 15.12e^{-2.133 \times 10^{-3}t},$$

380

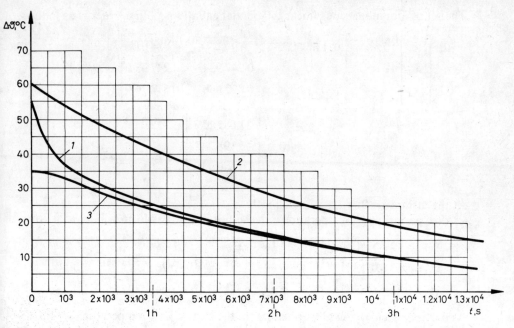

Fig. 6.11. Cooling of a 250 MVA transformer from its steady-state temperature: *1* — winding; *2* — core, *3* — oil

$$x_2 = 63.63e^{-0.113 \times 10^{-3}t} - 3.89e^{-0.404 \times 10^{-3}t} + 0.26e^{-2.133 \times 10^{-3}t},$$

$$x_3 = 34.54e^{-0.113 \times 10^{-3}t} + 2.45e^{-0.404 \times 10^{-3}t} - 1.99e^{-2.133 \times 10^{-3}t}.$$

The cooling of the 250 MVA transformer from its steady-state temperature is shown in Fig. 6.11.

Let us calculate how the transformer will warm up under its rated load, starting from instant $t = 0$. According to the foregoing data:

$$P_1 = 520 \times 10^3 \text{ W}, \quad P_2 = 300 \times 10^3 \text{ W}, \quad P_3 = 80 \times 10^3 \text{ W}.$$

With these data:

$$\frac{P_1}{a_1} = \frac{520 \times 10^3 \text{ s}^{-1} \text{ °C}}{13.8 \times 10^6} = 37.681\ 159 \times 10^{-3} a^{-1} \text{ °C},$$

$$\frac{P_2}{a_2} = \frac{300 \times 10^3 \text{ s}^{-1} \text{ °C}}{48.4 \times 10^6} = 6.198\ 347 \times 10^{-3} \text{ s}^{-1} \text{ °C},$$

$$\frac{P_3}{a_3} = \frac{80 \times 10^3 \text{ s}^{-1} \text{ °C}}{122.9 \times 10^6} = 0.650\ 936 \times 10^{-3} \text{ s}^{-1} \text{ °C},$$

$$f(\tau) = \begin{bmatrix} 37.681\ 159 \\ 6.198\ 347 \\ 0.650\ 936 \end{bmatrix} \times 10^{-3}.$$

381

The set of inhomogeneous linear differential equations of first order is as follows:

$$
\begin{bmatrix} \dot{x}_1 \\ \dot{x}_2 \\ \dot{x}_3 \end{bmatrix} - 10^{-3} \begin{bmatrix} -1.884\,058 & 0 & 1.884\,058 \\ 0 & -0.247\,934 & 0.247\,934 \\ 0.211\,554 & 0.097\,640 & -0.518\,308 \end{bmatrix} \begin{bmatrix} x_1 \\ x_2 \\ x_3 \end{bmatrix} =
$$

$$
= 10^{-3} \begin{bmatrix} 37.681\,159 \\ 6.198.347 \\ 0.650\,936 \end{bmatrix}.
$$

At the instant $t=0$: $x_{10}=x_{20}=x_{30}=0$.

After performing the operations, the solution in scalar form will be as follows:

$$
x_1 = -36.75e^{-0.113 \times 10^{-3}t} - 3.13e^{-0.404 \times 10^{-3}t} - 15.12e^{-2.133 \times 10^{-3}t} + 55,
$$

$$
x_2 = -63.63e^{-0.113 \times 10^{-3}t} - 3.89e^{-0.404 \times 10^{-3}t} - 0.26e^{-2.133 \times 10^{-3}t} + 60,
$$

$$
x_3 = -34.54e^{-0.113 \times 10^{-3}t} - 2.45e^{-0.404 \times 10^{-3}t} - 1.99e^{-2.133 \times 10^{-3}t} + 35.
$$

As it can be seen, in the limit when $t=\infty$, the steady-state temperature rise values will be:

$$
x_{1\infty} = 55\,°C; \quad x_{2\infty} = 60\,°C; \quad x_{3\infty} = 35\,°C.
$$

The warming up of the transformer is shown in Fig. 6.12.

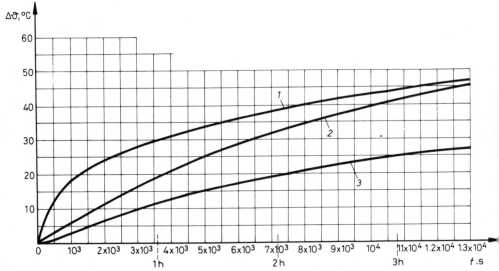

Fig. 6.12. Warming-up of a 250 MVA transformer under its rated load: 1 — winding, 2 — core; 3 — oil

382

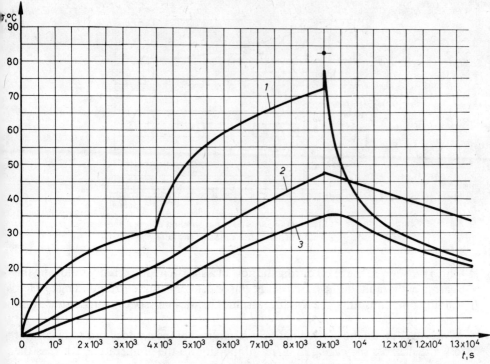

Fig. 6.13. Warming-up and cooling of a 250 MVA transformer under varying load: *1* — winding; *2* — core; *3* — oil

The calculation method presented above is suitable for computing the temperature rise conditions of transformers operated with a load of arbitrary magnitude and duration.

In Fig. 6.13 the temperature rise conditions of the transformer dealt with in the example have been plotted for the following load schedule.

After switching on, the transformer is loaded with its rated load for a period of 4×10^3 s. Then, a period of 40% overload operation follows, lasting 5×10^3 s. This overloaded operation is followed by a symmetrical three-phase short-circuit lasting for 3s, at the end of which the transformer is disconnected from the network, and a cooling-down period follows. During this cooling period both the motor-driven pumps and the fans are kept in operation. Figure 6.13 has been plotted on the basis of solutions corresponding to the given initial conditions of the set of differential equations described.

6.1.6 The short-circuit temperature rise

Under the effect of a short-circuit current flowing through the transformer, the loss developing in the winding causes the heat content of the winding and interturn insulation to increase. Due to the short duration of short-circuits, practically no

383

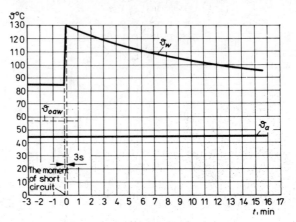

Fig. 6.14. Temperature rise vs. time curve of a transformer winding during a short-circuit and cooling of the same after clearing of the fault

heat is transferred to the oil during the flow of a short-circuit current. Following the clearing of the fault, the heat transfer begins and, after a time corresponding to a few winding time constants, the winding assumes its normal service temperature again (Fig. 6.14).

The loss developing in the winding is composed of the d.c. loss and eddy-current loss, the d.c. loss increasing in direct proportion to the temperature, whereas the eddy-current loss is inversely proportional to it.

The loss arising during a short circuit in a conductor of length l is converted into an increase of the heat content of the conductor without transferring it to the oil.

$$\left[\left(\frac{I}{A_c}A_c\right)^2\frac{1}{\sigma_{75}}\frac{235\,°C+\vartheta_w+\varDelta\vartheta}{(235+75)\,°C}\frac{l}{A_c}+\right.$$
$$\left.+\frac{e_{75}}{100}\left(\frac{I}{A_c}A_r\right)^2\frac{1}{\sigma_{75}}\frac{(235+75)\,°C}{235\,°C+\vartheta_w+\varDelta\vartheta}\frac{l}{A_c}\right]dt=$$
$$=(A_cl_c\rho_cc_c+A_pl_p\rho_pc_p)d\varDelta\vartheta=A_cl\,(\rho_cc_c+k\rho_pc_p),\qquad(6.56)$$

where I is the short-circuit current: A_c is the cross-sectional area of the conductor: A_p is the cross-sectional area of interturn insulation of the conductor; $k=A_p/A_c$ is the ratio of the two cross-sectional areas; $J=I/A_c$ is the short-circuit current density; σ_{75} is the specific electrical conductance of the material of the copper conductor at 75 °C; ϑ_w is the average temperature of the winding at incidence of the short-circuit; l is the length of conductor; ρ_c is the specific density of copper; ρ_p is the specific density of insulating material; e_{75} is the eddy-current loss expressed as a percentage of the d.c. loss, at 75 °C; $\varDelta\vartheta$ is the short-circuit temperature rise with respect to ϑ_w; and $d\varDelta\vartheta$ is the temperature change during the period dt.

Making substitutions:

$$\frac{J^2}{\sigma_{75}}\left(\frac{235\,°C+\vartheta_w}{310\,°C}+\frac{\varDelta\vartheta}{310\,°C}+\frac{e_{75}3.1\,°C}{235\,°C+\vartheta_w+\varDelta\vartheta}\right)dt=(\rho_cc_c+k\rho_pc_p)d\varDelta\vartheta\,.\qquad(6.57)$$

384

Separating the variables:

$$dt = \frac{\sigma_{75}}{J^2}(\rho_c c_c + k\rho_p c_p)\frac{1}{\dfrac{235\,°C + 9_w}{310\,°C} + \dfrac{\Delta 9}{310\,°C} + \dfrac{e_{75}3.1\,°C}{235\,°C + 9_w + \Delta 9}}\,d\Delta 9. \tag{6.58}$$

Removing the complex fraction from the right side of the equation:

$$dt = \frac{310\sigma_{75}(\rho_c c_c + k\rho_p c_p)\,°C}{J^2} \cdot \frac{\Delta 9 + 9_w + 235\,°C}{\Delta 9^2 + 2(9_w + 235\,°C)\Delta 9 + (9_w + 235\,°C)^2 + e_{75}960\,°C^2}\,d\Delta 9. \tag{6.59}$$

Introducing the following notations:

$$\Delta 9 = x; \quad d\Delta 9 = dx;$$

$$\frac{310\sigma_{75}(\rho_c c_c + k\rho_p c_p)\,°C}{J^2} = a_1, \tag{6.60}$$

where the dimension of a_1 is s,

$$9_w + 235\,°C = a_2 \tag{6.61}$$

and

$$e_{75} \times 960\,°C^2 = a_3. \tag{6.62}$$

With the above notations the differential equation of transient temperature rise is

$$dt = \frac{a_1 x + a_1 a_2}{x^2 + 2a_2 x + a_2^2 + a_3}\,dx. \tag{6.63}$$

Factoring out the quotient $a_1/2$ from the right side of the equation:

$$dt = \frac{a_1}{2}\frac{2x + 2a_2}{x^2 + 2a_2 x + a_2^2 + a_3}. \tag{6.64}$$

Integrating both sides:

$$\int dt = \frac{a_1}{2}\int \frac{2x + 2a_2}{x^2 + 2a_2 x + a_2^2 + a_3}\,dx.$$

Since the numerator of the fraction on the right side is the derivative of the denominator, the final result of integration is

$$t = \frac{a_1}{2}\ln(x^2 + 2a_2 x + a_2^2 + a_3) + C. \tag{6.65}$$

If $t = 0$, then $x = 0$. The integration constant is

$$C = -\frac{a_1}{2}\ln(a_2^2 + a_3). \tag{6.66}$$

Substituting this value into the equation obtained after integration:

$$t = \frac{a_1}{2}\ln\frac{x^2 + 2a_2 x + a_2^2 + a_3}{a_2^2 + a_3}. \tag{6.67}$$

Writing both sides as exponents

$$e^t = \left(\frac{x^2 + 2a_2 x + a_2^2 + a_3}{a_2^2 + a_3}\right)^{\frac{a_1}{2}}, \tag{6.68}$$

$$x^2 + 2a_2 x + (a_2^2 + a_3)(1 - e^{2t/a_1}) = 0,$$

$$x = -a_2 + \sqrt{a_2^2 - (a_2^2 + a_3)(1 - e^{2t/a_1})^2},$$ (6.69)

$$x = \sqrt{e^{2t/a_1}(a_2^2 + a_3) - a_3} - a_2,$$ (6.70)

$$\Delta \vartheta = \sqrt{e^{2t/a_1}[(\vartheta_w + 235\ ^\circ\mathrm{C})^2 + e_{75} 960\ ^\circ\mathrm{C}] - e_{75} 960\ ^\circ\mathrm{C}^2} - (\vartheta_w + 235\ ^\circ\mathrm{C}),$$ (6.71)

$$\Delta \vartheta = 31\ ^\circ\mathrm{C} \sqrt{e^{2t/a_1}\left[\left(\frac{\vartheta_w + 235\ ^\circ\mathrm{C}}{31\ ^\circ\mathrm{C}}\right)^2 + e_{75}\right] - e_{75}} - (\vartheta_w + 235\ ^\circ\mathrm{C}),$$ (6.72)

where

$$a_1 = \frac{310\ ^\circ\mathrm{C}\sigma_{75}(\rho_c c_c + k\rho_p c_p)}{J^2}.$$ (6.73)

As stated in the introduction, the differential equation has been set up by considering the eddy-current loss and also the heat stored in the insulating material.

If the eddy-current loss is disregarded, then the terms containing σ_{75} cancel out from the equation obtained for $\Delta \vartheta$. If the heat stored in the insulant is also disregarded, a_1 will be as follows:

$$a_1 = \frac{310\ ^\circ\mathrm{C}\sigma_{75}\rho_c c_c}{J^2}.$$ (6.74)

The physical quantities which appear in the expression for a_1 are:
   For electrolytic copper:

$$\rho_c = 8900\ \mathrm{kg\,m^{-3}}, \qquad \sigma_{75} = \frac{10^6}{0.0217}\ \mathrm{A\,V^{-1}\,m^{-1}},$$

$$c_c = 401\ \mathrm{W\,s\,kg^{-1}\,^\circ C^{-1}} \qquad \text{(in the range 100 to 200\ ^\circ\mathrm{C}).}$$

For aluminium:

$$\rho_a = 2700\ \mathrm{kg\,m^{-3}}, \qquad \sigma_{75} = \frac{10^6}{0.036}\ \mathrm{A\,V^{-1}\,m^{-1}},$$

$$c_a = 961\ \mathrm{W\,s\,kg^{-1}\,^\circ C^{-1}} \qquad \text{(in the range 100 to 200\ ^\circ\mathrm{C}).}$$

For oil-paper:

$$\rho_p = 600\ \mathrm{kg\,m^{-3}}: \quad c_p = 1600\ \mathrm{W\,s\,kg^{-1}\,^\circ C^{-1}}.$$

In the case of copper winding material, evaluating the heat stored to the insulation, the exponent of e is

$$\frac{2t}{a_1} = \frac{1}{2.55 + k\,0.685}\left(\frac{J}{10^8}\right)^2 t,$$ (6.75)

where the dimension of the expression preceding that in brackets is $\mathrm{m^4\,s^{-1}\,A^{-2}}$.
   Neglecting the heat stored in the insulation

$$\frac{2t}{a_1} = 0.392\left(\frac{J}{10^8}\right)^2 t,$$ (6.76)

where the dimension of the number standing before the bracket is $\mathrm{m^4\,s^{-1}\,A^{-2}}$.

386

In the case of aluminium winding material, considering again the heat stored in the insulation, the exponent of e is

$$\frac{2t}{a_1} = \frac{1}{1.115 + k0.413} \left(\frac{J}{10^8}\right)^2 t, \tag{6.77}$$

where the dimension of the expression preceding the bracket is $\mathrm{m^4\,s^{-1}\,A^{-2}}$.

Neglecting the heat stored in the insulation:

$$\frac{2t}{a_1} = 0.897 \left(\frac{J}{10^8}\right)^2 t, \tag{6.78}$$

where the dimension of the number preceding the bracket is $\mathrm{m^4\,s^{-1}\,A^{-2}}$.

The factor k can take into account not only the ratios of cross-sectional areas, but also the delay with which the heat is transferred from the winding material to the interturn insulation. If, due to this delay, the value of k is halved, the error committed adds to the safety margin.

Example 6.5. Let us assume a transformer of $\varepsilon = 10\%$ impedance voltage. When loaded with its rated current, the current density in the H.V. winding is $J' = 3.5\ \mathrm{A\ mm^{-2}}$. Let us determine the temperature rise of the winding under the effect of a three-phase terminal fault, if the transformer, fed from a supply of infinite short-circuit capacity, is disconnected after a delay of 3 s. Let the average temperature of the winding, at the incidence of the fault, be $\vartheta_w = 85\ ^\circ\mathrm{C}$. Let the eddy-current loss referred to 75 °C be 15% of the d.c. loss: $e_{75} = 15\%$. Let the heat stored in the interturn insulation be neglected. The winding is made of copper.

The short-circuit current density is

$$J = \frac{100}{\varepsilon} J' \times 10^6 = 35 \times 10^6\ \mathrm{A\ m^{-2}}.$$

Duration of the short-circuit is $t = 3$ s. The exponent of e is

$$\frac{2t}{a_1} = 0.392 \left(\frac{35 \times 10^6}{10^8}\right)^2 3 = 0.145; \quad e^{0.145} = 1.156;$$

$\left(\dfrac{85 + 235}{31}\right)^2 = 107$. With these data, the temperature rise developing during the period of 3 s is

$$\Delta\vartheta = \left(31\sqrt{1.156\,(107 + 15) - 15} - 320\right)\ ^\circ\mathrm{C} = 28\ ^\circ\mathrm{C}.$$

Thus, at the instant of clearing the fault, the average temperature of the winding will be $(85 + 28)\ ^\circ\mathrm{C}$ $= 113\ ^\circ\mathrm{C}$.

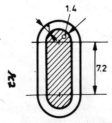

Fig. 6.15. Dimensions of insulated conductor

Example 6.6. A 250 MVA three-phase 400/132 kV autotransformer of $\varepsilon = 12\%$ impedance voltage is connected on its 400 kV side to a system having a short-circuit capacity of 20 000 MVA. The average temperature of the 400 kV winding of the transformer under the effect of a 150% transient overload is 130 °C. During this overload operation a three-phase terminal fault lasting for 2 s occurs on the 132 kV side. The 400 kV disc-type windings of the transformer are wound of the conductor material shown in Fig. 6.15.

In this example we want to calculate the final temperature attained by the winding at the instant the fault is cleared, the energy fed into the winding during the fault, the volume expansion of the windings, and the flow velocity of the oil in the 3-in. I.D. pipe connecting the tank with the conservator, brought about by the oil displaced by the increased volume of the expanded winding, if the tank is assumed to be rigid.

Consider first the combined impedance of the network and transformer, referred to 250 MVA.

$$\varepsilon' = 12 + \frac{250 \times 100}{20\,000} = 13.25\%.$$

In the 400 kV winding of the transformer, the current density is $J' = 3.3$ A mm^{-2} at its 250 MVA rated load. The short-circuit current density is

$$J = 3.3 \times 10^6 \frac{100}{13.25} \text{ A m}^{-2} = 24.9 \times 10^6 \text{ A m}^{-2}.$$

The ratio of eddy-current loss to d.c. loss at the rated load and 75 °C is $e_{75} = 16\%$.
With the data of Fig. 6.15 the cross-sectional area of insulation is

$$A_p = 1(2 \times 7.2 + 3.8\pi) \times 10^{-6} \text{ m}^2 = 26.3 \times 10^{-6} \text{ m}^2.$$

and of copper is

$$A_c = (7.2 \times 2.8 + 1.4^2\pi) \times 10^{-6} \text{ m}^2 = 26.33 \times 10^{-6} \text{ m}^2.$$

$$k = \frac{A_p}{A_c} = \frac{26.3 \times 10^{-6}}{26.33 \times 10^{-6}} \approx 1.$$

Exponent of e:

$$\frac{2t}{a_1} = \frac{1}{2.55 + 1 \times 0.685}\left(\frac{24.9 \times 10^6}{10^8}\right)^2 2 = 0.0383; \quad e^{0.0383} = 1.039; \quad \left(\frac{130 + 235}{31}\right)^2 = 139.$$

With these data the temperature rise developing during 2 s is $\Delta\vartheta = (31\sqrt{1.039(139 + 16)} - 16 - 365$ °C $= = 8$ °C. Thus, at the instant of clearing the fault, the average temperature of the winding is $(130 + 8)$ °C $= = 138$ °C. The amount of energy fed into the transformer during the fault:

Mass of the windig $m_c = 27 \times 10^3$ kg. Specific heat of copper $c_c = 401$ W s kg^{-1} °C^{-1}. Mass of interturn insulation $m_p = 1.82 \times 10^3$ kg. Specific heat of paper $c_p = 1600$ W s kg^{-1} °C^{-1}. Energy fed into the winding during the fault:

$$Q = \Delta\vartheta(m_c c_c + m_p c_p) = [8(27 \times 10^3 \times 401 + 1.82 \times 10^3 \times 1600)]\text{W s} =$$
$$= 110 \times 10^6 \text{ W s} = 30.6 \text{ kW h}.$$

Change of volume of the winding due to the short-circuit: The volumetric expansion factor of copper $\beta_c = 4 \times 10^{-5}$ °C^{-1}. Change of volume of the winding

$$\Delta V_c = V_c \beta_c \Delta\vartheta = \frac{27\,000}{8900} \times 4 \times 10^{-5} \times 8 \text{ m}^3 = 0.97 \times 10^{-3} \text{ m}^3.$$

Change of volume of the insulation, due to short-circuit: Let the coefficient of volumetric thermal expansion of the insulation be equal to that of the oil:

$$\beta_p = 8 \times 10^{-4} \text{ °C}^{-1}.$$

Change of volume of insulation:

$$\Delta V_p = V_p \beta_p \Delta\vartheta = \frac{1820}{600} \times 8 \times 10^{-4} \times 8 \text{ m}^3 = 19.7 \times 10^{-3} \text{ m}^3.$$

Total increase of volume during the short-circuit:

$$\Delta V = \Delta V_c + \Delta V_p = 20.67 \times 10^{-3} \text{ m}^3.$$

388

Cross-sectional area of the 3-in pipe of 82.6 mm I.D. is

$$\frac{(82.6 \times 10^{-3})^2 \pi}{4} \, m^2 = 5350 \times 10^{-6} \, m^2 = 5.35 \times 10^{-3} \, m^2.$$

Oil velocity in the pipe, considering that the change of volume comes about through a period of 2 s.

$$w = \frac{20.67 \times 10^{-3}}{2 \times 5.35 \times 10^{-3}} = 1.93 \, m \, s^{-1}.$$

As demonstrated by the example, even the very small temperature change of 8 °C can bring about a relatively large change of volume and a high oil velocity in the pipe connecting the transformer with the conservator and containing the Buchholz relay. This oil velocity would be capable of causing operation of the Buchholz relay. The movement of the winding caused by a short-circuit gives rise to pressure surges in the oil contributing to the operation of the Buchholz relay. In practice, it may happen that the Buchholz relay does not trip because of the elastic expansion of the tank walls.

6.1.7 Load carrying capacity of transformers

Two characteristic modes of operation can be distinguished with respect to the aging of insulation, as follows:

1. Normal operation during which the deterioration under varying conditions of load and ambient temperature is normal.
2. Operation under increased load, which may just be permitted when necessary without risking the soundness of the transformer, but is associated with a higher rate of deterioration than normal.

With the exception of generator-transformers, the load imposed on transformers varies between a lower and a higher level within the day and during the year. In order to avoid the necessity of assigning rigid load limits or of putting standby units in service for the periods of increased load, the thermal inertia of transformers and the relation between deterioration and temperature are utilized, since these permit, even with natural cooling, operation of the transformer with loads exceeding its rating for limited periods of time without modifying the thermal output of the transformer.

Transformers with forced cooling are equipped with automatic cooling control by which the thermal inertia of the transformer can be better utilized and whose heat dissipation is increased or decreased automatically as a function of the load. Obviously, the highest permissible load must not exceed the maximum value specified by the manufacturer. By fully utilizing the thermal inertia and/or by the use of automatic cooling, it is possible to achieve 20 to 25 years of useful thermal service life for transformers, even with varying loads that occasionally exceed their maximum continuous ratings.

6.1.7.1 Temperature rise of active and inactive parts under the effect of loads varying with time

If at a given instant the temperatures of active and inactive parts and the load vs. time function are known, then the temperatures valid for any instant can be determined by calculation.

Temperature rise values belonging to the rated output S_n are taken as a basis.

For natural oil flow the average temperature rise of the winding, calculated from the results of resistance measurement is $\Delta\vartheta_w = 65$ °C. Maximum temperature rise of the oil $\Delta\vartheta_{om} = 55$ °C. Average temperature rise of the oil in the winding $\Delta\vartheta_{oaw} = 44$ °C. Temperature difference between average temperature rise of the winding and that of the oil $\Delta\vartheta_{w-o} = 21$ °C. Temperature rise of the hot spot is

$$\Delta\vartheta_c = \Delta\vartheta_{om} + 1.1\Delta\vartheta_{w-o}, \tag{6.79}$$

$$= (55 + 1.1 \times 21)\,°C = 78\,°C \quad \text{(see Fig. 6.2).}$$

For forced oil flow the difference between the temperatures of the oil at the inlet and outlet of the coolers usually lies between 5 and 8 °C. Let us assume a temperature difference $\Delta\vartheta_{co} = 6$ °C.

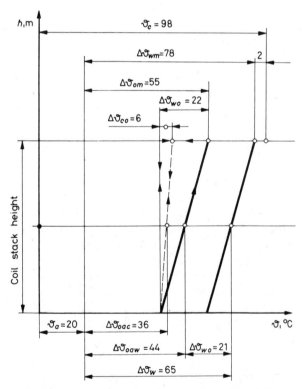

Fig. 6.16. Temperatures and temperature rises of winding and oil considered in the case of forced oil flow

All temperature rise figures are identical with those given for the case of natural oil flow, except that because of the mixing of cold oil ascending in the tank through the shunt path with hot oil leaving the winding, the temperature rise $\Delta\vartheta_{om}$ of the oil entering the coolers will not equal 55 °C that would correspond to the maximum temperature of the oil leaving the winding but only reach $\Delta\vartheta_{om} = 39$ °C, while the temperature rise of the oil entering the tank at the bottom of the winding is invariably $(39-6)$ °C $= 33$ °C. The maximum temperature of the oil leaving the winding, in the case of forced oil flow, cannot be measured directly (see Fig. 6.16). In fact, the temperature distribution will be more favourable than that shown since the value of $\Delta\vartheta_{w-o}$ will be a few degrees lower when changing over to forced oil flow. This is also indicated by the fact that, in the case of forced oil flow, the difference $\Delta\vartheta_{w-o}$ between average oil and average winding temperatures varies with the 0.9th power of heat loss developing in the winding, as opposed to the 0.8th power of loss in the formula for natural oil flow. As shown by the temperature pattern of Fig. 6.16, if the heat flux per unit transfer area is taken equal to its permissible level in the case of natural oil flow, i.e., if according to the example, $\Delta\vartheta_{w-o} = 21$ °C is chosen, then the average temperature rise of the winding and of the hot-spot remain 65 and 78 °C, respectively, equal to those in the case of natural oil flow. If, on the other hand, the heat flux per unit transfer area is increased so as to obtain a temperature rise of $\Delta\vartheta_{om} = 55$ °C for the mixed oil entering the cooler, then the 98 °C temperature assumed for the hot-spot will be exceeded considerably. Consequently, a transformer provided with forced oil cooling cannot be loaded on the basis of the design temperature of the oil entering the coolers.

6.1.7.2 Steady-state temperature rise figures under the effect of loads differing from the rated load

Let the following notations be introduced:
Assume the ratio between actual load and maximum continuous rating to be

$$K = \frac{S}{S_n}.$$
(6.80)

Let the ratio between the load loss at rated load and the no-load loss at rated no-load voltage be

$$d = \frac{P_l}{P_0}.$$
(6.81)

Let us indicate the various temperature rise values developing under the rated load by attaching n to the suffix of the respective symbols.
The maximum temperature rise of the oil is

$$\Delta\vartheta_{om} = \Delta\vartheta_{omn}\left(\frac{1+dK^2}{1+d}\right)^m.$$
(6.82)

The exponent appearing in the formula is $m = 0.8$ in the case of natural oil and air cooling and $m = 1.0$ in the case of natural oil flow and forced air cooling.

In IEC "Loading Guide for Oil-Immersed Transformers" the following values are considered: $m=0.9$, $d=5$, $\Delta\vartheta_{omn}=55\,°C$ for natural oil flow and $\Delta\vartheta_{omn}=40\,°C$ for forced oil flow, for reasons of simplification and unification. It is mentioned in IEC 76 that, when calculating with these values, the error introduced in the temperature rise is within ± 2 per cent. The use of actual values occurring in practice, instead of $d=5$, in the case of high loads, has but little influence on the final result.

The hot-spot temperature rise is

$$\Delta\vartheta_c = \Delta\vartheta_{omn}\left(\frac{1+dK^2}{1+d}\right)^m + 1.1\Delta\vartheta_{(w-o)n}K^{2n}. \tag{6.83}$$

The value of n for natural oil flow is $n=0.8$, for forced oil flow $n=0.9$ and for forced-directed oil flow $n=1.0$.

The average temperature rise of the oil is

$$\Delta\vartheta_{oaw} = \Delta\vartheta_{(oaw)n}\left(\frac{1+dK^2}{1+d}\right)^m. \tag{6.84}$$

The average temperature of the winding is

$$\Delta\vartheta_w = \Delta\vartheta_{wn}\left(\frac{1+dK^2}{1+d}\right)^m + \Delta\vartheta_{(w-o)n}K^{2n}. \tag{6.85}$$

Example 6.7. Let us consider a transformer of maximum continuous rating $S_n=40$ MVA provided with natural oil flow and natural air cooling. Let further data be $P_l=160$ kW, $P_0=50$ kW,

$$\Delta\vartheta_{wn}=\upsilon 5\,°C, \quad \Delta\vartheta_{(w-o)n}=25\,°C, \quad \Delta\vartheta_{omn}=48\,°C.$$

Let us calculate the hot-spot temperature and the maximum oil temperature for a load of $S=56$ MVA and an ambient temperature of $\vartheta_a=22\,°C$.

$$d=\frac{P_l}{P_0}=\frac{160}{50}=3.2, \quad K=\frac{S}{S_n}=1.4,$$

$$m=0.8, \quad n=0.8.$$

The temperature rise of the hot-spot is

$$\Delta\vartheta_c = \left[48\left(\frac{1+3.2\times 1.4^2}{1+3.2}\right)^{0.8} + 1.1\times 25\times 1.4^{2\times 0.8}\right]°C = (65+47)°C = 112\,°C.$$

Thus the hot-spot temperature is

$$\vartheta_c = \vartheta_a + \Delta\vartheta_c = (22+112)°C = 134\,°C.$$

The maximum oil temperature rise, as already obtained is

$$\Delta\vartheta_{om} = 65\,°C$$

and the maximum oil temperature

$$\vartheta_{om} = \vartheta_a + \Delta\vartheta_{om} = (22+65)°C = 87\,°C.$$

Temperature rise figures with load varying in time

As has already been shown, a transformer cooling or warming behaves as a three-mass system. It has also been stated that the exponential law is applicable to a cooling or warming body only if its heat dissipation is proportional to its temperature rise.

In practical calculations, in order to shorten and simplify the work, the transformer is considered as a two-mass body whose cooling and warming follows an exponential law. Only two time constants are used, viz. that of the difference between the temperatures of the winding and the oil, and that of the temperature rise of the oil. In the following, the former will be more shortly called the time constant of the winding and the latter the time constant of the transformer. Usual values of the time constant of the winding lie in the range 3 to 15 min, whereas those of transformers with natural oil flow are between 4 and 8 h, and those with forced air flow lie between 1 and 3 h.

If the steady-state temperature rise values of a transformer are known at a given instant, then when changing to another load, the temperature rise values for the period of transition are obtained by adding to the temperature those applying at instant $t=0$ the product of the difference between the steady-state temperature rise corresponding to the load and the temperature rise pertaining to instant $t=0$, multiplied by $(1-e^{-t/\tau})$.

Example 6.8. Let the time constant of the transformer dealt with in the preceding example be $\tau_{tr}=4$ h, and that of its winding $\tau_w=10$ min. At instant $t=0$, let the temperature of the transformer be equal to the ambient temperature. Assume $\vartheta_a=28$ °C. Let the transformer be loaded with 30 MVA for 2 h, then with 50 MVA for 3 h.

It is required to find the temperature of the hottest spot of the winding at the end of the eighth minute after applying the 50 MVA load, and the maximum oil temperature at the end of the fifth hour of the loaded period.

As the first step, the steady-state temperature rise values pertaining to the loads of $S=30$ and 50 MVA are determined. For $S=30$ MVA; $K=\dfrac{30}{40}=0.75$. Temperature rise of the hot-spot:

$$\Delta\vartheta_c=\left[48\left(\frac{1+3.2\times0.75^2}{1+3.2}\right)^{0.8}+1.1\times25\times0.75^{2\times0.8}\right]°C=(34.8+17.2)°C=52\ °C.$$

Maximum temperature rise of the oil: $\Delta\vartheta_{om}=34.8$ °C. For $S=50$ MVA; $K=\dfrac{50}{40}=1.25$. Temperature rise of the hot spot:

$$\Delta\vartheta_c=\left[48\left(\frac{1+3.2\times1.25^2}{1+3.2}\right)^{0.8}+1.1\times25\times1.25^{2\times0.8}\right]°C=(64+33)°C=97\ °C.$$

Maximum oil temperature rise: $\Delta\vartheta_{om}=64$ °C. At the end of the 2nd hour: $(1-e^{-2/4})=0.394$. Temperature rise of the hot-spot:

$$\Delta\vartheta_c=52\times0.394°C=20.4\ °C.$$

Maximum oil temperature rise:

$$\Delta\vartheta_{om}=34.8\times0.394°C=13.7\ °C.$$

The temperature rise of the hottest spot at the end of the eighth minute after this instant, with $(1-e^{-8/10})=0.545$, is

$$\Delta\vartheta_c=[34.8+17.2+(33-17.2)0.545]°C=60.6\ °C.$$

Temperature of the hot-spot:
$$\vartheta_c = (28 + 60.6)°C = 88.6 °C.$$

Maximum temperature rise of the oil at the end of the fifth hour after the commencement of loading with $(1 - e^{-3/4}) = 0.528$:
$$\Delta\vartheta_{om} = [34.8 + (64 - 34.8)0.528]°C = 50.2 °C.$$

Maximum oil temperature rise
$$\vartheta_{om} = (28 + 50.2)°C = 78.2 °C.$$

6.1.7.3 Tables of permissible loads [84]

The tables in the International Standard referred to are suitable for quick determination of the duration of permissible loads for transformers. In the tables the steady-state condition of the transformer under the lower load preceding a higher load is taken as the parameter on which the calculation of the temperature rise valid for instant $t = 0$ is based. The date of the tables were calculated by using the method demonstrated in the foregoing. The highest temperature permitted to develop at the hot-spot is limited to 140 °C, and the highest permissible overload is 1.5 times the rated output.

6.1.7.4 Other factors imposing a limit on the load

As stated in the introduction, the temperature of the active parts of a transformer with forced cooling is kept within specified limits by the automatic control of cooling. This automatic control is adapted to the cooling system of the transformer.

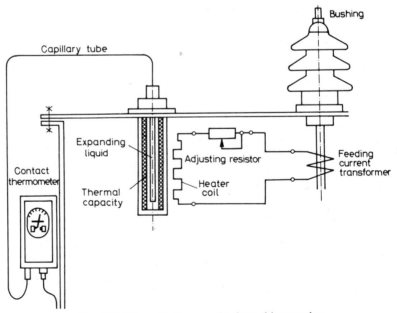

Fig. 6.17. Schematic diagram of a thermal image relay

By monitoring the temperature, it starts or stops the motor-driven fans, depending on the thermal condition of the transformer. For the purpose of temperature sensing, thermal images combined with contact thermometers are generally used. These thermometers start the auxiliaries and put the standby cooling equipment into operation depending on function of the assumed temperature of the hot-spot. With transformers provided with compact coolers it is common practice to shut down the fans at $+10\,°C$ or lower temperatures, when the transformer is used in cold condition. This is necessary because under the effect of heavy air flow the high-viscosity oil thickens in the oil pipes of compact coolers to such an extent that, even with the oil pumps in operation, oil is brought into circulation only very slowly.

Further tasks of the control equipment of automatic cooling are to disconnect the transformer when the hot-spot reaches a temperature of $140\,°C$ and to block the operation of the tap changer if the load current exceeds its rated value by 40 or 50%.

The principle of the thermal image is shown in Fig. 6.17. The thermometer-type sensor of the thermal image is immersed into the hottest oil in the tank and is surrounded by a heater coil carrying a current proportion of the load. Once the thermal block and the matching resistor have been adjusted, the thermal relay copies the temperature of the hottest spot (hot-spot) of the winding. The cooling auxiliaries are controlled by the contact thermometer accommodated in the associated control box.

6.2 Cooling of the winding

6.2.1 Losses developing in the winding

Let one or both windings of an untapped two-winding transformer consist of disc-type coils. In any loaded condition losses of different magnitude develop in discs of identical geometry constituting the winding stack. Each disc is assumed to contain an equal number of turns. The loss generated not only varies from disc to disc, but is of different magnitude in different layers of the disc-coils.

The different magnitudes of losses result from variation of stray flux density along the stack, from the axially varying temperature of the oil flowing around the discs and from the variation of temperature radially within the winding. Parts of the cooling surfaces are covered by the spacers placed between the disc-coils, so that heat is transferred by conduction from the parts covered up by the spacers to the cooled parts of the winding. This causes slight temperature gradient in the winding material.

6.2.1.1 Effect of stray flux

In a cylindrical winding arrangement the direction of the stray magnetic field in the middle of the winding stack is, to a good approximation, parallel to the longitudinal axis of the stack. Its radial distribution is approximately linear, i.e. it is highest in the part adjacent to the leakage channel, zero at the outer mantle of the winding, and it is deflected at the winding ends. The local value of stray flux density varies in magnitude and direction from point to point and, thus, it has a radial component as well.

6.2.1.2 Effect of cooling oil

Another reason why the losses in the various discs differ from each other is the difference in their temperatures. The oil streaming upwards in the winding stack from below is warmed up by the heat transferred from the coils. Thus, the oil in contact with a disc arranged at a higher level is of higher temperature. The oil temperature increases gradually while ascending along the winding stack. Since a temperature gradient must be present for heat transfer, the discs at a higher level would be warmer even if they were able to transfer the loss arising in them to the oil at the same temperature drop as those located at a lower level. The resistivity of winding material increases with temperature, so the d.c. loss will be higher in a coil located at a higher level. Although the eddy-current loss will decrease due to growing resistivity, the overall loss will, nevertheless, be higher at a higher temperature.

6.2.1.3 Effect of eddy-current loss on the overall loss developing in the layers of disc-type windings

The losses developing in the various layers within a disc will differ due to the varying local stray flux densities. If, e.g., the average eddy-current loss is 15%, the magnitude of the local flux density in the middle of the winding in the layer adjacent to the leakage channel will be close to 45%. This means that within a coil the overall loss varies in the ratio of 1 to 1.45.

Let us now examine the effect of the eddy-current loss on the overall loss developing in the different layers, by means of an example.

Example 6.9. Let us consider a 250 MVA regulating transformer having a voltage ratio 400/126 kV +11.5%–12.5%. The 400 kV winding tapped in the middle consists of 2 × 50 disc-coils. Each disc contains 32 turns of winding material having bare/insulated dimensions of $10 \times 2.8/12 \times 4.8$ mm^2. There are two 6 mm wide cooling ducts in each coil (see Fig. 6.18). The radial size of each coil is $(32 \times 4.8 + 12)$ mm = = 165.6 mm. The coils are separated from each other by spacers having an average thickness of 7.25 mm. The current flowing in each conductor is 90.25 A. Excitation of the leakage channel:

$$\frac{32 \times 90.25}{19.25 \times 10^{-3}} \text{ A m}^{-1} = 150 \times 10^3 \text{ A m}^{-1}.$$

The maximum stray flux density is: $B_{ma} = 1.257 \times 10^6 \sqrt{2} \times 150 \times 10^3$ V s m^{-2} = 0.265 V s m^{-2}.

In the coils located in the middle of the winding stack the stray flux direction is axial with no radial component, and along the radius its magnitude drops linearly to zero on reaching the outer mantle of the winding.

The local specific eddy-current loss can be calculated from the following formula:

$$v_e = C\sigma\rho^{-1}f^2 B_m^2 d^2 , \tag{6.86}$$

where C is a constant equal to 1.65 s^{-4}; σ is the conductivity in A^{-1} V^{-1} m^{-1}, ρ is the specific density in

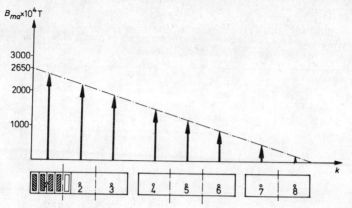

Fig. 6.18. Variation of axial leakage flux density in a disc located in the middle of a winding stack

$\mathrm{kg\,m^{-3}}$, f is the frequency in $\mathrm{s^{-1}}$, B_m is the maximum stray flux density in $\mathrm{V\,s\,m^{-2}}$ and d is the conductor dimension perpendicular to the direction of the stray flux density in m.

Let the conductivity at 75 °C and the specific density of copper be

$$\sigma_{75} = \frac{10^6}{0.0217}\,\mathrm{A\,V^{-1}\,m^{-1}}$$

and

$$\rho_c = 8900\,\mathrm{kg\,m^{-3}}.$$

With these data, the specific eddy-current loss is obtained from the formula:

$$v_e = C_1 f^2 B_m^2 d^2. \tag{6.87}$$

The value of constant C_1 appearing in the formula is $8.55 \times 10^3\,\mathrm{m^2\,kg^{-1}\,A\,V^{-1}}$. The specific d.c. loss can be calculated in the following way:

$$v_{dc} = \sigma^{-1}\rho^{-1}J^2, \tag{6.88}$$

where σ is the conductivity in $\mathrm{A\,V^{-1}\,m^{-1}}$, ρ is the specific density in $\mathrm{kg\,m^{-3}}$ and J is the current density in $\mathrm{A\,m^{-2}}$. Substituting the values of σ_{75} and ρ_C given above into Eqn. (6.88), the following formula results for the d.c. loss:

$$v_{dc} = C_2 J^2, \tag{6.89}$$

where $C_2 = 2.44 \times 10^{-12}\,\mathrm{m^4\,kg^{-1}\,V\,A^{-1}}$.

The maximum eddy-current loss develops in the coils located in the middle section of the winding stack, in the coil layers adjacent to the leakage channel. The maximum specific eddy-current loss is

$$v_e = 8.55 \times 10^3 \times 50^2 \times 0.265^2 \times (2.8 \times 10^{-3})^2\,\mathrm{W\,kg^{-1}} = 11.8\,\mathrm{W\,kg^{-1}}.$$

The specific d.c. loss at $J = 3.3 \times 10^6\,\mathrm{A\,m^{-2}}$ current density is

$$v_{dc} = \frac{2.44}{10^{12}}(3.3 \times 10^6)^2\,\mathrm{W\,kg^{-1}} = 26.4\,\mathrm{W\,kg^{-1}}.$$

The ratio of maximum specific eddy-current loss to the specific d.c. loss is

$$\frac{v_e}{v_{dc}} = \frac{11.8}{26.4} = 0.447.$$

This result shows that in the layer lying closest to the leakage channel the local eddy-current loss is 44.7%, i.e. the loss per kg developing in that layer is 1.447 times higher than the d.c. loss.

Table 6.6

Maximum local value of axial leakage flux density and magnitude of specific losses in the various layers of an eight-layer disc coil located in the middle of a winding stack

Serial No. of layer	1	2	3	4	5	6	7	8
B_{ma}, 10^4 T	2494.1	2182.4	1870.6	1558.8	1247	935.3	623.5	311.8
v_e, W kg^{-1}	10.4	8	5.9	4.1	2.6	1.47	0.655	0.163
v_{dc}, W kg^{-1}					26.4			
$v_e + v_{dc}$, W kg^{-1}	36.8	34.4	32.3	30.5	29	27.87	27	26.56
$\frac{v_e}{v_{dc}} \times 100$, %	39.8	30.3	22.3	15.55	9.85	5.57	2.48	0.617

Table 6.7

Maximum local value of leakage flux density, its axial and radial component and magnitude of specific losses in the various layers of an eight-layer disc coil located at the end of a winding stack

Serial No. of layer	1	2	3	4	5	6	7	8
B_{ma}, 10^4 T	1370	1165	975	790	602	417	226	25
B_{mr}, 10^4 T	355	512	620	693	733	743	720	650
B_m, 10^4 T	1410	1270	1155	1050	950	853	757	651
v_{ea}, W kg^{-1}	3.15	2.27	1.6	1.05	0.61	0.293	0.086	0.00105
v_{er}, W kg^{-1}	2.7	5.6	8.25	10.25	11.7	11.8	11.1	9
v_e, W kg^{-1}	5.85	7.87	9.85	11.3	12.31	12.093	11.186	9.001
v_{dc}, W kg^{-1}					26.4			
$v_e + v_{dc}$, W kg^{-1}	32.25	34.27	36.25	37.7	38.71	38.49	37.56	35.4
$\frac{v_e}{v_{dc}} \times 100$, %	22.2	29.8	37.4	42.8	46.7	45.8	42.4	34.1

In Table 6.6 the following values are given for eight points of the winding: maximum axial flux density B_{ma}, specific local eddy-current loss v_e, total specific loss $v_e + v_{dc}$ and percentage of local eddy-current loss $(v_e/v_{dc}) \times 100$.

Table 6.7 shows the total stray flux density, and its axial and radial components, passing through the disc at one end of the winding stack, and the associated specific local eddy-current losses. For the sake of completeness it should be noted that the summation of loss components is an approximation only (see Fig. 6.19). In Table 6.7, the specific losses appear in the same sequence as in Table 6.6.

The components of stray flux density have been obtained from the computer program run for the purpose of determining the short-circuit forces arising in the transformer of the example. The program takes into account the effect of the limb of the core. The division of the disc-coil into eight sections has also resulted from computer processing. Such division into sections has been found to be sufficient for determining the local short-circuit forces.

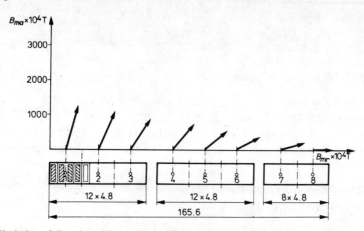

Fig. 6.19. Variation of direction and magnitude of leakage flux density in the end disc of a winding stack

From the data of Tables 6.6 and 6.7 the following conclusions can be drawn. The average eddy-current loss of the end disc is 37.5%, the average overall loss is 36.3 W kg^{-1}. The maximum deviation from this average value is found in the 5th group in a positive sense in 6.5%, and in the 1st group in a negative sense in -11%. The average eddy-current loss of the middle coil is 15.8%, the average overall loss is 30.6 W kg^{-1}. From this latter average value the maximum deviation in positive sense is in the 1st group $+20\%$ and in negative sense in the 8th group -13%.

In the end disc the loss per kg is $\dfrac{36.3-30.6}{30.6} \times 100 = 18.65\%$ higher than in the middle coil.

6.2.1.4 Effect of temperature difference of the oil

Let us assume a difference of 20 °C between the temperatures of the oil entering the winding at its bottom and leaving it at the top. This means that the temperature rise of the disc located at the bottom of the winding takes place with respect to an oil temperature 20 °C lower than that of a disc at the top. Let us further assume, for an ambient temperature of $\vartheta_a = 20$ °C, an average temperature rise of $\varDelta\vartheta_{oaw} = 45$ °C above the ambient temperature for the oil flowing in the winding of the coil, viz. an oil temperature of $\vartheta_{oaw} = (20+45)$ °C $= 65$ °C. The temperature of the oil entering the winding at its bottom is 10 °C lower than 65 °C, viz. $\vartheta_{o1} = 55$ °C, whereas the temperature of the oil leaving the winding at the top is 20 °C higher than 55 °C, viz. $\vartheta_{02} = 75$ °C. Let the sum of the heat drops across the interturn insulation and the drops across its surface be $\Delta\vartheta_{w-o} = \Delta\vartheta_p + \Delta\vartheta_s = 25$ °C, assumed to be equal to a first approximation for both the bottom and top discs. With these data the average copper temperature of the bottom disc is $\vartheta_{w1} = (55+25) = 80$ °C, and that of the top disc is $\vartheta_{w2} = (75+25) = 100$ °C.

In the formula for specific d.c. loss, the value of factor $C_2 = 2.44 \times 10^{-12}$ m kg^{-1} V A^{-1} should be increased in the ratio $\dfrac{235+80}{235+75} = \dfrac{315}{310}$ in the case of 80 °C, and in the

ratio of $\dfrac{235+100}{235+75} = \dfrac{335}{310}$ in the case of 100 °C. Thus for calculating the specific d.c.

loss developing at 80 °C the formula $v_{dc} = \dfrac{2.48}{10^{12}} J^2$, and for that developing at 100 °C

the formula $v_{dc} = \dfrac{2.64}{10^{12}} J^2$, should be used. The dimension of the first factors

appearing on the right-hand side of the formulae is m^4 kg^{-1} V A^{-1}.

In the formula for specific eddy-current loss, factor $C_1 = 8.55 \times 10^3$ m^2 kg^{-1} VA^{-1} should be reduced by the ratio 310/315 in the case of 80 °C, and by the ratio 310/335 in the case of 100 °C. Hence, for calculating the specific eddy-current loss valid at 80 °C the formula

$$v_e = 8.42 \times 10^3 f^2 B_m^2 d^2 \tag{6.90}$$

and for that valid at 100 °C the formula

$$v_e = 7.9 \times 10^3 f^2 B_m^2 d^2 \tag{6.91}$$

should be used. In both, the dimension of the first factor is m^2 kg^{-1} A V^{-1}.

Continuing Example 6.9 by using the preceding formulae, the specific losses developing are $(26.8 + 9.75)$ W kg^{-1} = 36.55 W kg^{-1} in the bottom disc at 80 °C, and $(28.55 + 9.17)$ W kg^{-1} = 37.72 W kg^{-1} in the top disc of 100 °C. Further, under the effect of a 20 °C temperature difference, the loss developing per kg in the top disc will be $\dfrac{37.72 - 36.55}{36.55} \times 100 = 3.2\%$ higher than in that at the bottom.

6.2.2 Cooling surfaces and heat dissipation of windings

The spacers placed between the disc coils cover 31 to 39% of the horizontal convection cooling surfaces of the coils. (See Figs 6.20 and 6.21). Heat is transferred from the parts of coils covered by the spacers to the exposed cooling surfaces by thermal conduction. Thus, by 31 to 39% more heat is dissipated through the uncovered sections than that developing there. Due to the increased specific thermal load imposed on these exposed surfaces, both the temperature drop across the insulation and the surface temperature drop increase.

Let us now examine what temperature drop will occur in the winding material in the section between the middle of the spacer and the exposed surface, if the loss

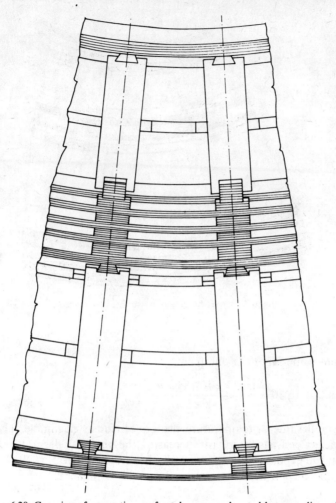

Fig. 6.20. Covering of convection surfaces by spacers located between disc-coils

developing under the spacers passes in both directions through the insulation material toward the exposed cooling surface.

With the data of Example 6.9 the loss generated in 1 m³ of winding material in the outermost coil is

$$p = (v_{dc} + v_e)\rho_c = 36.3 \times 8900 \text{ W m}^{-3} = 0.323 \times 10^6 \text{ W m}^{-3}. \qquad (6.92)$$

Let the spacer width be $b = 55 \times 10^{-3}$ m, and the thermal conductivity of the copper winding material be

$$\lambda_c = 390 \text{ W m}^{-1} \, {}^{\circ}\text{C}^{-1}. \qquad (6.93)$$

Substituting this value into the known formula of

$$\Delta \vartheta = \frac{pb^2}{8\lambda_c},$$

26

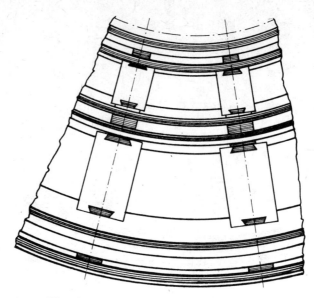

Fig. 6.21. Covering of cooling surfaces by spacers

the following value for $\Delta\vartheta$ is obtained:

$$\Delta\vartheta = \frac{0.323 \times 10^6 (55 \times 10^{-3})^2}{8 \times 390}\,°C = 0.313\,°C.$$

Such a small value of temperature drop is obviously negligible. The covering effect of spacers consists mainly of increasing the heat flux per unit transfer area imposed on the exposed surfaces, whereby both the temperature drop across the insulation and the surface temperature drop will be higher.

6.2.2.1 Disc-coil windings without intermediate axial cooling ducts

Two basic arrengements are distinguished. One is shown in Fig. 6.22. It is characterized by the presence of two axial oil ducts of thicknesses d_1 and d_2 respectively, along the inner and outer mantles of the winding, and confined by insulating cylinders.

The other arrangement is that illustrated in Fig. 6.23, having an axial, vertical oil duct of thickness d_1 only along the inner mantle, confined by an insulating cylinder, whereas the outer mantle of the winding is surrounded by a practically infinite oil space.

The disc coils are kept at a distance s from each other by insulating spacers. In the radial direction, the width of axial ducts is ensured by strips also made of insulating material.

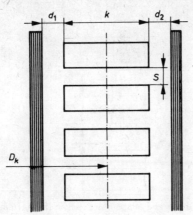

Fig. 6.22. Disc-coil without intermediate axial cooling duct, with oil duct confined from both sides by insulating cylinders

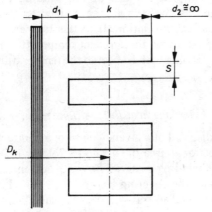

Fig. 6.23. Disc-coil without intermediate axial cooling duct, with oil duct confined on one side by an insulating cylinder

6.2.2.2 Simplification in the calculation
of temperature rise for disc coils

In order to simplify the temperature rise calculation for disc coils, the curvature of the coils is disregarded, the calculation takes into account the d.c. loss as increased by the average value of the eddy-current loss and the radial heat flow is neglected. The disc coils are considered as if an equal amount of loss were developing in each of them, their average temperature rise above the mean oil temperature is investigated, and their surface area as reduced by the part covered by the spacers is taken as the convection surface. The surface temperature drop is given as a function of heat flux per unit transfer area. The formula for surface temperature drop is calculated for a

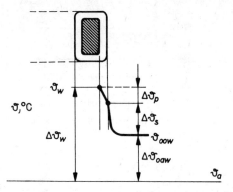

Fig. 6.24. Temperature drops and temperature rises in the surroundings of disc-coils

given average oil temperature, and for other oil temperatures a conversion key is given. The temperature drop across the insulation is calculated with a special formula. The average temperature rise of the winding above the average oil temperature is obtained as the sum of the surface temperature drop and the temperature drop across the insulation. If to this temperature drop the average temperature rise of the oil above the ambient temperature is added, the average temperature rise of the winding with respect to the temperature of the surroundings is obtained (see Fig. 6.24).

6.2.2.3 Average heat flux per unit transfer area

Let us investigate a disc coil of n layers wound of a conductor material having a bare/insulated cross-sectional area of $[a \times b/(a + 2\delta) \times (b + 2\delta)]$ (see Fig. 6.25). Let the conductance of the winding material be σ, its current density J, the average eddy-current loss $e\%$ and the covering factor 35%.

The average heat flux per unit transfer area q is calculated from the ratio of the loss developing in the winding of length l to the convection surface of the same winding corrected by the covering factor.

$$q = \frac{P}{F} = \frac{\left(1 + \dfrac{e}{100}\right)\dfrac{J^2}{\sigma}\,nabl}{[2n(a + 2\delta) + 2(b + 2\delta)]l} \times$$

$$\times \frac{1}{1 - 0.35} = \left(1 + \frac{e}{100}\right)\frac{J^2}{\sigma}K', \tag{6.94}$$

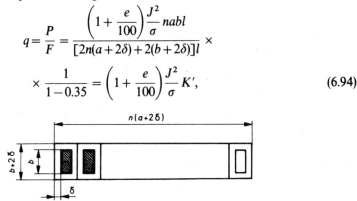

Fig. 6.25. Notations used in the calculation of average heat flux per unit transfer area of a coil

404

where

$$K' = 0.77 \frac{nab}{na + b + 2\delta(1 + n)}. \tag{6.95}$$

Let us substitute σ by the conductance of copper at 75 °C,

$$\sigma_{75} = \frac{10^6}{0.0217} \, \text{A V}^{-1} \, \text{m}^{-1}.$$

Let us introduce the factor

$$K = 0.0167 \times 10^{-6} \frac{nab}{na + b + 2\delta(1 + n)}. \tag{6.96}$$

The dimension of coefficient 0.0167×10^6 appearing in the formula is m A^{-1} V. Hence, the average heat flux per unit transfer area will be

$$q = \left(1 + \frac{e}{100}\right) J^2 K. \tag{6.97}$$

Example 6.10. Using the values $a = 3 \times 10^{-3}$ m, $b = 12 \times 10^{-3}$ m, $n = 12$, $\delta = 1 \times 10^{-3}$ m, $e = 4.5\%$, and $J = 3 \times 10^6$ A m^{-2}, let us calculate the average heat flux per unit transfer area.

$$K = 0.0167 \times 10^{-6} \frac{12 \times 3 \times 10^{-3} \times 12 \times 10^{-3}}{12 \times 3 \times 10^{-3} + 12 \times 10^{-3} + 2 \times 1 \times 10^{-3}(1 + 12)} \, \text{m}^2 \, \text{A}^{-1} \text{V} =$$

$$= 97.5 \times 10^{-12} \, \text{m}^2 \, \text{A}^{-1} \, \text{V}.$$

$$q = \left[\left(1 + \frac{4.5}{100}\right) 9 \times 10^{12} \times 97.5 \times 10^{-12}\right] \text{W m}^{-2} = 920 \, \text{W m}^{-2}.$$

The result obtained for q is the average heat flux per unit transfer area valid for 75 °C. For a temperature other than 75 °C, the first term of the coefficient in brackets on the right-hand side of the expression is to be multiplied by the quotient $(235 \, °C + \vartheta)/(235 + 75) \, °C$, and its second term by the quotient $(235 + 75) \, °C/235 \, °C + \vartheta$, where ϑ is the temperature for which the value of q is to be calculated.

6.2.2.4 Average surface heat transfer coefficient of the winding

Let us consider the disc coil winding of Fig. 6.26. From the vertical inner surface $h_1 l$ and vertical outer surface of equal area a power P_1 is transferred to the oil with a surface heat transfer coefficient of α_1. From the horizontal lower surface $b_2 l$ and horizontal upper surface of equal area a power P_2 is transferred to the oil with a surface heat transfer coefficient of α_2. The total temperature drop is the sum of the temperature drop developing across the insulation of thickness δ and thermal conductance λ and the surface temperature drop across the convective film — boundary layer—between insulation and oil. Let the total temperature drop be $\Delta\vartheta$, and the overall conductance for heat transfer referred to outer surfaces of insulation be k_1 and k_2. Finally, let us neglect the radial heat flow between the layers. The following equations may be written for the powers transferred to the oil.

$$P_1 = 2k_1 h_1 l \Delta \vartheta, \tag{6.98}$$

$$P_2 = 2k_2 h_2 l \Delta \vartheta. \tag{6.99}$$

Power P dissipated by the piece of winding of length l is the sum of losses dissipated through the vertical and horizontal lateral surfaces:

$$P = P_1 + P_2. \tag{6.100}$$

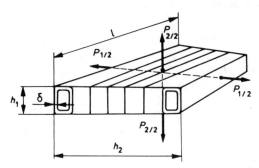

Fig. 6.26. Notations used in the calculation of average surface heat transfer coefficient of a coil

Let the overall conductance for heat transfer pertaining to the entire surface, i.e. to $2(h_1 + h_2)l$, be denoted by k, and the surface heat transfer coefficient by α. With these symbols

$$P = P_1 + P_2 = 2k(h_1 + h_2)l \Delta \vartheta. \tag{6.101}$$

Let the outer surface of the insulation be identical with the convection surface, and the quotient of heat conduction coefficient of insulation and its thickness be denoted by a:

$$a = \frac{\lambda}{\delta}. \tag{6.102}$$

The overall conductances for heat transfer are

$$k_1 = \frac{\alpha_1 a}{\alpha_1 + a}, \quad k_2 = \frac{\alpha_2 a}{\alpha_2 + a}, \quad k = \frac{\alpha a}{\alpha + a}. \tag{6.103}$$

Substituting the above expressions for heat transfer coefficients into the equation of transmitted heat:

$$P = P_1 + P_2 = 2 \frac{\alpha a}{\alpha + a}(h_1 + h_2)l \Delta \vartheta =$$

$$= 2 \frac{\alpha_1 a}{\alpha_1 + a} h_1 l \Delta \vartheta + 2 \frac{\alpha_2 a}{\alpha_2 + a} h_2 l \Delta \vartheta. \tag{6.104}$$

Expressing α from the above and introducing the notation $b = h_2/h_1$, the following equation is obtained:

$$\alpha = \frac{\alpha_1\alpha_2(1+b) + a(\alpha_1 + \alpha_2 b)}{\alpha_1 b + \alpha_2 + a(1+b)}. \tag{6.105}$$

If the value of a is re-substituted into the above expression of α, the following relation is obtained for the average surface heat transfer coefficient:

$$\alpha = \frac{\delta\alpha_1\alpha_2(1+b) + (\alpha_1 + \alpha_2 b)\lambda}{\delta(\alpha_1 b + \alpha_2) + (1+b)\lambda}. \tag{6.106}$$

For an uninsulated winding $\delta = 0$

$$\alpha = \frac{\alpha_1 + \alpha_2 b}{1 + b}. \tag{6.107}$$

6.2.3 Oil flow along the cooling surfaces of the winding and the surface heat transfer coefficient

The oil flow exhibits boundary layer properties along the cooling surfaces of the winding. This means that along the cylindrical inner and outer surfaces of the disc-coil, thin boundary layers are formed. These boundary layers break away from the

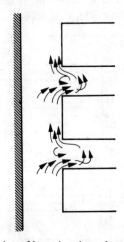

Fig. 6.27. Separation and disintegration of boundary layer formed along the mantle of the disc-coil

inner and outer upper edges of the lower disc, disintegrateing in the horizontal duct, and regenerate at the bottom surface of the upper disc and again pass in the form of thin boundary layers along the inner and outer mantles of the disc. This process repeats as many times as there are disc coils contained in the winding (see Fig. 6.27).

The intermediate disintegration of boundary layers taking place in the ducts between the coils does not mean the beginning of turbulent flow, it only promotes mixing of oil particles of different temperatures.

6.2.3.1 Depth of penetration of oil flow
into the horizontal ducts

From the vertical ducts the oil flow only enters the horizontal gaps if there are small values of k or large values of s and low heat flux per unit transfer area q. If oil flow penetration into the horizontal ducts occurs, the path of the oil flow increases.

6.2.3.2 The flow resistance imposed by the winding
and the length of flow path

As stated above, with large values of q and k and small values of s, the oil flow does not penetrate into the horizontal ducts, therefore in such cases the flow resistance imposed by the winding is not affected by the geometrical dimensions. With small values of q and k and large dimension s the situation is different, and the resistance to flow caused by the winding increases due to lengthening of the flow path.

6.2.3.3 Horizontal flow in the radial direction

Due to minor manufacturing asymmetries and to radial variation of loss, oil flows with alternately varying outward and inward directions have also been observed occasionally. These abnormal oil flows, as shunt paths, reduce the flow resistance caused by the winding (see. Fig. 6.28).

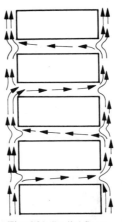

Fig. 6.28. Radial flows

6.2.3.4 Thickness of the boundary layer
in vertical axial ducts

The thickness of the laminar boundary layer increases with height in a vertical duct. According to Eckert [27], the thickness of the boundary layer in the unconfined half space in the laminar region, for $Gr_m Pr_m < 2 \times 10^9$, is

$$\delta = 3.93 l Pr^{-0.5}(0.952 + Pr)^{0.25} Gr^{-0.25}. \tag{6.108}$$

In the laminar region the above formula is used for calculating the thickness of boundary layers developing in vertical oil ducts. In the formula l is to be substituted by the distance of the point in question measured from the bottom of the winding, including the size of the horizontal oil ducts, up to $s=20$ mm. Above $s=20$ mm the turning of the boundary layer into the horizontal ducts should also be taken into account. Dimension l will thereby be larger than before, but the increase of the thickness of the boundary layer due to the increased dimension l is negligible.

The length of the laminar region measured from the bottom of the winding is generally rather small. Proceeding upwards along the axial length of the winding the laminar region is followed by a non-laminar zone and, then, by the turbulent zone. It is especially the non-laminar zone which is characteristic of the major part of the winding. In this zone, an increased duct thickness would reduce the flow resistance of the winding, but because of the resistance of the lower laminar zone connected in series with it, the flow resistance of the winding is decisively determined by the flow resistance of the laminar zone.

Example 6.11. Let us calculate the laminar boundary layer thickness at a height of $l=0.2$ m measured from the lower end of a winding stack 2.4 m high. The winding stack constitutes the outer winding, viz. it is not surrounded by an insulating cylinder. Let the ambient temperature be $\vartheta_a=20\,°\text{C}$, the temperature rise of the oil entering the tank at its bottom $\Delta\vartheta_{oiw}=35\,°\text{C}$ above ambient temperature, the temperature rise of the oil leaving the winding $\Delta\vartheta_{oow}=55\,°\text{C}$ above ambient temperature, and the surface temperature drop of the winding $\Delta\vartheta_s=15\,°\text{C}$. With these data the mean temperature of the boundary layer of the oil—the film average temperature—at the bottom of the winding is

$$\vartheta'_m=\vartheta_a+\Delta\vartheta_{oiw}+\frac{\Delta\vartheta_s}{2}=\left(20+35+\frac{15}{2}\right)°\text{C}=62.5\,°\text{C}.$$

After ascending to a height of 2.4 m, the temperature of the oil flowing in the winding like the mean temperature of the boundary layer of the oil, will be higher by $\Delta\vartheta_{oow}-\Delta\vartheta_{oiw}=(55-35)°\text{C}=20\,°\text{C}$. After covering a distance $l=0.2$ m, the mean temperature of the boundary layer will be

$$20\times\frac{0.2}{2.4}\,°\text{C}=1.6\,°\text{C}$$

higher than at the bottom of the winding. Accordingly, at level $l=0.2$ above the bottom of the winding, the mean temperature of the boundary layer of the oil, i.e. the film average temperature, is

$$\vartheta_m=(62.5+1.6)\,°\text{C}=64.16\,°\text{C}.$$

The physical constants of the oil are determined for this temperature. From Table 6.2

$$\beta_{64.16}=7.9\times10^{-4}\,°\text{C}^{-1},\qquad Pr_{64.16}=84.58,$$

$$\nu_{64.16}=6.25\times10^{-16}\,\text{m}^2\,\text{s}^{-1}.$$

The *Gr* number

$$Gr=\frac{\beta g l^3}{\nu^2}\Delta\vartheta_s \tag{6.109}$$

$$=\frac{7.9\times10^{-4}\times9.81\times0.008}{(6.25\times10^{-6})^2}15=2.42\times10^6,$$

$$Gr^{-0.25}=2.35\times10^{-2},\qquad Pr^{-0.5}=0.109,\qquad (0.952+Pr)^{0.25}=3.041.$$

Substituting the above values into formula (6.108)

$$\delta = 3.93 \times 0.2 \times 0.109 \times 3.041 \times 2.53 \times 10^{-2} = 6.59 \times 10^{-3} \text{ m}.$$

Now, let us check whether we are still within the region of laminar flow:

$$Pr \times Gr = 84.58 \times 2.42 \times 10^6 = 2.05 \times 10^8.$$

This value being lower than 2×10^9, the condition of laminar flow is satisfied.

Due to the finite width of the duct, the thickness of the boundary layer developing in the vertical duct will be less than this value.

6.2.3.5 Influence of the vertical duct width on the oil flow resistance

The boundary layer thickness obtainable from Eckert's formula, denoted by δ_{sp} in the following, is equal to that gap size through which, assuming an oil space unrestricted from one side, at least 90% of the oil of the flow stream passes (see Fig. 6.29), where the distance from the heated wall is indicated by x.

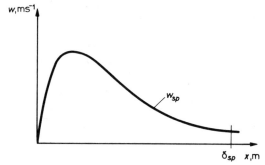

Fig. 6.29. Velocity distribution in the boundary-layer flow developing along the heat transfer surface located in an unconfined semi-space. δ_{sp} is the boundary-layer thickness

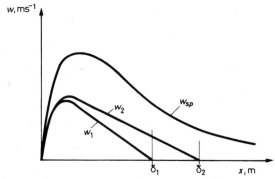

Fig. 6.30. The change of velocity maximum caused by restricting the unconfined semi-space to oil-duct width

410

If the flow developing freely and spontaneously is cut off, i.e. the infinite half of the space is restricted to an oil duct equal to or narrower than the gap size obtainable from Eckert's formula, then a considerable distortion of the velocity profile is to be expected. This situation is shown in Fig. 6.30, with two velocity profiles developing for gap widths δ_1 and δ_2. It can be stated as a general rule that any duct size below the thickness of the boundary layer developing spontaneously increases the flow resistance of the oil, decreases the mass flow, and increases the temperature difference $\Delta\vartheta_{wo}$ between the oil entering and leaving the winding stack.

From the preceding paragraphs it also follows that a vertical duct size larger than the width of the spontaneous boundary layer does not significantly improve the flow conditions in the laminar region. A vertical duct size smaller than half of the Eckert width for the spontaneous boundary layer significantly increases the flow resistance of the winding. For constructional reasons, no such small ducts are employed in practice.

6.2.3.6 Flow velocity profile of the oil in a vertical duct

In the preceding sections the effect of duct width on the oil flow has been demonstrated by variation of the velocity profile (see Fig. 6.30).

According to Schuh [121] the velocity profile in the boundary layer of a flow taking place in an unrestricted half space develops as shown in Fig. 6.31. In the

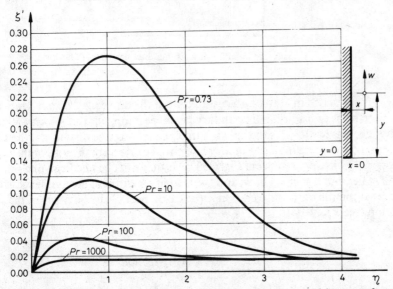

Fig. 6.31. Velocity profiles of boundary layer flow developing along a vertical heat transfer surface located in an unconfined half-space, according to Schuh [121]. Coordinates of an arbitrary point are x and y, in m. Vertical velocity is w, in ms^{-1}. Along the horizontal axis the values of $\eta = \dfrac{v}{x}(Gr_x)^{0.25}$ and

along the vertical axis the values of $\zeta' = \dfrac{w}{2\sqrt{gx}} \dfrac{1}{\sqrt{\beta\Delta\vartheta_s}}$ are to be found.

Local Gr number is $Gr_x = \dfrac{\beta g x^3 \Delta\vartheta_s}{v^2}$

dimensionless number η appearing on the abscissa axis the local Grashof number is

$$Gr_x = \frac{\beta g x^3 \Delta \vartheta_s}{v^2}.$$

(6.110)

From Fig. 6.31, velocity w can be determined for a point situated at distance x in the radial direction and at height y above the bottom of the winding.

Reverting to Fig. 6.30, the impeding effect of any finite duct size is to be attributed not only to the fact that the flow velocity must necessarily be zero at the wall of the oil duct, but also because, this wall being unheated, the viscosity of the oil is higher

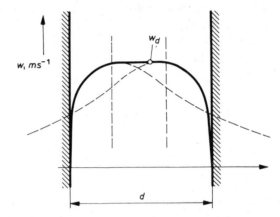

Fig. 6.32. Velocity profile of flowing oil in a duct heated from both sides

along this surface than at the surface adjacent to the disc coils. In an oil duct heated from both sides two adjoining boundary layers develop, but the velocity maximum will not decrease as long as the duct width exceeds twice the distance of the velocity maximum developing in the boundary layer of one side from the wall of the disc coil; (see Fig. 6.32). As apparent from Fig. 6.33, a vertical duct heated from one side, is not entirely filled out by the oil flow, and the velocity maximum always develops along the heated mantle of the disc coil. In some cases, part of the oil flows in the reverse direction (see Fig. 6.34).

6.2.3.7 Flow cross-section

In a wide vertical oil duct heated from one side only, the velocity profile is not quadratic and the effective flow cross-section area is much smaller than the available geometrical cross-sectional area. If the mean flow duct width is calculated as δ_{sp} 0.56, then the flow is hydraulically equivalent to a laminar flow having a velocity profile of parabolic shape. The formula for the pipe friction of fluids can be used and the resulting pressure drop can be calculated.

412

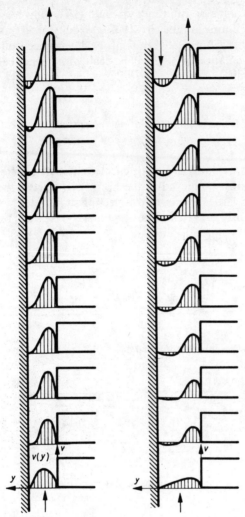

Fig. 6.33. Velocity distribution of the oil in a narrow duct heated from one side

Fig. 6.34. Thickening of laminar boundary layer in a wide oil duct [52]

6.2.3.8 Difference between temperatures of the oil entering and leaving the winding

The value of $\Delta \vartheta_{wo}$ depends on the amount of loss developing in the active parts of the transformer, on the buoyancy acting in the entire cooling circuit and on the hydraulic resistance of the cooling circuit. The fraction of the resistance of the cooling circuit caused by the winding is composed of the resistance due to liquid friction, of the resistance resulting from the outlet impetus of the oil and of the resistance corresponding to the accelerating energy required by rearrangement of

413

the velocity profile of the oil entering the winding. The flow losses of oil are covered by buoyancy of the entire cooling circuit. The velocity of oil flowing in the winding, and therefore also the value of $\Delta \vartheta_{wo}$ can only be determined if all factors listed above are known. The calculation of $\Delta \vartheta_{wo}$ will be dealt with further below, with the design of the cooling circuit.

6.2.3.9 Methods for calculating the surface temperature drop of disc-type coil windings

Two calculation methods are distinguished. The first method is followed if a definite oil flow can be achieved in the horizontal ducts either by asymmetrical arrangement of the vertical ducts, or by selecting small values for k and large values for s (see Figs 6.35, 6.37, and 6.39). The other method is used if the heat is transferred

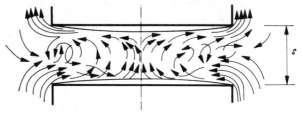

Fig. 6.35. Definite oil flow in wide horizontal ducts, with low heat flux per unit transfer area [52]

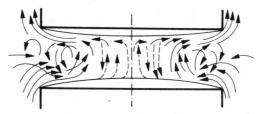

Fig. 6.36. In wide ducts and with high heat flux per unit transfer area penetration depth decreases and a stagnant oil layer develops in the middle of the section [52]

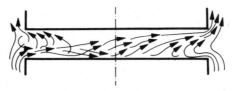

Fig. 6.37. Flow rate of oil through narrow gaps due to the differential pressure in the vertical ducts beside a horizontal gap [52]

Fig. 6.38. Stagnant oil in a narrow horizontal duct [52]

from the horizontal ducts through a practically unmoving oil layer by conduction to the oil flowing in the vertical ducts (see Figs 6.36 and 6.38).

In the first case an average surface heat transfer coefficient may be construed, calculated from the relation

$$Nu_m = c(Gr \times Pr)_m^n. \tag{6.111}$$

This relation may be set up by using data obtained from temperature rise measurements. This method can also be used in cases, where the width of the disc coil between two vertical oil ducts is less than 35 mm and the thickness of the horizontal duct is 8 per cent or more of the radial size of the winding. The surface heat transfer coefficients of horizontal surfaces are generally different, those of the lower surfaces being higher than those of the upper ones, and both being lower than that of a vertical surface. It is only worthwhile refining the calculation by distinguishing between the coefficients for upper and lower disc-coil surfaces (instead of calculating with an average surface heat transfer coefficient), if, for the

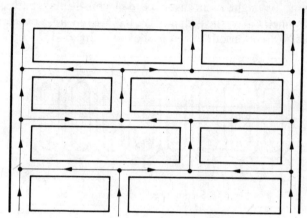

Fig. 6.39. Asymmetrically arranged vertical oil ducts

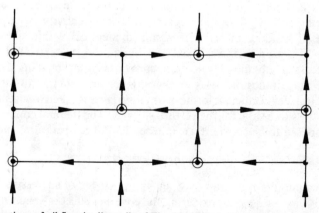

Fig. 6.40. Directions of oil flow in disc-coils of Fig. 6.39. Points of mixing are indicated by circles

415

arrangement of Fig. 6.39, the calculation of oil velocities and the pattern of temperature distribution are thereby also made possible (see Fig. 6.40). The surface heat transfer coefficient is, of course, a function of the local flow velocity developed, thermal load, Pr number and other physical characteristics of the oil.

The second calculation method is employed in cases when there is no definite oil flow present in the horizontal ducts, and the heat is transmitted from them principally by conduction to the oil flowing in the vertical ducts. In the meeting points of vertical and horizontal ducts the oil of the horizontal duct delivers its heat content to the colder oil flowing in the vertical duct by way of mixing. In the horizontal ducts an effective turbulent thermal conductivity is taken. The value of this thermal conductivity is higher than the usual value of conductivity valid for oil in the static state [54].

6.2.4 Temperature differences inside the winding

Let us assume an eight-layer disc-type coil constituting part of a winding containing asymmetrical vertical ducts, i.e. there is oil flow present also in the horizontal ducts. A section of the coil is shown in Fig. 6.41.

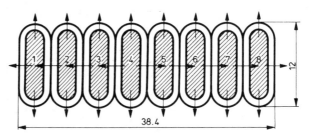

Fig. 6.41. Section of an eight-layer disc-coil

Neglecting the curvature of the winding, the coverage of horizontal surfaces by the spacers is assumed to be 40%. Let us neglect the temperature drop across the winding material in view of the high thermal conductivity of copper, i.e. the temperature of winding material is assumed identical within each layer. The temperature of the oil in the vertical and horizontal ducts around the winding section beyond the boundary layer, is also considered uniform. A distinction is made between the magnitudes of losses developing per unit volume of the layers. The surface heat transfer coefficient for horizontal surfaces is taken as already known to be 0.8 times the value valid for the vertical surfaces. The thermal conductivity of oil-paper is supposed to be constant, i.e. independent of temperature, and to take the value

$$\lambda_p = 0.1395 \text{ W m}^{-1} {}^\circ\text{C}^{-1}.$$

Let the magnitude of overall loss per kg $v_{dc} + v_e = v$ of successive layers of the winding be $v_1, v_2, \ldots, v_8$. Considering the covering effect of spacers, these values are increased by 66.7%, i.e. the above specific values are multiplied by $1/1 - 0.4$.

416

If the specific loss per kg is now further multiplied by the density of copper, $\rho_c = 8900 \text{ kg m}^{-3}$, then the values of p, i.e. the magnitudes of loss developing in each unit volume of the winding material are obtained:

$$p_1 = \frac{1}{1-0.4} v_1 \, 8900 \text{ kg m}^{-3} = 1.48 \times 10^4 \text{ kg m}^{-3} v_1 \, ,$$

$$p_2 = 1.667 \, v_2 \, 8900 \text{ kg m}^{-3} = 1.48 \times 10^4 \text{ kg m}^{-3} v_2 \, ,$$

$$\vdots$$

$$p_k = 1.667 \, v_k \, 8900 \text{ kg m}^{-3} = 1.48 \times 10^4 \text{ kg m}^{-3} v_k \, ,$$

$$\vdots$$

$$p_8 = 1.667 \, v_8 \, 8900 \text{ kg m}^{-3} = 1.48 \times 10^4 \text{ kg m}^{-3} v_8 . \tag{6.112}$$

Let the coil investigated be that part of the larger radial size winding which lies between two vertical oil ducts, closest to the leakage channel, i.e.

$$v_1 > v_2 \ldots > v_8 .$$

Denote the losses developing in the piece of conductor of length l perpendicular to the plane of the diagram by $P_1, P_2, \ldots P_8$, the dimensions of the conductor section by a and b, the cross-sectional area of conductor by A. With these notations, the loss developing in the kth piece of conductor of length l is

$$P_k = Alp_k = ablp_k = 1.48 \times 10^4 ablv_k , \tag{6.113}$$

where the dimension of factor 1.48×10^4 is kg m^{-3}.

Losses $P_1, P_2, \ldots P_8$ of different magnitudes developing in the various layers leave the disc-type coils in the directions indicated by arrows in Fig. 6.41. As it can be seen, part of the heat flows radially between the layers. The loss developing in the winding has to leave the disc coil across the outer surface. The power balance is

$$\sum_{k=1}^{8} P_k = \sum_{k=1}^{8} \alpha_k A_k \, \Delta \vartheta_{sk} . \tag{6.114}$$

In the equation of power balance A_k is the convection surface of the kth layer, α_k is the surface heat transfer coefficient of the kth layer and $\Delta \vartheta_{sk}$ is the surface temperature drop in the boundary layer of the kth layer, this boundary layer being in direct contact with the oil.

The losses of different magnitudes developing in the various layers leave through the paths of minimum temperature drop. The heat leaving a layer passes, first, through an interturn insulation of thickness δ and of heat conductance λ_p, at the expense of a temperature drop $\Delta \vartheta_p$. Denoting the surface area of the paper layer perpendicular to the direction of heat flow by A_p, and its thermal resistance by R_p, then the temperature drop across the paper layer, assuming a heat flux per unit transfer area q, will be

$$\Delta \vartheta_p = q A_p R_p = q A_p \frac{\delta}{A_p \lambda_p} = q \frac{\delta}{\lambda_p} . \tag{6.115}$$

From the outer surface of the paper layer, the heat either reaches the oil flowing in the duct through thermal resistance R_c at the expense of surface temperature drop

$\Delta\vartheta_s$, or flows into the adjacent layer across the interturn insulation of the latter. The effect of radial heat flow on the temperature rise of the various layers is demonstrated by an example.

If the convection surface of the layer is denoted by A_c, the heat flux per unit transfer area by q and the surface heat transfer coefficient by α, then the temperature drop across the surface will be.

$$\Delta\vartheta_s = qA_cR_C = qA_c\frac{1}{A_c\alpha} = \frac{q}{\alpha}.\qquad(6.116)$$

Example 6.12 Calculate the temperature rise of turns of an eight-layer disc-type coil winding, relative to the oil temperature which is assumed to remain constant (see Fig. 6.41). The following specific overall losses develop in the various layers differing in their magnitude due to the different eddy-current loss components.

Denote the specific losses developing in the various layers by $v_1, v_2, \ldots, v_8$.

$$v_1 = 30.90 \text{ W kg}^{-1},$$
$$v_2 = 28.90 \text{ W kg}^{-1},$$
$$v_3 = 27.20 \text{ W kg}^{-1},$$
$$v_4 = 25.70 \text{ W kg}^{-1},$$
$$v_5 = 24.40 \text{ W kg}^{-1},$$
$$v_6 = 23.10 \text{ W kg}^{-1},$$
$$v_7 = 22.70 \text{ W kg}^{-1},$$
$$v_8 = 22.30 \text{ W kg}^{-1}.$$

Loss developing in a conductor of 1 m length: Let the cross-sectional area of a conductor be $26.3 \times 10^{-6} \text{ m}^2$ and the density of copper 8900 kg m^{-3}. Mass of a $1-m$ length of conductor:

$$26.3 \times 10^{-6} \times 1 \times 8900 \text{ kg} = 0.234 \text{ kg}.$$

Let the spacer covering be 40%, i.e. an area belonging to a conductor length of 1 m must dissipate the loss developing in a conductor of $1/1 - 0.4 = 1.66$ m length.

Multiplying the loss developing in 1 kg of conductor material by 0.234×1.6 kg gives the amount of loss that is to be transferred to the oil across the surface area belonging to a conductor length of 1 m. Let these losses be denoted by $P_1, P_2, \ldots, P_8$.

$$P_1 = 12.05 \text{ W}, P_2 = 11.28 \text{ W}, P_3 = 10.60 \text{ W}, P_4 = 10.00 \text{ W},$$
$$P_5 = 9.50 \text{ W}, P_6 = 9.10 \text{ W}, P_7 = 8.85 \text{ W}, P_8 = 8.70 \text{ W}.$$

Heat transfer surfaces for 1 m conductor length (see Fig. 6.42).

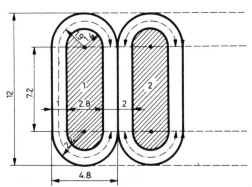

Fig. 6.42. Dimensions of conductors of an eight-layer disc-coil

Convection surface area of the outermost layer:

$$A_{c1} = 1 \times (4.8\pi + 7.2) \times 10^{-3} \, m^2 = 22.25 \times 10^{-3} \, m^2 .$$

Convection surface area of any intermediate layer:

$$A_{c2} = 1 \times 4.8\pi \times 10^{-3} \, m^2 = 15.05 \times 10^{-3} \, m^2 .$$

Surface area of the outermost layer for the mean thickness of the paper layer:

$$A_{p1} = 1 \times (3.8\pi + 7.2) \times 10^{-3} \, m^2 = 19.12 \times 10^{-3} \, m^2 .$$

Surface area of any intermediate layer for the mean thickness of the paper layer:

$$A_{p2} = 1 \times 3.8\pi \times 10^{-3} \, m^2 = 11.92 \times 10^{-3} \, m^2 .$$

Heat transfer area between any two layers:

$$A_t = 1 \times 7.2 \times 10^{-3} \, m^2 = 7.2 \times 10^{-3} \, m^2 .$$

Thickness of insulation layer between any layer and oil:

$$\delta = 1 \times 10^{-3} \, m .$$

Thickness of insulation between any two layers:

$$2\delta = 2 \times 1 \times 10^{-3} = 2 \times 10^{-3} \, m .$$

The surface heat transfer coefficient can be expressed as an exponential function of heat flux per unit transfer area. Let the surface heat transfer coefficient be

$$\alpha = 2.1 q^{0.5} , \tag{6.117}$$

where q is the heat flux per unit transfer area and the dimension of factor 2.1 is $W^{0.5} \, m^{-1} \, °C^{-1}$. For simplicity let us assume that the surface heat transfer coefficient is identical for the horizontal and the vertical surfaces. Denote by symbol P^*, the loss dissipated by the outermost layer through convection. Dividing this loss by the convection surface of the layer, the heat flux per unit transfer area is obtained. Denote the surface heat transfer coefficient pertaining to this thermal load by α^*:

$$\alpha^* = 2.1 \sqrt{\frac{P^*}{22.25 \times 10^{-3}}} = 14 \sqrt{P^*} .$$

Denote by P^{**} the loss dissipated by any intermediate layer through convection. Its heat transfer coefficient is

$$\alpha^{**} = 2.1 \sqrt{\frac{P^{**}}{15.05 \times 10^{-3}}} = 17.1 \sqrt{P^{**}} ,$$

where in the expressions for α^* and α^{**} the dimension of the first factor is, in both cases, $W^{0.5} \, m^{-2} \, °C^{-1}$.

Thermal conductivity of insulating paper is

$$\lambda_p = 0.1395 \, W \, m^{-1} \, °C^{-1} .$$

Thermal resistances: Generally, the thermal resistance between any layer and the oil is

$$R_{c+p} = \frac{1}{A_c\alpha} + \frac{\delta}{A_p\lambda_p} = \frac{A_p\lambda_p + \delta A_c\alpha}{A_c\alpha A_p\lambda_p}. \tag{6.118}$$

The thermal resistance of paper insulation between any two layers is

$$R_p = \frac{\delta'}{A_t\lambda_p}, \tag{6.119}$$

where δ' is the thickness of insulating paper between the two layers.

The thermal resistance of the outermost layer up to the oil, is

$$R^* = \frac{19.12 \times 10^{-3} \times 0.1395 + 1 \times 10^{-3} \times 22.25 \times 10^{-3} \times 14 \times \sqrt{P^*}\ W^{-0.5}}{22.25 \times 10^{-3} \times 14 \times \sqrt{P^*}\ 19.12 \times 10^{-3} \times 0.1395\ W^{-0.5}}\ ^\circ\mathrm{C\,W^{-1}}$$

$$= \frac{2.67\ W^{0.5} + 0.311\ \sqrt{P^*}}{0.835\ \sqrt{P^*}}\ ^\circ\mathrm{C\,W^{-1}} = \left(\frac{3.20\ W^{0.5}}{\sqrt{P^*}} + 0.37\right)^\circ\mathrm{C\,W^{-1}}.$$

The thermal resistance of any intermediate layer up to the oil is:

$$R^{**} = \frac{11.92 \times 10^{-3} \times 0.1395 + 1 \times 10^{-3} \times 15.05 \times 10^{-3} \times 17.1\ \sqrt{P^{**}}\ W^{-0.5}}{15.05 \times 10^{-3} \times 17.1\ \sqrt{P^{**}}\ 11.92 \times 10^{-3} \times 0.1395\ W^{-0.5}}\ ^\circ\mathrm{C\,W^{-1}}$$

$$= \frac{1.665\ W^{0.5} + 0.2575\ \sqrt{P^{**}}}{0.428\ \sqrt{P^{**}}}\ ^\circ\mathrm{C\,W^{-1}} = \left(\frac{3.88\ W^{0.5}}{\sqrt{P^{**}}} + 0.6\right)^\circ\mathrm{C\,W^{-1}}.$$

The thermal resistance of insulation between two layers is

$$R_p = \frac{2 \times 10^{-3}}{7.2 \times 10^{-3} \times 0.1395}\ ^\circ\mathrm{C\,W^{-1}} = 1.99\ ^\circ\mathrm{C\,W^{-1}}.$$

The product of thermal resistance and power passing through it gives the temperature drop, with the dimension of $^\circ$C. The temperature drop of any layer with respect to the oil temperature, assumed to remain constant, is equal to the product of the thermal resistance R_{c+p} of the layer and the power passing through it (see Fig. 6.43).

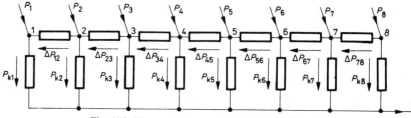

Fig. 6.43. Thermal scheme of an eight-layer disc-coil

420

The direction of heat flow is indicated by arrows in the diagram. As shown, the reference directions of heat flow between adjacent layers have been assumed unidirectional, although we already know that, from the 5th layer on, the direction of radial heat flow will probably be from left to right, instead of that assumed. In the calculated results the correct flow direction will be shown unequivocally by the signs of transferred power.

As shown by Fig. 6.43 eight nodes correspond to the eight layers, the nodes being marked with serial numbers from 1 to 8. The losses P_1, P_2, ..., P_8 developing in the various layers are those given numerically above.

The outermost layer on the left-hand side obtains a power ΔP_{12} through resistance R_p. For example, the loss developing in the conductor arranged in the 6th layer is P_5, and the 5th layer obtains a power ΔP_{56} from the 6th layer and transfers ΔP_{45} to the 4th layer.

From the various layers the following powers are transferred to the oil:

1st layer: $P_1 + \Delta P_{12} = 12.05\ \text{W} + P_{12}$,
2nd layer: $P_2 + \Delta P_{23} - \Delta P_{12} = 11.28\ \text{W} + \Delta P_{23} - \Delta P_{12}$,
3rd layer: $P_3 + \Delta P_{34} - \Delta P_{23} = 10.60\ \text{W} + \Delta P_{34} - \Delta P_{23}$,
4th layer: $P_4 + \Delta P_{45} - \Delta P_{34} = 10.00\ \text{W} + \Delta P_{45} - \Delta P_{34}$,
5th layer: $P_5 + \Delta P_{56} - \Delta P_{45} = 9.50\ \text{W} + \Delta P_{56} - \Delta P_{45}$,
6th layer: $P_6 + \Delta P_{67} - \Delta P_{56} = 9.10\ \text{W} + \Delta P_{67} - \Delta P_{56}$,
7th layer: $P_7 + \Delta P_{78} - \Delta P_{67} = 8.85\ \text{W} + \Delta P_{78} - \Delta P_{67}$,
8th layer: $P_8 + 0 - \Delta P_{78} = 8.70\ \text{W} - \Delta P_{78}$.

Let us write the temperature drops of layers with respect to the oil, denoted by $\Delta \vartheta_1$, $\Delta \vartheta_2$, ... $\Delta \vartheta_8$.

$$\Delta \vartheta_1 = \left(\frac{3.20\,°\text{C W}^{-0.5}}{\sqrt{12.05\,\text{W} + \Delta P_{12}}} + 0.375\,°\text{C W}^{-1} \right)\left(12.05\ \text{W} + \Delta P_{12} \right) =$$

$$= 3.20\,°\text{C W}^{-0.5}\sqrt{12.05\,\text{W} + \Delta P_{12}} + 4.51\,°\text{C} + 0.375\,°\text{C W}^{-1}\Delta P_{12},$$

$$\Delta \vartheta_2 = 3.88\,°\text{C W}^{-0.5}\sqrt{11.28\,\text{W} + \Delta P_{23} - \Delta P_{12}} + 6.77\,°\text{C} +$$
$$+ 0.6\,°\text{C W}^{-1}(\Delta P_{23} - \Delta P_{12}),$$

$$\Delta \vartheta_3 = 3.88\,°\text{C W}^{-0.5}\sqrt{10.60\,\text{W} + \Delta P_{34} - \Delta P_{23}} + 6.35\,°\text{C} +$$
$$+ 0.6\,°\text{C W}^{-1}(\Delta P_{34} - \Delta P_{23}),$$

$$\Delta \vartheta_4 = 3.88\,°\text{C W}^{-0.5}\sqrt{10.00\,\text{W} + \Delta P_{45} - \Delta P_{34}} + 6.00\,°\text{C} +$$
$$+ 0.6\,°\text{C W}^{-1}(\Delta P_{45} - \Delta P_{34}),$$

$$\Delta \vartheta_5 = 3.88\,°\text{C W}^{-0.5}\sqrt{9.50\,\text{W} + \Delta P_{56} - \Delta P_{45}} + 5.70\,°\text{C} +$$
$$+ 0.6\,°\text{C W}^{-1}(\Delta P_{56} - \Delta P_{45}),$$

$$\Delta \vartheta_6 = 3.88\,°\text{C W}^{-0.5}\sqrt{9.10\,\text{W} + \Delta P_{67} - \Delta P_{56}} + 5.46\,°\text{C} +$$
$$+ 0.6\,°\text{C W}^{-1}(\Delta P_{67} - \Delta P_{56}),$$

$$\Delta \vartheta_7 = 3.88\,°\text{C W}^{-0.5}\sqrt{8.85\,\text{W} + \Delta P_{78} - \Delta P_{67}} + 5.31\,°\text{C} +$$
$$+ 0.6\,°\text{C W}^{-1}(\Delta P_{78} - \Delta P_{67}),$$

$$\Delta \vartheta_8 = 3.20\,°\text{C W}^{-0.5}\sqrt{8.70\,\text{W} - \Delta P_{78}} + 3.26\,°\text{C} -$$
$$- 0.375\,°\text{C W}^{-1}\,\Delta P_{78}.$$

421

For the temperature drops between adjacent layers the following can be written:

$$\Delta\vartheta_2 - \Delta\vartheta_1 = 1.99\,°\text{C W}^{-1}\,\Delta P_{12},$$

$$\Delta\vartheta_3 - \Delta\vartheta_2 = 1.99\,°\text{C W}^{-1}\,\Delta P_{23},$$

$$\Delta\vartheta_4 - \Delta\vartheta_3 = 1.99\,°\text{C W}^{-1}\,\Delta P_{34},$$

$$\Delta\vartheta_5 - \Delta\vartheta_4 = 1.99\,°\text{C W}^{-1}\,\Delta P_{45},$$

$$\Delta\vartheta_6 - \Delta\vartheta_5 = 1.99\,°\text{C W}^{-1}\,\Delta P_{56},$$

$$\Delta\vartheta_7 - \Delta\vartheta_6 = 1.99\,°\text{C W}^{-1}\,\Delta P_{67},$$

$$\Delta\vartheta_8 - \Delta\vartheta_7 = 1.99\,°\text{C W}^{-1}\,\Delta P_{78}.$$

Substituting the various values of $\Delta\vartheta$, 7 equations are obtained with 7 unknown quantities. Changing over to simpler mathematical notation, the equations will be the following: instead of $\Delta\vartheta_1, \Delta\vartheta_2, \ldots, \Delta\vartheta_8$, write $y_1, y_2, \ldots, y_8$; and instead of $\Delta P_{12}, \Delta P_{23}, \ldots, \Delta P_{78}$, write $x_{12}, x_{23}, \ldots, x_{78}$:

$$3.88\sqrt{11.28 + x_{23} - x_{12}} - 3.20\sqrt{12.05 + x_{12}} + 2.26 + 0.6x_{23} - 2.965x_{12} = 0,$$

$$3.88(\sqrt{10.60 + x_{34} - x_{23}} - \sqrt{11.28 + x_{23} - x_{12}}) - 0.42 + 0.6(x_{12} + x_{34}) - 3.19x_{23} = 0,$$

$$3.88(\sqrt{10 + x_{45} - x_{34}} - \sqrt{10.6 + x_{34} - x_{23}}) - 0.35 + 0.6(x_{23} + x_{45}) - 3.19x_{34} = 0,$$

$$3.88(\sqrt{9.5 + x_{56} - x_{45}} - \sqrt{10 + x_{45} - x_{34}}) - 0.3 + 0.6(x_{34} + x_{56}) - 3.19x_{45} = 0,$$

$$3.88(\sqrt{9.1 + x_{67} - x_{56}} - \sqrt{9.5 + x_{56} - x_{45}}) - 0.24 + 0.6(x_{45} + x_{67}) - 3.19x_{56} = 0,$$

$$3.88(\sqrt{8.85 + x_{78} - x_{67}} - \sqrt{9.1 + x_{67} - x_{56}}) - 0.15 + 0.6(x_{56} + x_{78}) - 3.19x_{67} = 0,$$

$$3.20\sqrt{8.7 - x_{78}} - 3.88\sqrt{8.85 + x_{78} - x_{67}} - 2.05 + 0.6x_{67} - 2.965x_{78} = 0.$$

Solving the equations, the following results are obtained: $x_{12} = 1.05456$, $x_{23} = 0.03620$, $x_{34} = -0.232\,38$, $x_{45} = -0.291\,60$, $x_{56} = -0.325\,36$, $x_{67} = -0.481\,47$, $x_{78} = -1.146\,17$. Reverting to the physical way of writing, the former numbers indicate the quantities of heat flowing between the layers in the radial direction, expressed in W: $P_{12} = 1.05456\ W$, $P_{23} = 0.036\,20\ W$, $\ldots$, $P_{78} = -1.146\,17\ W$. Re-substituting the values obtained for $x_{12}, x_{23}, \ldots, x_{78}$ into the original equations, the following values are obtained for $y_1, y_2, \ldots, y_8$, i.e. for the temperature drops: $\Delta\vartheta_1 = 16.49°\text{C}$, $\Delta\vartheta_2 = 18.59\,°\text{C}$, $\Delta\vartheta_3 = 18.66\,°\text{C}$, $\Delta\vartheta_4 = 18.20\,°\text{C}$, $\Delta\vartheta_5 = 17.62\,°\text{C}$, $\Delta\vartheta_6 = 16.97\,°\text{C}$, $\Delta\vartheta_7 = 16.01\,°\text{C}$, $\Delta\vartheta_8 = 13.73\,°\text{C}$.

Example 6.13. Let us calculate the temperature rise above the oil temperature, assumed to remain constant, in the disc coil dealt with in the preceding example, if equal losses develop in all layers and the radial heat flow is left out of consideration.

Let the loss to be transferred through the surface of the winding of 1 m length be equal to the sum of the losses obtained in the preceding example:

$$P = P_1 + P_2 + \ldots + P_8 = (12.05 + 11.28 + 10.6 + 10 + 9.5 + 9.1 + 8.85 + 8.7)\ W = 80.08\ W.$$

The convection surface area of the winding of 1 m length is

$$A_c = (8 \times 4.8\pi + 2 \times 7.2)\,10^{-3} \times 1\ \text{m}^2 = 0.1349\ \text{m}^2.$$

The area pertaining to the mean thickness of insulation is

$$A_p = (8 \times 3.8\pi + 2 \times 7.2)\,10^{-3} \times 1\ \text{m}^2 = 0.1099\ \text{m}^2.$$

The thickness of insulation is

$$\delta = 1 \times 10^{-3}\ \text{m}.$$

The thermal conductivity of insulating paper is

$$\lambda_p = 0.1395\ \text{W m}^{-1}\,°\text{C}^{-1}.$$

The surface heat transfer coefficient is

$$\alpha = 2.1 \sqrt{\frac{P}{A_c}} = 2.1 \sqrt{\frac{80.08}{0.1349}} \ \text{W M}^{-2}\,°\text{C}^{-1} = 55.3 \ \text{W m}^{-2}\,°\text{C}^{-1}.$$

The thermal resistance is

$$R_{c+p} = \frac{A_p \lambda_p + \delta A_c \alpha}{A_c \alpha A_p \lambda_p} = \frac{0.1099 \times 0.1395 + 1 \times 10^{-3} \times 0.1349 \times 55.3}{0.1349 \times 55.3 \times 0.1099 \times 0.1395} \,°\text{C W}^{-1} = 0.199 \,°\text{C W}^{-1}.$$

The temperature drop is

$$\Delta \vartheta = R_{c+p} P = 0.199 \times 80.08 = 15.95 \,°\text{C}.$$

Collating the two examples, it will be seen that the hot-spot temperature of the winding is higher by $(18.66 - 15.95)\,°\text{C} = 2.71\,°\text{C}$ than the average temperature of the disc.

6.2.5 Calculation of differences between mean winding and oil temperatures of disc-type coil with natural oil flow

6.2.5.1 Measurements of temperature rise on disc-type coil windings and evaluation of results

The following method shows how measurements of temperature rise can be utilized for setting up an equation suitable for calculating temperature rise conditions of disc-coils. The temperature rise data have been taken over from a table appearing on page 307 of Reference [11], transcribed into the system of measures used in this book.

The thermocouple for measuring the winding temperature was embedded into the second disc from the top in its middle layer, under the spacers. The temperature of the surrounding oil was monitored by a thermocouple placed at the height of the measured disc at a distance 40 mm in the radial direction from the outer mantle of the disc. The insulated size of the disc was $h_1 = 10$ mm, $h_2 = 50$ mm. (The width of the oil duct between the discs of the winding stack and that of the inner oil duct was 6 mm. No insulating cylinder surrounded the outer mantle of the winding, leaving it exposed to the bulk of oil.

The results of measurements and the similarity factors calculated by the author and relating to the mean boundary-layer (film average) temperature of the oil, are contained in Table 6.8. In calculating the temperature across the paper insulation the increase of thermal conductivity of the paper with temperature was taken into consideration. As linear dimension appearing in the Gr number the height of the disc, $h_1 = 10 \times 10^{-3}$ m was chosen. By plotting the Nu numbers as functions of $10^4\,Gr \times Pr$ in a log–log coordinate system, the diagram shown in Fig. 6.44 was obtained.

The straight line found to match best the points calculated from the measuring points was drawn, and two points lying wide apart on the straight line selected. A power–function relation exists between the coordinates of the two points.

Table 6.8

Results of temperature rise measurements performed on a winding consisting of disc coils and similarity numbers calculated from the results

No.	Temperature of surrounding oil ϑ_o, °C	Max. winding temperature ϑ_{wm}, °C	Temp. drop in paper $\Delta\vartheta_p$, °C	Surface temperature drop $\Delta\vartheta_s$, °C	Mean boundary layer temperature ϑ_m, °C	Surface heat flux per unit transfer area q, W m⁻²	Average surface heat-transfer coefficient α, W m⁻² °C⁻¹	Nu_m	Pr_m	Gr_m	$(Gr \times Pr)_m \times 10^{-4}$
1	24.7	38.5	3.2	10.6	30.0	474	44.7	3.42	265.0	171.2	4.54
2	23.1	54.2	9.3	21.8	34.0	1515	69.5	5.33	226.6	497.0	11.25
3	25.3	71.8	12.7	33.8	42.2	2910	94.5	7.28	166.8	1 540.0	25.85
4	27.2	106.0	25.2	53.6	54.0	6600	123.2	9.6	111.8	5930.0	66.30
5	34.4	47.0	2.0	10.6	39.7	481	45.3	3.49	180.0	398.0	7.17
6	31.7	60.0	9.1	19.2	41.3	1395	72.7	5.6	172.2	792.0	13.62
7	34.1	80.2	13.3	32.8	50.5	2910	88.7	6.87	127.0	2650.0	33.70
8	34.0	110.9	24.9	52.0	60.0	6470	124.5	9.7	93.9	5930.0	55.75
9	44.6	56.6	4.0	8.0	48.6	508	63.5	4.91	133.4	587.0	7.89
10	44.0	70.0	6.8	19.2	53.6	1480	77.2	5.99	113.0	1 985.0	22.40
11	44.1	87.7	13.2	30.4	59.3	3070	101.0	7.88	95.9	4 680.0	45.00
12	45.3	119.3	23.4	50.6	70.6	6880	136.0	10.67	73.0	14 100.0	103.00
13	74.5	85.6	2.9	8.2	78.6	560	68.2	5.37	60.5	3480.0	21.05
14	73.0	97.0	6.6	17.4	81.7	1610	92.4	7.29	57.0	8 190.0	46.70
15	75.1	115.4	12.5	27.8	89.0	3340	120.2	9.52	50.5	17 100.0	86.30
16	76.3	148.8	25.1	47.4	100.0	7450	157.0	12.52	43.5	41 300.0	180.00
17	98.4	108.5	1.9	8.2	102.5	600	73.2				
18	98.9	121.6	6.7	16.0	106.9	1720	107.2	Physical properties of oil			
19	100.2	139.2	12.2	26.8	113.6	3550	132.8	are given in Table 6.2.			
20	100.9	168.8	19.2	48.2	125.0	7820	162.5	up to 100 °C only.			

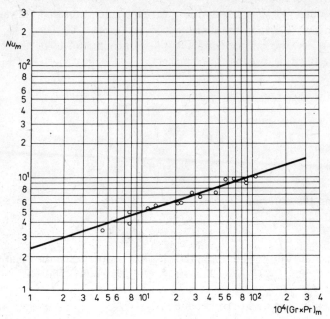

Fig. 6.44. Relation $Nu_m = 0.1095(Gr, Pr)_m^{0.333}$

The two points are $Gr \times Pr = 1 \times 10^4$ to which $Nu = 2.3$ corresponds and $Gr \times Pr = = 300 \times 10^4$ to which $Nu = 15$ corresponds. If a power–function relation exists between the points, then

$$c(10^4)^n = 2.3,$$
$$c(300 \times 10^4)^n = 15.$$

Solving the above pair of equations, the results are $c = 0.1095$ and $n = 0.333$, from which the equation of similarity criteria obtained from the measurements is,

$$Nu_m = 0.1095(Gr \times Pr)_m^{0.333}, \tag{6.120}$$

where subscript m means that the temperature-dependent physical properties of the oil have to be substituted by their values prevailing at the mean boundary-layer temperature.

Let us express the value of α in explicit form from the preceding equation:

$$\alpha = Nu_m \frac{\lambda}{h_1}, \tag{6.121}$$

thus

$$\alpha = 0.1095 Pr^{0.333} \left(\frac{\beta h_1^3 g \Delta \vartheta_s}{v^2} \right)^{0.333} \frac{\lambda}{h_1},$$

or

$$\alpha = 0.1095 Pr^{0.333} \lambda \left(\frac{\beta g}{v^2} \right)^{0.333} \Delta \vartheta_s^{0.333}. \tag{6.122}$$

425

As seen, the linear dimension has disappeared which means that the value of α does not depend on the size of the disc.

Let us express the surface temperature drop by the terms of heat flux per unit transfer area and heat transfer coefficient:

$$\Delta\vartheta_s = q\frac{1}{\alpha} = q\frac{1}{0.1095Pr^{0.333}\lambda\left(\dfrac{\beta g}{\nu^2}\right)^{0.333}\Delta\vartheta_s^{\,0.333}}. \tag{6.123}$$

From which

$$\Delta\vartheta_s = 5.3q^{0.75}(Pr\beta g)^{-0.25}\lambda^{-0.75}\nu^{0.5}. \tag{6.124}$$

Substituting the criterion of $Pr = \dfrac{\nu c\rho}{\lambda}$,

$$\Delta\vartheta_s = 5.3q^{0.75}\nu^{0.25}(c\rho\beta g)^{-0.25}\lambda^{-0.5}. \tag{6.125}$$

Let us substitute into the equation the physical properties of the oil at temperature $\vartheta_m = 70\,°C$, $Pr = 74$, $\beta = 7.95\times10^{-4}\,°C^{-1}$, $\lambda = 0.1276\,W\,m^{-1}\,°C^{-1}$, and $\nu = 5.4\times10^{-6}\,m^2\,s^{-1}$.

After substitution and rearrangement,

$$\Delta\vartheta_s = 0.0662q^{0.75} = \frac{1}{15.1}q^{0.75} \tag{6.126}$$

is obtained.

The dimension of the factor 0.0662 appearing in the formula is $m^{1.5}\,°C\,W^{-0.75}$.

Example 6.14 With $q = 6880\,W\,m^{-2}$ and $\vartheta_m = 70\,°C$,

$$\Delta\vartheta_s = 0.0662\times6880^{0.75}\,°C = 49.5\,°C.$$

To a rather good approximation, we have obtained the measured data shown in the 12th row of Table 6.8.

6.2.5.2 Surface heat transfer coefficient and temperature drop in the case of a vertical surface cooled in an unlimited half-space

The surface heat transfer coefficient of the winding can be calculated, in this case also, from the power function (6.111):

$$Nu_m = c(Gr\times Pr)_m^n.$$

The temperature-dependent physical characteristics of the oil should be substituted into the equation, using their values at the mean boundary-layer temperature of the oil, as indicated by subscript m. Constant c and power n depend on the magnitude of the product $(Gr\times Pr)_m$.

In practical cases the product $(Gr\times Pr)_m$ falls in the range of 5×10^2 to 2×10^7. For that range $c = 0.54$ and $n = 0.25$.

426

The surface heat transfer coefficient in explicit form is

$$\alpha = 0.54(\beta g Pr_m)^{0.25} \frac{\lambda_m}{v_m^{0.5}} \left(\frac{\Delta \vartheta_s}{l}\right)^{0.25}. \tag{6.127}$$

where l is the dimension of the heated surface in the direction of flow. Let us introduce the factor of

$$A = 0.54(\beta g Pr_m)^{0.25} \frac{\lambda_m}{v_m^{0.5}}. \tag{6.128}$$

With it, the following expression is obtained for the heat transfer coefficient:

$$\alpha = A\left(\frac{\Delta \vartheta_s}{l}\right)^{0.25}. \tag{6.129}$$

Let us now calculate the value of A for oil at a temperature $\vartheta_m = 70\,°C$. The physical characteristics of the oil are then

$$\vartheta_m = 70\,°C$$

$$\beta = 7.95 \times 10^{-4}\,°C^{-1},\ Pr = 74,\ \lambda = 0.1276\ W\ m^{-1}\,°C^{-1},$$
$$v = 5.4 \times 10^{-6}\ m^2\ s^{-1}.$$

$$A = 0.54(7.95 \times 10^{-4} \times 9.81 \times 74)^{0.25} \frac{0.1276}{(5.4 \times 10^{-6})^{0.5}}\ W\ m^{-1.75}\,°C^{-1.25} =$$
$$= 25.8\ W\ m^{-1.75}\,°C^{-1.25}.$$

The surface heat transfer coefficient is

$$\alpha = 25.8\left(\frac{\Delta \vartheta_s}{l}\right)^{0.25}, \tag{6.130}$$

where the dimension of the factor 25.8 is $W\ m^{-1.75}\,°C^{-1.25}$, $\Delta \vartheta_s$ is the surface temperature drop and l is the dimension of the heated surface in the direction of flow.

As known, power P transferred across surface A can be expressed in terms of heat flux per unit transfer area q, surface heat transfer coefficient α and surface temperature drop $\Delta \vartheta_s$:

$$q = \frac{P}{A} = \alpha \Delta \vartheta_s = 25.8\left(\frac{\Delta \vartheta_s}{l}\right)^{0.25} \Delta \vartheta_s =$$

$$= 25.8 \frac{\Delta \vartheta_s^{1.25}}{l^{0.25}}, \tag{6.131}$$

where the dimension of factor 25.8 is again $W\ m^{-1.75}\,°C^{-1.25}$. Let us express the value of $\Delta \vartheta_s$ from the above equation:

$$\Delta \vartheta_s = q^{0.8} \frac{l^{0.2}}{13.5}, \tag{6.132}$$

where the dimension of the denominator 13.5 is $W^{0.8}\ m^{-1.4}\,°C^{-1}$.

Presenting the heat transfer coefficient as a function of the heat flux per unit transfer area

$$\alpha = \frac{13.5}{l^{0.2}} q^{0.2} ,$$
(6.133)

is obtained, in which the dimension of factor 13.5 is again $W^{0.8} m^{-1.4} {}^{\circ}C^{-1}$.

Let us select a characteristic size of l and let height h_1 of the disc be 15×10^{-3} m. Let us substitute this characteristic size into the formula for surface temperature drop. After the substitution and rearranging, the following formula is obtained for the surface temperature drop:

$$\Delta \vartheta_s = \frac{1}{31.4} q^{0.8} .$$
(6.134)

The dimension of factor $1/31.4$ appearing in the formula is $m^{1.6} {}^{\circ}C W^{-0.8}$.

6.2.5.3 Equations for calculating the surface temperature drop of various winding arrangements

The scale model measurements described in the foregoing sections have shown that the flow in vertical ducts resembles a boundary-layer pattern, similar to the free flow pattern of a liquid which develops along a heated vertical plane in an unlimited half-space. It has also been explained that if the width of the vertical duct heated from one side exceeds the thickness of the boundary layer which develops spontaneously as is always the case due to technological reasons, then, however large the duct is made the reduction of flow velocity will remain the same, relative to the velocity of free flow that would develop in an infinite halfspace. This phenomenon has been explained by the cutting-off of the velocity profile. It has also been shown that if the duct is heated from both sides and its width is at least as great as the thickness of the spontaneous boundary layer, no reduction of velocity will take place.

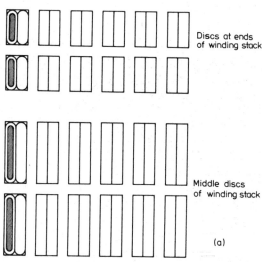

Discs at ends of winding stack

Middle discs of winding stack

(a)

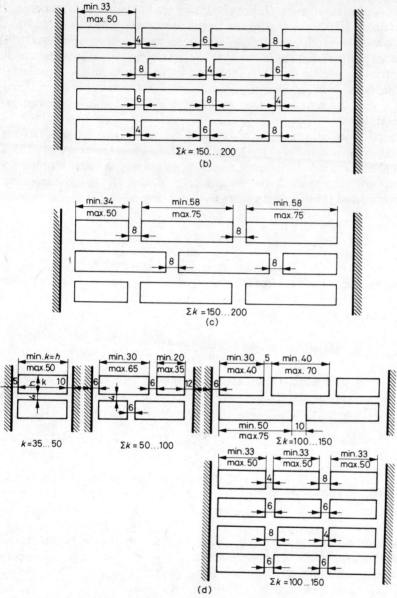

Fig. 6.45. Winding sections of large transformers

The windings of large transformers contain two ducts heated from one side and at least one duct heated from two sides. All these vertical oil ducts are asymmetrically displaced with respect to each other after every disc (see Fig. 6.45). These asymmetrically arranged vertical ducts make for definite oil flow in the horizontal ducts.

If only two lateral oil ducts are provided, each heated from one of its sides, and the width of horizontal ducts up to 50 mm radial disc size is smaller than 8% of the disc size, then no definite oil flow will develop in the horizontal ducts, and the oil will fail to enter the horizontal ducts. In this case, the heat is transferred to the lateral ducts by turbulent convection from the upper and lower heating surfaces bordering the horizontal ducts. In large power transformers the use of such winding arrangements and duct sizes are always avoided because no definite oil flow develops in the horizontal ducts.

It can be stated as a rule that the cooling system of large power transformers should be so constructed as to ensure oil flow of boundary-layer-type in all cooling ducts and also the development of a definite oil velocity in the horizontal ducts.

For a given boundary-layer temperature of the oil, as corroborated by experimental data for all winding arrangements of Fig. 6.46, the surface temperature drop is

$$\Delta \vartheta_s = \frac{1}{B} q^n , \qquad (6.135)$$

where the dimension of B depends on the exponent $m^{2n} \, °C \, W^{-n}$.

For the surface facing the free half space, with $n = 0.8$, for the value of the above exponent $B = 31.4 \, m^{1.6} \, °C \, W^{-0.8}$ was obtained.

The surface temperature drop (6.134) is

$$\Delta \vartheta_s = \frac{1}{31.4} q^{0.8} .$$

For a disc coil dissipating its heat across one of its mantle surface into unconfined space, from experimental data $n = 0.75$ and $B = 15.1 \, m^{1.5} \, °C \, W^{-0.75}$ have been obtained. The surface temperature drop (Eqn 6. 126) is given by

$$\Delta \vartheta_s = \frac{1}{15.1} q^{0.75} .$$

The drawings (a) to (f) of Fig. 6.46 illustrate cross-sections of tapped windings. Based on the foregoing, the highest heat transfer coefficient can be measured on the inner surfaces of arrangement (c), where the disc is in contact with a duct heated from two sides.

Assuming an identical specific loss referred to the unit of mass, the lowest surface temperature drop arises in arrangement (d), if only because the heat flux per unit transfer area is the lowest in that case. For arrangements (a) to (f) the use of values $n = 0.75$ and $B = 25. \, m^{1.5} \, °C \, W^{-0.75}$ seems appropriate.

For cases (g) and (h) the use of values $B = 15.1 \, m^{1.5} \, °C \, W^{-0.75}$ and $n = 0.75$, and for cases (i) and (j), $n = 0.75$, but $B = 20 \, m^{1.5} \, °C \, W^{-0.75}$ seems to be justified.

The formulae used above are valid for oils having a mean boundary-layer temperature of $\vartheta_m = 70 \, °C$ and for disc coil windings of axial size $h_1 = 15 \times 10^{-3} \, m$. In the case of the helical coils illustrated by drawings (a) to (c) the characteristic size is the full height of the winding, this being $l = 2 \, m$ here. When calculating with the valid formulae, relatively low values are obtained for the surface temperature drops, if the heat fluxes per unit transfer area are not increased. Normally, from the middle

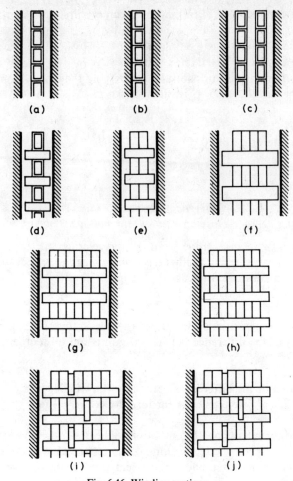

Fig. 6.46. Winding sections

of the winding upwards, the flow is associated with the mixing of hotter and colder partions of the oil, and heat transfer improves, and this effect is taken into account by the respective formulae.

6.2.5.4 Temperature drop across the insulation

The temperature drop taking place across the insulation is calculated from the formula (6.115) as

$$\Delta\vartheta_p = q\frac{\delta}{\lambda_p},$$

where δ is the thickness of one side of interturn insulation, λ_p is the thermal conductivity of insulating paper and q is the heat flux per unit transfer area.

The thermal conductance of oil-paper is temperature dependent:

$$\lambda_p = 0.0465 \, \vartheta_p^{0.279},\tag{6.136}$$

where the dimension of the factor 0.0465 is $\mathrm{W\,m^{-1}\,^\circ C^{-1.279}}$ and ϑ_p is the temperature of the paper. For the sake of unifying the calculations, the thermal conductance pertaining to 55 °C is often used, i.e.

$$\lambda_p = 0.1395 \, \mathrm{W\,m^{-1}\,^\circ C^{-1}}.\tag{6.137}$$

For a paper temperature of 70 °C,

$$\lambda_p = 0.15 \, \mathrm{W\,m^{-1}\,^\circ C^{-1}}.$$

6.2.5.5 *Difference between average winding and average oil temperatures; the hot-spot temperature*

The difference between average winding and average oil temperatures, sometimes called the mean winding gradient, is obtained as the sum of surface temperature drop and temperature drop across the insulation (6.7):

$$\Delta\vartheta_{w-o} = \Delta\vartheta_s + \Delta\vartheta_p.$$

The hot-spot temperature is the sum of the maximum oil temperature and the local maximum of the difference between winding and oil temperatures.

6.2.6 Calculation of the differences between mean winding and oil temperatures of disc-type coils for forced-directed oil flow

With forced directed oil flow the conditions of heat transfer change substantially. Because of the high speed of oil flow along the surface of windings under the effect of the external compelling force, the surface heat transfer is improved.

In Fig. 6.47, temperature drops are plotted for three different oil flow velocities for equal surface heat flux densities. The temperature drop across the insulation is

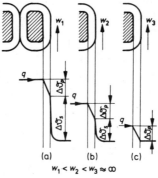

Fig. 6.47. Temperature drop in the winding insulation and at the boundary layer between insulation and boundary layer, with equal heat flux per unit transfer area but different oil flow velocities

432

identical in all the three cases, but the surface temperature drop denoted by $\Delta\vartheta_s$ decreases with increasing oil velocity and falls to zero at infinitely high oil velocity. With higher heat flux densities the temperature drop across the insulating paper will, of course, also rise but the dependence of surface temperature drop on oil velocity remains unchanged. According to experiments carried out at the GANZ

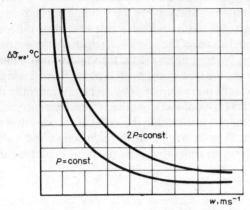

Fig. 6.48. Relation between oil flow velocity and temperature of the oil entering and leaving the winding, with forced-directed oil flow

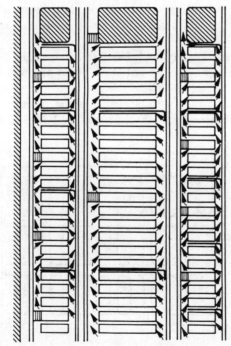

Fig. 6.49. Oil paths connected hydraulically in parallel with forced-directed oil flow

Electric Works, in conformity with the technical literature, the surface temperature drop decreases to such an extent at oil velocity values of a few dm s^{-1} that it is not worthwhile increasing the oil velocity further, for this would bring about an unnecessary increase of pumping work without perceptively reducing the resultant thermal resistance.

For dissipating a given amount of heat loss, the required difference between the temperatures of oil entering and leaving the winding depends hyperbolically on the oil velocity (see Fig. 6.48). With an identical temperature difference $\Delta\vartheta_{wo}$ between inlet and outlet temperatures of the oil passing through the winding, dissipation of a doubled loss requires twice as much oil, i.e. the application of doubled flow velocity.

In order to influence the flow of oil, several parallel ducts are generally provided. In such parallel ducts the oil velocities developing are generally different (Fig. 6.49).

The number of ducts connected in parallel hydraulically is usually so determined as to make the oil flow proportional to the losses of the windings to be cooled and to permit the required flow velocities to develop along the cooling surfaces. The adjustment of oil flows is, generally, performed on prototype units.

Methods of influencing the direction of oil flow within the winding

Forced-directed oil cooling, like any other, is most effective if the largest possible surface area of the parts to be cooled is kept in contact with the oil flowing at a predetermined velocity. In order to ensure the required flow velocities of the oil everywhere along the cooling surfaces, it is sometimes essential to have flow directions opposed to those that would develop under conditions of gravitational flow.

Two methods are used within the winding to influence the direction of flow of the cooling oil, differing in whether or not baffle plates are used to achieve the above aim.

Within the arrangements without baffle plates, there are symmetrical and asymmetrical variants (see Figs 6.50, 6.51, and 6.52).

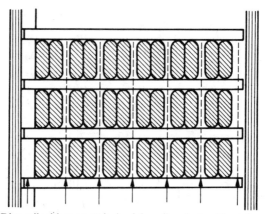

Fig. 6.50. Disc-coil with symmetrical axial cooling duct, without flow deflection

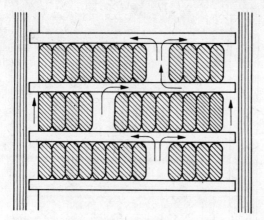

Fig. 6.51. Disc-coil with asymmetrical deflected-flow axial cooling duct

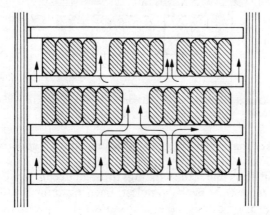

Fig. 6.52. Disc-coil with symmetrical deflected-flow axial cooling duct

Drawbacks of the arrangements employing baffle plates are the manufacturing difficulties, the additional cost of baffle plates, the difficulties of sealing baffle plates against oil leakages and the liability of parts to damage due to shrinking of the windings. Their advantages are the possibility of varying the number of parallel ducts within wide limits; the easy adjustment of hydraulic characteristics, the good space factors and high surface thermal conductivities (Fig. 6.53).

The inlet of the cooling oil may either be at the bottom only, which is the simplest and most generally adopted arrangement, or at the bottom and also at the middle of the winding stack. Though the second arrangement is more complicated, yet it offers the advantage of providing the lower part of the upper half of the winding with cold oil. From the upper part of the lower half of the winding stack the warmed-up oil is led out through ducts confined by insulating cylinders. To make the oil enter the winding stack, the openings between the transformer and the tank are shut off with

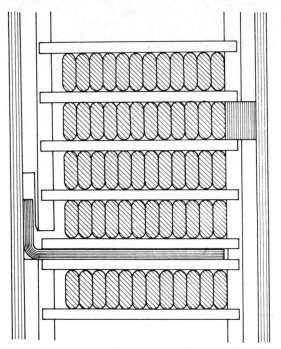

Fig. 6.53. Disc-coil with three parallel connected oil paths, with guiding elements

insulating plates at the level of the bottom spacers, and the cold oil is forced into the shut-off region from outside (see Fig. 6.54).

The cooling oil flows under the windings through openings provided in the frame girders. This arrangement is the simplest when traditional tanks are used. The bottom beam can also be designed as a box girder. In this case the inside of the box girder will be the space kept under pressure, from where the oil is distributed for feeding under the windings. In such cases, the flexible piping is connected after lowering the transformer into the tank (see Fig. 6.55).

The most up-to-date and easy-to-assemble design is that of transformers provided with bell-type tanks (see Fig. 6.56). Checking of the sealings of oil ducts and distributing baffles is the easiest with this type of construction.

In the two latter cases, the preliminary drying of the transformer is performed with hot air blown through the oil ducts.

Surface heat transfer coefficient

According to measurements carried out by the author, in the case of forced-directed oil flow, the surface heat transfer coefficient can be calculated from the following equation

$$Nu_m = 0.62 Re_m^{0.64} Pr_m^{0.333} \sqrt{\frac{d_h}{L}}, \qquad (6.138)$$

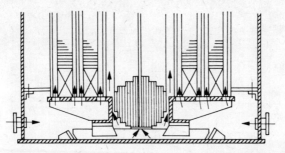

Fig. 6.54. Bottom shut-off plate type cooling oil inlet of traditional oil tank

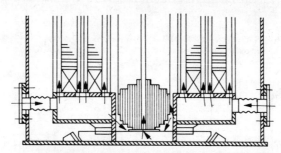

Fig. 6.55. Flexible tube type cooling oil inlet of traditional oil tank

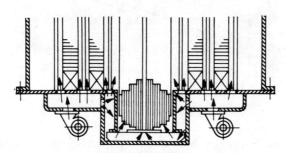

Fig. 6.56. Cooling oil inlet of bell-type tank

where the characteristics of oil measured at the mean boundary-layer temperature ϑ_m should be substituted in the formula: d_h is the hydraulic diameter of the oil duct and L is its length. The validity of the formula has been verified by measurements performed in the following ranges: $Re_m = 160$ to 820, $Pr_m = 40$ to 120 and $d_h/L = 0.0905$.

Some of the results of measuring are shown in Figs 6.57 and 6.58. Let us express the value of α^{-1} from the preceding equation of criteria:

$$\frac{1}{\alpha} = \frac{1}{0.62} d_h^{-0.14} L^{0.5} \lambda^{-2/3} v^{0.31} (c\rho)^{-1/3} w^{-0.64}.$$ (6.139)

437

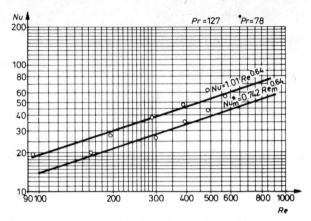

Fig. 6.57. Results of measurements on a model coil. Variation of Nu number as a function of Re number. The upper line refers to the mean oil temperature $Pr = 127$, the lower line to the mean boundary-layer temperature, $Pr_m = 78$

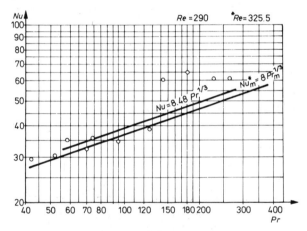

Fig. 6.58. Results of measurements on a model coil. Variation of Nu number as a function of Pr number. The upper line refers to the mean oil temperature $Re = 290$, the lower line to the mean boundary-layer temperature, $Re_m = 325.5$

Physical constants of the oil at temperature

$$\vartheta_m = 70\,°C:$$
$$\lambda = 0.1276 \text{ W m}^{-1}\,°C^{-1},$$
$$v = 5.4 \times 10^{-6} \text{ m}^2 \text{ s}^{-1},$$
$$c = 2060 \text{ W s kg}^{-1}\,°C^{-1},$$
$$\rho = 852 \text{ kg m}^{-3},$$
$$\frac{1}{\alpha} = 1.175 \times 10^{-3} L^{0.5} d_h^{-0.14} w^{-0.64}. \tag{6.140}$$

The dimension of factor 1.175×10^{-3} appearing in the formula is $\text{m}^{2.28}\,\text{s}^{-0.64}\,°C\,\text{W}^{-1}$.

The surface temperature drop

$$\Delta\vartheta_s = \frac{q}{\alpha} = q1.175 \times 10^{-3} L^{0.5} d_h^{-0.14} w^{-0.64}. \tag{6.141}$$

Here, the dimension of factor 1.175×10^{-3} is $m^{2.28} s^{-0.64} {}^\circ C\, W^{-1}$.

Example 6.15. In a winding cooled by forced-directed oil flow, the oil flows with a velocity $w = 0.2$ ms^{-1}, in 5×100 mm horizontal channels of length $L = 120$ mm. What will the surface temperature drop be when the surface heat flux density is $q = 2000$ W m^{-2}? The hydraulic diameter is

$$d_h = \frac{4A}{c} = \frac{4 \times 5 \times 100 \times 10^{-6}}{210 \times 10^{-3}}\, m = 9.52 \times 10^{-3}\, m,$$

$$d_h^{-0.14} = (9.52 \times 10^{-3})^{-0.14}\, m^{-0.14} = 1.92\, m^{-0.14},$$

$$L = 0.12\, m;\ L^{0.5} = 0.12^{0.5} m^{0.5} = 0.346\, m^{0.5},$$

$$w^{-0.64} = 0.2^{-0.64}\, m^{-0.64} s^{0.64} = 2.8\, m^{-0.64} s^{0.64}.$$

Therefore

$$\Delta\vartheta_s = 2000 \times 1.175 \times 10^{-3} \times 0.346 \times 1.92 \times 2.8\ {}^\circ C = 4.36\ {}^\circ C.$$

As it is apparent from the above, even at very low oil velocities very small surface temperature drops are obtained.

6.3 Radiator cooling

6.3.1 The physics of natural oil cooling

6.3.1.1 The calculation of pressure difference

An outline diagram of the cooling circuit is shown in Fig. 6.59. To facilitate understanding the route of the oil heated up in the active parts of the transformer and cooled down in the coolers is indicated by a single broken line, and the heat transfer is explained in the following by reference to the phenomena taking place in the winding.

Along the vertical length $A - B$ of the winding, an amount of heat P is transferred to the oil entering at point A at the bottom of the winding. The heat transferred through the winding surfaces causes a temperature rise $\Delta\vartheta_{wo}$ in the oil, whose mean specific heat is c and which has a mass flow of Φ:

$$P = c\Phi\Delta\vartheta_{wo}. \tag{6.142}$$

Since both the density of the oil

$$\rho = \rho_0 \frac{1}{1 + \beta\vartheta} = \rho_0(1 - \beta\vartheta) \tag{6.143}$$

and its specific gravity

$$\gamma = \gamma_0(1 - \beta\vartheta), \tag{6.144}$$

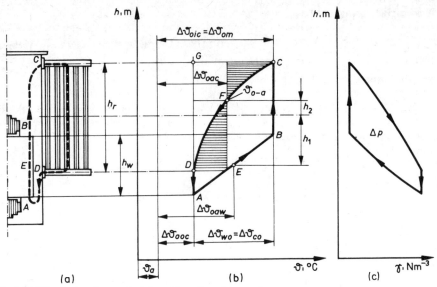

Fig. 6.59. The physics of natural oil flow: (a) outline of the cooling circuit; (b) temperature of flowing oil in the ϑ–h coordinate system; (c) the process of warming and cooling of the oil in the γ–h coordinate system

decrease with rising temperature, the warmed oil flows upwards and is replaced by cold oil from below.

The warm oil enters the cooler at point C, its heat P is dissipated, the oil cools down, its specific gravity increases, and it leaves the cooler at point D. Then, the oil thus cooled again enters the winding at point A, and the process keeps repeating.

Plotting the process in the ϑ–h coordinate system, a closed curve is obtained for the steady-state condition (see Fig. 6.59).

If the function

$$\gamma = f(\vartheta) \tag{6.145}$$

of the oil is known, then the process of warming and cooling of the oil can be drawn up in the γ–h coordinate system as well (see Fig. 6.59). In the steady-state condition there are different temperatures at every level of the closed cooling circuit, hence different values of specific gravity are found for the oil and the specific gravity is a function of height:

$$\gamma = f(h). \tag{6.146}$$

The phenomenon is identical with the case of gravity-circulated hot-water heating. Area $ABCDA$ is proportional to the buoyancy keeping the oil in circulation, or in other words, to the pressure resulting from the difference between specific gravities of hot and cold oil. Thus, the mass flow of the oil Φ is maintained in the cooling circuit by the pressure difference

$$\Delta p = \oint \gamma(h)\, dh. \tag{6.147}$$

440

The pressure difference Δp balances the resistance of flow. If the slight temperature dependence of specific heat is disregarded then to dissipate the same amount of heat loss, the product $\Phi \Delta \vartheta_{wo}$ must be constant. When designing a transformer, efforts are made to keep the temperature rise $\Delta \vartheta_{wo}$ of the oil flowing through the active parts as low as possible. In order to minimize the difference between the maximum temperature rise of the oil specified by the standards and the mean temperature rise of the oil, $\Delta \vartheta_{oac}$. By this, to dissipate a given amount of heat, the average temperature rise $\Delta \vartheta_{o-a}$ of the radiator can be increased, resulting in lower weight and cost of the cooler. A lower value of $\Delta \vartheta_{wo}$ is obtained if Φ is high. And a high Φ can be achieved either by reducing the loss of flow of the oil, or by increasing the value of Δp.

Determination of pressure

$$\Delta p = \oint \gamma(h) \, dh.$$

The pressure is determined from the temperature loop, by using the equation

$$\Delta p = \gamma_k \beta_k \oint \vartheta(h) \, d\vartheta. \tag{6.148}$$

Factor $\gamma_k \beta_k$ is the change of specific gravity of the oil per degree centigrade change of temperature at the weighted mean temperature ϑ_k. The dimension of the said factor is $\mathrm{N\,m^{-3}\,{}^\circ C^{-1}}$.

A simplified diagram of the temperature loop is given in Fig. 6.59. Along section A–B the oil is heated up in the winding of height h_w, and along section C–D it is cooled in the radiator bank of length h_r. In the $\vartheta - h$ coordinate system, section A–B has been plotted as being linear, assuming that along each unit length of the winding an identical amount of heat $\mathrm{W\,m^{-1}}$ is transferred to the oil, and the specific heat of oil does not change with temperature.

The characteristic point of section A–B is point E, which is at level $h_w/2$ of the winding, and where the temperature of the oil corresponds to the mean temperature rise $\Delta \vartheta_{oaw}$ of the oil with respect to the inlet temperature ϑ_a of the cooling air. The shape of curve C–D corresponds to the temperature pattern of a condenser cooler, i.e. to that pattern which develops when the flow capacity of one medium taking part in the heat exchange process—air in the present case—is assumed to be infinite. The characteristic point of that section of the curve is point F, which denotes the logarithmic mean temperature difference $\Delta \vartheta_{o-a}$ relative to that of the cooling air which is assumed to be constant in the cooler:

$$\Delta \vartheta_{o-a} = \frac{\Delta \vartheta_{co}}{\ln \dfrac{1}{1 - \dfrac{\Delta \vartheta_{co}}{\Delta \vartheta_{oic}}}}. \tag{6.149}$$

Point F is height h_2 above the middle height of the radiator. The difference between the middle level of the radiator and that of the winding is denoted by h_1. The level difference between points E and F is

$$h = h_1 + h_2. \tag{6.150}$$

441

The area of the heat loop, from Fig. 6.59, is

$$T = (h_1 + h_2)\Delta\vartheta_{co}. \tag{6.151}$$

Dimension h_2 is obtained as follows. Area $CDGC$ is

$$T_1 = h_r(\Delta\vartheta_{o-a} - \Delta\vartheta_{ooc}). \tag{6.152}$$

Since

$$\Delta\vartheta_{ooc} = \Delta\vartheta_{oic} - \Delta\vartheta_{co},$$

$$T_1 = h_r(\Delta\vartheta_{o-a} + \Delta\vartheta_{co} - \Delta\vartheta_{oic}). \tag{6.153}$$

Let us transform this area into a rectangle of equal surface area, but having a base equal to $\Delta\vartheta_{co}$ and a height of $\left(\dfrac{h_r}{2} - h_2\right)$.

$$T_1 = \Delta\vartheta_{co}\left(\frac{h_r}{2} - h_2\right) = h_r(\Delta\vartheta_{o-a} + \Delta\vartheta_{co} - \Delta\vartheta_{oic}). \tag{6.154}$$

Thus

$$h_2 = h_r\left(\frac{\Delta\vartheta_{oic}}{\Delta\vartheta_{co}} - \frac{\Delta\vartheta_{o-a}}{\Delta\vartheta_{co}} - 0.5\right). \tag{6.155}$$

The area of the temperature loop is

$$T = \left[h_1 + h_r\left(\frac{\Delta\vartheta_{oic}}{\Delta\vartheta_{co}} - \frac{\Delta\vartheta_{o-a}}{\Delta\vartheta_{co}} - 0.5\right)\right]\Delta\vartheta_{co} \tag{6.156}$$

or

$$T = h_1\Delta\vartheta_{co} + h_r(\Delta\vartheta_{oic} - \Delta\vartheta_{o-a} - 0.5\Delta\vartheta_{co}). \tag{6.157}$$

The pressure

$$\Delta p = \gamma_a\beta_a T. \tag{6.158}$$

For a mean oil temperature of $\vartheta_{oa} = 65\,°\text{C}$, $\gamma_{65} = \rho_{65}g = 855 \times 9.81\ \text{N m}^{-3} = 8380\ \text{N m}^{-3}$, $\beta_{65} = 7.9 \times 10^{-4}\ °\text{C}^{-1}$, $\gamma_{65}\beta_{65} = 6.62\ \text{N m}^{-3}\ °\text{C}^{-1}$.

Example 6.16. Let us consider a 40 MVA 120/11 kV natural oil flow, forced/natural air cooled transformer. In the case of natural air cooling the transformer may be loaded with a maximum 70% of its rated current. In this case the load loss is $0.7^2 \times 100 = 49\%$ of the value pertaining to the rated current. The transformer is provided with radiator-type cooling. The radiator groups are mounted on the side of the tank, with the motordriven fans accomodated below them.

From the constructional drawings $h_1 = 0.643$ m and $h_r = 1.965$ m. From the temperature rise measurements performed with natural cooling and at 70% of rated current the following data have been obtained:

$$\vartheta_a = 20\,°\text{C}, \quad \vartheta_{oic} = 80\,°\text{C}, \quad \vartheta_{ooc} = 60\,°\text{C}.$$

Let us draw up the temperature diagram and specific gravity diagram of the transformer oil, and determine the value of pressure required for maintaining the oil flow (Fig. 6.60).

442

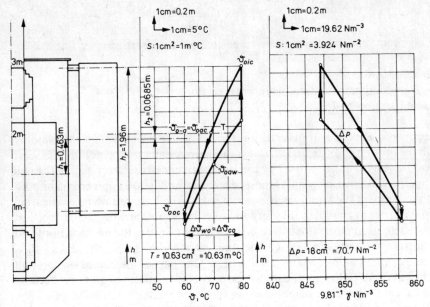

Fig. 6.60. Determination of pressure maintaining the flow of oil in the cooling circuit of a 40 MVA transformer provided with natural oil flow

From the temperature rise data $\Delta \vartheta_{oic} = 60\,°C$, $\Delta \vartheta_{ooc} = 40\,°C$, and $\Delta \vartheta_{co} = 20\,°C$. Let us calculate the logarithmic-mean temperature difference $\Delta \vartheta_{o-a}$:

$$\Delta \vartheta_{o-a} = \frac{\Delta \vartheta_{co}}{\ln \dfrac{1}{1 - \dfrac{\Delta \vartheta_{co}}{\Delta \vartheta_{oic}}}} = \frac{20}{\ln \dfrac{1}{1 - \dfrac{20}{60}}}\,°C = 49.3\,°C.$$

To draw the diagram, dimension h_2 is also required from

$$h_2 = h_r \left(\frac{\Delta \vartheta_{oic}}{\Delta \vartheta_{co}} - \frac{\Delta \vartheta_{o-a}}{\Delta \vartheta_{co}} - 0.5 \right),$$

whence

$$h_2 = 1.96 \left(\frac{60}{20} - \frac{49.3}{20} - 0.5 \right)\,m = 0.0685\,m.$$

The area of the temperature loop is

$$T = h_1 \Delta \vartheta_{co} + h_r (\Delta \vartheta_{oic} - \Delta \vartheta_{o-a} - 0.5 \Delta \vartheta_{co}) =$$
$$= 0.463 \times 20\,m\,°C + 1.965(60 - 49.3 - 0.5 \times 20)\,m\,°C = 9.26\,m\,°C + 1.37\,m\,°C = 10.63\,m\,°C.$$

For

$$\vartheta_{oa} = 70\,°C: \quad \gamma_{70} = 852 \times 9.81\,N\,m^{-3} = 8360\,N\,m^{-3},$$
$$\beta_{70} = 7.95 \times 10^{-4}\,°C^{-1}, \quad \gamma_{70}\beta_{70} = 8360 \times 7.95 \times 10^{-4}\,N\,m^{-3}\,°C^{-1} = 6.65\,N\,m^{-3}\,°C^{-1}.$$

The magnitude of pressure required is

$$\Delta p = \gamma_{70}\beta_{70}T = 6.65 \times 10.63\,N\,m^{-2} = 70.7\,N\,m^{-2}.$$

443

6.3.1.2 The oil flow path

The oil circuit may be divided into three characteristic sections (see Fig. 6.59(a))

1. Heated section $A–B$
2. with good approximation, isothermal sections $B–C$ and $D–A$
3. cooled section $C–D$.

In section $A–B$ heat is transferred to the oil. The cold oil enters the winding through the bottom end spacer, where it comes into contact with winding surfaces at temperatures higher than its own by the surface temperature drop $\Delta\vartheta_s$, it absorbs heat from the windings, its pecific gravity decreases, and it flows upwards. The path of oil flow between the windings is determined by the horizontal and vertical oil ducts marked out by the spacer blocks and laths. By suitable arrangement of vertical oil ducts a definite, unidirectional flow can also be obtained in the oil ducts bordered by the bottom and top planes of the disc coils. As has been shown, the flow of oil in the windings follows a boundary-layer pattern. One part of the heated oil cools off along the tank walls. This shunt path will be neglected in our investigations, because in practical cases, the heat dissipated along such shunt paths amounts to only a few % of that transferred to the surroundings through the coolers. The shunt oil path is shown in Fig. 6.61.

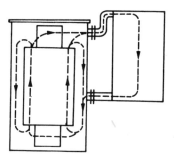

Fig. 6.61. The main cooling circuit and the shunt cooling circuit

The flat oil ducts running between the windings have four specific features:

(a) Along the flow path, the oil is heated by the windings constituting the two longitudinal sides of the ducts with an approximately equal heat flux per unit length of the oil path.

(b) The disc coils arranged above each other are surrounded by oil of a temperature increasing with height, and their surface are at temperatures exceeding the surface temperature drop $\Delta\vartheta_s$ of the oil surrounding them.

(c) The mean temperature of the boundary layer between winding insulation and oil is higher by $\Delta\vartheta_s/2$ than the temperature of the oil surrounding the winding, i.e. of that flowing in the ducts. The oil mass of the boundary layer represents a considerable part of the mass of oil within the winding, thus the mean temperature

444

of the boundary layer significantly influences the temperature attained by the flowing oil.

(d) In spite of the boundary-layer character of the oil flow the temperature of the oil leaving the winding at point B can be considered identical and homogeneous with good approximation due to the articulated arrangement of oil ducts and their frequent changes of direction.

The oil flowing in the vertical cooling ducts of the core flows in a duct heated from two sides. Although the specific losses developing in the corners of the core and in the limbs are very different, the temperatures of these duct walls are almost equal along the path of flow, due to the good heat conductance parallel to the plane of the sheets. The flow is also of boundary-layer character here.

6.3.1.3 Mass flow, temperature rise and pressure difference

The oil of mass flow Φ_0 and mean specific heat c_0, is heated through a temperature rise of $\Delta\vartheta_{wo}$ by the thermal power P transferred across convection surface A_c of the winding at a heat flux per unit transfer area of q.

$$\Delta\vartheta_{wo} = \frac{qA_c}{c_0\Phi_0}. \tag{6.159}$$

Let us denote quotient $\dfrac{A_c}{c_0}$ by C_1:

$$\Delta\vartheta_{wo} = C_1 \frac{q}{\Phi_0}. \tag{6.160}$$

The pressure acting in the gravitational cooling system as shown, with substitution of $h = h_1 + h_2$, is approximately

$$\Delta p = h\gamma_o\beta\Delta\vartheta_{wo}. \tag{6.161}$$

Let us introduce the notation $C_2 = h\gamma_0\beta$, with which

$$\Delta p = C_2\Delta\vartheta_{wo}. \tag{6.162}$$

Substituting the expression for $\Delta\vartheta_{wo}$ into the equation for Δp and putting $C_1 C_2$ equal to C_3,

$$\Delta p = C_3 \frac{q}{\Phi_0}. \tag{6.163}$$

This pressure overcomes the hydraulic resistance of the streaming oil. With laminar flow the pressure drop resulting from the flow loss is

$$\Delta p = C_4\Phi_0 \tag{6.164}$$

and the oil flow is

$$\Phi_0 = C_5\sqrt{q}, \tag{6.165}$$

445

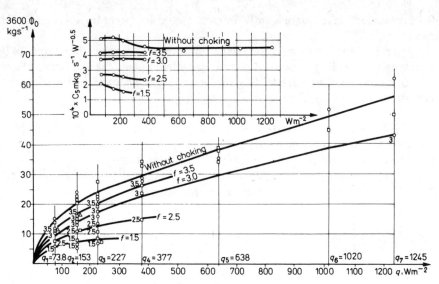

Fig. 6.62. Relation $\Phi_0 = C_5\sqrt{q}$ measured experimentally [50]

where $C_5 = \sqrt{C_3/C_4}$. The temperature rise of the oil is

$$\Delta\vartheta_{wo} = C_6\sqrt{q}, \qquad (6.166)$$

where $C_6 = C_1/C_5$, and the oil pressure is

$$\Delta p = C_7\sqrt{q}, \qquad (6.167)$$

where $C_7 = C_2 C_6$.

The existence and validity of the above relations have been verified. A set of curves plotted on the basis of measured data is shown in Fig. 6.62. The first set of curves represents the relation $\Phi_0 = C_5\sqrt{q}$. The measurements have been performed on a model winding. For the purpose of varying the resistance of the flow circuit a gate valve has been installed. The number of revolutions made with the hand wheel of the gate valve marked with f in the figures, has been recorded as measure of throttling and which was used as parameter. Within the range of value, of q occurring in practice, it can be seen that factor C_5 appears to be constant.

Figure 6.63 shows the relation $\Delta\vartheta_{wo} = C_6\sqrt{q}$ with the flow resistance, i.e. throttling, as parameter. In Fig. 6.64 the relation $\Delta p = C_7\sqrt{q}$ is represented, also on the basis of experimental data. Here, also throttling is used as a parameter in the function q.

From the three sets of curves it will be seen that in practical cases factors C_5, C_6 and C_7 can be taken as constant, provided a reasonable parameter for throttling is used, and in the vicinity of $q = 1000$ W m^{-2}.

The measurements have been performed with the three different winding arrangements (see Fig. 6.65). For the measuring points marked O the characteristic

446

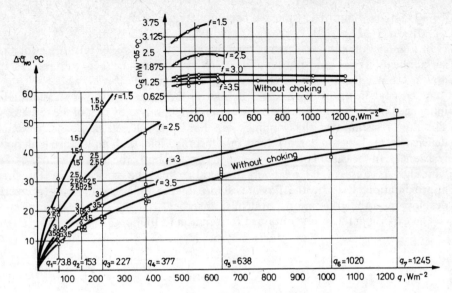

Fig. 6.63. Relation $\Delta\vartheta_{wo} = C_6\sqrt{q}$ measured experimentally [50]

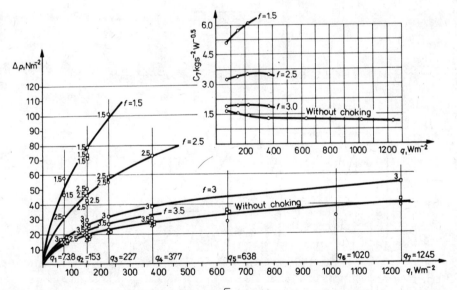

Fig. 6.64. Relation $\Delta p = C_7\sqrt{q}$ measured experimentally [50]

values of the arrangement were: $s = 20$ mm, $k = 50$ mm, $o = 15$ mm; for the measuring points marked x the respective values were: $s = 4$ mm, $k = 40$ mm, $o = 5$ mm, and for those marked Δ: $s = 10$ mm, $k = 30$ mm, $o = 10$ mm.

In Fig. 6.66 the pressure area is shown as a function of heat flux per unit transfer area for the winding arrangement having characteristic dimensions: $s = 40$ mm,

$k = 40$ mm, $o = 5$ mm. The values of heat flux per unit transfer area were $q_1 = 73.8$, $q_2 = 163$, $q_3 = 227$, $q_4 = 377$, $q_5 = 638$, $q_6 = 1020$, $q_7 = 1245$ W m^{-2}.

In Fig. 6.67 the pressure area is represented as a function of throttling, for the surface heat flux per unit transfer area 227 W m^{-2}. As can be seen, with increasing flow resistance the pressure area and $\Delta\vartheta_{wo}$ also increase.

As clearly shown by the last set of curves, the value of $\Delta\vartheta_{wo}$ is substantially influenced by the flow resistance of the system and more so by that of the radiators. The flow resistance of the windings can hardly be altered but constructional deficiencies such as too narrow ducts or a throttled inlet path may cause high flow resistances. In the case of properly designed constructions, the flow resistance of the windings, because of its relatively low value, may be neglected in the first approximation. It is by the flow resistance of the radiators that the pressure developing in the gravitational system is absorbed.

The experimental results obtained for relation (6.160)

$$\Delta\vartheta_{wo} = C_1 \frac{q}{\Phi_0}$$

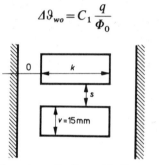

Fig. 6.65. Dimensions of coils used in the experiments

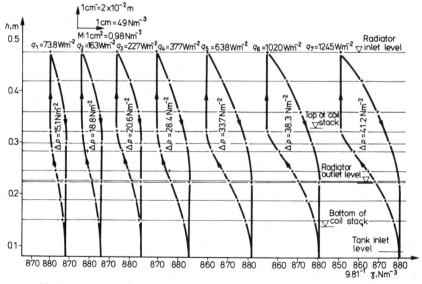

Fig. 6.66. Pressure area as a function of heat flux per unit transfer area [50]

448

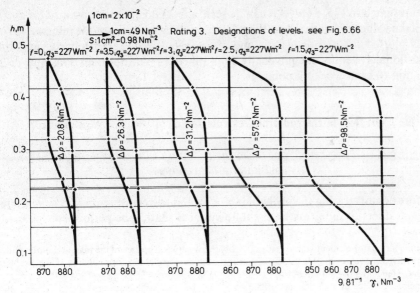

Fig. 6.67. Pressure area with constant surface heat load, as a function of throttling [50]

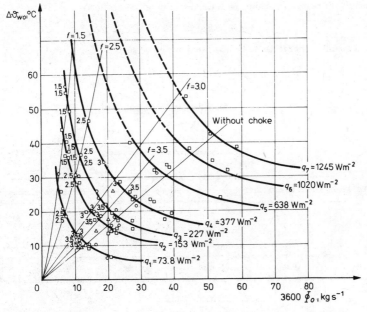

Fig. 6.68. Experimental results obtained for relation $\Delta \vartheta_{wo} = C_1 \dfrac{q}{\Phi_0}$ [50]

are represented in Fig. 6.68. As it can be seen, a working point is obtained for every pair of the values of q and of throttling. Each working point lies at the intersection of curves plotted for a $q=$ constant and $f=$ constant value, respectively, and the gravitational system is unequivocally determined by the coordinates Φ_0 and $\Delta\vartheta_{wo}$ belonging to each working point.

6.3.2 Oil flow in the radiators, calculation of oil-side temperature drop

6.3.2.1 Flow velocity of oil in the radiators with natural oil flow and natural air cooling

Before demonstrating the probable velocity and temperature profiles developing in the oil ducts of radiators let the following calculation be performed.

Assume a radiator element with a spacing of 2440 mm between centres of the inlet and outlet stubs. There are five oil ducts of double trapezoidal shape, each having a cross-sectional area of 3.54 cm² (see Fig. 6.69). Hence the overall inlet/outlet area is $A_0 = 5 \times 3.54$ cm² $= 17.71$ cm² $= 17.71 \times 10^{-4}$ m². The circumference of each duct is $c = 8.06 \times 10^{-2}$ m. The hydraulic diameter is

$$d_h = \frac{4A}{c} = \frac{4 \times 3.54 \times 10^{-4}}{8.06 \times 10^{-2}}\,\text{m} = 1.76 \times 10^{-2}\,\text{m}.$$

Overall surface area of the oil ducts is $A_{co} = 5 \times 8.06 \times 10^{-2} \times 2.44$ m² $= 0.983$ m². Air-side surface of each element is $A_{ca} = 1.276$ m². Assume $\vartheta_a = 20\,°C$, air-side specific heat flux with convection $q_a = 220$ W m^{-2}, $\Delta\vartheta_{co} = 20\,°C$, $\Delta\vartheta_{oic} = 60\,°C$, $\Delta\vartheta_{ooc} = 40\,°C$.

The quantity of heat dissipated by each element through convection is

$$P_a = q_a A_{ca} = 220 \times 1.276\,\text{W} = 280.72\,\text{W}.$$

An equal quantity of heat has to be dissipated on the oil side:

$$P_o = P_a = q_o A_{co} = \frac{A_{ca}}{A_{co}} q_a A_{co} = \frac{1.276}{0.983} \times 220 \times 0.983\,\text{W} =$$

$$= 285.57 \times 0.983\,\text{W} = 280.72\,\text{W}.$$

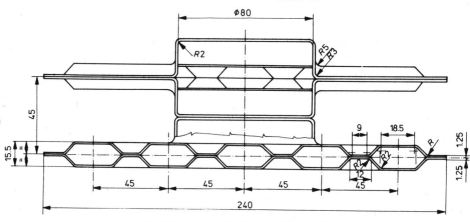

Fig. 6.69. Section of radiators of Table 6.9

As shown, the oil-side heat flux per unit transfer area is the product of the air-side heat flux per unit transfer area and the surface ratio.

The difference between heat contents of inlet and outlet oil entering and leaving the radiator is equal to the quantity of heat transferred through the oil-side surface to the radiator wall and through that to the cooling air. At temperature

$$\vartheta_{oac} = \vartheta_o + \varDelta\vartheta_{oac} + \frac{\varDelta\vartheta_{co}}{2}(20+40+10) \; °C = 70 \; °C$$

let the oil density be $\rho = 852 \; kg \; m^{-3}$, and the specific heat $c = 2060 \; W \; s \; kg^{-1} \; °C^{-1}$. Let the oil velocity in the radiator ducts be denoted by w. On the basis of power balance we may write:

$$P_0 = P_a = w A_0 \rho c, \varDelta\vartheta_{co}.$$

Thus, the required oil velocity is

$$w = \frac{P_o}{A_0 \rho c \varDelta\vartheta_{co}} = \frac{280.72}{17.71 \times 10^{-4} \times 852 \times 2060 \times 20} \; m \; s^{-1} =$$

$$= 0.452 \times 10^{-2} \; m \; s^{-1} \approx 0.5 \; cm \; s^{-1}.$$

The outermost radiator element transferring heat through its half surface also by radiation causes an increase of the air-side average power density by $q = q_r/2$, whereby q_r the heat flux per unit transfer area dissipated by radiation is indicated. In the present case the maximum average heat flux per unit transfer area will be $(220 + 350/2) \; W \; m^{-2} = 395 \; W \; m^{-2}$. This value is 80% higher than the heat flux per unit transfer area referred to convection only, so the oil velocity ought to increase to 0.81 cm s^{-1} if $\varDelta\vartheta_{co}$ remains unchanged. Of course, this higher heat flux per unit transfer area requires an increased oil flow in the outermost element. The increased oil flow of the outermost element has hardly any influence on the values of $\varDelta\vartheta_{o-a}$ and $\varDelta\vartheta_{co}$ of the entire transformer. If radiating plates are placed between the radiator elements, then about 40% more heat is dissipated by the inner surface of the outermost radiator than that which would result from the convective heat flux per unit transfer area. This would mean that the maximum heat flux per unit transfer area would increase by an additional 88/2 W m^{-2}, i.e. q_{max} would be as high as 439 W m^{-2}. This value is 100% higher than the original convective heat flux per unit transfer area i.e. the conceivable maximum oil velocity is 0.904 cm s^{-1}.

As it is apparent from the above, the oil flows through the radiators with a very low velocity. Let us calculate the oil-side heat-transfer coefficient and surface temperature drop with an oil velocity of 0.45 cm s^{-1}.

6.3.2.2 Oil-side heat transfer coefficient and surface temperature drop

The motion of the oil in the oil ducts is laminar. As already seen, the oil is kept in circulation by the pressure represented by the area of the closed-circuit curve of specific gravity vs. height difference plotted for the oil particles taking part in the heat exchange. The motion of the oil may, therefore, be considered as a forced flow to which the relation known from [90] is applicable:

$$Nu_m = 0.85 \times 0.74 Re_m^{0.2}(Gr \times Pr)_m^{0.1} Pr_m^{0.2}. \tag{6.168}$$

By means of the above formula let us calculate the value of the oil-side heat transfer coefficient. The physical characteristic values of the oil at the mean boundary-layer temperature will be substituted into the formula.

Let the oil-side temperature drop be assumed to be $\Delta\vartheta_{os}=4\,°C$. With this value, the mean boundary-layer temperature is

$$\vartheta_m=\vartheta_{oac}-\frac{\Delta\vartheta_{os}}{2}=\left(70-\frac{4}{2}\right)°C=68\,°C.$$

The physical characteristics of the oil valid for this temperature are taken from Table 6.1.

$v_m=5.652\times10^{-6}\,m^2\,s^{-1}$,
$\beta_m=7.932\times10^{-4}\,°C^{-1}$, $\quad Pr_m=77.13$,
$\lambda_m=0.127\,744\,W\,m^{-1}\,°C^{-1}$, $\quad w=0.45\times10^{-2}\,m\,s^{-1}$,
$d_h=1.76\times10^{-2}\,m$.

Then

$$Pr_m^{0.2}=77.13^{0.2}=2.385,$$

$$d_h^3=(1.76\times10^{-2})^3\,m^3=5.452\times10^{-6}\,m^3,$$

$$v_m^2=(5.652\times10^{-6})^2\,m^4\,s^{-2}=31.945\times10^{-12}\,m^4\,s^{-2}.$$

The Re number is:

$$Re_m=\frac{wd_h}{v_m}=\frac{0.45\times10^{-2}\times1.76\times10^{-2}}{5.652\times10^{-6}}=14.01,$$

$$Re_m^{0.2}=14.01^{0.2}=1.696.$$

The Gr number is

$$Gr_m=\frac{\beta_m g d_h^3\Delta\vartheta_{os}}{v_m^2}=\frac{7.932\times10^{-4}\times9.81\times5.452\times10^{-6}\times4}{31.945\times10^{-12}}=5311,$$

$$(Gr\times Pr)_m^{0.1}=(5311\times77.13)^{0.1}=409\,637^{0.1}=3.64.$$

The Nu number is

$$Nu_m=0.85\times0.74Re_m^{0.2}(Gr\times Pr)_m^{0.1}Pr_m^{0.2}$$

$$=0.85\times0.74\times1.696\times3.64\times2.385=9.26.$$

The heat transfer coefficient is

$$\alpha=Nu_m\frac{\lambda_m}{d_h}=9.26\frac{0.127\,744}{1.76\times10^{-2}}\,W\,m^{-2}\,°C^{-1}=67.22\,W\,m^{-2}\,°C^{-1}.$$

For the purpose of checking, the heat flux per unit transfer area is calculated:

$$q=\alpha\Delta\vartheta_{os}=67.22\times4\,W\,m^{-2}=269\,W\,m^{-2}.$$

For the case of natural oil flow and natural air cooling taking $\vartheta_a=20\,°C$, $\Delta\vartheta_{co}=20\,°C$, $\Delta\vartheta_{oac}=70\,°C$ and a radiator length of 2 m, the probable shape of velocity curves, temperature profiles and isotherms are shown in Figs 6.70(a), (b) and (c).

At the inlet point of the radiator the velocity curve is parabolic and the temperature "curve" is a straight line, since no heat has been dissipated at this point yet. In the course of the downward flow of the oil the velocity profile develops continuously from parabolic through a roughly trapezoid shape into a W-form

452

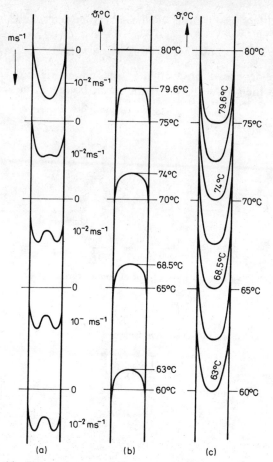

Fig. 6.70. (a) Flow velocities (b) temperatures and (c) temperature isotherms of the oil cooled in a radiator

curve. The straight line of the temperature first becomes a trapezoid and finally goes over into a semicircle. While plotting the curves, the viscosity of the oil, the strong dependence of viscosity on temperature and the poor thermal conductivity of the oil have been taken into consideration.

From knowledge of the following quantities: thermal power P dissipated by each radiator element, the oil-side cross-sectional area of flow A_o, the oil-side surface A_{co} of each radiator element, the density ρ at the average temperature of the oil flowing in the radiator and its specific heat c, the oil-side heat transfer coefficient α, oil-side surface temperature drop $\Delta\vartheta_{os}$, and difference $\Delta\vartheta_{co}$ between inlet and outlet temperatures of the oil entering and leaving the radiator, the power balance equation can be set up: from the equality of thermal power dissipated by a radiator element and the reduction of oil-side heat content:

$$P = \alpha A_{co} \Delta\vartheta_{os} = w A_0 \rho c \Delta\vartheta_{co}, \tag{6.169}$$

where w is the velocity of the oil flowing in the radiator.

The oil-side surface heat transfer coefficient can be determined from the equation (6.168):

$$Nu_m = 0.85 \times 0.74 Re_m^{0.2}(Gr \times Pr)_m^{0.1} Pr_m^{0.2} .$$

The temperature-dependent physical quantities appearing in the formula are to be substituted with their values valid at the mean boundary-layer temperature ϑ_m.

The explicit form of α is

$$\alpha = 0.85 \times 0.74 d_h^{-0.5}(\rho_c)^{0.3}\left(\frac{\beta g}{\nu}\right)^{0.1} \lambda^{0.7} w^{0.2} \Delta\vartheta_{os}^{0.1} . \tag{6.170}$$

From the power balance:

$$\alpha = \frac{P}{A_{co}\Delta\vartheta_{os}} \quad \text{and} \quad w = \frac{P}{A_0\rho c \Delta\vartheta_{co}} .$$

$$\Delta\vartheta_{os}^{1.1} = \frac{A_o^{0.2}\Delta\vartheta_{co}^{0.2}d_h^{0.5}}{0.85 \times 0.74 A_{co}g^{0.1}}\left(\frac{\nu}{\rho c \beta}\right)^{0.1}\frac{1}{\lambda^{0.7}}P^{0.8} . \tag{6.171}$$

K is the product of factors which are independent of temperature:

$$K = \frac{A_o^{0.2}\Delta\vartheta_{co}^{0.2}d_h^{0.5}}{0.85 \times 0.74 A_{co}g^{0.1}} \tag{6.172}$$

and B is the product of temperature-dependent factors:

$$B = \left(\frac{\nu}{\rho c \beta}\right)^{0.1}\frac{1}{\lambda^{0.7}} . \tag{6.173}$$

The oil-side surface temperature drop is

$$\Delta\vartheta_{os}^{1.1} = KBP^{0.8} , \tag{6.174}$$

from which

$$\Delta\vartheta_{os} = (KB)^{0.91}P^{0.727} . \tag{6.175}$$

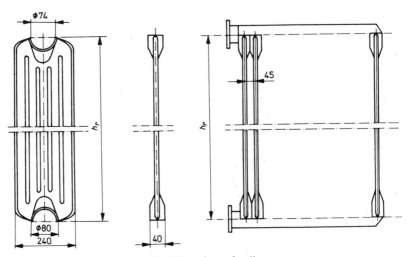

Fig. 6.71. Dimensions of radiators

Table 6.9

Radiator properties of one element, hydraulic characteristics

Length h_r, m	Mass without oil m_r, kg	Volume of oil filling V_0, 10^{-3} m³	Flow cross-section of oil A_0, m²	Oil-side convection surface A_{co}, m²	Air-side convection surface A_{ca}, m²	Hydraulic diameter of one oil duct d_{ho}, m	Cross-sectional area of air duct between two elements A_a, m²	Hydraulic diameter of one air duct between two elements d_{ha}, m	Width b, m	Spacing between centre lines of elements s, m
0.915	4.32	1.620		0.369	0.478					
1.190	5.60	2.107		0.480	0.622					
1.640	7.72	2.904	1.771×10^{-3}	0.661	0.858	1.76×10^{-2}	8.43×10^{-3}	6.44×10^{-2}	0.240	0.045
1.965	9.25	3.480		0.792	1.028					
2.440	11.50	4.321		0.983	1.276					

The outline dimensions of radiators of lengths falling in the range of $h_r = 0.915$ to 2.44 m are given in Fig. 6.71, and their further geometrical characteristics are contained in Table 6.9. Since the five radiator types differ in length only, the value of K valid for the various types of radiators is

$$K = \frac{A_o^{0.2} d_h^{0.5}}{0.85 \times 0.74 g^{0.1}} \frac{\Delta \vartheta_{co}^{0.2}}{A_{co}} = \frac{(17.71 \times 10^{-4})^{0.2} \times (1.76 \times 10^{-2})^{0.5}}{0.85 \times 0.74 \times 9.81^{0.1}} \times$$

$$\times \frac{\Delta \vartheta_{co}^{0.2}}{A_{co}} = 47.27 \times 10^{-3} \frac{\Delta \vartheta_{co}^{0.2}}{A_{co}}.$$

Let the difference between inlet and outlet temperatures of the oil be 20 °C, i.e. $\Delta \vartheta_{co} = 20$ °C, then

$$K_{20} = 86.06 \times 10^{-3} \frac{1}{A_{co}}.$$

Table 6.10

Values of factor B as a function of mean boundary-layer temperature

Mean boundary-layer temperature ϑ_m, °C	B $\dfrac{m^{1.2}\ ^\circ C^{0.9}}{W^{-0.8}\ s^{-0.2}}$
40	0.668
50	0.645
60	0.623
70	0.610
80	0.596
90	0.587
100	0.582

Table 6.11

Factors used for calculating the oil-side surface temperature drop of a radiator element of length $h_r = 0.915$ m; and the values of surface temperature drop for powers $P = 100$ W and $P = 500$ W, as a function of mean boundary layer temperature

Mean boundary-layer temperature ϑ_m, °C	KB °C$^{1.1}$ W$^{-0.8}$	$(KB)^{0.91}$ °C W$^{-0.727}$	Power P, W	Temperature drop $\Delta \vartheta_{os}$, °C	Power P, W	Temperature drop $\Delta \vartheta_{os}$, °C
40	0.1558	0.1842		5.24		16.88
50	0.1504	0.1784		5.07		16.35
60	0.1453	0.1728		4.92		15.84
70	0.1423	0.1695	100	4.82	500	15.54
80	0.1390	0.1660		4.72		15.21
90	0.1369	0.1637		4.66		15.08
100	0.1357	0.1624		4.62		14.88

Table 6.12

Factors used for calculating the oil-side surface temperature drop of a radiator element of length $h_r = 2.44$ m; and the values of surface temperature drop for powers $P = 100$ W and $P = 1000$ W, as a function of mean boundary layer temperature

Mean boundary-layer temperature ϑ_m, °C	KB °C$^{1.1}$ W$^{-0.8}$	$(KB)^{0.91}$ °C W$^{-0.727}$	Power P, W	Temperature drop $\Delta\vartheta_{os}$, °C	Power P, W	Temperature drop $\Delta\vartheta_{os}$, °C
40	0.0585	0.0755		2.15		11.45
50	0.0565	0.0731		2.08		11.09
60	0.0545	0.0709		2.02		10.74
70	0.0534	0.0695	100	1.98	1000	10.54
80	0.0522	0.0681		1.94		10.33
90	0.0514	0.0671		1.91		10.18
100	0.0510	0.0666		1.89		10.10

The values of B for the range of mean boundary-layer temperatures $\vartheta_m = 40$ to 100 °C are compiled in Table 6.10. Values of KB and $(KB)^{0.91}$ pertaining to two radiator lengths and the values of $\Delta\vartheta_{os}$, each calculated for two different values of P as a function of ϑ_m are contained in Tables 6.11 and 6.12. The diagrams of Fig. 6.72 and 6.73 have been plotted on the basis of data of the tables.

6.3.2.3 Effect of oil velocity on the oil-side surface temperature drop

The data of Tables 6.11 and 6.12, and the diagrams given in Figs 6.72 and 6.73, showing the thermal powers dissipated by radiator elements of two differing lengths, as a function of oil-side surface temperature drop, refer to the case of $\Delta\vartheta_{co} = 20$ °C.

This means that, for identical values of P, the oil velocity pertaining to the radiator element of length $h_r = 2.44$ m is 2.67 times higher than that developing in the radiator of length $h_r = 0.915$. Obviously, whether this oil velocity really develops or not depends on the buoyancy acting in the cooling circuit and the pressure drops occurring in it. This fact can be noticed in the diagram relating to radiator length $h_r = 2.44$ m, showing a considerably smaller $\Delta\vartheta_{os}$ associated with the same P than in the case of $h_r = 0.915$.

It can be seen from the formula of the oil-side surface temperature drop that, when the oil velocity is doubled, i.e. $\Delta\vartheta_{co} = 10$ °C is chosen instead of $\Delta\vartheta'_{co} = 20$ °C, the following value is obtained as the new surface temperature drop:

$$\Delta\vartheta'_{os} = \Delta\vartheta_{os}\left(\frac{10}{20}\right)^{0.182} = \Delta\vartheta_{os} \times 0.88 .$$

This shows that doubling the oil velocity reduces the already small value of $\Delta\vartheta_{os}$ (a few °C only) not by half but to 88%. By doubling the heat dissipation of one element, the surface temperature drop is increased to $2^{0.727} = 1.655$ times its original value.

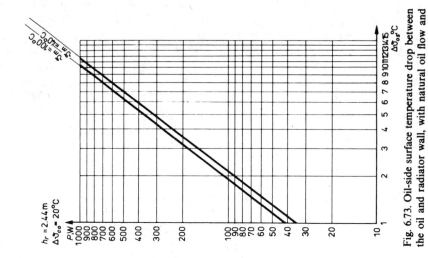

$h_r = 2.44$ m
$\Delta\vartheta_{co} = 20°C$

Fig. 6.73. Oil-side surface temperature drop between the oil and radiator wall, with natural oil flow and $\Delta\vartheta_{co} = 20\,°C$, $h_r = 2.44$ m

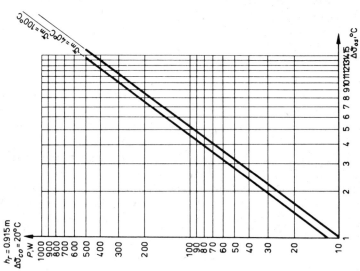

$h_r = 0.915$ m
$\Delta\vartheta_{co} = 20°C$

Fig. 6.72. Oil-side surface temperature drop between the oil and radiator wall, with natural oil flow and $\Delta\vartheta_{co} = 20\,°C$, $h_r = 0.915$ m

458

By increasing the flow velocity of the oil (e.g. by pumping) alone i.e. without improving the heat dissipation on the air side, no substantial increase in the amount of heat dissipated is to be expected. The reason why an increased flow velocity of the oil is justified with forced air cooling is that otherwise, due to the flow resistance of the radiator tubes, an increased difference between inlet and outlet oil temperature would be required to bring about a transfer of heat flux required by the increased heat flux per unit transfer area.

6.3.3 Pressure drop in the radiators

6.3.3.1 Average axial viscosity

The introduction of this concept is required for calculating the pressure drop in radiators.

The average viscosity of the oil flowing in a radiator of length h_r is defined as

$$v_{al} = \frac{1}{h_r} \int_0^{h_r} v(h)\, dh \tag{6.176}$$

is understood. The independent variable h is the distance measured from the upper end of the radiator.

The function $v(h)$ is made up of function $v(\vartheta)$ characterizing the oil and of function $\vartheta(h)$:

$$v(h) = v[\vartheta(h)],$$

thus, v is indirectly a function of h.

In the expression of average viscosity let us change over from integration limits of length to those of temperature limits:

$$v_{al} = \frac{1}{h_r} \int_{\vartheta_{ooc}}^{\vartheta_{oic}} v(\vartheta)\, \frac{dh}{d\vartheta}\, d\vartheta. \tag{6.177}$$

Function ϑ_h corresponding to the condenser-type temperature pattern is

$$\vartheta = \vartheta_a + \Delta\vartheta_{oic} \left[1 - \frac{\Delta\vartheta_{co}}{\Delta\vartheta_{oic}}\right]^{\frac{h}{h_r}}. \tag{6.178}$$

This last expression of average viscosity prescribes the differentiation of the inverse of function $\vartheta(h)$. It is simpler to perform the differentiation in function $\vartheta(h)$, and subsequently to calculate the reciprocal of the result:

$$\frac{dh}{d\vartheta} = \frac{1}{\dfrac{d\vartheta}{dh}}.$$

459

The differential is

$$\frac{d\vartheta}{dh} = \Delta\vartheta_{oic}\frac{1}{h_r}\left[1 - \frac{\Delta\vartheta_{co}}{\Delta\vartheta_{oic}}\right]^{\frac{h}{h_r}}\ln\left(1 - \frac{\Delta\vartheta_{co}}{\Delta\vartheta_{oic}}\right).$$ (6.179)

The average viscosity is

$$v_{al} = \frac{\Delta\vartheta_{o-a}}{\Delta\vartheta_{co}}\int_{\vartheta_{ooc}}^{\vartheta_{oic}}\frac{v(\vartheta)}{\vartheta - \vartheta_a}\,d\vartheta,$$ (6.180)

where from Eqn. (6.149)

$$\Delta\vartheta_{o-a} = \frac{\Delta\vartheta_{co}}{\ln\dfrac{1}{1 - \dfrac{\Delta\vartheta_{co}}{\Delta\vartheta_{oic}}}}.$$

If the value of function $v(\vartheta)$, in the range of $30 < \vartheta < 110\,°C$, is

$$v = 6.3 \times 10^{-3}\vartheta^{-1.66},$$ (6.181)

then

$$v_{al} = \frac{6.3}{10^3}\frac{\Delta\vartheta_{o-a}}{\Delta\vartheta_{co}}\int_{\vartheta_{ooc}}^{\vartheta_{oic}}\frac{1}{\vartheta^{2.66} - \vartheta_a\vartheta^{1.66}}\,d\vartheta,$$ (6.182)

where the dimension of factor 6.3×10^{-3} is $m^2\,s^{-1}\,°C^{1.66}$. The definite integral can most simply be calculated by plotting, for a given value of ϑ_a, the curve

$$\frac{v(\vartheta)}{\vartheta - \vartheta_a},$$

and by measuring the area under the curve between temperature limits ϑ_{ooc} and ϑ_{oic} with a planimeter.

Table 6.13

Values of function $\dfrac{v(\vartheta)}{\vartheta - \vartheta_a}$ appearing in the formula of axial viscosity for $\vartheta_a = 10,\ 20,\ 30\,°C$, as a function of temperature

Tempera-ture ϑ, °C	$\dfrac{v(\vartheta)}{\vartheta - 10\,°C} \times 10^6$ $m^2\,s^{-1}\,°C^{-1}$	$\dfrac{v(\vartheta)}{\vartheta - 20\,°C} \times 10^6$ $m^2\,s^{-1}\,°C^{-1}$	$\dfrac{v(\vartheta)}{\vartheta - 30\,°C} \times 10^6$ $m^2\,s^{-1}\,°C^{-1}$
25	1.8	–	–
30	1.05	2.1	–
35	0.68	1.132	3.4
40	0.467	0.7	1.4
45	–	–	0.765
50	0.243	0.323	0.485
60	0.14	0.175	0.233
70	0.09	0.118	0.135
80	0.06	0.07	0.084
90	0.0437	0.05	0.0583
100	0.0333	0.0376	0.0428

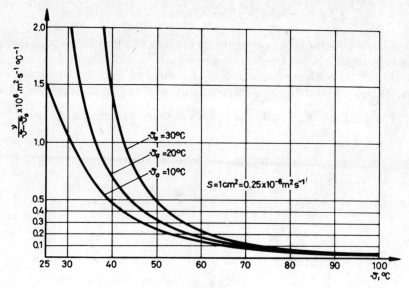

Fig. 6.74. The values of function $\dfrac{v(\vartheta)}{\vartheta - \vartheta_a}$ for three different air temperatures

In order to facilitate the calculations that follow the function $\dfrac{v(\vartheta)}{\vartheta - \vartheta_a}$ is calculated for the cases of $\vartheta_a = 10, 20$ and $30\,^{\circ}\mathrm{C}$, utilizing the data of Table 6.1. The data of Table 6.13 have been used for plotting the curves of Fig. 6.74.

Example 6.17. Calculate the value of viscosity for the radiator banks of the preceding example taking $\varDelta\vartheta_{o-a} = 49.3\,^{\circ}\mathrm{C}$, $\varDelta\vartheta_{co} = 20\,^{\circ}\mathrm{C}$, $\vartheta_{oic} = 80\,^{\circ}\mathrm{C}$, $\vartheta_{ooc} = 60\,^{\circ}\mathrm{C}$ and $\vartheta_a = 20\,^{\circ}\mathrm{C}$.

By planimetering in the diagram of Fig. 6.74 the section of area under the curve between 60 and 80 $^{\circ}\mathrm{C}$ corresponding to $\vartheta_a = 20\,^{\circ}\mathrm{C}$ is obtained as 8.95 cm^2. Considering that temperature scale of the curve is 1 cm = 2.5 $^{\circ}\mathrm{C}$, and the scale of function $\dfrac{v(\vartheta)}{\vartheta - 20\,^{\circ}\mathrm{C}}$ is 1 cm = 0.1×10^{-6} m^2 s^{-1} $^{\circ}\mathrm{C}^{-1}$, i.e. the scale of the area marked S is 1 cm$^2 = 0.25 \times 10^{-6}$ m^2 s^{-1}, the value of the definite integral

$$\int_{60}^{80} \frac{v(\vartheta)}{\vartheta - 20\,^{\circ}\mathrm{C}}$$

is $0.25 \times 10^{-6} \times 8.95$ m^2 s$^{-1} = 2.24 \times 10^{-6}$ m^2 s^{-1}.

With this result, the value of average viscosity is

$$v_{al} = \frac{\varDelta\vartheta_{o-a}}{\varDelta\vartheta_{co}} \int_{\vartheta_{ooc}}^{\vartheta_{oic}} \frac{v(\vartheta)}{\vartheta - \vartheta_a} \, d\vartheta = \frac{49.3}{20} \, 2.24 \times 10^{-6} \ \mathrm{m^2\,s^{-1}} =$$

$$= 5.52 \times 10^{-6} \ \mathrm{m^2\,s^{-1}}.$$

461

6.3.3.2 Calculation of pressure drop in the radiators

Consider n radiator groups mounted on the tank of a transformer. Assume each radiator group to contain z elements arranged one behind another. Let the total oil mass flow of the cooling system be Φ_0, the mean density of the oil flowing in the radiators be ρ_0, and the axial average viscosity of the oil be v_{al}. Thus, the oil mass flow of one element is $\dfrac{\Phi_o}{nz}$, and the volume flow of one element $\psi_0 = \dfrac{\Phi_0}{\rho_0 nz}$.

Based on measurements, the pressure drop of radiator-type cooling is

$$\Delta p_r = C v_{al} \frac{\Phi_0}{\rho_0 nz} + Bz^2 \left(\frac{\Phi_0}{\rho_0 nz}\right)^2 = C v_{al}\psi_0 +$$
$$+ Bz^2\psi_0^2 . \tag{6.183}$$

For the type of radiators shown in Fig. 6.71, from measurements show

$$C = h_r 48.5 \times 10^9 , \tag{6.184}$$

where h_r is the length of the radiator, and the dimension of factor 48.5×10^9 is m^{-7} kg.

$$B = 5.63 \times 10^6 \ m^{-7} \ kg . \tag{6.185}$$

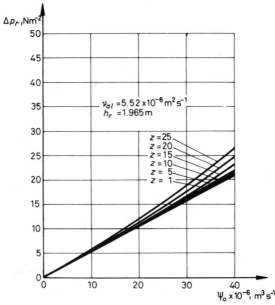

Fig. 6.75. Pressure drop in radiators containing different numbers of elements; $h_r = 1.965$ m; $v_{al} = 5.52 \times 10^{-4} \ m^2 \ s^{-1}$

The value of Bz^2 for a few different numbers of elements:

$$z=1, \quad B \times 1^2 = 5.63 \times 10^6 \, \text{m}^{-7} \, \text{kg},$$
$$z=5, \quad B \times 5^2 = 0.141 \times 10^9 \, \text{m}^{-7} \, \text{kg},$$
$$z=10, \quad B \times 10^2 = 0.563 \times 10^9 \, \text{m}^{-7} \, \text{kg},$$
$$z=15, \quad B \times 15^2 = 1.267 \times 10^9 \, \text{m}^{-7} \, \text{kg},$$
$$z=20, \quad B \times 20^2 = 2.252 \times 10^9 \, \text{m}^{-7} \, \text{kg},$$
$$z=25, \quad B \times 25^2 = 3.519 \times 10^9 \, \text{m}^{-7} \, \text{kg}.$$

Let the height of radiators be $h_r = 1.965$ m, and select the value of average viscosity to be equal to that obtained in Example 6.17, i.e. $v_{al} = 5.52 \times 10^{-6}$ m^2 s^{-1}. Product Cv is

$$Cv = 0.526 \times 10^6 \, \text{kg} \, \text{m}^{-4} \, \text{s}^{-1}.$$

Let us represent Δp_r as a function of

$$x = \frac{\Phi_o}{\rho_0 nz} \times 10^{-6} = \psi_0 \times 10^{-6}. \tag{6.186}$$

The equations needed for calculating the curves are

$$z=1; \quad \Delta p_r = 0.526 + 5.63 \times 10^{-6} x^2,$$
$$z=5; \quad \Delta p_r = 0.526 + 0.141 \times 10^{-3} x^2,$$
$$z=10; \quad \Delta p_r = 0.526 + 0.563 \times 10^{-3} x^2,$$
$$z=15; \quad \Delta p_r = 0.526 + 1.267 \times 10^{-3} x^2,$$
$$z=20; \quad \Delta p_r = 0.526 + 2.252 \times 10^{-3} x^2,$$
$$z=25; \quad \Delta p_r = 0.526 + 3.519 \times 10^{-3} x^2.$$

Curves $\Delta p_r = f(x)$ pertaining to the different numbers of elements are plotted in Fig. 6.75. In the six formulae the dimension of factor 0.526 is kg m^{-4} s^{-1}, and that of the multiplying factors of x^2 is m^{-7} kg.

6.3.4 Pressure drop in the oil ducts of the winding

In order to be able to apply known formula of flow resistance for the case of laminar flow applicable, it is appropriate to introduce the concept of reduced duct width δ_r which is independent of the geometrical dimension of duct width and depends on the spontaneous boundary-layer thickness δ_{sp}. The calculation of δ_{sp} has also been demonstrated by an example. According to measurements for a vertical duct width of $d = 3.5$ mm, $\delta_r = 0.5 \, \delta_{sp}$, and for $d = 10$ mm $\delta_r = 0.57$ were found to be correct choices, whereas for $d > 3$ mm and $\delta_{sp} \leq 1.25$ the value

$$\delta_r = 0.56 \delta_{sp} \tag{6.187}$$

seemed to be appropriate.

About 35% of the horizontal surfaces of disc coils are covered by spacers and about 20% of the vertical surfaces are covered by strips. Take the width of a vertical flow duct as equal to the uncovered length of the arc between two spacers along the circumference of the disc (Fig. 6.76). Let the width of each of the ducts located along the inner surface of each coil, of equal number to the spacers, be l_1, and that of each of the ducts arranged along the outer mantle and also equal in number to that of the spacers be l_2. The length of each vertical duct is equal to the height of the winding h_w.

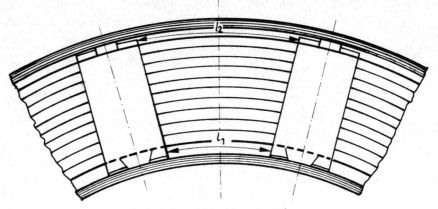

Fig. 6.76. Width of flow channel

The hydraulic diameters of the duct are

$$d_{h1} = 4\frac{\delta_r l_1}{2(\delta_r + l_1)} = 2\frac{\delta_r l_1}{\delta_r + l_1},$$ (6.188)

$$d_{h2} = 2\frac{\delta_r l_2}{\delta_r + l_2}.$$ (6.189)

Let the equivalent hydraulic diameter be

$$d_h = \frac{d_{h1} + d_{h2}}{2}.$$ (6.190)

In the following, the equivalent hydraulic diameter is used. It would make no sense to calculate separately the pressures prevailing in the inner and outer oil ducts, because the two vertical ducts are coupled with each other after each coil. The pressure is therefore approximately uniform at every height level, because if there were a difference, it would immediately be equalized by the radial flow developing between successive spacers.

Let the loss developing in the winding be P_w, the difference between inlet and outlet oil temperatures $\Delta\vartheta_{wo}$, the mean oil density ρ_o and its specific heat c_o.

The oil volume flow is

$$\psi_o = \frac{P_w}{\rho_o c_o \Delta\vartheta_{wo}}.$$ (6.191)

Denote the number of spacers placed around the periphery by n. The cross-sectional area available for the flow of oil is

$$A_o = n(l_1 + l_2)\delta_r.$$ (6.192)

The measurements referred to above have shown that the oil flow developing in vertical ducts of the winding cannot be approximated by the parabolic velocity

464

Table 6.14

Hydraulic diameters and the value of A for ducts of different cross-
sectional areas

Shape of cross-sectional area	Hydraulic diameter d_h, m	A
circle of diameter d	d	64
square of side length a	a	57
equilateral triangle of side length a	0.58a	53
ring of width a	2a'	96
rectangle of sides a and b		
$\frac{a}{b} = 0$	2a	96
$\frac{a}{b} = 0.10$	1.81a	85
$\frac{a}{b} = 0.20$	1.67a	76
$\frac{a}{b} = 0.25$	1.60a	73
$\frac{a}{b} = 0.33$	1.50a	69
$\frac{a}{b} = 0.50$	1.30a	62

distribution characteristic of laminar flow between planes, because the flow here is
of boundary-layer character (see Section 6.2.3.4). The flow resistance of the winding
in the laminar section, calculated as if the flow filled the available duct section
entirely, is much lower than that obtained by measurements. As mentioned before
the adoption of the reduced duct width δ, enables us to apply the known laminar
flow formula for this flow of boundary-layer character.

Because of its parabolic velocity distribution, the maximum oil velocity is 3/2 of
the mean value. Let the maximum oil velocity be denoted by w:

$$w = \frac{3\Psi_o}{2A_0}.$$ (6.193)

The pressure drop of a liquid flowing in a duct of hydraulic diameter d_h and length
h_w is

$$\Delta p_w = \frac{A}{Re} \frac{h_w}{d_h} \frac{w^2}{2} \rho_0,$$ (6.194)

where A is a number depending on the shape of the duct, its value being in the range
85 to 95 in practical cases (Table 6.14).

Substituting the value $Re = \frac{wd_h}{v}$, where v is the average viscosity of flowing oil:

$$\Delta p_w = \frac{3}{4} A \frac{h_w}{d_h^2 A_0} \frac{P_w}{c_o \Delta \vartheta_{wo}} v.$$ (6.195)

6.3.5 Air flow along the radiators, calculation of air-side temperature drop

6.3.5.1 The condenser-type temperature pattern

When calculating the temperature variation of media flowing in heat exchangers, the changing value on Nu, i.e. the variation of surface heat transfer coefficient α along the surface is generally not taken into account, and instead of the local number Nu an average value of Nu or α is used in the calculation.

In most cases it is left out of consideration that α is a fractional-power function of surface temperature drop. Correspondingly, the differential equation giving the temperature of a cooling or warming liquid, from which the reduction of heat content or temperature drop due to heat dissipation taking place along a path of unit width and length dh can be determined, is the following:

$$\alpha \Delta \vartheta \, dh = \Phi_o c_p d\vartheta. \tag{6.196}$$

Separating the variables:

$$\frac{d\vartheta}{\Delta \vartheta} = \frac{\alpha}{\Phi_o c_p} dh = a \, dh, \tag{6.197}$$

where

$$a = \frac{\alpha}{\Phi_o c_p}. \tag{6.198}$$

Integrating both sides:

$$ln\Delta\vartheta = ah + b. \tag{6.199}$$

If $h=0$, then $\Delta\vartheta = \Delta\vartheta_{oic}$, and if $h=h_r$, then $\Delta\vartheta = \Delta\vartheta_{ooc}$. Substituting the above,

$$b = ln\vartheta_{oic} \tag{6.200}$$

and

$$ln\Delta\vartheta_{ooc} = ah_r + ln\Delta\vartheta_{oic}$$

are obtained, from which

$$a = \frac{ln\Delta\vartheta_{ooc} - ln\Delta\vartheta_{oic}}{h_r}. \tag{6.201}$$

Re-substituting the integration constants

$$ln\Delta\vartheta = (ln\Delta\vartheta_{ooc} - ln\Delta\vartheta_{oic})\frac{h}{h_r} + ln\Delta\vartheta_{oic}. \tag{6.202}$$

Introducing the relation $\Delta\vartheta_{ooc} = \Delta\vartheta_{oic} - \Delta\vartheta_{co}$, it may be written:

$$ln\Delta\vartheta_{ooc} - ln\Delta\vartheta_{oic} = ln\frac{\Delta\vartheta_{ooc}}{\Delta\vartheta_{oic}} = ln\frac{\Delta\vartheta_{oic} - \Delta\vartheta_{co}}{\Delta\vartheta_{oic}} = ln\left(1 - \frac{\Delta\vartheta_{co}}{\Delta\vartheta_{oic}}\right),$$

$$ln\Delta\vartheta = ln\left(1 - \frac{\Delta\vartheta_{co}}{\Delta\vartheta_{oic}}\right)\frac{h}{h_r} + ln\Delta\vartheta_{oic}, \tag{6.203}$$

i.e. the surface temperature drop will be

$$\Delta\vartheta = \Delta\vartheta_{oic}\left(1 - \frac{\Delta\vartheta_{co}}{\Delta\vartheta_{oic}}\right)^{\frac{h}{h_r}} \tag{6.204}$$

or

$$\Delta\vartheta = \Delta\vartheta_{oic} \left(\frac{1}{1 - \dfrac{\Delta\vartheta_{co}}{\Delta\vartheta_{oic}}} \right)^{-\frac{h}{h_r}} \tag{6.205}$$

This temperature equation is true if α does not depend on the surface temperature drop $\Delta\vartheta$.

In the preceding sections it has been demonstrated that the heat dissipation on the air side of natural oil flow, natural air cooled radiators is dominated by the air-side heat transfer coefficient and that on the oil side by the magnitude of volumetric flow. In the following, it will be seen that with radiators shorter than 1 m, the temperature of the air may be assumed constant on the air side beyond the

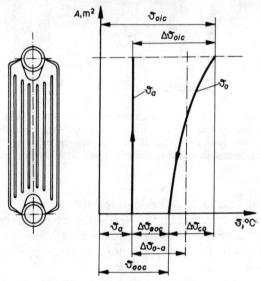

Fig. 6.77. The condenser-type temperature pattern

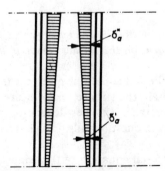

Fig. 6.78. In the case of short radiators the boundary layers do not come into contact with each other, not even at the upper part of the radiator

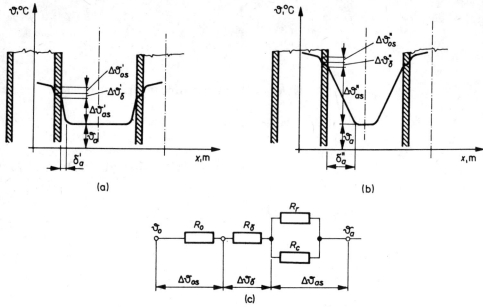

Fig. 6.79. Thermal gradients in the air duct between two radiators and in the radiator wall, for the case of a condenser-type temperature pattern. The temperature gradients developing at the bottom level, (a) and those developing at the top level, are shown in (b). The diagram (c) shows the temperature drops schematically. R_c is the thermal resistance to convection and R_r the thermal resistance to radiation

boundary layer. Even the radiator elements located at the top are so far apart that the boundary layers do not come into contact with each other, i.e. a condenser-type temperature pattern develops as if the air had a very high flow-stream capacity rate. The formation of a condenser-type temperature pattern depends on the air-side flow conditions, the oil-side flow having no influence on it (see Figs 6.77, 6.78, and 6.79).

In deducing of the temperature equation corresponding to the condenser type temperature pattern, the air-side heat transfer coefficient has been assumed constant and independent of surface temperature drop.

6.3.5.2 Temperature pattern in the case of heat transfer coefficient dependent on the surface temperature drop

In practical cases, the criterion equations given for the air-side Nu number and thus for the air-side α contain the surface temperature drops raised to either the 0.125th, or to the 0.25th or the 1/3 power depending on the range in which the product $Gr \times Pr$ of the flow in question falls. Let us consider the case of $\alpha = a\Delta\vartheta^{1/3}$ formula (6.30).

For that case, the differential equation of the temperature of the cooling oil is

$$a\Delta\vartheta^{\frac{1}{3}}\,\Delta\vartheta dh = a\Delta\vartheta^{\frac{4}{3}}\,dh = \Phi c_p d\vartheta\,. \qquad (6.206)$$

468

Separating the variables

$$\frac{d\vartheta}{\varDelta\vartheta^{\frac{4}{3}}} = \frac{a}{\varPhi c_p}\,dh = g\,dh,$$ (6.207)

where

$$g = \frac{a}{\varPhi c_p}.$$ (6.208)

Integrating Eqn. (6.207):

$$\int \frac{d\vartheta}{\varDelta\vartheta^{\frac{4}{3}}} = g\int dh,$$ (6.209)

the following result is obtained:

$$-3\varDelta\vartheta^{-\frac{1}{3}} = gh + k,$$ (6.210)

where if

$$h = 0,\quad \varDelta\vartheta = \varDelta\vartheta_{oic}$$

and if

$$h = h_r,\quad \varDelta\vartheta = \varDelta\vartheta_{ooc}.$$

Thus the integration constant

$$k = -3\varDelta\vartheta_{oic}^{-\frac{1}{3}},$$ (6.211)

and

$$-3\varDelta\vartheta_{ooc}^{-\frac{1}{3}} = gh_r - 3\varDelta\vartheta_{oic}^{-\frac{1}{3}},$$

so that

$$g = \frac{3}{h_r}\left(\varDelta\vartheta_{oic}^{-\frac{1}{3}} - \varDelta\vartheta_{ooc}^{-\frac{1}{3}}\right).$$ (6.212)

Re-substituting

$$\varDelta\vartheta^{-\frac{1}{3}} = \left(\varDelta\vartheta_{ooc}^{-\frac{1}{3}} - \varDelta\vartheta_{oic}^{-\frac{1}{3}}\right)\frac{h}{h_r} + \varDelta\vartheta_{oic}^{-\frac{1}{3}}.$$ (6.213)

The surface temperature drop:

$$\varDelta\vartheta = \frac{1}{\left[\left(\varDelta\vartheta_{ooc}^{-\frac{1}{3}} - \varDelta\vartheta_{oic}^{-\frac{1}{3}}\right)\dfrac{h}{h_r} + \varDelta\vartheta_{oic}^{-\frac{1}{3}}\right]^3}.$$ (6.214)

If

$$\varDelta\vartheta_{ooc} = \varDelta\vartheta_{oic} - \varDelta\vartheta_{co},$$

then

$$\varDelta\vartheta = \frac{1}{\left\{\left[\left(\varDelta\vartheta_{oic} - \varDelta\vartheta_{co}\right)^{-\frac{1}{3}} - \varDelta\vartheta_{oic}^{-\frac{1}{3}}\right]\dfrac{h}{h_r} + \varDelta\vartheta_{oic}^{-\frac{1}{3}}\right\}^3}.$$ (6.215)

If

$$\varDelta\vartheta_{ooc} = m\varDelta\vartheta_{oic},$$

then

$$\varDelta\vartheta_{ooc}^{-\frac{1}{3}} - \varDelta\vartheta_{oic}^{-\frac{1}{3}} = m^{-\frac{1}{3}}\varDelta\vartheta_{oic}^{-\frac{1}{3}} - \varDelta\vartheta_{oic}^{-\frac{1}{3}} = \varDelta\vartheta_{oic}^{-\frac{1}{3}}(m^{-\frac{1}{3}} - 1),$$ (6.216)

from which

$$\varDelta\vartheta = \frac{\varDelta\vartheta_{oic}}{\left[\left(m^{-\frac{1}{3}} - 1\right)\dfrac{h}{h_r} + 1\right]^3}.$$ (6.217)

The average temperature difference, $\varDelta\vartheta'_{o-a}$, is obtained by integrating Eqn. (6.217) between limits $h = 0$ and $h = h_r$, and dividing the result by h_r.

$$\varDelta\vartheta'_{o-a} = \frac{1}{h_r}\int_{h=0}^{h=h_r} \frac{\varDelta\vartheta_{oic}}{\left[\left(m^{-\frac{1}{3}} - 1\right)\dfrac{h}{h_r} + 1\right]^3}\,dh.$$ (6.218)

Table 6.15

$$\text{Values of function of } \Delta\vartheta = \frac{60\,°C}{\left(0.145\dfrac{h}{h_r}+1\right)^3}\,°C \text{ for different values of } \frac{h}{h_r}$$

$\dfrac{h}{h_r}$	0	0.1	0.2	0.3	0.4	0.5	0.6	0.7	0.8	0.9
$\Delta\vartheta$, °C	60	57.4	55	52.7	50.7	48.3	46.8	44.9	43.1	41.6

The integration may be performed most simply by plotting the function for a given value of m and $\Delta\vartheta_{oic}$, then using a planimeter and dividing the area k obtained under the curve by h_r.

Example 6.18. Assuming $\Delta\vartheta_{oic}=60\,°C$ and $\Delta\vartheta_{ooc}=40\,°C$, determine the value of $\Delta\vartheta'_{o-a}$. From the above data

$$m=\frac{\Delta\vartheta_{ooc}}{\Delta\vartheta_{oic}}=\frac{40}{60}=\frac{2}{3},\qquad m^{-\frac{1}{3}}=\sqrt[3]{\frac{3}{2}}=1.145.$$

As a first step, the function $\Delta\vartheta$ is written:

$$\Delta\vartheta=\frac{\Delta\vartheta_{oic}}{\left[\left(m^{-\frac{1}{3}}-1\right)\dfrac{h}{h_r}+1\right]^3}=\frac{60}{\left(0.145\dfrac{h}{h_r}+1\right)^3}\,°C.$$

Next, $\Delta\vartheta$ is calculated for ten values of h/h_r. The results of the calculation are compiled in Table 6.15.

Finally the area under curve $\Delta\vartheta$ is determined by adding up the values of $\Delta\vartheta$ taken from every second column of the table and doubling the sum obtained. The resulting product divided by 10 gives the required $\Delta\vartheta'_{o-a}$.

$$\Delta\vartheta'_{o-a}=\frac{57.4+52.8+48.3+44.9+41.6}{10}\,2\,°C=48.98\,°C\approx49\,°C.$$

As seen, the result does not differ much either from the logarithmic-mean temperature difference (49.3 °C) or from the mean average temperature rise (50 °C).

6.3.5.3 Effect of radiator length on air-side heat transfer coefficient

If the radiator is shorter than the critical length, then the surface heat transfer coefficient is calculated from the similarity equation (6.111):

$$Nu_m=c(Gr\times Pr)^n_m$$

valid for the flow in an unconfined space. Factor c and exponent n depend on the value of product $Gr\times Pr$ valid for the temperature ϑ_m of the boundary layer. In practical cases, the value of product $Gr\times Pr$ falls in the range 2×10^7 to 1×10^{13}, for which

$$c=0.135\quad\text{and}\quad n=1/3$$

470

are valid. In such a case, no geometrical factor appears in the explicit expression for the surface heat transfer coefficient, and the flow is considered as turbulent.

The surface heat transfer coefficient in explicit form is

$$\alpha = 0.135 (\beta g Pr_m)^{1/3} \frac{\lambda_m}{v_m^{2/3}} \Delta \vartheta^{1/3}. \tag{6.219}$$

In the case of radiators shorter than the critical length the logarithmic-mean temperature difference calculated in the preceding sections is applicable. The logarithmic-mean temperature difference has been calculated with the assumption that the increasing thickness of the boundary layer along the surface of radiator elements in the direction of flow does not become so thick as to come into contact with the adjacent boundary layer even at the top of the radiator bank, i.e. that the condenser-type temperature pattern is applicable, which supposes the presence of an air core at ambient temperature of infinitely high flow-stream capacity.

6.3.5.4 Critical radiator length

For a given radiator pitch and as a function of mean boundary layer temperature and surface heat flux per unit transfer area it is always possible to determine a critical dimension in the direction of flow at which the boundary layers on the two sides just make contact with each other. This critical dimension is termed the critical radiator length. In the section of the flow channel above the point where the laminar boundary layers of the two sides make contact, on each side, a thin laminar boundary layer develops adhering closely to the radiator wall, and inside the channel the air ascends with turbulent flow, causing thereby an increase of α.

In the case of radiator elements longer than the critical length, the condenser-type temperature pattern is no longer valid, and radiator cooling changes into a counterflow type of heat exchanger. In reality, in spite of the reduced temperature difference, due to the specific dissipated heat of radiators longer than critical will increase slightly over that calculated with the α value valid for the laminar flow of the condenser-type temperature patterns, because of the higher value of α.

Now suppose the critical radiator length is to be calculated for a given arrangement. Let the centre lines of radiator elements be spaced 45 mm apart and the average thickness of an element be 10 mm. This means that if the laminar boundary layer on one side reaches the width $d = (45 - 10)/2 \text{ mm} = 17.5 \text{ mm}$, the two boundary layers come into contact with each other.

As known, the thickness of the laminar boundary layer at distance x from the leading edge of the plate, i.e. in our case from the lower edge of the radiator, is

$$\delta_a = 5.83 \sqrt{\frac{xv}{w_0}}, \tag{6.220}$$

where v is the viscosity of air and w_0 is the flow velocity of air outside the boundary layer.

Substituting $\delta_a = 17.5 \times 10^{-3} \text{ m}$, the formula gives that distance x', at which the two boundary layers just contact each other.

Table 6.16

Distance of boundary layers coming into contact with each other, measured from the bottom of radiator, for different mean boundary-layer temperatures and air velocities

Mean boundary-layer temperature ϑ_m, °C	Viscosity m² s⁻¹	Air velocity w_o, m s⁻¹			
		0.5	1.0	1.5	2.0
		Spacing at which boundary layers come into contact x', m			
20	15.7×10^{-6}	0.287	0.573	0.860	1.150
30	16.61×10^{-6}	0.271	0.542	0.815	1.080
40	17.6×10^{-6}	0.256	0.512	0.767	1.023
50	18.6×10^{-6}	0.242	0.484	0.727	0.967
60	19.6×10^{-6}	0.230	0.460	0.688	0.917

The value obtained is

$$x' = 9 \times 10^{-6} \frac{w_0}{v},$$ (6.221)

where the dimension of factor 9×10^{-6} is m². The values of x' for different air velocities w_0 and ϑ_m mean boundary-layer temperature are contained in Table 6.16.

As can be seen, the boundary layers come into contact a short distance above the bottom of the radiator in the case of low air velocities and high temperatures.

6.3.5.5 Maximum thickness of laminar boundary layer, thickness of turbulent boundary layer

If the air flows along a heated vertical plane with a velocity w_0, then at some distance x_{cr} from the bottom leading edge the laminar boundary layer changes over into a turbulent layer developing over a thin laminar layer. The maximum laminar boundary layer thickness is the boundary layer thickness pertaining to value x_{cr}. The turbulent boundary layer develops at the instant when the value of

$$Re_{cr} = \frac{w_0 x_{cr}}{v} = 4.85 \times 10^5$$ (6.222)

is reached. The laminar boundary layer of maximum thickness is calculated from the formula of the laminar boundary layer thickness:

$$\delta_a = \frac{5.83x}{\sqrt{Re}}.$$ (6.223)

By substituting the value of Re_{cr}:

$$x_{cr} = 4.85 \times 10^5 \frac{v}{w_0}$$ (6.224)

Table 6.17

Ratio of distance of turbulent boundary-layer from bottom of radiator to maximum laminar boundary-layer thickness in the function of mean boundary-layer temperature

Mean boundary-layer temperature m, °C	Air velocity w_o, ms^{-1}			
	0.5	1.0	1.5	2.0
	distance x_{cr}, m/thickness δ_{cr}, m			
20	15.25	7.6	5.07	3.81
	0.1275	0.0637	0.0425	0.0318
30	16.15	8.03	5.36	4
	0.135	0.0675	0.045	0.0338
40	17.6	8.5	5.68	4.19
	0.143	0.0715	0.0477	0.0358
50	18.1	9	6	4.43
	0.151	0.0756	0.0502	0.0377
60	19.05	9.47	6.32	4.67
	0.159	0.0795	0.053	0.0397

and

$$\delta_{cr} = 4.06 \times 10^3 \frac{v}{w_0}. \tag{6.225}$$

Values of x_{cr} for a few air velocities and mean boundary-layer temperatures are given in Table 6.17.

Collating the data of Tables 6.16 and 6.17 it can be seen that turbulent flow sets in-between the radiators already at very short distances above the bottom of the radiators, due to the contact taking place between the laminary boundary layers, whereas in the case of a vertical heated wall facing a free half space, the turbulent flow appears after several metres of radiator length.

The thickness of turbulent flow starts to build up at the distance x_{cr} above the edge of the plate. With x denoting the distance measured from x_{cr} taken as origin of the coordinate system, the thickness of the turbulent boundary layer is

$$\delta_t = \frac{0.37x}{Re^{0.2}} = 0.37 \sqrt[5]{\frac{x^4 v}{w_0}}. \tag{6.226}$$

6.3.5.6 Character of air flow with natural air cooling

Whether the flow of air ascending between the radiators is laminar or turbulent depends on the value of $Re = w d_h/v$.

It is known that the flow becomes turbulent for values $Re \geq 2320$.

For calculating the Re number, the hydraulic diameter has to be known. The hydraulic diameter of the channels formed by 240 mm wide radiator elements

473

Table 6.18

Values of the factor d_h/v as a function
of air temperature

Air temperature ϑ_a, °C	Viscosity v, m² s⁻¹	$\dfrac{d_h}{v}$, s m⁻¹
20	15.7×10^{-6}	4102
30	16.61×10^{-6}	3877
40	17.6×10^{-6}	3659
50	18.6×10^{-6}	3462

spaced 45 mm apart and the side coverplates according to Table 6.9, is

$$d_h = \frac{4A}{c} = 6.44 \times 10^{-2}\ \text{m}\,.$$

The air viscosity v appearing in the Re number and the values of d_h/v are given in Table 6.18. For example, for $\vartheta_a = 20\,^\circ\text{C}$, $Re = 4102w$.

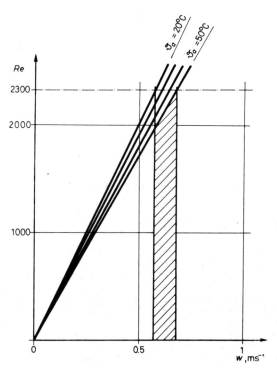

Fig. 6.80. Values of the Re number as a function of air velocity, for different air temperatures

It can be seen that at an air temperature of 20 °C, the critical value $Re_{cr}=2320$ is reached at a relatively low air velocity of $w=0.57\ \mathrm{m\ s^{-1}}$. Re numbers for different values of ϑ_a are shown as a function of air velocity w in Fig. 6.80. It is apparent from the diagram that for air temperatures encountered in practice, the flow is regarded as turbulent when $w=0.67\ \mathrm{m\ s^{-1}}$. This result has otherwise been reached in discussing the critical radiator lengths.

6.3.5.7 Air-side surface heat transfer coefficient with radiators exceeding the critical length

The formulae available for calculation of surface heat transfer coefficients, relating either to laminar or to turbulent flow, are generally not valid for the transition zone ranging from $Re=2000$ to $10\,000$. It would be possible to use the Nu values obtainable from the diagram of Fig. 6.81 based on the experiments of Ilyin [90]. However, since the diagram is a result of heat transfer measurements made on equipment differing from radiators, the heat transfer coefficient should be calculated, for safety reasons, from the equation:

$$Nu_m=0.85 \times 0.74\ Re_m^{0.2} \times (Gr \times Pr)_m^{0.1}\ Pr_m^{0.2}$$

relating to laminar flow.

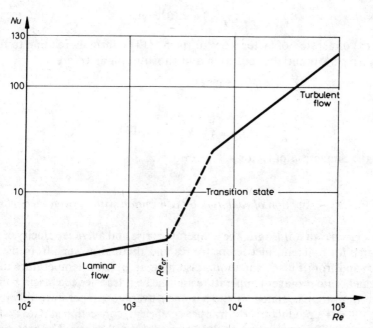

Fig. 6.81. Heat transfer for transient flow, according to the experiments of Ilyin [90]. The comparison refers to the values of $Pr=1$ and $Gr\,Pr=1$

6.3.5.8 Pressure drop due to wall friction

For calculating wall friction, two formulae are used in practice. The first is valid in the case of laminar flow (see Eqn. 6.194):

$$\Delta p_f = \frac{A}{Re}\frac{h_r}{d_h}\rho\frac{w^2}{2}, \tag{6.227}$$

where the value of A for the case of $a/b = 0.035/0.24 = 0.146$, according to Table 6.14, is $A \approx 80$.

The second formula is valid for turbulent flow:

$$\Delta p_f = \frac{0.3164}{Re^{0.25}}\frac{h_r}{d_h}\rho\frac{w^2}{2}. \tag{6.228}$$

For safety reasons, wall friction is calculated from the formula given for the case of turbulent flow.

The formula for frictional pressure drop which is valid in the case of laminar flow, after substituting $Re = \dfrac{wd_h}{v}$, is

$$\Delta p_f = \frac{40}{d_h^2}h_r\rho vw. \tag{6.229}$$

Substituting into the formula the value $d_h = 6.44 \times 10^{-2}$ and performing the multiplication indicated:

$$\Delta p_f = 9.6 \times 10^3\, h_r\rho vw, \tag{6.230}$$

where the dimension of factor 9.6×10^3 is m^{-2}. The formula relating to turbulent flow, after performing the reduction and substitution as before, is

$$\Delta p_f = \frac{0.3164}{2d_h^{1.25}}h_r\rho v^{0.25}w^{1.75}, \tag{6.231}$$

$$\Delta p_f = 4.88 h_r\rho\, v^{0.25}w^{1.75}, \tag{6.232}$$

where the dimension of factor 4.88 is $m^{-1.25}$.

6.3.5.9 Heat dissipation of radiators by convention with natural air and oil flow

For a given radiator length, the temperature rise and average velocity of the air is calculated for different inlet temperatures and thermal powers, from the thermal balance and from Euler's equation. Assuming an average temperature difference, the mean boundary-layer temperature and the heat transfer coefficient valid at this temperature are calculated. If the assumed average temperature difference was correct, then the product of heat transfer coefficient, convection surface and average temperature difference is equal to the assumed thermal power. The heat dissipation of radiators are then plotted in a coordinate system where the abscissae are the temperature rise values of the radiator wall, $\Delta\vartheta''_{o-a}$. The temperature of ambient air is chosen as parameter.

The temperature rise of the radiator wall is selected as one axis because in calculating the temperature rise of the transformer winding, the former rise is obtained as a quantity which determines the size of the cooling equipment.

6.3.5.10 Calculation of heat transfer diagrams

Symbols (see Fig. 6.82)

ϑ_1 inlet temperature of air entering the cooler,

$\Delta\vartheta$ temperature rise of air in the cooler,

$\vartheta_1 + \dfrac{\Delta\vartheta}{2}$ average temperature of air in the cooler,

$\Delta\vartheta'_{o-a}$ average temperature rise of the radiator wall above average temperature of the air flowing in the cooler,

$\vartheta_m = \vartheta_1 + \dfrac{\Delta\vartheta}{2} + \dfrac{\Delta\vartheta'_{o-a}}{2}$ mean boundary layer temperature of the air flowing in the cooler,

$\Delta\vartheta''_{o-a} = \vartheta_1 + \dfrac{\Delta\vartheta}{2} + \Delta\vartheta'_{o-a} - \vartheta_1 = \dfrac{\Delta\vartheta}{2} + \Delta\vartheta'_{o-a}$ average temperature rise of the radiator wall with respect to that of the inlet cooling air,

Δp the air-side pressure difference in the cooler which can be calculated from the difference of specific gravities of cold and hot air,

h_r the radiator length. The heat dissipation diagrams have been determined for the following 5 different radiator lengths:
$h_{r1} = 0.915$ m, $h_{r2} = 1.19$ m, $h_{r3} = 1.64$ m,
$h_{r4} = 1.965$ m, $h_{r5} = 2.44$ m,

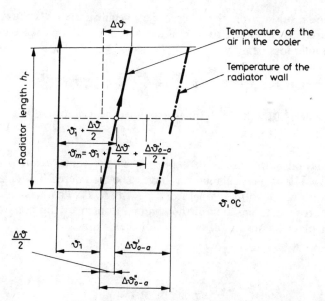

Fig. 6.82. Temperatures and temperature rises in the calculation of heat dissipation of radiators

g	9.81 m s^{-2}, gravitational acceleration,
d_h	the hydraulic diameter, $d_h = 6.44 \times 10^{-2}$ m in the radiators under examination,
A_a	the cross-sectional area of air flow between two radiators. $A_a = 8.43 \times 10^{-3}$ m^2,
A_{ca}	the convection surface of one radiator element. The convection surfaces pertaining to the five different radiator lengths:

$$A_{ca1} = 0.478 \ m^2, \ A_{ca2} = 0.622 \ m^2, \ A_{ca3} = 0.858 \ m^2,$$
$$A_{ca4} = 1.028 \ m^2, \ A_{ca5} = 1.276 \ m^2,$$

P	the thermal power dissipated through convection by one radiator element,
w_a	the flow velocity of the air in the cooler,
α	the air-side heat transfer coefficient,
ρ_1	the density of air entering the cooler at temperature ϑ_1,
ρ_2	the density of air leaving the cooler at temperature $\vartheta_1 + \Delta\vartheta$.
c_p	the specific heat of air at the average temperature $\vartheta_1 + \dfrac{\Delta\vartheta}{2}$ of the air flowing in the cooler,
$\rho_a = \dfrac{\rho_1 + \rho_2}{2}$	the mean air density in the cooler at average temperature $\vartheta_1 + \dfrac{\Delta\vartheta}{2}$ of the air flowing in the cooler,
ν_a	the kinematic viscosity of air at the average temperature $\vartheta_1 + \dfrac{\Delta\vartheta}{2}$ of the air flowing in the cooler,
ϑ_2	the temperature of the air leaving the cooler, $\vartheta_2 = \vartheta_1 + \Delta\vartheta$,
T_1	the thermodynamic temperature of the air entering the cooler, $T_1 = \vartheta_1 + 273 \ °C$,
T_2	the thermodynamic temperature of the air leaving the cooler, $T_2 = \vartheta_2 + 273 °C = \vartheta_1 + \Delta\vartheta + 273 °C$.

Pairs of equations for determining the temperature rise $\Delta\vartheta$
of the air and air velocity w_a

The pressure resulting from the difference of specific gravities is equal to the sum of the pressure corresponding to the outlet loss and of the pressure drop due to friction. The inlet loss being very low is neglected:

$$\Delta p = \frac{\rho_1 - \rho_2}{2} h_r g = \rho_a \frac{w_a^2}{2} + \frac{0.3164}{2 d_h^{1.25}} h_r \rho_a \nu_a^{0.25} w_a^{1.75} \ . \tag{6.233}$$

The heat transferred by the radiator is expanded heating of the air:

$$P = c_p \rho_a w_a \, A_a \Delta\vartheta \ . \tag{6.234}$$

Let the mean and outlet values be expressed by the values relating to the inlet conditions. Two basic physical relations have to be considered. The density of air is inversely proportional to its temperature: $\rho_2 = \rho_1 T_1/T_2$. The mass of air passing through the cross-sectional area of the channel formed by the radiators and the lateral cover plate is constant at all levels of the channel:

$$A_a w_1 \rho_1 = A_a w_2 \rho_2 \tag{6.235}$$

from which

$$w_2 = w_1 \frac{\rho_1}{\rho_2} = w_1 \frac{T_2}{T_1} \ . \tag{6.236}$$

478

Further,

$$\frac{\rho_1-\rho_2}{2}=\frac{\rho_1}{2}\left(1-\frac{T_1}{T_2}\right)=\frac{\rho_1}{2}\left(1-\frac{273\,°C+\vartheta_1}{273\,°C+\vartheta_1+\varDelta\vartheta}\right), \qquad (6.237)$$

$$\rho_a=\frac{\rho_1+\rho_2}{2}=\frac{\rho_1}{2}\left(1+\frac{T_1}{T_2}\right)=\frac{\rho_1}{2}\left(1+\frac{273\,°C+\vartheta_1}{273\,°C+\vartheta_1+\varDelta\vartheta}\right) \qquad (6.238)$$

and

$$w_a=\frac{w_1+w_2}{2}=\frac{w_1}{2}\left(1+\frac{T_2}{T_1}\right)=\frac{w_1}{2}\left(1+\frac{273\,°C+\vartheta_1+\varDelta\vartheta}{273\,°C+\vartheta_1}\right). \qquad (6.239)$$

At normal atmospheric pressure, the density of dry air at temperature ϑ_1 is

$$\rho_1=1.252\frac{273\,°C}{273\,°C+\vartheta_1}\ \text{kg}\,\text{m}^{-3}. \qquad (6.240)$$

The specific heat of air at temperature $\left(\vartheta_1+\dfrac{\varDelta\vartheta}{2}\right)$ is

$$c_p=1006.67\ \text{W}\,\text{s}\,\text{kg}^{-1}\,°C^{-1}+0.133\ \text{W}\,\text{s}\,\text{kg}^{-1}\,°C^{-2}\left(\vartheta_1+\frac{\varDelta\vartheta}{2}\right). \qquad (6.241)$$

The viscosity of air at temperature $\left(\vartheta_1+\dfrac{\varDelta\vartheta}{2}\right)$ is

$$v_a=15.7\times10^{-6}\ \text{m}^2\,\text{s}^{-1}+0.1\times10^{-6}\ \text{m}^2\,\text{s}^{-1}\,°C^{-1}\left(\vartheta_1+\frac{\varDelta\vartheta}{2}-20\,°C\right). \qquad (6.242)$$

Example 6.19. Let us determine, for given values of h_r and ϑ_1, the values of $\varDelta\vartheta$ and w_1 belonging together and valid in the case of different magnitudes of P.

(a) If $h_{r2}=1.19\,\text{m}$, $\quad\vartheta_1=30\,°C$, $\quad P=200\,\text{W}$:

$$\frac{1.13}{2}\left(\frac{\varDelta\vartheta}{303\,°C+\varDelta\vartheta}\right)1.19\times9.81=\frac{1.13}{2}\left(\frac{2\times303\,°C+\varDelta\vartheta}{303\,°C+\varDelta\vartheta}\right)\times\frac{w_1^2}{4\times2}\left(\frac{2\times303\,°C+\varDelta\vartheta}{303\,°C}\right)^2\text{s}^2\,\text{m}^{-2}+$$

$$+\frac{0.3164}{2(6.44\times10^{-2})^{1.25}}\times1.19\frac{1.13}{2}\left(\frac{2\times303\,°C+\varDelta\vartheta}{303\,°C+\varDelta\vartheta}\right)\left\{\left[15.7+\frac{0.1}{°C}\left(10\,°C+\varDelta\vartheta\right)\right]\times\right.$$

$$\left.\times10^{-6}\right\}^{0.25}\left[\frac{w_1}{2}\frac{2\times303\,°C+\varDelta\vartheta}{303\,°C}\right]^{1.75}\text{m}^{-1.75}\,\text{s}^{1.75}.$$

$$200=\left[1006.67+\frac{0.133}{°C}\left(30\,°C+\frac{\varDelta\vartheta}{2}\right)\right]\varDelta\vartheta\frac{1.13}{2}\times$$

$$\times\left(\frac{2\times303\,°C+\varDelta\vartheta}{303\,°C+\varDelta\vartheta}\right)\frac{w_1}{2}\left(\frac{2\times303\,°C+\varDelta\vartheta}{303\,°C}\right)8.43\times10^{-3}\,\text{m}^{-1}\,\text{s}\,°C^{-1}.$$

From the above pair of equations, for the given values of ϑ_1, h_r and P, the values of $\varDelta\vartheta$ and w_1 belonging together have been obtained.

(b) Let us now substitute the value of P given under (a) and the value of w_1 obtained from the above set of equations into the heat transfer equation relating to the air side. This air-side heat transfer equation is

$$P=\alpha A_{ca}\varDelta\vartheta'_{o-a}. \qquad (6.243)$$

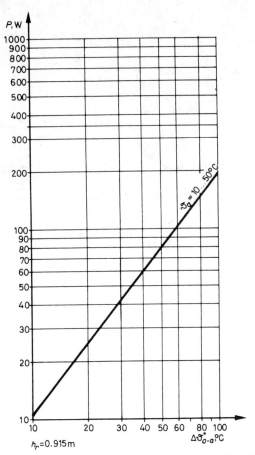

Fig. 6.83. Heat dissipation by convection of radiators of length $h_r = 0.91$ m. Natural air cooling

Substituting the value of surface heat transfer coefficient valid for laminar flow, the following equation is obtained:

$$P = 0.85 \times 0.74 d_h^{-0.5} g^{0.1} A_{ca} (\rho_m c_{pm})^{0.3} \times \left(\frac{\beta_m}{\nu_m}\right)^{0.1} \lambda_m^{0.7} w_a^{0.2} (\Delta\vartheta'_{o-a})^{1.1} . \qquad (6.244)$$

Here, the temperature-dependent physical quantities should be substituted by their values valid at the mean boundary-layer temperature ϑ_m.

The further course of the calculation is as follows: Select a value for $\Delta\vartheta'_{o-a}$ tentatively and substitute it into the equation. The mean boundary-layer temperature is then obtained from ϑ_1 chosen further above, $\Delta\vartheta$ that has resulted from the calculation and $\Delta\vartheta'_{o-a}$ chosen now, viz.:

$$\vartheta_m = \vartheta_1 + \frac{\Delta\vartheta}{2} + \frac{\Delta\vartheta'_{o-a}}{2} .$$

From the equation relating to the heat transfer the values of P and $\Delta\vartheta'_{o-a}$ belonging together are obtained. As the final step, the values of P are plotted in the function of $\Delta\vartheta''_{o-a} = \Delta\vartheta'_{o-a} + \dfrac{\Delta\vartheta}{2} .$

480

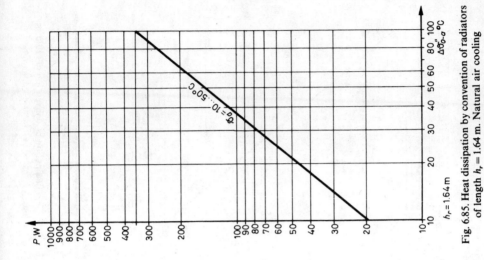

Fig. 6.85. Heat dissipation by convention of radiators
of length $h_r = 1.64$ m. Natural air cooling

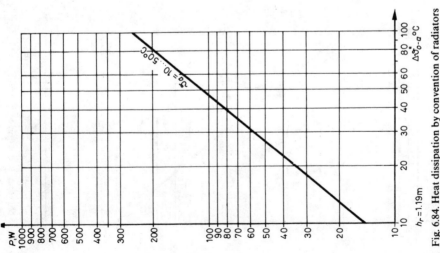

Fig. 6.84. Heat dissipation by convention of radiators
of length $h_r = 1.19$ m. Natural air cooling

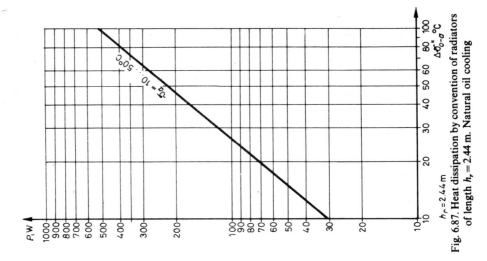

Fig. 6.87. Heat dissipation by convention of radiators of length $h_r = 2.44$ m. Natural oil cooling

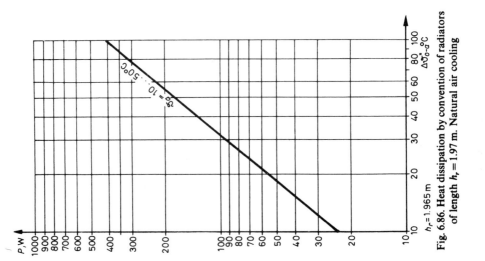

Fig. 6.86. Heat dissipation by convention of radiators of length $h_r = 1.97$ m. Natural air cooling

482

From the first pair of equations the results $\Delta\vartheta = 29.9\,°C$ and $w_1 = 0.69\ \mathrm{m\ s^{-1}}$ have been obtained. Let us choose $\Delta\vartheta'_{o-a} = 65.9\,°C$. With this value

$$\vartheta_m = \vartheta_1 + \frac{\Delta\vartheta}{2} + \frac{\Delta\vartheta'_{o-a}}{2} = \left(30 + \frac{29.9}{2} + \frac{65.9}{2}\right)°C = 77.9\,°C.$$

The convection surface of the radiator element of $h_{r2} = 1.19\ \mathrm{m}$ is, as already seen, $A_{ca2} = 0.622\ \mathrm{m^2}$. The heat transfer equation is

$$200 = 0.85 \times 0.74(6.44 \times 10^{-2})^{-0.5} \times 9.81^{0.1} \times 0.622 \times (0.974 \times 1019)^{0.3} \times$$

$$\times \left(\frac{\frac{1}{351}}{21.5 \times 10^{-6}}\right)^{0.1} \times (2.92 \times 10^{-2})^{0.7} \left(\frac{0.69}{2} \times \frac{656}{303}\right)^{0.2} \times 65.9^{1.1}$$

If the quantity on the right side of the equation is equal or close to 200, then the value of $\Delta\vartheta'_{o-a}$ has been selected properly. Otherwise, a trial should be made with a new value of $\Delta\vartheta'_{o-a}$. Let us assume that the first attempt has proved correct. Thereby the required pair of values is found:

$$\Delta\vartheta''_{o-a} = \Delta\vartheta'_{o-a} + \frac{\Delta\vartheta}{2} = \left(65.9 + \frac{29.9}{2}\right)°C = 80.76\,°C$$

and with this average temperature rise the heat dissipated by the radiator is equal to 200 W.

Results obtained by the method described and processed by a computer are shown in Figs 6.83 to 6.87, which apply to the ϑ_1 range of 10 to 50 °C occurring in practice.

6.3.5.11 Fin efficiency

The radiators represented in Fig. 6.71 contain 5 oil pipes. Each oil pipe shown in Fig. 6.69 has two fins. Let us determine the fin efficiency for one fin.

Dimensions according to the diagram are $\delta = 2.5 \times 10^{-3}\ \mathrm{m}$, $l = 6 \times 10^{-3}\ \mathrm{m}$. The radiator is made of cold-rolled deep-drawn steel plates having a thermal conductivity of

$$\lambda = 45.4\ \mathrm{W\ m^{-1}\,°C^{-1}}.$$

The fin efficiency is

$$\eta_f = \frac{\mathrm{th}(ml)}{ml}, \tag{6.245}$$

where

$$m = \sqrt{\frac{2\alpha_a}{\lambda\delta}}, \tag{6.246}$$

with α_a being the air-side heat transfer coefficient.

Substituting the fin data into the equation for m:

$$m = \alpha_a^{0.5} \sqrt{\frac{2}{45.4 \times 2.5 \times 10^{-3}}} = \alpha_a^{0.5}\,4.2,$$

where the dimension of factor 4.2 is $\mathrm{W^{-0.5}\,°C^{0.5}}$.

$$ml = \alpha_a^{0.5}\,4.2 \times 6 \times 10^{-3} = \alpha_a^{0.5}0.0252,$$

where the dimension of factor 0.0252 is $\mathrm{m\ W^{-0.5}\,°C}$.

31*

Example 6.20. Let α_a be equal to 36 W m^{-2} °C^{-1} and with this value $ml = 36^{0.5} \times 0.0252 = 0.151$. The fin efficiency will be

$$\eta_f = \frac{\text{th}\,(ml)}{ml} = \frac{\text{th}\,0.151}{0.151} = \frac{0.1493}{0.151} \approx 1.$$

The fin efficiency is close to unity even for relatively high values of α. This means that application of fins is a very efficient way to increase the air-side surface area.

It is not worthwhile splitting up the air-side surface of radiators into one part multiplied by the fin efficiency, and into another not so multiplied because the fin efficiency in the case of radiators is, as shown, very close to unity.

6.3.5.12 Overall thermal conductance

Let the air-side surface area of the radiator be A_a, that of the oil-side A_o, the area pertaining to the half wall thickness of the radiator A_δ, the oil-side heat transfer coefficient α_o, the air side heat transfer coefficient α_a, the heat conductance of radiator material λ and the wall thickness of the radiator δ. The reciprocal of the overall thermal conductance k_{A_a} calculated for the air-side convection surface area is

$$\frac{1}{k_{A_a}} = \frac{1}{\alpha_o \dfrac{A_o}{A_a}} + \frac{1}{\lambda \dfrac{A_\delta}{\delta A_a}} + \frac{1}{\alpha_a}. \tag{6.247}$$

The radiators shown in Fig. 6.71 differ only in their lengths. Let the length of the radiator be denoted by h_r. For any radiator (see Table 6.9):

$$A_o = 0.403 h_r,$$
$$A_a = 0.523 h_r.$$

The dimension of the multiplying factors of h_r is m.

$$\frac{A_o}{A_a} = \frac{0.403 h_r}{0.523 h_r} = 0.77.$$

According to Fig. 6.69

$$A_\delta \approx A_o = 0.77 A_a, \quad \text{so} \quad \frac{A_\delta}{A_a} = 0.77.$$

The wall thickness of the radiator is $\delta = 1.25 \times 10^{-3}$ m and its heat conductance is $\lambda = 45.4$ W m^{-1} °C^{-1}. With these values the denominator of the second term in the expression of the reciprocal of the overall thermal conductance is

$$\frac{\lambda}{\delta} \frac{A_\delta}{A_a} = \frac{45.4}{1.25 \times 10^{-3}} 0.77 \text{ W m}^{-2}\text{°C}^{-1} = 2.797 \times 10^4 \text{ W m}^{-2}\text{°C}^{-1}.$$

Hence, the second term is

$$\frac{1}{\lambda \dfrac{A_\delta}{\delta A_a}} = \frac{1}{2.797 \times 10^4} \text{ W}^{-1}\text{m}^2\text{°C} \approx 0$$

484

and it will be neglected further on. The overall thermal conductance related to the air-side convection surface, for the radiators of Fig. 6.71, is thus

$$\frac{1}{k_{A_a}} = \frac{1}{0.77\alpha_o} + \frac{1}{\alpha_a}.$$ (6.248)

Let the power transferred by one radiator through convection be denoted by P. The entire temperature drop, if the logarithmic-mean temperature difference is $\Delta\vartheta_{o-a}$, is

$$\Delta\vartheta_{o-a} = \frac{P}{A_a}\frac{1}{k_{A_a}} = \frac{q_a}{k_{A_a}},$$ (6.249)

where q_a is the heat flux per unit transfer area related to the air-side surface area. The entire temperature drop consists of two parts, i.e. of the surface temperature drop on the oil side:

$$\Delta\vartheta_{os} = \frac{q_a}{0.77\alpha_0},$$ (6.250)

and of that on the air side:

$$\Delta\vartheta_{as} = \frac{q_a}{\alpha_a}.$$ (6.251)

Thus the entire temperature drop will be

$$\Delta\vartheta_{o-a} = \Delta\vartheta_{os} + \Delta\vartheta_{as}.$$ (6.252)

6.3.6 Heat transfer of forced air cooled radiators

6.3.6.1 Air-side heat transfer coefficient

In the case of forced air cooling the similarity equation

$$Nu_f = 0.23 Re_f^{0.8} \times Pr_f^{0.4}\varepsilon_1$$ (6.253)

may be used for calculating the surface heat transfer coefficient. It is valid for the forced turbulent flow of liquids or gases, e.g. air in ducts. In the formula the physical characteristics should be substituted by values referred to the mean temperature ϑ_f. ε_1 is a correction factor depending on the ratio length of duct length/hydraulic diameter and on the value of Re_f. The numerical values of correction factor ε_1 are contained in Table 6.19. In the table the duct length is taken equal to the radiator

Table 6.19

Values of factor ε_1 in the formula for heat transfer coefficient, as a function of the ratio radiator length/hydraulic diameter l_r/d_h, in the range of $Re_f = 1 \times 10^4 - 2 \times 10^4$

l_r/d_h, m/m	1	2	5	10	15	20	30	40	50
ε_1	1.65	1.50	1.34	1.23	1.117	1.133	1.07	1.03	1.00

length, h_r. With cooling by ventilated air the usual air velocities are always higher than 1m s^{-1}, and the hydraulic diameter of the flow duct is of a value to make Re_f exceed 2200, so that the flow qualifies as turbulent.

From the similarity equation, the heat transfer coefficient in explicit form is as follows:

$$\alpha_a = 0.023 \, \lambda^{0.6} \left(\frac{c_p \rho}{v}\right)^{0.4} \frac{\varepsilon_1}{d_h^{0.2}} \, w^{0.8} . \qquad (6.254)$$

6.3.6.2 Hydraulic diameter

For ventilation from below, the hydraulic diameter d_h is the hydraulic diameter of the flat flow duct formed by two adjoining radiator elements and a lateral cover plate.

For ventilation from the side, the hydraulic diameter is the free cross-sectional hydraulic diameter which is the free cross-sectional area A_a available for the flow of air between two radiator elements, divided by the convection surface area A_c belonging to A_a, multiplied by four times the path length L of the flow:

$$d_h = 4L \frac{A_a}{A_c} . \qquad (6.255)$$

Where several radiator elements are located one behind another, both convection area and flow path length must of course be multiplied accordingly.

6.3.6.3 Forced air cooled radiators as oil–air heat exchangers

The flow of oil may be natural, i.e. a gravitational flow, or a forced flow kept in circulation by pumping.

Consider the first case, and suppose a constant rate of loss is to be dissipated. With growing air flow velocity, i.e. with increasing air-side heat transfer coefficient, the average temperature rise of the radiator drops rapidly at first, then the rate of decrease becomes steadily slower, until a steady-state temperature rise value is attained which corresponds to the increased heat transfer coefficient. Together with the reduced average temperature rise of the radiator, the area enclosed by the rising-falling temperature curve plotted in the γ–h coordinate system expands, i.e. the circulating pressure and with it the buoyancy, increases. The pressure rise is due partly to the increasing difference between the heights of the centres of heating and cooling as a consequence of the ascending centre of heating in the radiators (caused by the improved cooling), and partly to the increased difference between temperatures of the oil entering and leaving the windings, defined by the oil flow velocity developing spontaneously in the windings, which also increases due to the higher oil viscosity associated with the decreased average oil temperature. The increased pressure area of the diagram is absorbed by the higher flow resistance of the oil kept in circulation at increased velocity.

With very high air velocities, i.e. with very large air-side heat transfer coefficients, the oil is caused to cool off in the uppermost sections of the radiators (Fig. 6.88). In such a marginal case it will be observed that the largest part of the radiator serves as an oil pipe, and a very small fraction of the surface is sufficient for the required dissipation of heat. The difference between the temperatures of the oil entering and leaving the winding develops spontaneously. The pressure area is always large enough to cover the pressure drop resulting from the flow loss of the oil.

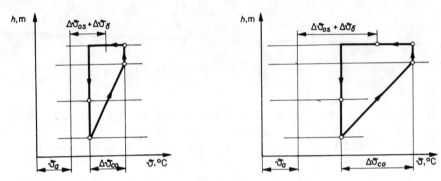

Fig. 6.88. Warming and cooling of oil with infinitely large air-side surface heat transfer coefficient, when dissipating thermal power P

Fig. 6.89. Effect of infinitely large air-side heat transfer coefficient on the $\vartheta - h$ curve when dissipating a thermal power $P' > P$

In the case examined the loss to be dissipated was assumed constant, and only the air-side heat transfer coefficient has been increased. Let us now examine the case when the thermal power to be dissipated is increased simultaneously with an increasing air-side heat transfer coefficient so as to keep the average temperature rise of the oil with respect to the temperature of the cooling air constant. With increasing power to be dissipated the temperature difference between the oil entering and leaving the winding will increase. This is one reason for growth of the pressure area, the other being the rise of the centre of heating to a higher level. The flow velocity of the circulating oil increases, and the larger pressure area is absorbed by the flow resistances. The viscosity of the oil does not change much in the circuit, since the average temperature rise of the oil has been kept constant. There is a theoretical upper limit imposed on the possible increase of thermal power. If an infinitely high air-side heat transfer coefficient is assumed, then the average temperature rise of the oil in the cooler is equal to the sum of the cooling air temperature, the oil-side temperature drop and the temperature drop occurring in the radiator wall (see Fig. 6.89). The radiators again play the part of an oil duct in this marginal case, since an infinitesimally small area at the top part of the radiator will be capable of dissipating the heat, given the assumed infinitely high heat transfer coefficient.

In the preceding discussion the marginal cases attained by increasing the air-side heat transfer coefficient have been investigated. As apparent also from Figs 6.88 and

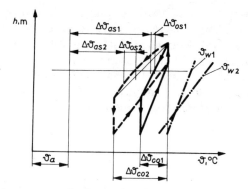

Fig. 6.90. $\vartheta - h$ curves occurring in practice with natural and forced air cooling

6.89, the pressure areas cannot be increased beyond all limits. This means that the oil velocity does not increase proportionally with improvement of the external cooling, but only to about 2.5 to 3 times the oil velocity obtainable with natural cooling, whatever the improvement of external cooling.

In practice, we content ourselves with increasing the air-side heat transfer to a level at which the heat dissipation obtained by forced ventilation is 2.3 to 3 times higher than that of natural air cooling, (see Fig. 6.90).

6.3.6.4 Forced oil-side flow

When applying forced oil-side flow, the flow will become turbulent in the radiators. The oil-side heat transfer coefficient will increase, and the surface temperature drop decreases. Let us assume that, by increasing the air velocity, we have succeeded in achieving a very high air-side surface heat transfer coefficient. By changing over to forced flow on the oil side, the oil-side temperature drop is also reduced. Assuming infinitely high heat transfer coefficients on both sides, the average temperature of the cooler will be the sum of the temperature drop in the radiator wall and the temperature of the air (see Fig. 6.91).

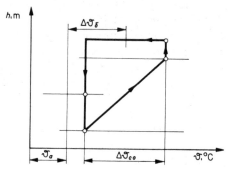

Fig. 6.91. Effect of infinitely large oil-side and air-side heat transfer coefficients on temperature drops

If forced flow of the oil is extended to the cooler only, i.e. the flow of oil heated in the winding is still brought about by gravitation, then the difference between the temperatures of the oil entering at the bottom and leaving at the top of the winding will with good approximation develop spontaneously i.e. the difference will increase with increasing heat flux per unit transfer area and vice-versa.

Pumping of the oil and reduction of the temperature of the oil entering the cooler go together, since the oil leaving the cooler becomes mixed with warmer oil leaving the winding. For safety reasons, the heat flux per unit transfer area of the winding should not be selected at a value higher than that considered permissible in the case of natural oil flow, because, as mentioned above, the difference between the temperatures of the oil entering and leaving the winding develops spontaneously, corresponding to the heat flux per unit transfer area and the flow resistance of the windings. In the case of radiator-type cooling, application of forced oil flow is only justifiable if, for some reason, the centre of cooling of the radiators cannot be raised, i.e. the pressure area cannot be increased. Forced oil flow, combined with good air-side heat transfer, allows a smaller-size cooler for a transformer of a given equal rating, than a system without oil-side forced flow.

If the oil flow is directed within the winding, the situation changes fundamentally. The temperature rise of the winding will no longer develop spontaneously, but it will become a function of the oil velocity.

6.3.6.5 Temperature difference between oil entering the winding at the bottom and leaving it at the top

Several times in the preceding discussion, mention has been made of the spontaneously developing difference between the temperatures of the oil entering at the bottom of the winding and leaving at its top. This concept means the temperature difference which develops when the hydraulic resistance of the cooling circuit attached to the winding is zero. In reality, for a given cooling arrangement, and in the case of a given heat flux per unit transfer area the temperature difference developing may be either smaller or bigger than that developing spontaneously. It is smaller if the pressure area is increased by raising the centre of cooling, and it is bigger if such an increase is insufficient to overcome the pressure increase caused by the hydraulic resistances extraneous to the windings.

For a transformer of given geometry, the characteristic curve $\Delta\vartheta_{co}=f(q)$ can be plotted as a function of the loss to be dissipated and with the average temperature rise of the cooler kept constant. The slope of the curve at any arbitrary point gives the increment $d\Delta\vartheta_{co}$ pertaining to a unit increase of thermal load. The dimension of $\dfrac{d\Delta\vartheta_{co}}{dq}$ is $°C\, m^2\, W^{-1}$, i.e. that of thermal resistance. During the measurement the average temperature rise of the cooler can be maintained at a constant value e.g. by intensifying the ventilation of the cooler.

6.3.6.6 Air-side heat transfer coefficient with forced air-cooling. Design data

In the equation given for the air-side heat transfer coefficient, the hydraulic diameter d_h and a multiplication factor ε_1 appear.

For the radiators of Fig. 6.71, with ventilation from below, the value of hydraulic diameter is $d_h = 6.44 \times 10^{-2}$ m. Calculation of the hydraulic diameter for this concrete case has already been performed. As seen from Table 6.9, five different radiator lengths are used in practice.

The values of ratios $l/d_h = h_r/d_h$ for the five radiator lengths and the relevant values of ε_1 are contained in Table 6.20.

In Table 6.21 the values of factors A and B facilitating calculation of the air-side heat transfer coefficient are given as a function of mean boundary-layer temperature ϑ_m of the air:

$$A = 0.023 \, \lambda^{0.6} \left(\frac{c_p \rho}{v} \right)^{0.4}, \tag{6.256}$$

$$B = \frac{A}{d_h^{0.2}} = \frac{A}{(6.44 \times 10^{-2})^{0.2} m^{0.2}} = 1.73 \, A \, \mathrm{m}^{-0.2}. \tag{6.257}$$

Table 6.20

Values of factor ε_1 for the radiators of Table 6.9 as a function of radiator length/hydraulic diameter h_r/d_h

h_r, m	0.915	1.19	1.64	1.965	2.44
h_r/d_h, m/m	14.2	18.5	25.5	30.5	37.9
ε_1	1.17	1.14	1.10	1.07	1.04

Table 6.21

Values of factors A and B in the formula for surface heat transfer coefficient as a function of mean boundary-layer temperature ϑ_m

Mean boundary-layer temperature ϑ_m, °C	A W $s^{0.8}$ $m^{-2.6}$ °C^{-1}	B W $s^{0.8}$ $m^{-2.8}$ °C^{-1}
10	3.65	6.325
20	3.55	6.150
30	3.49	6.050
40	3.42	5.925
50	3.35	5.800
60	3.29	5.700
70	3.24	5.620
80	3.19	5.520
90	3.13	5.420
100	3.09	5.360

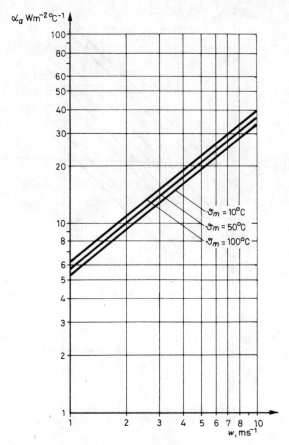

Fig. 6.92. Air-side surface heat transfer coefficient with forced air cooling

With forced air cooling using air blown from below, for the radiator arrangement shown in Fig. 6.71, the air-side heat transfer coefficient is

$$\alpha_a = B\varepsilon_1 w^{0.8} , \tag{6.258}$$

where w is the flow velocity of air at the half height of the radiator, i.e. the mean air velocity. The values of α_a, for the case of $\varepsilon_1 = 1$ and with the mean boundary layer temperature of the air used as parameters, are given in Fig. 6.92, as a function of air velocity. The value of α_a pertaining to ϑ_m and w read from the diagram should be multiplied by the factor ε_1 for the radiator length concerned and taken from Table 6.20.

6.3.6.7 Air-side surface temperature drop for the case of forced air cooling

In the preceding paragraph the value of air-side heat transfer coefficient has been determined as

$$\alpha_a = B\varepsilon_1 w^{0.8} .$$

491

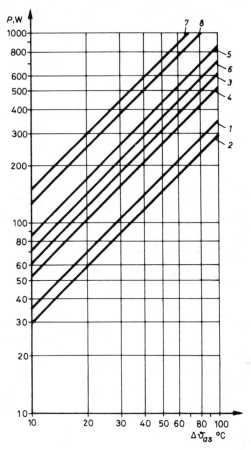

Fig. 6.93. Magnitude of air-side surface temperature drop in forced air cooled radiators ventilated from below (see Table 6.22)

The air-side surface temperature drop is given by Eqn. (6.251) as

$$\Delta \vartheta_{as} = \frac{q_a}{\alpha_a}.$$

Let us calculate and plot on a log–log chart the thermal power dissipated by convection, as a function of surface temperature drop, for two different radiator lengths, for two air velocities and two mean boundary layer temperatures as parameters. Let one of the radiators be of length $h_{r1} = 0.915$ m, and the other of $h_{r5} = 2.44$ m. From the relation $A_{ca} = 0.523$ m h_r, the air-side surface area of the shorter radiator is $A_{ca1} = 0.478$ m^2 and that of the longer is $A_{ca5} = 1.276$ m^2. Assume, for the first case, $\vartheta_m = 10\ ^{\circ}$C and, for the second, $\vartheta_m = 100\ ^{\circ}$C. Let the air velocity be, in the first case, $w = 1$ m^{-1} and, in the second case, $w = 2$ m s^{-1}. The diagrams plotted from the results of calculation are given in Fig. 6.93 and the values of parameters in Table 6.22.

492

Table 6.22

Air-side surface temperature drop in forced air cooled
radiators ventilated from below.
Characteristic figures relating to diagrams of Fig. 6.93

No.	Radiator length h_r, m	Air velocity w, m s^{-1}	Mean boundary-layer temperature ϑ_m, °C
1	0.915	1	10
2	0.915	1	100
3	2.44	1	10
4	2.44	1	100
5	0.915	2	10
6	0.915	2	100
7	2.44	2	10
8	2.44	2	100

6.3.6.8 Temperature rise of the air in forced air cooled radiators

Air entering at the bottom of the radiators with inlet temperature ϑ_1, inlet velocity w_1, density ρ_1 and specific heat c_p, acquires by convection a quantity of power P, warms up, and leaves at the top of the radiators with outlet temperature ϑ_2, density ρ_2 and velocity w_2. Let the difference between inlet and outlet temperatures of the air be denoted by $\vartheta_2 - \vartheta_1 = \Delta\vartheta$, the cross-sectional area between the two radiator elements and lateral cover plates by A_a, the mean density by ρ_a and the mean air velocity by w_a. Let us, further, suppose that the mass of air flowing through any cross-sectional area of the duct per unit time is constant (see relation (6.235)):

$$w_1\rho_1 = w_a\rho_a = w_2\rho_2 .$$

From the energy balance,

$$P = c_p\rho_a w_a A_a \Delta\vartheta = c_p\rho_1 w_1 A_a \Delta\vartheta . \tag{6.259}$$

At normal atmospheric pressure the density of dry air is

$$\rho_1 = 1.252 \frac{273\ °C}{273\ °C + \vartheta_1}\ \text{kg m}^{-3} .$$

The specific heat of air at temperature $\vartheta_1 + \dfrac{\Delta\vartheta}{2}$ is

$$c_p = 1006.67 + 0.133\left(\vartheta_1 + \frac{\Delta\vartheta}{2}\right) .$$

In the formula the dimension of the first term is W s kg^{-1} °C^{-1}, and that of factor 0.133 W s kg^{-1} °C^{-2}. For the values of ϑ_1 in the range 20 to 40 °C $c_p \approx 1012$ W s kg^{-1} °C^{-1}.

The cross-sectional area of the flow duct, in the case of the radiators of Fig. 6.71, is $A_a = 8.43 \times 10^{-3} \, m^2$. Let us substitute into the expression for P the air and flow duct data:

$$P = \left[1006.67 \text{ W s kg}^{-1}\,°C^{-1} + 0.133 \text{ W s kg}^{-1}\,°C^{-2} \left(\vartheta_1 + \frac{\varDelta\vartheta}{2} \right) \right] \times$$

$$\times 1.252 \text{ kg m}^{-3} \frac{273\,°C}{273\,°C + \vartheta_1} \times 8.43 \times 10^{-3}\,m^2\, w_1\, \varDelta\vartheta.$$

Replacing the expression in square brackets by $c_p = 1012$ W s kg^{-1} valid for the range 20 to 40 °C, after rearrangement

$$P = 2916 \frac{\varDelta\vartheta}{273\,°C + \vartheta_1}\, w_1 \tag{6.260}$$

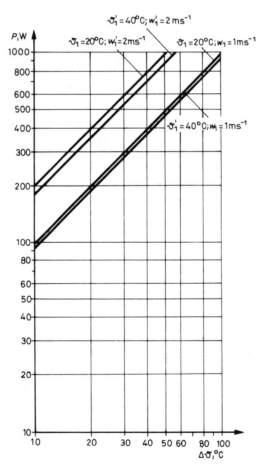

Fig. 6.94. Warming of air in forced air cooled radiators

is obtained, where P is the heat transferred through convection by one radiator element, ϑ_1 is the inlet air temperature, $\Delta\vartheta$ is the temperature rise of the air due to the heat dissipated through the radiators, and w_1 is the flow velocity of air between the radiators. The dimension of factor 2916 is $W\,s\,m^{-1}$.

Temperature rise figures calculated from the above equation are shown in Fig. 6.94, for the cases $w_1 = 1\,m\,s^{-1}$, $w_1' = 2\,m\,s^{-1}$, $\vartheta_1 = 20\,°C$ and $\vartheta_1' = 40\,°C$.

6.3.7 Heat dissipation through radiation

6.3.7.1 The Stefan–Boltzmann law

According to the Stefan–Boltzmann law, the quantity of heat flux of a black body of temperature T_1, K, delivered by radiation to surroundings of temperature T_2, K, per unit area is

$$q_r = 5.7 \times 10^{-8}\varepsilon(T_1^4 - T_2^4). \tag{6.261}$$

The factor 5.7×10^{-8} is the radiation constant of the ideal black body; its dimension is $W\,m^{-2}\,K^{-4}$.

A multiplying coefficient ε appearing in the formula as the degree of blackness or relative radiation power, being the ratio of the specific heat power of the body in question to that of the ideal black body, with both bodies being at equal

Table 6.23

Total degree of blackness of some materials

Material	Total degree of blackness
Silver, polished	0.02
Zinc, polished	0.05
Aluminium, polished	0.08
Nickel	0.12
Copper	0.15
Cast iron	0.25
Aluminium paint	0.55
Brass, polished	0.60
Copper, oxidized	0.60
Steel, oxidized	0.70
Bronze paint	0.80
Black varnish paint	0.90
Plaster, rough, calcareous concrete	0.91
White varnish paint	0.95
Green paint	0.95
Grey paint	0.95
Soot	0.95
Ideal black body	1.00

temperature. Thus, ε is a dimensionless number with a value between 0 and 1. The numerical value of ε depends on the material, surface quality, colour and temperature of the body concerned. Degrees of blackness of a few materials are given in Table 6.23 for the temperature range 0 to 100 °C encountered under normal operating condition of transformers. As it is apparent from the data of the table, the power radiated by surfaces coated with aluminium paint is 42 per cent less per unit area than for those provided with the usual green or grey coating, assuming the same temperature difference between radiating surface and ambient atmosphere.

For engineering calculations the equation may be simplified by multiplying the difference in brackets by a factor 10^{-8}, and since the equation is mainly applied for calculating the thermal power radiated by radiators and tanks to their surroundings of temperature ϑ_a, instead of T_1 the value 273 °C $+\vartheta_a+\Delta\vartheta$ is substituted and instead of T_2 the value 273 °C $+\vartheta_a$. Thus, the Stefan–Boltzmann law will go over into the following modified form:

$$q_r = 5.7\varepsilon\left[\left(\frac{273\,°C + \vartheta_a + \Delta\vartheta}{100}\right)^4 - \left(\frac{273\,°C + \vartheta_a}{100}\right)^4\right]. \tag{6.262}$$

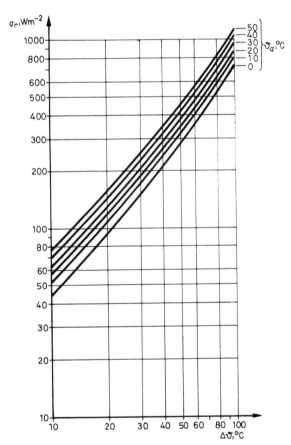

Fig. 6.95. Power density radiated as a function of temperature rise at different ambient temperatures

Table 6.24

Power density q_r radiated into surroundings of temperature ϑ_a
as a function of temperature rise $\Delta\vartheta$ with $\varepsilon = 0.95$

Temperature of surroundings ϑ_a, °C	Temperature rise $\Delta\vartheta$, °C									
	10	20	30	40	50	60	70	80	90	100
	Radiated power density q_r, W m^{-2}									
0	43.5	95.2	152.5	215.8	285.6	361.9	445.6	536.8	636.3	744.3
10	51.6	109.0	172.3	242.0	318.4	402.0	493.3	592.8	700.7	817.6
20	57.3	120.6	190.4	266.7	350.3	441.6	541.1	649.1	765.9	892.5
30	63.2	113.0	209.3	292.9	384.2	483.8	591.7	708.5	835.1	971.8
40	69.7	146.0	229.7	320.9	420.5	528.4	645.2	771.8	908.5	1055.7
50	76.3	159.9	251.1	350.7	458.7	575.5	702.0	838.8	986.0	1144.1

The dimension of factor 5.7 is W m^{-2} °K^{-4}. The numerical values of q_r for $\varepsilon = 0.95$ and for $\vartheta_a = 0$, 10, 20, 30, 40, 50 °C are shown as a function of $\Delta\vartheta$ by Fig. 6.95 and Table 6.24.

6.3.7.2 The radiating surface

The surface to be considered as radiating is that of the external covering of the transformer, independent of the degree of articulation of this surface. The area corresponding to the vertical projection of the transformer radiates with the maximum oil temperature, ϑ_{om}, whereas the four enclosing side walls of the transformer radiate with the average oil temperature, ϑ_{oac}. The bottom of the transformer is not taken into account as a radiating surface. Strictly speaking, the area of the radiating surface is equal to that of the external enclosure only in the case of an ideal or close-to-ideal black body. This condition is approximated by the green or machine-grey colour but, e.g. in the case of an aluminium coating, the surface of radiators should be considered with a reduced area, due to the radiation and reflection of surfaces facing each other.

6.3.7.3 Heat exchange by radiation between parallel surfaces

Assume two parallel surfaces are located in a distance from each other to allow neglect of heat radiated sidewise. The power per unit area transmitted from surface 1 of absolute temperature T_1 K to surface 2 of absolute temperature T_2 K by means of radiation is

$$q_{12} = 5.7 \times 10^{-8} \varepsilon_n (T_1^4 - T_2^4), \tag{6.263}$$

where the dimension of factor 5.7×10^{-8} is W m^{-2} K^{-4} and ε_n is the reduced degree of blackness, calculated from the degrees of blackness ε_1 and ε_2 of surfaces 1 and 2 by means of the relation

$$\varepsilon_n = \frac{1}{\dfrac{1}{\varepsilon_1} + \dfrac{1}{\varepsilon_2} - 1}. \tag{6.264}$$

6.3.7.4 Heat exchange by radiation for an enclosed body

If a body of radiating surface A_1, absolute temperature T_1 K and degree of blackness ε_1 is surrounded by a body of internal surface area A_2, absolute temperature T_2 K and degree of blackness ε_2, then the power transmitted by means of radiation per unit area of the radiating body is given by Eqn. (6.263) as

$$q_{12} = 5.7 \times 10^{-8} \varepsilon_n (T_1^4 - T_2^4),$$

where the dimension of factor 5.7×10^{-8} is $W \, m^{-2} \, K^{-4}$. In this case, the reduced degree of blackness is

$$\varepsilon_n = \frac{1}{\dfrac{1}{\varepsilon_1} + \dfrac{A_1}{A_2}\left(\dfrac{1}{\varepsilon_2} - 1\right)}. \tag{6.265}$$

This kind of heat exchange by radiation takes place in the case of a transformer installed in a chamber.

6.3.7.5 Effect of solar radiation

Consider a transformer of temperature T_1 K, radiation area A_1, and of degree of blackness ε_1 operating at a site having an ambient temperature T_2 K. A surface area A_0 of the transformer is exposed to direct sunshine, and the absorption capacity of that surface A_0 for solar radiation is a_r. Let the radiation capability of the sun be denoted by q_r. The power density transferred from the transformer to its surroundings by radiation from its surface A_1 is

$$q_{12} = 5.7 \times 10^{-8} \varepsilon_1 (T_1^4 - T_2^4) - a_r \frac{A_0}{A_1} q_r, \tag{6.266}$$

where the dimension of factor 5.7×10^{-8} is $W \, m^{-2} \, K^{-4}$. The absorptivity referred to solar radiation, denoted above by a_r and depending on the material, colour and smoothness of the surface A_0, is a dimensionless number less than unity which gives

Table 6.25

Absorptivity of solar radiation by various materials

Material	Absorptivity a_r
Aluminium, polished	0.26
Steel, polished	0.45
Steel, oxidized, rusty	0.74
Copper, polished	0.26
White paint	0.12–0.26
Grey, green paints	0.90–0.97
Black paint	0.97–0.99
Zinc plated steel, new	0.66
Zinc plated steel, contaminated	0.89

Table 6.26

Irradiation power density of the sun on a clear summer day
in latitude 40°

Hour of the day, h	Irradiation q_r, W m^{-2}			
	Vertical surface, facing			
	East	South	West	Horizontal surface
6	227	–	–	46.5
9	611	81.3	–	675.0
12	–	244.5	–	947.0
15	–	81.3	610	674.0
18	–	–	547	46.5

the ratio of power absorbed by an area of transformer surface to the power of solar radiation power incident on the said surface. Values for the absorptivity of solar radiation are given in Table 6.25 for a few materials. Solar power density means the thermal power transferred by solar radiation to a unit area. The dimension of solar power density, q_r, is W m^{-2}. Figures of solar power density for latitude 40° on a clear summer day are contained in Table 6.26.

6.3.7.6 Mean angle factor. Factor of irradiation

Let two areas A_1 and A_2 of finite extension be given. Select a reference area A_p, which can relate to either A_1 or A_2.

Of the entire energy radiated into the half space by an infinitesimal area dA_1 of surface A_1, the part incident on surface A_2 is as follows:

$$\varphi' = \int_{A_2} \frac{\cos \varphi_1 \cos \varphi_2}{\pi r^2} dA_2, \tag{6.267}$$

where r is the distance between infinitesimal area dA_1 and infinitesimal area dA_2 of surface A_2; φ_1 and φ_2 are the angles between the normal vectors of areas dA_1, dA_2 and r respectively.

The angle factor is the surface integral of φ' taken over surface A_1, divided by the reference surface A_p.

$$\varphi_{12} = \frac{1}{A_p} \int_{A_1} \varphi' dA_1. \tag{6.268}$$

In the case of large surfaces arranged close to each other, the angle factor is unity.

Considering angle factor φ_{12}, the radiated power density transferred from surface 1 to surface 2 is

$$q_{12} = 5.7 \varepsilon_n \varphi_{12} \left[\left(\frac{T_1}{100} \right)^4 - \left(\frac{T_2}{100} \right)^4 \right], \tag{6.269}$$

where the dimension of factor 5.7 is W m^{-2} K^{-4}.

6.3.7.7 Radiation screen placed between parallel surfaces

In some cases reduction of heat transmission by radiation is necessary. By placing a radiation screen between surfaces the specific heat dissipation will be reduced to q_e:

$$q_e = \frac{\dfrac{\varepsilon_{1e}\varepsilon_{e2}}{\varepsilon_{1e}+\varepsilon_{e2}}}{\varepsilon_{12}} q_{12}, \qquad (6.270)$$

where q_{12} is the heat dissipation without screen, ε_{1e} is the reduced degree of blackness between surface 1 and the screen, ε_{e2} is that between the screen and surface 2 and ε_{12} is the reduced degree of blackness between surface 1 and 2. If $\varepsilon_{12}=\varepsilon_{1e}=\varepsilon_{e2}$, then

$$q_e = \frac{1}{2} q_{12}. \qquad (6.271)$$

6.3.7.8 Examples for heat radiation calculations

Example 6.2.1. Assuming $\varepsilon_1 = 0.95$, $\varepsilon_2 = 0.8$, $\varepsilon'_{1e} = 0.2$, $\varepsilon'_{e2} = 0.6$, $q_{12} = 350 \text{ W m}^{-2}$, calculate to what value will the heat transmitted by radiation from surface 1 to surface 2 decrease, when using the screen.

$$\varepsilon_{12} = \frac{1}{\dfrac{1}{0.95}+\dfrac{1}{0.8}-1} = 0.767,$$

$$\varepsilon_{1e} = \frac{1}{\dfrac{1}{0.95}+\dfrac{1}{0.2}-1} = 0.247,$$

$$\varepsilon_{e2} = \frac{1}{\dfrac{1}{0.6}+\dfrac{1}{0.8}-1} = 0.521,$$

$$q_e = \frac{\dfrac{0.247\times0.521}{0.247+0.521}}{0.767} q_{12} = 0.218 q_{12} = 0.218 \times 350 \text{ W m}^{-2} = 76.3 \text{ W m}^{-2}.$$

Example 6.2.2. Obtain the formula for determining the reduced degree of blackness when a radiation screen is employed.

Notation: q_{1e} is the power density transmitted by radiation from surface 1 to screen e, q_{e2} is the power density transmitted by radiation from screen e to surface 2, ε_{1e} is the reduced degree of blackness between surface 1 and screen e, ε_{e2} is the reduced degree of blackness between screen e and surface 2, ε_{12} is the reduced degree of blackness between surfaces 1 and 2, without screen.

The power density transmitted by radiation from surface 1 to screen e is

$$q_{1e} = \varepsilon_{1e} \times 5.7 \times \left[\left(\frac{T_1}{100}\right)^4 - \left(\frac{T_e}{100}\right)^4\right],$$

where the dimension of factor 5.7 is $\text{W m}^{-2}\text{ K}^{-4}$.

The power density transmitted by radiation from screen e to surface 2 is

$$q_{e2} = \varepsilon_{e2} \times 5.7 \times \left[\left(\frac{T_e}{100}\right)^4 - \left(\frac{T_2}{100}\right)^4\right],$$

where the dimension of factor 5.7 is $\text{W m}^{-2}\text{ K}^{-4}$.

500

If no heat is dissipated by convection from the screen, then

$$q_{e1} = q_{e2} = q_e.$$

Taking $\left(\dfrac{T_e}{100}\right)^4$ from the first equation and substituting it into the second:

$$-\frac{q_{e1}}{\varepsilon_{e1}5.7} + \left(\frac{T_1}{100}\right)^4 = \left(\frac{T_e}{100}\right)^4,$$

where the dimension of factor 5.7 is $W\,m^{-2}\,K^{-4}$.

$$q_{e2} = \varepsilon_{e2}5.7\left[\left(\frac{T_1}{100}\right)^4 - \frac{q_{1e}}{\varepsilon_{e1}5.7} - \left(\frac{T_2}{100}\right)^4\right],$$

where the dimension of factor 5.7 is $W\,m^{-2}\,K^4$.

After rearrangement, and considering that

$$q_{1e} = q_{e2},$$

$$q_e = q_{1e} = q_{e2} = \frac{\varepsilon_{1e}\varepsilon_{e2}}{\varepsilon_{1e} + \varepsilon_{e2}}5.7\left[\left(\frac{T_1}{100}\right)^4 - \left(\frac{T_2}{100}\right)^4\right], \qquad (6.272)$$

where the dimension of factor 5.7 is $W\,m^{-2}\,K^{-4}$.

If, without the screen, the power density from surface 1 to surface 2 is q_{12}, then by Eqn. (6.270)

$$q_e = \frac{\dfrac{\varepsilon_{1e}\varepsilon_{e2}}{\varepsilon_{1e} + \varepsilon_{e2}}}{\varepsilon_{12}}q_{12}.$$

Example 6.23. The 40 MVA transformer dealt with in Example 6.16, is installed with its longitudinal axis in an east-west direction, in an outdoor substation. On each of the shorter sides of the tank, perpendicular to the said longitudinal axis, there is a fire-protecting wall spaced 1 m from the tank.

On a clear summer day, at noon, the ambient air temperature is $\vartheta_a = 30\,°C$; the sun shines parallel to the protecting walls, irradiating one longitudinal side and the top of the transformer.

Calculate the amount of power dissipated by the transformer by radiation, taking into account the effect of sunshine.

Let us introduce the following simplifications. The two shorter sides of the transformer radiate on to the protective walls of temperature $35\,°C$ having a grade of blackness $\varepsilon = 0.91$. Let the angle factor be equal to unity, i.e. the wall be considered as of infinite extension and no heat be radiated into the surroundings of $30\,°C$ by the short sides. Most of the power is radiated by the two longer sides into the $30\,°C$ surroundings and a smaller fraction of it on to the protecting walls. Let us neglect the radiation on to the walls and take the angle factor equal to 1, but assume the grade of blackness is equal to the reduced grade of blackness calculated from the grades of blackness of wall and transformer

$$\varepsilon_n = \frac{1}{\dfrac{1}{0.91} + \dfrac{1}{0.95} - 1} = 0.87.$$

The top surface of the transformer radiates into empty space assumed to have an effective temperature of $\vartheta_{sp} = -53\,°C$. The absorption capacity referred to solar radiation is $a_r = 0.97$. The quantity of power absorbed by the transformer from solar radiation is as follows (see Table 6.26). The area of top covering surface of the transformer is $A_{ru} = 15.899\ m^2$, further, with $q_{ru} = 947\ W\,m^{-2}$.

$P_{ru} = a_r q_{ru} A_{ru} = 0.97 \times 947 \times 15.899\ W = 14\,600\ W$. The lateral surface area of the transformer facing south is

$$A_{rs} = (4.7 + 0.85 \times 2.08 + 4.54 \times 0.92)\ m^2 = 15.72\ m^2, \quad q_{rs} = 244.5\ W\,m^{-2},$$

$$P_{rs} = a_r q_{rs} A_{rs} + 0.97 \times 244.5 \times 15.72\ W = 3730\ W.$$

501

The overall power from the sun is

$$P_s = (14\,600 + 3730)\text{ W} = 18\,330\text{ W}.$$

The quantity of power dissipated by the transformer by way of radiation: the top covering surface radiates with an absolute temperature corresponding to

$$\Delta\vartheta_{om} = 60\,°\text{C}, \quad T = (273 + 30 + 60)\text{ K} = 363\text{ K},$$

$\dfrac{T_1}{100} = 3.63$ the absolute temperature corresponding to the effective temperature $\vartheta_{sp} = -53\,°\text{C}$ of empty space is $T_2 = (273 - 53)\text{ K} = 220\text{ K}$,

$$\frac{T_2}{100} = 2.2, \quad \varepsilon = 0.95.$$

Thus

$$q_{12} = 5.7 \times 0.95(3.63^4 - 2.2^4)\text{ W m}^{-2} = 812\text{ Wm}^{-2},$$
$$P_{ru} = q_{12}A_{ru} = 812 \times 15.899\text{ W} = 12\,900\text{ W}.$$

The power dissipated by radiation from the bank of radiators arranged on the two longitudinal sides is calculated as follows. Using the diagram shown in Fig. 6.95, with $\Delta\vartheta''_{o-a} = 50\,°\text{C}$, $\vartheta_a = 30\,°\text{C}$ and grade of blackness $\varepsilon = 0.95$ $q'_{rr} = 385\text{ W m}^{-2}$. As mentioned above, this value is reduced by the ratio 0.87/0.95 because of partial reflection by the protective wall:

$$q_{rr} = \frac{0.87}{0.95}\,385\text{ W m}^{-2} = 352\text{ W m}^{-2}.$$

The lateral projection area of the two radiator banks on the longitudinal sides and one radiator bank on the short side is

$$A_{rr} = 2(4.7 + 0.85)2.08\text{ m}^2 = 23.1\text{ m}^2.$$

The power dissipated by radiation is $P_{rr} = q_{rr}A_{rr} = 352 \times 23.1 = 8100\text{ W}$. The lower part of the lateral tank wall radiates with a temperature of $\Delta\vartheta = 30\,°\text{C}$. With $\Delta\vartheta = 30\,°\text{C}$ and $\vartheta_a = 30\,°\text{C}$, $q'_{ru} = 208\text{ Wm}^{-1}$.

The surface area is $A_{ru} = 2 \times 0.92 \times 4.54\text{ m}^2 = 8.33\text{ m}^2$. The power dissipated by radiation $P_{ru} = \dfrac{\varepsilon_r}{\varepsilon}q'_{ru}A_{ru} =$

$= \dfrac{0.87}{0.95} \times 208 \times 8.33\text{ W} = 1585\text{ W}$. On the short side the power density of radiation from the surface radiating with $\Delta\vartheta''_{o-a} = 50\,°\text{C}$ is $q_{ru} = 5.7 \times 0.87 \times (3.53^4 - 3.08^4)\text{ W m}^{-2} = 324\text{ W m}^{-2}$. Here, the temperature rise of $\Delta\vartheta = 35\,°\text{C}$ and the reduced grade of blackness, 0.87, of the protective wall have already been taken into account. The surface area is $(1.3 + 1.56 + 1.5)2.08\text{ m}^2 = 9.07\text{ m}^2$. The power radiated $P_{ru} = q_{ru}A_{ru} = 324 \times 9.07\text{ W} = 2940\text{ W}$. The lower strips radiate at $\Delta\vartheta = 30\,°\text{C}$.

The power density of radiation is $q_{rl} = 5.7 \times 0.87 \times (3.33^4 - 3.08^4)\text{ W m}^{-2} = 163\text{ W m}^{-2}$. The surface area is $A_{rl} = 2 \times 1.56 \times 0.92\text{ m}^2 = 2.87\text{ m}^2$. The power radiated is $P_{rl} = q_{rl}A_{rl} = 163 \times 2.87\text{ W} = 467\text{ W}$.

The overall quantity of power dissipated by the transformer by way of radiation is

$$P = (12\,900 + 8100 + 1585 + 2940 + 467) = 25\,992\text{ W}.$$

The difference between the power dissipated by radiation and transferred by solar radiation is

$$(25\,992 - 18\,330)\text{ W} = 7662\text{ W}.$$

Therefore, due to absorbed solar radiation, the amount of power dissipated into the surroundings by way of radiation has been reduced from 25 992 W to 7662 W.

Example 6.24. Assume a bank of radiators consisting of 10 elements each of 1 m length and coated, as usual with machine-grey paint. Let the radiator bank be arranged in such a way as to have further radiator banks accommodated on its right and left sides, so that only the side surface area of the outermost radiator elements and the vertical projection onto the horizontal plane of the upper envelope

502

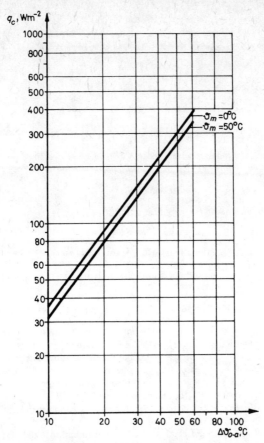

Fig. 6.96. Power density dissipation by laminar convection as a function of logarithmic-mean temperature difference, in the case of air having different mean boundary-layer temperatures

surface should be taken into account from the radiation point of view. The spacing between inlet and outlet studs of radiator elements is 915 mm, the convection surface of each element being 48 dm^2 = 0.48 m^2 and that of the entire bank of 10 elements A_{cr} = 4.8 m^2. The pitch of the radiators is 45 mm, their width 240 mm. The thickness of each element is 15 mm. Let a steel plate coated with a machine-grey paint be inserted between each pair of adjacent radiator elements in the middle of the pitch, with the width of the plates equal to that of the radiator elements, and their length such as to bear up against the top and bottom collecting tubes. Find to what extent the inserted steel plates will increase the heat dissipation capacity of the bank, with ambient temperature $\vartheta_a = 20\,°C$, average radiator temperature rise $\Delta\vartheta'_{o-a} = 50\,°C$ and $\Delta\vartheta_{oic} = 60\,°C$.

Power transferred by convection without inserted steel plates:

According to Fig. 6.96, at mean boundary-layer temperature of

$$\vartheta_m = \vartheta_a + \frac{\Delta\vartheta'_{o-a}}{2} = \left(20 + \frac{50}{2}\right)°C = 45\,°C \quad \text{and} \quad \Delta\vartheta'_{o-a} = 50\,°C,$$

$$q_k = 270 \text{ W m}^{-2}.$$

Power dissipated by the radiator bank:

$$P_c = q_c A_{cr} = 270 \times 4.8 = 1.32 \times 10^3 \text{ W}.$$

503

Power transferred by radiation without inserted steel plates:

The upper envelope surface of the radiator bank radiates with a power density corresponding to $\Delta\vartheta_{oic} = 60\ °C$. From the diagram of Fig. 6.95, with $\vartheta_a = 20\ °C$ and $\Delta\vartheta = 60\ °C$:

$$q'_r = 441\ W\ m^{-2}.$$

The upper envelope surface is

$$A'_r = (9 \times 0.045 + 0.02)0.24\ m^2 = 0.102\ m^2,$$

so

$$P'_r = q'_r A'_r = 441 \times 0.102\ W = 45\ W.$$

The lateral envelope surface radiates with a power density corresponding to $\Delta\vartheta_{o-a} = 50\ °C$. From Fig. 6.95 with $\vartheta_a = 20\ °C$ and $\Delta\vartheta = 50\ °C$,

$$q''_r = 350\ W\ m^{-2}.$$

The lateral envelope surface, taking the length of radiator to 0.9 m, is

$$A''_r = 0.9 \times 0.24\ m^2 = 0.216\ m^2,$$

$$P''_r = q''_r A''_r = 350 \times 0.216\ W = 75.6\ W.$$

The power transferred by convection and radiation, without that of the top and bottom collecting tubes, is

$$P = P_c + P'_r + P''_r = (1.32 \times 10^3 + 45 + 75.6)\ W = 1441\ W.$$

Calculation of power dissipation with inserted steel plates: Let us assume that the thickness of boundary layers along the surfaces of both the radiator and the steel plate, on the air side, is not so great as to make the layer come into contact with its counterpart, either at the top or at the bottom, i.e. in the case of both the radiator and steel plate the condenser-type temperature pattern may be reckoned with, the flow pattern remaining undisturbed by the steel plate. If this assumption is right, then the conditions of heat transfer by convection may be applied to the diagram shown in Fig. 6.96, which is based on assuming the air-side heat transfer coefficient to vary with the 1/3rd power of surface temperature drop.

Heat balance with steel plates inserted. Heat is transferred by radiation to the steel plates, the latter are warmed up and the heat thus absorbed is transmitted to the air by convection.

Let us denote by T_1 the absolute temperature corresponding to the average temperature rise of the radiator given by $\Delta\vartheta_{o-a}$, and by T_2 the absolute temperature of the steel plate showing a temperature rise $\Delta\vartheta_{i1}$. Thus,

$$T_1 = (273\ °C + \vartheta_a + \Delta\vartheta_{o-a}),$$

and

$$T_2 = (273\ °C + \vartheta_a + \Delta\vartheta_{i1}).$$

The quantity of power density radiated from the surface of the radiator to the steel plate:

$$q_{12} = 5.7\varepsilon_n \left[\left(\frac{T_1}{100} \right)^4 - \left(\frac{T_2}{100} \right)^4 \right],$$

where the dimension of factor 5.7 is $W\ m^{-2}\ K^4$. The power density transferred by convection to the surroundings is

$$q_{2c} = \alpha\Delta\vartheta_{i1}.$$

The heat balance equation is

$$q_{12} = q_{2k} = 5.7\varepsilon_n \left[\left(\frac{T_1}{100} \right)^4 - \left(\frac{T_2}{100} \right)^4 \right] = \alpha\Delta\vartheta_{i1},$$

where the dimension of factor 5.7 is again $W\ m^{-2}\ K^{-4}$.

Reverting to the data of the example, let the grade of blackness of both the radiator and steel plate be 0.95. The reduced grade of blackness

$$\varepsilon_n = \frac{1}{\dfrac{1}{0.95} + \dfrac{1}{0.95} - 1} = 0.905,$$

$$T_1 = (273 + 20 + 50) \text{ K} = 343 \text{ K},$$

$$\frac{T_1}{100} = 3.43 \text{ K}.$$

Let us assume a temperature rise of $\Delta\vartheta_{i1} = 30\ °C$ of the steel plate above the ambient air temperature of $\vartheta_a = 20\ °C$.

$$T_2 = (273 + 20 + 30) \text{ K} = 323 \text{ K},$$

$$\frac{T_2}{100} = 3.23 \text{ K},$$

$$q_{12} = 5.7 \times 0.905(3.43^4 - 3.23^4)\ \text{W m}^{-2} = 152.5\ \text{W m}^{-2}.$$

From the diagram of Fig. 6.96, for a mean boundary-layer temperature of

$$\vartheta_m = \vartheta_a + \frac{\Delta\vartheta_{i1}}{2} = \left(20 + \frac{30}{2}\right)\ °C = 35\ °C$$

and a temperature rise of $\Delta\vartheta_{i1} = 30\ °C$,

$$q_{ci} = 140\ \text{W m}^{-2}.$$

As will be seen, too low a value has been selected for the temperature of the steel plate. Let us try the value $\Delta\vartheta_{i1} = 31\ °C$.

$$q_{12} = 5.7 \times 0.905(3.43^4 - 3.24^4)\ \text{W m}^{-2} = 145\ \text{W m}^{-2},$$

$$q_{ci} = 145\ \text{W m}^{-2}.$$

Let the size of steel plate be 260×850 mm giving a surface area of

$$A'_{i1} = 2 \times 0.85 \times 0.26\ \text{m}^2 = 0.442\ \text{m}^2.$$

The overall surface area of the 9 steel plates is

$$A_{i1} = 9 \times 0.442\ \text{m}^2 = 3.97\ \text{m}^2.$$

The power dissipated from the steel plates by convection:

$$P_{i1} = q_{ci} A_{i1} = 145 \times 3.97\ \text{W} = 575\ \text{W}.$$

Overall power dissipation of the radiator bank equipped with inserted steel plates

$$P_t = (1441 + 575)\ \text{W} = 2016\ \text{W}.$$

Increase of heat dissipation achieved by the inserted steel plates:

$$\frac{2016 - 1441}{1441}\ 100 = 40.7\%.$$

This increase of dissipated power is possible if the oil side is capable of delivering the required power.

Example 6.25. The external dimensions and surface areas of an air-cooled 40 MVA transformer equipped with fitted-on radiators are as follows. Cover: $1.72 \times 4.95\ \text{m}^2 = 8.514\ \text{m}^2$, tank sides: $2 \times (1.56 + 4.54) \times 3\ \text{m}^2 = 36.6\ \text{m}^2$.

The radiator banks are fitted to the two longitudinal sides and to one of the shorter sides. The vertical projection on the horizontal plane of the rows of radiators mounted on the two longitudinal sides is $2 \times 4.7 \times 0.65 \text{ m}^2 = 6.11 \text{ m}^2$. The vertical projection on the horizontal plane of the group radiators mounted on one of the short sides is $1.5 \times 0.85 \text{ m}^2 = 1.275 \text{ m}^2$. The envelope surface of the outer side of the rows of the radiator bank is $2\left(0.65 + 4.7 + 0.85 + \dfrac{1.5}{2}\right)2.08 \text{ m}^2 = 31.616 \text{ m}^2$. On the two longitudinal sides the number of radiator elements is 2×17 groups each with 9 elements $= 306$. On one short side there are 6 groups with 12 elements in each group $= 72$ radiator elements. The overall number of radiator elements is thus 378. The spacing of inlet and outlet stubs is 1965 mm, the surface area of one element is 1.028 m^2. Overall radiator surface area is thus $378 \times 1.028 \text{ m}^2 = 388.6 \text{ m}^2$. Let us calculate the power dissipated by the transformer through radiation, if $\vartheta_a = 20\,°\text{C}$, $\varDelta\vartheta_{om} = 60\,°\text{C}$, and $\varDelta\vartheta_{o-a} = 50\,°\text{C}$. The transformer is painted machine-grey, with $\varepsilon = 0.95$.

The temperature of the top cover of the tank and the upper end of radiators corresponds to that of the maximum temperature rise of the oil. From Table 6.26, with $\varDelta\vartheta_{om} = 60\,°\text{C}$, the power density transferred by radiation toward the surroundings of temperature $\vartheta_a = 20\,°\text{C}$ is

$$q_{rm} = 441.6 \text{ W m}^{-2}.$$

The area of the radiating surface is the sum of the top cover surface and the vertical projection of radiators on the horizontal plane:

$$A_{rm} = (8.514 + 6.11 + 1.275) \text{ m}^2 = 15.899 \text{ m}^2.$$

Power transferred by radiation

$$P_{rm} = q_{rm} A_{rm} = 441.6 \times 15.899 \text{ W} = 7 \times 10^3 \text{ W}.$$

The temperature rise of the lateral surfaces of the radiators and that of the free side of the tank is that corresponding to the logarithmic-mean temperature difference. From the table, with $\varDelta\vartheta_{o-a} = 50\,°\text{C}$, the power density transferred by radiation towards the surroundings of temperature $\vartheta_a = 20\,°\text{C}$ is

$$q_{ra} = 350.3 \text{ W m}^{-2}.$$

The area of the radiating surface is the sum of the outer lateral envelope surface of the rows of radiator groups and part of the free short side corresponding to the radiator length:

$$A_{ra} = (31.616 + 1.56)2.08 \text{ m}^2 = 34.856 \text{ m}^2.$$

Power transferred by radiation

$$P_{ra} = q_{ra} A_{ra} = 350.3 \times 34.856 \text{ W} = 12.2 \times 10^3 \text{ W}.$$

The average temperature rise of the tank sides below the bottom end of the radiators is assessed to be $\varDelta\vartheta_{rt} = 30\,°\text{C}$, considering that $\varDelta\vartheta_{ooc} = 40\,°\text{C}$. From Table 6.26, with $\varDelta\vartheta_{rt} = 30\,°\text{C}$, the power density transferred by radiation to the surroundings of ambient temperature $\vartheta_a = 20\,°\text{C}$ is

$$q_{rt} = 190.4 \text{ W m}^{-2}.$$

Area of radiating surface

$$A_{rt} = 36.6 \frac{3 - 2.08}{3} \text{ m}^2 = 11.2 \text{ m}^2.$$

Power transferred by radiation

$$P_{rt} = q_{rt} A_{rt} = 190.4 \times 11.2 \text{ W} = 2.13 \times 10^3 \text{ W}.$$

Overall power dissipated by the transformer is

$$P_r = (7 + 12.2 + 2.13) \text{ kW} = 21.33 \text{ kW}.$$

6.3.8 Design of radiator cooling

6.3.8.1 Design sequence

In the case of natural oil flow and natural air flow, the design of radiator-type cooling is performed in the steps described below.

1. Determine the sum of all losses arising in the transformer considering the relevant tolerances. According to the majority of national standards, the sum of all losses of a transformer may exceed the sum of the guaranteed no-load and load losses by 10%. This increased value is taken as the overall design loss from the point of view of cooling.

2. Calculate the increments between average winding temperature and average oil temperature for the different windings. The values of these increments are generally different and the maximum temperature drop or increment $\Delta\vartheta_{w-o}$ is taken as the design value.

3. Subtract the maximum temperature drop $\Delta\vartheta_{w-o}$ from the value permitted by the standards as the average temperature rise for the winding. The temperature increment thus obtained is the average temperature rise of the oil above the ambient temperature.

$$\Delta\vartheta_{oaw} \approx \Delta\vartheta_{oac} = 65\,°C - \Delta\vartheta_{w-o}.$$

4. Determine the dimensions of the tank and, by assessing the power to be dissipated by radiation, the number of radiator groups and of elements in the groups. Draw up an outline sketch of the transformer.

5. Estimate the difference between inlet and outlet temperatures of the oil flowing through the winding, i.e. the value of $\Delta\vartheta_{wo}$, this value being also the difference between outlet and inlet oil temperatures of the radiator, $\Delta\vartheta_{wo} = \Delta\vartheta_{co}$.

6. Add half the difference between temperatures of the oil entering and leaving the radiator to the average temperature rise of the oil in the radiator, this sum being the maximum temperature rise of the oil:

$$\Delta\vartheta_{om} = \Delta\vartheta_{oac} + \frac{\Delta\vartheta_{co}}{2}.$$

Check whether the value of $\Delta\vartheta_{om}$ thus obtained is lower than the maximum temperature rise, $\Delta\vartheta_{om} = 60\,°C$, specified as upper limit for the oil by the standard.

7. Assuming $20\,°C$ as ambient temperature, determine the maximum oil temperature and average temperature of the radiator wall.

$$\vartheta_{om} = \vartheta_a + \vartheta_{om},$$

$$\vartheta_{oac} = \vartheta_a + \frac{\Delta\vartheta_{ca}}{2} + \Delta\vartheta_{o-a} - \Delta\vartheta_{os},$$

where $\Delta\vartheta_{os}$ is the oil-side temperature drop of the radiator, and $\Delta\vartheta_{ca}$ is the temperature rise of air in the radiator.

8. Multiply by the value of q_r relating to the maximum oil temperature the area of the top projection of radiators and that of the top cover to obtain that part of the losses dissipated through the projected top surfaces by way of radiation.

507

9. Multiply by q_r relating to the average radiator temperature the area of the outer surface of the enveloping mantle of the transformer to obtain that the part of the losses dissipated through the mantle surfaces by way of radiation.

10. Multiply the area of the top cover by the value of q_c pertaining to the maximum oil temperature, and multiply the surface area of the mantle of the tank by the value of q_c pertaining to the average temperature of the radiator to obtain those of the loss dissipated through the tank cover and wall, respectively, by way of convection. The values of q_c should be taken from the diagram shown in Fig. 6.96, bearing in mind that different values of q_c apply to the horizontal and vertical surfaces.

11. Substract from the overall design loss the part of the loss dissipated by way of radiation and that dissipated through the surface of the tank by convection. It is the remaining part of the loss that has to be dissipated by the radiators by way of convection.

12. Check, considering the diagrams of Figs 6.83 to 6.87, whether the radiators will be capable of dissipating the respective proportions of the loss at the given values of ϑ_a and $\varDelta\vartheta_{o-a}$. If so, proceed with the calculation.

13. For the assumed geometry of winding height, radiator length and difference of level between thermal centre points of windings and radiators, calculate the pressure area.

14. Using the assumed value of pressure difference between inlet and outlet temperatures of the oil entering and leaving the windings, calculate the oil flow required for the designed winding structure:

$$\Phi_o = \frac{P_w}{c_o \varDelta\vartheta_{co}},$$

where P_w is the loss of the designed winding.

15. Calculate for the designed winding the pressure drop $\varDelta p_w$ occurring in it.

16. Calculate the oil flow which corresponds to the difference between overall loss and the loss dissipated through tank and top cover by way of convection and through the projected area of top cover by way of radiation.

17. Calculate for the oil flow found according to the preceding item, the pressure drop $\varDelta p_r$ taking place in the radiators.

18. Check whether relation

$$\varDelta p_w + \varDelta p_r < \varDelta p$$

is satisfied. If an excessive deviation is found, repeat the calculation with a modified $\varDelta\vartheta_{wo}$.

6.3.8.2 Heat transfer efficiency of heat exchangers

The heat transfer efficiency of a heat exchanger is a dimensionless number equal to the ratio of the thermal power transferred in the heat exchanger to the thermodynamically maximum thermal power that can be transmitted in a counterflow heat exchanger through an infinite heat transfer surface.

508

Denote the warmer medium, i.e. the oil in the present case, by subscript w, and the colder medium, i.e. the air, by subscript c (see Fig. 6.97). Denote mass flows: Φ_w, Φ_c; average specific heats: c_w, c_c; flow-stream capacity rates: $C_w = \Phi_w c_w$, $C_c = \Phi_c c_c$; temperature of media: ϑ_{win}, ϑ_{wou}, ϑ_{cin}, ϑ_{cou}.

The difference between inlet temperatures of the two media is $\Delta_0 = \vartheta_{win} - \vartheta_{cin}$ (see Fig. 6.97), mean temperature difference is $\Delta\vartheta_{o-a}$.

The thermal power transferred in the heat exchanger is P; the thermal power that can be transferred in an infinite-surface heat exchanger is P_{max}, the heat transfer

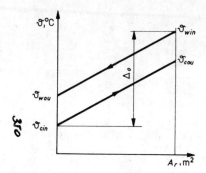

Fig. 6.97. Temperature conditions of media flowing in the heat exchanger

efficiencies of the heat exchanger is ε. The flow-stream capacity rates of the two media being generally different, let the flow-stream capacity rate of the medium with the lower value be denoted by C_{min}. According to the definition

$$\varepsilon = \frac{P}{P_{max}} = \frac{C_w(\vartheta_{win} - \vartheta_{wou})}{C_{min}(\vartheta_{win} - \vartheta_{cin})} = \frac{C_c(\vartheta_{cou} - \vartheta_{cin})}{C_{min}(\vartheta_{win} - \vartheta_{cin})}. \tag{6.273}$$

As it can be seen from the above equation,

$$P_{max} = C_{min}(\vartheta_{win} - \vartheta_{cin}) = C_{min}\Delta_o. \tag{6.274}$$

Denoting the heat exchanging surface by A, and the pertinent overall thermal conductance by k, the relation for ε wil be

$$\varepsilon = \frac{kA}{C_{min}} \frac{\Delta\vartheta_{o-a}}{\Delta_o}. \tag{6.275}$$

Let us introduce N to indicate the number of heat transfer units of the heat exchanger, this number being also dimensionless:

$$N = \frac{kA}{C_{min}}. \tag{6.276}$$

Introducing the ratio of flow-stream capacity rates K, also this being a dimensionless number:

$$K = \frac{C_{min}}{C_{max}}. \tag{6.277}$$

509

The thermodynamic properties of the heat exchanger of a given kind are completely determined by the latter three dimensionless numbers.

Example 6.26. A radiator bank of length $h_r = 2.44$ m dissipates 300 W thermal power at $\Delta\vartheta_{o-a} = 45$ °C. The temperature of the air while passing through the radiators rises by $\Delta\vartheta_a = 31$ °C. Inlet air temperature is $\vartheta_{ain} = 20$ °C, inlet oil temperature $\vartheta_{oin} = 90$ °C, outlet temperature of oil leaving the cooler is $\vartheta_{oou} = 70$ °C. Air-side convection surface area of radiator is $A = 1.276$ m^2. Calculate the heat transfer efficiency of cooling by the given radiator.

Mass flow on the air side:

$$\Phi_a = \frac{P}{c_a \Delta\vartheta_a} = \frac{300}{1016 \times 31} \text{ kg s}^{-1} = 9.6 \times 10^{-3} \text{ kg s}^{-1}.$$

Mass flow on the oil side with $\Delta\vartheta_o = 20$ °C:

$$\Phi_o = \frac{300}{2100 \times 20} \text{ kg s}^{-1} = 7.15 \times 10^{-3} \text{ kg s}^{-1}.$$

The values of products Φc are

$$\Phi_a c_a = 9.6 \times 10^{-3} \times 1016 \text{ W °C}^{-1} = 9.7 \text{ W °C}^{-1},$$
$$\Phi_o c_o = 7.15 \times 10^{-3} \times 2100 \text{ W °C}^{-1} = 15 \text{ W °C}^{-1}.$$

Maximum thermal power that can be transferred through an infinite area of heat transfer surface:

$$P_{max} = C_{min} \Delta_o = 9.7(90 - 20) \text{ W} = 680 \text{ W}.$$

Heat transfer efficiency of radiator cooling of the present example:

$$\varepsilon = \frac{P}{P_{max}} = \frac{300}{680} = 0.44, \text{ i.e. } 44\%.$$

In the present example the air side heat transfer coefficient is

$$\alpha = \frac{P}{\Delta\vartheta_{o-a} A} = \frac{300}{45 \times 1.276} \text{ W m}^{-2} \text{°C}^{-1} = 5.22 \text{ W m}^{-2} \text{°C}^{-1}.$$

6.4 Compact heat exchangers

Some large power transformers, mainly those of a size and mass approaching the limit for transport by rail, are provided with heat exchangers of relatively small space requirement and high heat transfer capability. Such oil–air or oil–water coolers are termed compact coolers, referring to their small space requirement. The motor-driven fans supplying air to the oil–air compact coolers are built integrally with the coolers. A typical cooler is shown in Fig. 6.98. The suction funnels have the double task of reducing the inlet pressure drop of air thus they are acting as confusors and of preventing re-suction of heated air.

A decisive point to be considered when selecting a motor-driven fan, beside its other specification data such as operating pressure, air delivery, efficiency, input power, r.p.m. and outdoor design, is its noise level. The noise level of a fan increases with peripheral speed, i.e. with the number of r.p.m. and blade diameter, and further with increasing operating pressure. High-pressure fans are provided with spin vanes before and with stator blading after the impeller, the latter for removing the tan-

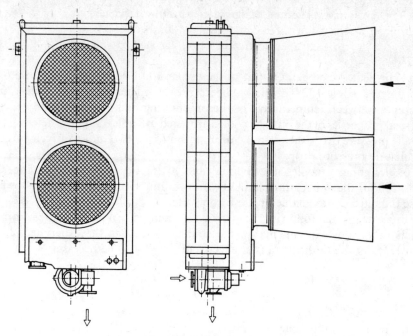

Fig. 6.98. Compact cooler

gential spin from the air. Transfer of high thermal power within small volumes can be achieved by dividing the heat transfer surface into small surface elements and by applying high flow velocities. In compact coolers of very small space requirement the power required for maintaining the flow is high.

Beyond a certain cooling capacity the capital cost and interest charges are so high in the case of radiator-type coolers used in conjunction with natural oil cooling that, although otherwise being the most reliable system and having almost no operating costs, it proves more economical to install compact coolers. Although these incur constant operating costs they have a much lower first cost, and thus also low interest charges.

The size of substation transformer bays is in some cases determined by the space requirement of bulky radiator banks provided with either natural or forced cooling. With the use of compact coolers fitted to the tank side, a smaller housing will be sufficient for a transformer.

The design of compact heat exchangers will be demonstrated on a concrete example by showing methods for determining the dimensions of active parts and the thermal and hydraulic characteristics of a compact oil–air cooler satisfying the following specification.

6.4.1 Thermal design of compact heat exchangers

6.4.1.1 Specification of an oil–air heat exchanger

Oil–air heat exchanger. Output of the cooler 100 kW, inlet temperature of air entering the cooler 32 °C, inlet temperature of oil entering the cooler 70 °C, difference between temperatures of the oil entering and leaving the cooler 5 °C. Physical properties of the oil to be cooled are listed in Table 6.2. Let us determine the oil-side pressure drop, the overall consumption of the two motor-driven fans, and the air-side pressure drop.

For designing a cooler, the thermal and hydraulic data of the arrangement envisaged for the application should be known. In the majority of cases, however, these data are not available or are incomplete.

It is shown in the following how the data required can be calculated from the results of various experiments. In the course of utilizing the experimental data knowledge of the fin heat transfer effectiveness will also be necessary.

6.4.1.2 Fin effectiveness

The fin heat transfer effectiveness is

$$\eta_f = \frac{\vartheta_f - \vartheta_2}{\vartheta_{w2} - \vartheta_f}, \tag{6.278}$$

(see Fig. 6.99), where ϑ_f is the average temperature of the fin, ϑ_2 is the temperature of the cooling air and ϑ_{w2} is the temperature of the fin root. The fin heat transfer effectiveness in the case of a pin-shaped fin of finite length is

$$\eta_f = \frac{\text{th}(ml)}{ml}, \tag{6.279}$$

where

$$m = 2\sqrt{\frac{\alpha_2}{\lambda d}} \tag{6.280}$$

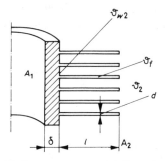

Fig. 6.99. Air-side fins, temperature and dimensional markings

512

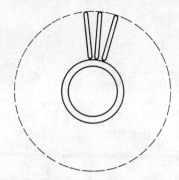

Fig. 6.100. Pin-type fins made of small diameter wire-spiral on the air side

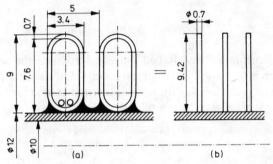

Fig. 6.101. (a) Longitudinal section of tube; (b) loops cut and straightened out

were l is the length of the pin-shaped fin, d is the pin diameter, λ is the heat conductance of the pin material, α_2 is the fin side surface heat transfer coefficient.

Example 6.27. Let a double-spiral cooling fin of the arrangement shown in Fig. 6.100 be given, wound of copper wire of the size given in Fig. 6.101(a). On cutting such a spiral perpendicular to the surface, fitting closely to the largest diameter, and straightening the curved pieces of wire thus obtained the pins shown in Fig. 6.101(b) will result.

Calculation of fin heat transfer effectiveness. From the diagram $d = 0.7 \times 10^{-3}$ m, $l = 9.42 \times 10^{-3}$ m, heat conductance of copper $\lambda = 391$ W m^{-1} °C^{-1},

$$m = 2\sqrt{\frac{\alpha_2}{\lambda d}} = 2\sqrt{\frac{\alpha_2}{391 \times 0.7 \times 10^{-3}}} \; \text{W}^{-0.5} \, °\text{C}^{0.5} = 3.83\sqrt{\alpha_2} \; \text{W}^{-0.5} \, °\text{C}^{0.5},$$

$$ml = 3.83\sqrt{\alpha_2} \times 9.42 \times 10^{-3} \; \text{m W}^{-0.5} \, °\text{C}^{0.5} = 36 \times 10^{-3}\sqrt{\alpha_2} \text{m W}^{-0.5} \, °\text{C}^{0.5}.$$

The fin heat transfer effectiveness is

$$\eta_f = \frac{\text{th} \, (36 \times 10^{-3}\sqrt{\alpha_2})}{36 \times 10^{-3}\sqrt{\alpha_2}},$$

where the dimension of factor 36×10^{-3} is $\text{W}^{-0.5} \, \text{m} \, °\text{C}^{0.5}$.

Let α_2 be equal to $\alpha_2 = 144$ W m^{-2} °C^{-1}.
Then

$$\eta_{f_{144}} = \frac{\text{th}\,(36 \times 10^{-3}\sqrt{144})}{36 \times 10^{-3}\sqrt{144}} = \frac{\text{th}\,0.432}{0.432} = \frac{0.40698}{0.432} = 0.94\,.$$

Or for $\alpha_2 = 196$ W m^{-2} °C^{-1},

$$\eta_{f_{196}} = \frac{\text{th}\,(36 \times 10^{-3}\sqrt{196})}{36 \times 10^{-3}\sqrt{196}} = \frac{\text{th}\,0.504}{0.503} = \frac{0.46525}{0.504} = 0.923.$$

As it can be seen, for a given geometrical arrangement the fin heat transfer effectiveness decreases with increasing heat transfer coefficient.

6.4.1.3 The overall thermal conductance, k

For a tube smooth inside and finned outside, for which the area of the inside surface of the tube is to a good approximation, equal to the area pertaining to the mean tube diameter (i.e. $A_1 = A_m$), the overall thermal conductance k_{A_2} referring to be finned outside surface area of the tube is given by

$$\frac{1}{k_{A_2}} = \frac{1}{\alpha_1 \dfrac{A_1}{A_2}} + \frac{\delta}{\lambda \dfrac{A_1}{A_2}} + \frac{1}{\alpha_2 \eta_f}, \qquad (6.281)$$

where δ is the wall thickness of the tube and λ is the heat conductance of the tube material.

Example 6.28. For the water–air cooler provided with double-spiral outer fins of the preceding example, with the horizontally staggered arrangement of tubes, the data furnished by Fig. 6.102 are available.

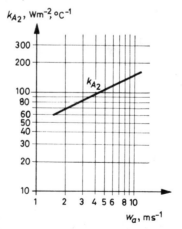

Fig. 6.102. Overall thermal conductance referred to the air side as a function of air velocity

514

The diagram gives the air-side heat transfer coefficient as a function of air velocity, for a water velocity of 3 m s^{-1}. The pipe material was brass, that of the spiral wire fins, copper. Determine the values of α_1, α_2 and η_f from the experimental date, for water velocity 3 m s^{-1}.

Determination of the value of α_1 begins with that of the Re number. It will be apparent from the diagram that when the heat transfer coefficient was determined, the tubes were arranged horizontally, the average water temperature was $\vartheta_{wa} = 20\,°C$, and the water velocity was $w_w = 3$ m s^{-1}. For water of 20 °C, $v_{20} = 1 \times 10^{-6}$ m^2 s^{-1}.

$$Re = w \frac{d}{v} = 3 \frac{10 \times 10^{-3}}{1 \times 10^{-6}} = 3 \times 10^4,$$

$$Re^{0.8} = (3 \times 10^4)^{0.8} = 3800.$$

This falls in the measuring range $Re > 2320$, i.e. the flow is turbulent throughout. For turbulent flow

$$Nu_f = 0.023\, Re^{0.8}\, Pr^{0.4}.$$

For water of 20 °C, $Pr = 7.06$, $Pr^{0.4} = 7.06^{0.4} = 2.18$. The heat conductance of water at 20 °C is

$$\lambda = 0.6 \text{ W m}^{-1}\,°C^{-1}, \quad \frac{\lambda}{d} = \frac{0.6}{10 \times 10^{-3}} \text{ W m}^{-2}\,°C^{-1} = 60 \text{ W m}^{-2}\,°C^{-1}.$$

From relation

$$Nu_f = \frac{\alpha d}{\lambda}: \quad \alpha = \frac{\lambda}{d} Nu_f.$$

Heat transfer coefficient α_1 pertaining to water velocity 3 m s^{-1} is

$$\alpha_1 = 60 \times 0.023 \times 3800 \times 2.18 \text{ W m}^{-2}\,°C^{-1} = 11\,450 \text{ W m}^{-2}\,°C^{-1}.$$

The surface areas of the tube of 1 m length are $A_1 = 0.0314$ m^2 and $A_2 = 0.295$ m^2.

$$\frac{A_1}{A_2} = \frac{0.0314}{0.295} = 0.1065, \quad \alpha_1 \frac{A_1}{A_2} = 11\,450 \times 0.1065 \text{ W m}^{-2}\,°C^{-1} =$$

$$= 1220 \text{ W m}^{-2}\,°C^{-1}.$$

The wall thickness is $\delta = 1 \times 10^{-3}$ m, the thermal conductivity of brass is $\lambda = 105$ W m$^{-1}\,°C^{-1}$,

$$\frac{\delta}{\lambda} = \frac{1 \times 10^{-3}}{105} \text{ W}^{-1} \text{ m}^2\,°C = \frac{1}{105 \times 10^3} \text{ W}^{-1} \text{ m}^2\,°C,$$

$$\frac{\delta}{\lambda \frac{A_1}{A_2}} = \frac{1}{11\,200} \text{ W}^{-1} \text{ m}^2\,°C.$$

$$\alpha_2 \eta_f = \alpha_2 \frac{\text{th}(36 \times 10^{-3} \sqrt{\alpha_2})}{36 \times 10^{-3} \sqrt{\alpha_2}} \text{ W m}^2\,°C^{-1} = 27.8\sqrt{\alpha_2}\, \text{th}(36 \times 10^{-3}\sqrt{\alpha_2}) \text{ W m}^{-2}\,°C^{-1},$$

where the dimension of factor 36×10^{-3} is W$^{-0.5}$ m °C$^{-0.5}$ and that of factor 27.8 is W$^{0.5}$ m^{-1} °C$^{-0.5}$. The reciprocal of the overall thermal conductance, referred to surface area A_2 for $w_w = 3$ m s^{-1} water velocity is

$$\frac{1}{k_{A_2}} = \left(\frac{1}{1220} + \frac{1}{11\,200} + \frac{1}{27.8\sqrt{\alpha_2}\, \text{th}\,(36.10^{-3}\sqrt{\alpha_2})} \right) \text{m}^2\,°C \text{ W}^{-1},$$

where the dimension of factor 36×10^{-3} is W$^{-0.5}$ m °C$^{-0.5}$ and that of factor 27.8 is W$^{0.5}$ m^{-1} °C$^{-0.5}$. Now calculate the values of k_{A_2} for the cases of $\alpha_2 = 81$ W m^{-2} °C^{-1} and $\alpha_2 = 169$ W m^{-2} °C^{-1}.

Noting that denominator $27.8\sqrt{81}$ th $(36 \times 10^{-3}\sqrt{81})$ W m^{-2} °C^{-1} = 78.3 W m^{-2} °C^{-1},

$$\frac{1}{k_{A_2}} = \left(\frac{1}{1220} + \frac{1}{11\,200} + \frac{1}{78.3}\right) \text{m}^2 \,°\text{C W}^{-1}$$

whence $k_{A_2} = 73.2$ W m^{-2} °C^{-1}, for $\alpha_2 = 81$ W m^{-2} °C^{-1}.

$$27.8\sqrt{169}\text{ th }(36 \times 10^{-3}\sqrt{169}) = 157.6 \text{ W m}^{-2}\,°\text{C}^{-1},$$

Again so for $\alpha_2 = 169$ W m^{-2} °C^{-1}

$$\frac{1}{k_{A_2}} = \left(\frac{1}{1220} + \frac{1}{11\,200} + \frac{1}{157.6}\right) \text{m}^2 \,°\text{C W}^{-1}$$

$$\text{and} \quad k_{A_2} = 138 \text{ W m}^{-2}\,°\text{C}^{-1}.$$

As read from the diagram of Fig. 6.102, the air velocity pertaining to $k_{A_2} = 73.2$ W m^{-2} °C^{-1} is 2.3 m s^{-1} and that pertaining to $k_{A_2} = 138$ W m^{-2} °C^{-1} is 8.9 m s^{-1}.

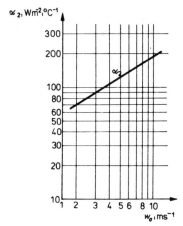

Fig. 6.103. Air-side heat transfer coefficient in the function of air velocity

The values of the related air speads and air-side heat transfer coefficients plotted on a log–log scale give the diagram shown in Fig. 6.103. There the data required for design of the air side are already obtained. Ít is, however worthwhile bringing the data obtained into a form satisfying the criteria of similarity.

6.4.1.4 Determination of the air-side equation, of the form $Nu = cRe^n Pr^{2/3}$

The air-side Re number. Denote, the smallest airside cross-sectional area of flow path by A_0, and the air-side cooling surface area, including fins, of a 1 m tube length by A (see Fig. 6.104). From the diagram: $A_c = 15.7 \times 10^{-3} \times 1$ m$^2 = 0.0157$ m^2 and $A = 0.295$ m^2, as has already been seen. Denote by L the distance between the leading edge of the first row of tubes and the leading edge of the imaginary row after the last row of tubes (see Fig. 6.105). By convention, the hydraulic diameter is

$$d_h = 4L\frac{A_c}{A}.$$

In our case $d_h = 4 \times 0.109 \dfrac{0.0157}{0.295}$ m $= 0.0232$ m .

The measurements were carried out with an air flow of 50 °C mean temperature. $v = 18.6 \times 10^{-6}$ m^2 s^{-1}. The Re number is

$$Re = \frac{d_h}{v} \qquad w_1 = \frac{0.0232}{18.6 \times 10^{-6}} \text{ m}^{-1}\text{s} \qquad w_1 = 1.247 \times 10^3 \text{ m}^{-1} \text{ s } w_1.$$

We now calculate the Re number pertaining to the two different air velocities.

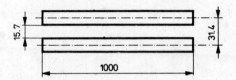

Fig. 6.104. Geometric data for calculation of hydraulic diameter

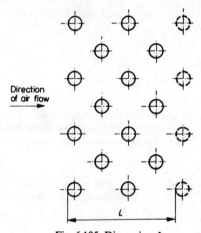

Fig. 6.105. Dimension L

For the case of $w_1 = 2.3$ m s^{-1},

$$Re = 1.247 \times 10^3 \times 2.3 = 2.87 \times 10^3,$$

and for the case of $w_1 = 8.9$ m s^{-1},

$$Re = 1.247 \times 10^3 \times 8.9 = 11.1 \times 10^3.$$

The thermal conductivity of air at 50 °C is $\lambda = 2.72 \times 10^{-2}$ W m^{-1} °C. The Nu number is

$$Nu = \frac{\alpha_2 d_h}{\lambda} = \frac{0.0232}{0.0272} \alpha_2 = 0.852 \, \alpha_2.$$

Next, calculate the Nu number pertaining to the two different values of α_2.

For
$$\alpha_2 = 81 \text{ W m}^{-2} \text{ °C}^{-1},$$
$$Nu = 0.852 \times 81 = 69,$$

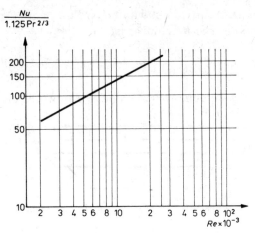

Fig. 6.106. Relation $Nu = 1.125\,Re^{0.545}\,Pr^{2/3}$

and for
$$\alpha_2 = 169\ \mathrm{W\ m^{-2}\,^\circ C^{-1}},$$
$$Nu = 0.852 \times 169 = 144.$$

The Pr number of air at 50 °C is $Pr = 0.722$, $Pr^{2/3} = 0.807$. In the equation of criteria, for the sake of uniformity, it is common practice to assume the form of variation of the Nu number as proportional to the (2/3)rd power of the Pr number. For this reason the (2/3)rd power will be calculated. Let the Nu number be a power function of the Pr number, and c be a proportionality factor:

$$Nu = cRe^n\,Pr^{2/3}. \tag{6.282}$$

To determine the values of unknown quantities c and n substitute the related Re and Nu numbers pertaining to $\alpha_2 = 81\ \mathrm{W\ m^{-2}\,^\circ C^{-1}}$ and to $\alpha_2 = 169\ \mathrm{W\ m^{-2}\,^\circ C^{-1}}$, into the above relation:

$$69 = c\ \ 2\,870^n \times 0.807,$$

$$144 = c\ \ 11\,100^n \times 0.807.$$

Solving the equations for the two unknowns, the following relation is obtained:

$$Nu = 1.125\,Re^{0.545} \times Pr^{2/3}. \tag{6.283}$$

The above equation obtained for the Nu number of the finned side is now, plotted in a log–log coordinate system, with the Re numbers on the abscissae axis and with quotient

$$\frac{Nu}{1.125\,Pr^{2/3}}$$

on the ordinate axis (see Fig. 6.106).

6.4.1.5 The fluid side heat transfer coefficient

The value of α_1 relating to the inner side of the tube and to the case of laminar flow can be calculated from the formula

$$Nu_m = k\,0.74Re_m^{0.2}(Gr \times Pr)_m^{0.1}\,Pr_m^{0.2}, \tag{6.284}$$

(see also Eqn. (6.188)). For tubes mounted vertically with a medium flowing downwards from above $k=0.85$, with a flow upwards from below (based on experimental data) $k=1.15$, and in the case of tubes arranged horizontally $k=1$. The physical properties of the flowing medium should be substituted into the equation using their values taken up at the mean boundary layer temperature, as indicated by subscript m. The value of α_1 in explicit form is

$$\alpha_1 = k\,0.74\lambda^{0.7}d_h^{-0.5}(\rho c_p)^{0.3}\left(\frac{\beta g}{\nu}\right)^{0.1}w^{0.2}\Delta\vartheta_{os}^{0.1}, \tag{6.285}$$

where $\Delta\vartheta_{os}$ is the temperature drop between oil and tube wall. The physical properties of the oil also have to be substituted into the equation for α_1 with their values taken at the mean boundary-layer temperature.

In the case of turbulent flow α_1 can be calculated just as before (see also Eqn. 6.253) from the formula

$$Nu_f = 0.023Re_f^{0.8}\,Pr_f^{0.4} \tag{8.286}$$

where the physical properties of the flowing oil should be substituted with their values taken at its average temperature. The value of α_1 in explicit form is

$$\alpha_1 = 0.023\frac{\lambda}{d_h}\left(\frac{wd_h}{\nu}\right)^{0.8}Pr^{0.4}. \tag{6.287}$$

6.4.1.6 Heat transfer effectiveness of internal fins
(see Fig. 6.107)

The heat transfer effectiveness of a straight fin of thermal conductivity λ and thickness δ_f, soldered into a tube of internal diameter d is

$$\eta_f = \frac{\mathrm{th}\left(m\dfrac{d}{2}\right)}{m\dfrac{d}{2}}, \tag{6.288}$$

where

$$m = \sqrt{\frac{2\alpha_1}{\lambda\delta_f}}. \tag{6.289}$$

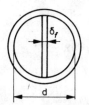

Fig. 6.107. Oil-side fin

The fin heat transfer effectiveness assumes a local decrease of fin temperatures over the whole fin surface area. The other parts of the surface are assumed to be at the temperature of the fin roots, i.e. having 100% heat transfer effectiveness.

Example 6.29. Let a fin be made of copper plate with $\lambda = 391$ W m^{-1} °C^{-1}, $d = 10 \times 10^{-3}$ m, and $\delta_f = 0.3 \times 10^{-3}$ m. Let α_1 be 80 W m^{-2} °C^{-1}. Calculate the fin heat transfer effectiveness and the value of quotient $\dfrac{1}{A_1 \alpha_1}$, °C W^{-1}, for both an unfinned and a finned surface. For the tube of 1 m length

$$A_1 = A_1' + A_1'' = d\pi \times 1 + 2d \times 1\eta_f.$$

The fin heat transfer effectiveness, with

$$m = \sqrt{\frac{2 \times 80}{391 \times 0.3 \times 10^{-3}}}\ \text{m}^{-1} = 37\ \text{m}^{-1},$$

$$\eta_f = \frac{\text{th}(37 \times 5 \times 10^{-3})}{37 \times 5 \times 10^{-3}} = \frac{\text{th}\,0.185}{0.185} = \frac{0.18291}{0.185} = 0.987.$$

$$\frac{1}{(A_1' + A_1''\eta_f)\alpha_1} = \frac{1}{(10 \times 10^{-3}\pi \times 1 + 2 \times 10 \times 10^{-3} \times 1 \times 0.987)80}\ \text{m}^2\,°\text{C W}^{-1} =$$

$$= \frac{1}{4.9}\ \text{m}^2\,°\text{C W}^{-1}$$

and for the unfinned tube:

$$\frac{1}{A_1\alpha_1} = \frac{1}{0.0314 \times 80}\ \text{m}^2\,°\text{C}^{-1} = \frac{1}{2.52}\ \text{m}^2\,°\text{C W}^{-1}.$$

6.4.1.7 Construction of a 100 kW oil–air cooler

Let us select for the oil-side the copper tube with inner fins shown in Fig. 6.107. On the air side the double spiral-wire cooling fins of Fig. 6.108 are soldered to the tube. The tube diameter is 10/12 mm, the inner fin is made of 0.3 mm thick copper plate, whereas the diameter of the air-side wire fins is 0.7 mm. Both inside and outside fins are fastened to the tube by soft soldering.

6.4.1.8 The oil side

The thermal power to be dissipated is $P = 100\,000$ W. The oil entering the cooler is at 70 °C, that leaving the cooler is at 65 °C. The specific heat of oil at 67.5 °C is $c = 2050$ W s kg^{-1} °C^{-1}. $\Delta\vartheta_{co} = (70 - 65)\,°\text{C} = 5\,°\text{C}$. The mass flow of the oil is

$$\Phi = \frac{P}{c\Delta\vartheta_{co}} = \frac{100\,000}{2050 \times 5}\ \text{kg s}^{-1} = 9.75\ \text{kg s}^{-1}.$$

The density of oil set at 67.5 °C is $\rho = 853.5$ kg m^{-3}. The oil volume passing through the cooler in a unit of time is

$$\psi = \frac{\Phi}{\rho} = \frac{9.75}{853.5}\ \text{m}^3\,\text{s}^{-1} = 0.01142\ \text{m}^3\,\text{s}^{-1}.$$

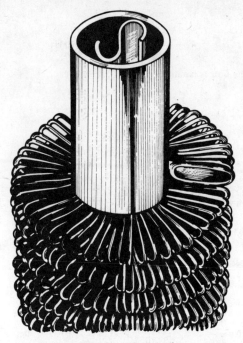

Fig. 6.108. Wire-finned tube

The cross-sectional area of the tube of $d = 10 \times 10^{-3}$ m internal diameter, with the 0.3×10^{-3} m thick fin taken into account, is

$$\left(\frac{10^2 \pi}{4} - 0.3 \times 10 \right) \text{mm}^2 = 75.5 \times 10^{-6} \, \text{m}^2 \,.$$

Let the reduction of the cross-sectional area for flow due to the soldering of the inner fin be 1.5 mm². Hence, the free cross-sectional area is $A_1 = 74 \times 10^{-6}$ m². Let the cooler be a double-pass unit on the oil side, each pass comprising 135 parallel tubes each 1.94 m long. The oil velocity in the tubes is

$$w_0 = \frac{\psi}{A_1} = \frac{0.01142}{135 \times 74 \times 10^{-6}} \, \text{m s}^{-1} = 1.142 \, \text{m s}^{-1} \,.$$

6.4.1.9 Heat transfer coefficient on the oil side

The hydraulic diameter of the tube of D-shape flow cross-section is

$$d_h = \frac{4A_1}{c} = \frac{4 \times 74}{2 \times 25.7} \, \text{mm} = 5.77 \, \text{mm} = 5.77 \times 10^{-3} \, \text{m} \,.$$

In the formula, the surface area wetted by the liquid is

$$c = \frac{d\pi}{2} + d = \left(\frac{10\pi}{2} + 10 \right) \text{mm} = 25.7 \, \text{mm} \,.$$

It is assumed that on the oil side, up to the inner wall of the tube, the heat drop is $\Delta \vartheta_{s1} = 10 \, °\text{C}$. Correspondingly, the mean boundary-layer temperature of the oil will be $\vartheta_m = (67.5 - 10/2) \, °\text{C} = 62.5 \, °\text{C}$. The physical properties of oil are calculated for this temperature.

521

The *Re* number is

$$Re = \frac{1.142 \times 5.77 \times 10^{-3}}{6.55 \times 10^{-6}} = 1008 \, .$$

This value means that the flow is laminar.

$$Re^{0.2} = 1008^{0.2} = 3.97 \, .$$

The *Pr* number is

$$Pr = 88.3; \quad Pr^{0.2} = 88.3^{0.2} = 2.45 \, .$$

The *Gr* number is

$$Gr = \frac{7.9 \times 10^{-4}(5.77 \times 10^{-3})^3 \times 9.81 \times 10}{(6.55 \times 10^{-6})^2} = 347$$

$$(Gr \times Pr)^{0.1} = (347 \times 88.3)^{0.1} = 2.8 \, .$$

The *Nu* number is

$$Nu = 0.74 \, Re^{0.2}(Gr \, Pr)^{0.1} \times Pr^{0.2} = 0.74 \times 3.97 \times 2.8 \times 2.45 = 20.2$$

$$\lambda = 0.12815 \text{ W m}^{-1}\,{}^{\circ}\text{C}^{-1} \, .$$

The heat transfer coefficient is

$$\alpha_1 = Nu \frac{\lambda}{d_h} = 20.2 \frac{0.12815}{5.77 \times 10^{-3}} \text{ W m}^{-2}\,{}^{\circ}\text{C}^{-1} = 477 \text{ W m}^{-2}\,{}^{\circ}\text{C}^{-1} \, .$$

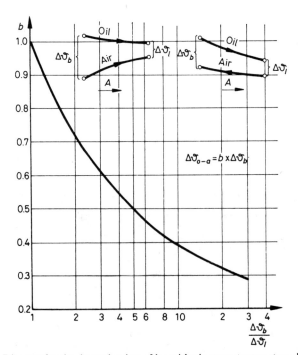

Fig. 6.109. Diagram for the determination of logarithmic-mean temperature differences

6.4.1.10 The logarithmic-mean temperature difference

The logarithmic-mean temperature difference between the temperatures of flowing media is calculated, in the case of both parallel-flow and counterflow heat exchangers, from the formula

$$\Delta \vartheta_{o-a} = \frac{\Delta \vartheta_b - \Delta \vartheta_l}{\ln \dfrac{\Delta \vartheta_b}{\Delta \vartheta_l}}. \tag{6.290}$$

The value $\Delta \vartheta_b$ and $\Delta \vartheta_l$ appearing in the above formula are the larger and smaller temperature difference between the media, respectively, interpreted according to the upper part of Fig. 6.109. The calculation of the logarithmic-mean temperature difference is facilitated by the use of the diagram of Fig. 6.109. The value of $\Delta \vartheta_b$ multiplied by b gives the value of $\Delta \vartheta_{o-a}$.

$$\Delta \vartheta_{o-a} = b \Delta \vartheta_b. \tag{6.291}$$

In the case of single-pass and multi-pass crossflow heat exchangers the logarithmic-mean temperature differences valid for counterflow and parallel-flow heat exchangers, respectively, are to be multiplied by correction factors ε. The correction factors valid for the double-pass parallel- and counterflow types of crossflow heat exchangers are given in Figs 6.110 and 6.111.

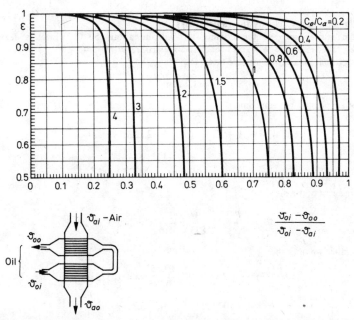

Fig. 6.110. Correction factor ε for cross-counterflow coolers. The parameter is the ratio C_0/C_a. The oil mixes in the oil chamber connecting the two oil paths

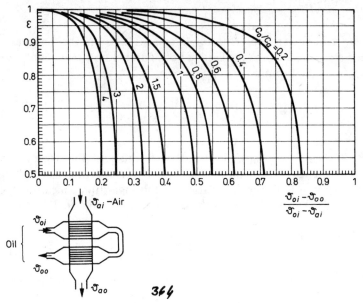

Fig. 6.111. Correction factor ε for crossflow coolers. The parameter is the ratio C_0/C_a. The oil mixes in the oil chamber connecting the two paths

6.4.1.11 The air side

Select for the air-side air velocity the value $3.2\ \mathrm{m\ s^{-1}}$.

The average temperature of the cooling air will be about $50\,^{\circ}\mathrm{C}$. For this temperature $\rho_a = 1.056\ \mathrm{kg\ m^{-3}}$, $c_p = 1016\ \mathrm{W\ s\ kg^{-1}\ ^{\circ}C^{-1}}$ and $\lambda_a = 0.0272\ \mathrm{W\ m^{-1}\ ^{\circ}C^{-1}}$.

The cross-sectional area of the air-side flow is

$$A_2 = (1.7/2)\ \mathrm{m^2}\ .$$

The volume of air passing per unit time is

$$\psi_a = w_a A_2 = 3.2\,\frac{1.7}{2}\ \mathrm{m^3\ s^{-1}} = 2.72\ \mathrm{m^3\ s^{-1}}\ .$$

The air flow

$$\Phi_a = \psi_a \rho_a = 2.72 \times 1.056\ \mathrm{kg\ s^{-1}} = 2.875\ \mathrm{kg\ s^{-1}}\ .$$

The temperature rise of the air

$$\Delta\vartheta_a = \frac{P}{\Phi_a c_p} = \frac{100\,000}{2.875 \times 1016}\ ^{\circ}\mathrm{C} = 34.2\ ^{\circ}\mathrm{C}\ .$$

For the Re number we note that

$$L = 0.27\ \mathrm{m},\quad d_h\,L = 4 \times 0.27\,\frac{0.0157}{0.295}\ \mathrm{m} = 0.0575\ \mathrm{m}\ ,$$

$$v = 18.6 \times 10^{-6}\ \mathrm{m^2\ s^{-1}},\quad w_a = 3.2\ \mathrm{m\ s^{-1}}\ ,$$

$$Re_f = \frac{3.2 \times 0.0575}{18.6 \times 10^{-6}} = 9900,\quad Re_f^{0.545} = 9900^{0.545} = 150\ .$$

The *Pr* number

$$Pr_f = 0.722, \quad Pr_f^{2/3} = 0.722^{2/3} = 0.807.$$

The *Nu* number

$$Nu_f = 1.125 \, Re_f^{0.545} \, Pr_f^{2/3} = 1.125 \times 150 \times 0.807 = 136.$$

The heat transfer coefficient

$$\alpha_2 = Nu_f \frac{\lambda_a}{d_h} = 135 \frac{0.0272}{0.0575} \text{ W m}^{-2}\,{}^\circ\text{C}^{-1} = 64.3 \text{ W m}^{-2}\,{}^\circ\text{C}^{-1}.$$

6.4.1.12 Heat transfer effectiveness of the fins

The oil-side fin heat transfer effectiveness:

$$m_1 = \sqrt{\frac{2 \times 447}{391 \times 0.3 \times 10^{-3}}} \text{ m}^{-1} = 87.3 \text{ m}^{-1},$$

$$\eta_{f1} = \frac{\text{th} \,(87.3 \times 5 \times 10^{-3})}{87.3 \times 5 \times 10^{-3}} = \frac{\text{th} \, 0.437}{0.437} = \frac{0.41114}{0.437} = 0.94.$$

The air-side fin heat transfer effectiveness:

$$m_2 = \sqrt{\frac{4 \times 64.3}{391 \times 0.7 \times 10^{-3}}} \text{ m}^{-1} = 30.6 \text{ m}^{-1},$$

$$\eta_{f2} = \frac{\text{th} \,(30.6 \times 9.42 \times 10^{-3})}{30.6 \times 9.42 \times 10^{-3}} = \frac{\text{th} \, 0.288}{0.288} = \frac{0.28026}{0.288} = 0.972.$$

6.4.1.13 The overall thermal conductance

Let the overall thermal conductance k be referred to the inner surface area A_1' of a smooth tube of 1 m length. Denote the overall thermal conductance referred to surface area A_1' by k_{A_1}, the area of the fin located in the tube by A_1'', the air-side surface area by A_2, the oil-side fin heat transfer effectiveness by η_{f1}, the oil-side heat transfer coefficient by α_1, that on the air side by α_2, the air-side fin heat transfer effectiveness by η_{f2}, the wall thickness of the tube by δ and its thermal conductivity by λ. In view of the small wall thickness, the surface area pertaining to the mean diameter of the tube A_m is taken as equal to A_1'. The reciprocal value of overall thermal conductance for heat transfer is

$$\frac{1}{k_{A_1'}} = \frac{1}{\left(1 + \dfrac{A_1''}{A_1'}\eta_{f1}\right)\alpha_1} + \frac{1}{\dfrac{\lambda}{\delta}} + \frac{1}{\dfrac{A_2}{A_1'}\eta_{f2}\alpha_2}. \tag{6.292}$$

Quantities appearing in the equation are $A_1' = d\pi 1 = 10 \times 10^{-3}\pi \times 1 \text{ m}^2 = 0.0314 \text{ m}^2$, $A_1'' = 2d \times 1 = 2 \times 10 \times 10^{-3} \times 1 \text{ m}^2 = 0.02 \text{ m}^2$, $A_2 = 0.295 \text{ m}^2$, $\delta = 1 \times 10^{-3} \text{ m}$, $\eta_{f1} = 0.94$, $\eta_{f2} = 0.972$, $\lambda = 391 \text{ W m}^{-1}\,{}^\circ\text{C}^{-1}$, $\alpha_1 = 447 \text{ W m}^{-2}\,{}^\circ\text{C}^{-1}$, $\alpha_2 = 64.3 \text{ W m}^{-2}\,{}^\circ\text{C}^{-1}$.

Substituting these values into the expression for the reciprocal of overall thermal conductance the first term on the right side of the equation will be

$$\frac{1}{\left(1+\dfrac{0.02}{0.0314}\,0.94\right)447}\ m^2\,{}^{\circ}C\ W^{-1}=\frac{1}{714}\,m^2\,{}^{\circ}C\ W^{-1}\,.$$

The second term on the right side of the equation is

$$\frac{1}{\dfrac{391}{1\times 10^{-3}}}\ m^2\,{}^{\circ}C\ W^{-1}=\frac{1}{391\,000}\,m^2\,{}^{\circ}C\ W^{-1}\,,$$

and the third term on the right side of the equation:

$$\frac{1}{\dfrac{0.295}{0.0314}\,0.972\times 64.3}\ m^2\,{}^{\circ}C\ W^{-1}=\frac{1}{587}\,m^2\,{}^{\circ}C\ W^{-1}\,.$$

The reciprocal of the overall conductance for heat transfer is

$$\frac{1}{k_{A_i'}}=\left(\frac{1}{714}+\frac{1}{391\,000}+\frac{1}{587}\right)m^2\,{}^{\circ}C\ W^{-1}\,.$$

Thus the overall conductance for the heat transfer will be

$$k_{A_i'}=322\ W\ m^{-2}\,{}^{\circ}C^{-1}\,.$$

6.4.1.14 The temperature drop

The cooler contains 270 tubes, each of length 1.94 m, dissipation capacity of the cooler is $P=100\,000$ W, the share of each tube being

$$P'=-\frac{100\,00}{270\times 1.94}\ W=191\ W\,.$$

The oil-side temperature drop

$$\varDelta\vartheta_{s1}=\frac{P'}{(A_1'+A_1''\eta_{f1})\alpha_1}=\frac{191}{(0.0314+0.02\times 0.94)\,447}\ {}^{\circ}C=8.48\ {}^{\circ}C\,.$$

The air-side temperature drop

$$\varDelta\vartheta_{s2}=\frac{P'}{A_2\eta_{f2}\alpha_2}=\frac{191}{0.295\times 0.972\times 64.3}\ {}^{\circ}C=10.32\ {}^{\circ}C\,.$$

The temperature drop in the tube wall

$$\varDelta\vartheta_{\delta}=\frac{P'}{\dfrac{\lambda}{\delta}}=\frac{191}{391\,000}\ {}^{\circ}C=0.000487\ {}^{\circ}C\,.$$

It is clear that the temperature drop across the tube wall may be neglected. The total temperature drop will, therefore, be the sum of the oil-side and air-side temperature drops:

$$\varDelta\vartheta_{o-a}=18.8\ {}^{\circ}C\,.$$

The same result is arrived at when calculating with the overall conductance for heat transfer $k_{A_i'}$:

$$\varDelta\vartheta_{o-a}=\frac{P}{k_{A_i'}A_1'\,270\times 1.94}=\frac{100\,000}{322\times 0.0314\times 270\times 1.94}\ {}^{\circ}C=18.8\ {}^{\circ}C\,.$$

526

6.4.1.15 The corrected logarithmic-mean temperature difference

The total temperature drop obtained in the preceding section is, at the same time, equal to the logarithmic-mean temperature difference multiplied by the correction factor ε: (see Eqns 6.149 and 6.290)

$$\Delta\vartheta_{o-a} = \varepsilon \frac{\Delta\vartheta_b - \Delta\vartheta_l}{\ln\dfrac{\Delta\vartheta_b}{\Delta\vartheta_l}}. \tag{6.293}$$

Assume that $\Delta\vartheta_b = 33\,°C$ and $\Delta\vartheta_l = 8.8\,°C$;

$$\frac{\Delta\vartheta_b}{\Delta\vartheta_l} = 3.75\,, \ b = 0.56\,.$$

The values of correction factors are shown by Figs. 6.110 and 6.111. We use Fig. 6.110 valid for cross-counterflow coolers. Select the value of ε to be 1, then

$$\Delta\vartheta_{o-a} = 1 \times b \times 33\,°C = 1 \times 0.56 \times 33\,°C = 18.5\,°C\,.$$

This value does not differ much from the total temperature drop. The real temperature rise of the oil will just reach the level as high at which the logarithmic-mean temperature difference is equal to the temperature drop pertaining to the heat transfer. The correction factor ε used in calculating the logarithmic-mean temperature difference has been taken equal to 1. As seen from Fig. 6.110, the value of the correction factor for the water ratio value $K = 0.1455$ is indeed equal to 1. The temperature variations of the oil and the air along the heat transfer surface are shown in Fig. 6.112.

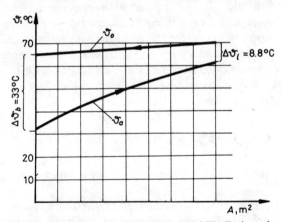

Fig. 6.112. Temperatures of media of a 100 kW oil–air cooler

6.4.1.16 Heat transfer efficiency of the heat exchanger

The flow-stream capacity rates:
For the oil side, $C_w = \Phi_w c_w = 9.75 \times 2050\ \text{W °C}^{-1} = 20\ 000\ \text{W °C}^{-1}$.
For the air side, $C_c = \Phi_c c_c = 2.875 \times 1015\ \text{W °C}^{-1} = 2915\ \text{W °C}^{-1}$.
Also

$$C_{max} = C_w\,, \ C_{min} = C_c\,.$$

527

The heat transfer efficiency

$$\varepsilon = \frac{P}{P_{max}} = \frac{2915(61.2-32)}{2915(70-32)} = \frac{29.2}{38} = 0.7675 \,.$$

Number of heat transfer units

$$kA = 322 \times 0.0314 \times 270 \times 1.94 \text{ W }^\circ\text{C}^{-1} = 5290 \text{ W }^\circ\text{C}^{-1} \,,$$

$$N = \frac{kA}{C_{min}} = \frac{5290}{2915} = 1.815 \,.$$

The flow-stream capacity rate ratio

$$K = \frac{C_{min}}{C_{max}} = \frac{2915}{20\,000} = 0.1455 \,.$$

6.4.2 Hydraulic design of compact heat exchangers

6.4.2.1 Oil-side static pressure drop

The oil-side static pressure drop consists of three parts, viz. the inlet and outlet pressure drops and that caused by the pipe friction.

The static inlet pressure drop is again composed of two parts. In the formula given below, the first term is the pressure drop arising from the reduction of the cross-sectional area of flow that can be calculated by means of Bernoulli's equation valid for ideal, friction-free fluids. The second part of the inlet pressure drop is the result of the irreversible free expansion occurring in all cases of sudden reduction of cross-sectional area of flow. The reason for this expansion is that at the inlet section of the tube, separation of the boundary layer takes place and, a rearrangement of the velocity profile occurs. Both phenomena are associated with loss of mechanical energy due to transformation into heat, with the consequential pressure drop.

The inlet pressure drop is

$$\Delta p_1 = \rho \frac{w^2}{2}(1-\sigma^2) + \rho \frac{w^2}{2} K_c = \frac{\rho w^2}{2}(1-\sigma^2 + K_c) \,. \tag{6.294}$$

The value of K_c valid for laminar flows can be taken from the diagram shown in Fig. 6.113.

The static outlet pressure rise also consists of two parts, as shown below. The first term is the pressure rise from the increase of the cross-sectional area of flow calculated by means of Bernoulli's equation valid for ideal, friction-free fluids. The second part is a pressure drop, being the consequence of the irreversible free expansion taking place at every sudden increase of the cross-sectional area of flow, due to boundary layer separation and to rearrangement of the velocity profile.

The outlet pressure rise is:

$$\Delta p_2 = \rho \frac{w^2}{2}(1-\sigma^2) - \frac{\rho w^2}{2} K_e = \frac{\rho w^2}{2}(1-\sigma^2 - K_e) \,. \tag{6.295}$$

The value of K_e for laminar flows can also be taken from the diagram of Fig. 6.113.

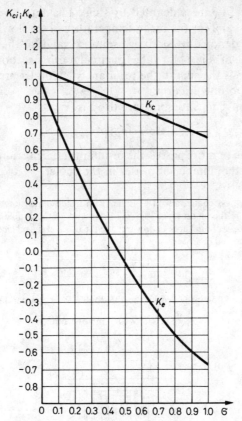

Fig. 6.113. Factors K_c and K_e in the range $4\left(\dfrac{L}{d_h}\right) Re \approx 1$ to ∞, for laminar flow

For one pass, the resultant static inlet and outlet pressure drop is given by the difference between the static inlet pressure drop and the static outlet pressure rise.

$$\Delta p' = \Delta p_1 - \Delta p_2 = \rho \frac{w^2}{2} (K_c + K_e). \qquad (6.296)$$

Pressure drop due to flow friction: The pressure drop caused by flow friction is

$$\Delta p_{fr} = \xi \frac{l}{d_h} \rho \frac{w^2}{2}. \qquad (6.297)$$

As the oil flow in the cooling tubes is not isothermal, the value of the friction coefficient ξ is calculated by Mihevev's formula.

$$\xi = \frac{A}{Re_f} \left(\frac{Pr_w}{Pr_f}\right)^{\frac{1}{3}} \left[1 + 0.22 \left(\frac{Cr_f Pr_f}{Re_f}\right)^{0.15}\right], \qquad (6.298)$$

where the similarity numbers denoted by suffix f have to be calculated with the physical quantities valid for the mean temperature of the flowing oil, whereas the similarity numbers denoted by suffix w use those valid for the temperature of the tube wall. The value of A is 64 for tubes of circular cross-section. The total oil-side static pressure drop is the sum of the resultant inlet–outlet pressure drop and the pressure drop due to flow friction.

Thus, the total oil-side static pressure drop is

$$\Delta p = \Delta p' + \Delta p_{fr}. \tag{6.299}$$

For coolers with two or more passes, the resultant static pressure drop is that of one pass multiplied by the number of passes in the cooler.

Assume, in a double-pass cooler, an oil chamber with inside dimensions of 290×890 mm in the direction perpendicular to the oil flow (see Fig. 6.114). Because of the double-pass arrangement of the cooler, the available cross-sectional area of flow outside the tubes in the oil chamber is half of the inner cross-sectional area of the oil chamber.

$$A_{ch} = \frac{0.29 \times 0.89}{2} \ \mathrm{m^2} = 0.129 \ \mathrm{m^2} .$$

The cross-sectional area of flow provided by the 135 double D-section parallel tubes is

$$A_t = 135 \times 74 \times 10^{-6} \ \mathrm{m^2} = 0.01 \ \mathrm{m^2} .$$

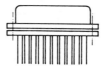

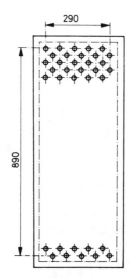

Fig. 6.114. Oil chamber dimensions

The ratio of flow sections

$$\sigma = \frac{A_t}{A_{ch}} = \frac{0.01}{0.129} = 0.0775, \ \sigma^2 = 0.006.$$

The value of the dimensionless ratio

$$\frac{4\dfrac{L}{d_h}}{Re}$$

for the mean temperature of the flowing oil, i.e. for 67.5 °C, is

$$\frac{4\dfrac{L}{d_h}}{Re} = \frac{4\dfrac{1.94}{5.77 \times 10^{-3}}}{\dfrac{1.142 \times 5.77 \times 10^{-3}}{5.75 \times 10^{-6}}} = \frac{1345}{1148} = 1.172.$$

From the diagram of Fig. 6.113: $K_c = 1.04$, $K_e = 0.8$.
The inlet pressure drop is

$$\Delta p_1 = \rho \frac{w^2}{2}(1 - \sigma^2 + K_c) \ \mathrm{N\,m^{-2}},$$

where

$$w = 1.142 \ \mathrm{m\,s^{-1}}, \ \frac{w^2}{2} = \frac{1.142^2}{2} \ \mathrm{m^2\,s^{-2}} = 0.66 \ \mathrm{m^2\,s^{-2}}$$

and

$$\rho = 853.5 \ \mathrm{kg\,m^{-3}}.$$

Thus

$$\Delta p_1 = 853.5 \times 0.66(1 - 0.006 + 1.04) \ \mathrm{N\,m^{-2}} = 1135 \ \mathrm{N\,m^{-2}}.$$

The outlet pressure rise is

$$\Delta p_2 = \rho \frac{w^2}{2}(1 - \sigma^2 - K_e) \ \mathrm{N\,m^{-2}},$$

$$\Delta p_2 = 853.5 \times 0.66(1 - 0.006 - 0.8) \ \mathrm{N\,m^{-2}} = 111 \ \mathrm{N\,m^{-2}}.$$

For one pass the resultant pressure drop is the difference between inlet pressure drop and outlet pressure rise. Since the cooler is a double-pass unit, the resultant pressure drop is twice that of one pass.

$$\Delta p' = 2(\Delta p_1 - \Delta p_2) = 2(1135 - 111) \ \mathrm{N\,m^{-2}} = 2048 \ \mathrm{N\,m^{-2}}.$$

To determine the pressure drop caused by flow friction, the quantities appearing in Miheyev's equation for the friction coefficient are first calculated. At the wall temperature of $(67.5 - 8.5) \,^{\circ}\mathrm{C} = 59 \,^{\circ}\mathrm{C}$, $Pr_w = 96.7$. At the temperature of the oil flow, i.e. at 67.5 °C,

$$Pr_f = 78.3, \ v_f = 5.75 \times 10^{-6} \ \mathrm{m^2\,s^{-1}},$$

$$\beta_f = 7.92 \times 10^{-4} \ ^{\circ}\mathrm{C^{-1}},$$

$$\Delta \vartheta_s = 8.5 \,^{\circ}\mathrm{C}, \ \rho_f = 853.5 \ \mathrm{kg\,m^{-3}}.$$

The Gr number

$$Gr_f = \frac{7.92 \times 10^{-4} \times (5.77 \times 10^{-3})^3 \times 9.81 \times 8.5}{(5.75 \times 10^{-6})^2} = 384.$$

34*

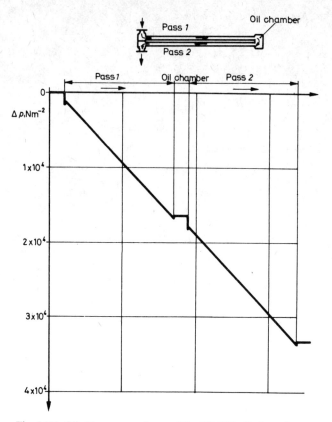

Fig. 6.115. Oil-side pressure drops of the 100 kW oil–air cooler

The *Re* number

$$Re_f = \frac{1.142 \times 5.77 \times 10^{-3}}{5.75 \times 10^{-6}} = 1145.$$

Thus

$$\xi = \frac{64}{1145}\left(\frac{96.7}{78.3}\right)^{1/3}\left[1 + 0.22\left(\frac{384 \times 78.3}{1145}\right)^{0.15}\right] = 0.0816.$$

For the two passes $l = 2 \times 1.94$ mm.
The pressure drop due to flow friction is given by relation (6.297)

$$\Delta P_{fr} = \xi\,\frac{l}{d_h}\,\rho\,\frac{w^2}{2}$$

$$\Delta p_{fr} = 0.0816\,\frac{2 \times 1.94}{5.77 \times 10^{-3}}\,853.5\,\frac{1.142^2}{2}\ \text{N m}^{-2} = 30\,500\ \text{N m}^{-2}.$$

The total oil-side pressure drop is

$$\Delta p = 2(\Delta p_1 - \Delta p_2) + \Delta p_{fr} = (2048 + 30\,500)\ \text{N m}^{-2} = 32\,548\ \text{N m}^{-2}.$$

The oil-side pressure drops are shown in Fig. 6.115.

532

6.4.2.2 The air-side static pressure drop

The air-side static pressure drop consists, in principle, of four parts. The first is the inlet pressure drop comprising the pressure drop resulting from the boundary-layer separation, characterized by the inlet loss factor, and the pressure drop resulting from the increase of velocity due to the reduced cross-sectional area of flow. The second part is due to the work used in accelerating the gas or air expanding under the effect of rising temperature and manifesting itself as a pressure drop. The velocity profile of the air entering the cooler with velocity w_1, assuming laminar flow in a tube of circular section, first goes over into many small bodies of air of paraboloid-shape velocity profiles with identical mean velocity w_1. The work required for this transformation of velocity profile is equal to the kinetic energy of the air flowing with identical velocity w_1 throughout the entire cross-sectional area of the flow. The second part of the accelerating work is represented by the energy required for transforming this body of air of paraboloid-shape velocity distribution and mean velocity w_1 into another body of air of a higher mean velocity w_2, but also having a paraboloid-shape velocity distribution. This explains the multiplying factor 2 of this part of the pressure drop in the formula.

The third part of the pressure drop results from the friction of the air flowing in the cooler. The coefficient of friction, fr valid for air-side fins of different shapes and based on experimental data, are given by various text books as a function of the air-side Re number defined in the foregoing sections.

The fourth part comprises the discharge loss characterized by K_e and the pressure rise contribution resulting from the decrease of velocity.

$$\Delta p = \frac{1}{2}\rho_1 w_1^2 \left[(K_c + 1 - \sigma^2) + 2\left(\frac{\rho_2}{\rho_1} - 1\right) + fr\frac{A}{A_c}\frac{\rho_a}{\rho_1} - (1 - \sigma^2 - K_e)\frac{\rho_2}{\rho_1} \right]$$

(6.300)

The four pressure components on the right-hand side of the formula are inlet, acceleration, friction, outlet, respectively. ρ_1 is the density of inlet air; w_1 is the velocity of inlet air before the cooler in section 1 (see Fig. 6.116), ρ_2 is the density of outlet air in section 2, calculated for the outlet temperature of the air. ρ_a is the density of air flowing in the cooler, referred to the mean temperature calculated by the logarithmic-mean temperature difference. When calculating the densities it should be remembered that ρ_a and ρ_2 refer to pressures p_a and p_2, respectively, where both pressures are lower than pressure p_1 of the inlet air.

For calculation of the air-side static pressure drop of the 100 kW oil–air cooler a relation $\Delta p = f(Re)$ obtained from experimental data is available, by which the total static pressure drop is provided as a function of the Re number of the air.

The mean temperature of the air flowing in the cooler is

$$\vartheta_a = (67.5 - 18.8)\,°C = 48.7\,°C,$$

$$v_a = 18.4 \times 10^{-6}\,m^2\,s^{-1}.$$

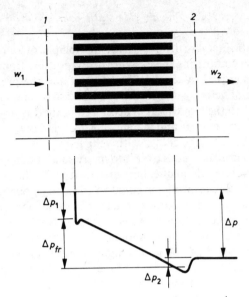

Fig. 6.116. Air-side pressure drop and pressure rise

As calculated further above, $d_h = 0.0575$ m

$$w_a = w_1 \frac{T_a}{T_1} \frac{p_1}{p_1 - \frac{p_a}{2}} = 3.04 \frac{273 + 50}{273 + 34} \frac{1}{1 - \frac{18 \times 10^{-4}}{2}} \text{ m s}^{-1} = 3.2 \text{ m s}^{-1}.$$

The Re number

$$Re = \frac{3.2 \times 0.0575}{18.47 \times 10^{-6}} = 10\,000.$$

The cooler contains 10 rows of cooling tubes in the direction of air flow, with $L = 0.27$ m.

$$d_h = 4 \times 0.27 \frac{0.0157}{0.295} \text{ m} = 0.0575 \text{ m}.$$

The formula for pressure drop obtained from experimental data is

$$\Delta p = 136 \times 10^{-8} \, Re^2 \text{ N m}^{-2}.$$

The air-side pressure drop is

$$\Delta p = 136 \times 10^{-8} \times 10\,000^2 \text{ N m}^2 = 136 \text{ N m}^2.$$

The next step of cooler design is to select, both for the oil side and air side, suitable pumps and fans capable of providing the required flows of oil and air at the given pressure drops.

6.4.2.3 Rating of the motor-driven pump and motor-driven fans

Power input of pump at the rated loads, is

$$P_p = \frac{1}{\eta_p \eta_m} \frac{\Delta p}{\rho_0} \Phi_0. \tag{6.301}$$

If the efficiency of the pump $\eta_p = 0.7$, motor efficiency $\eta_m = 0.75$, $\Delta p = 32\,548$ N m^{-2}, $\rho_c = 853.5$ kg m^{-3} and $\Phi_0 = 9.75$ kg s^{-1}, then

$$P_p = \frac{1}{0.7 \times 0.75} \frac{32\,548}{853.5} 9.75 \text{ W} = 707 \text{ W}.$$

The total power input of the two motor-driven fans at the rated load is

$$P_f = \frac{1}{\eta_f \eta_m} \frac{\Delta p}{\rho_a} \Phi_a. \tag{6.302}$$

Let the efficiency of the fan part of the motor-driven fan be $\eta_f = 0.6$ and that of the driving motor be $\eta_m = 0.7$. The estimated pressure drop resulting from the inlet loss of the air entering the fans is $\Delta p' = 54$ N m^{-2}. The pressure drop in the cooler is $\Delta p'' = 136$ N m^{-2}. The total pressure drop is

$$\Delta p = \Delta p' + \Delta p'' = (54 + 136) \text{ N m}^{-2} = 190 \text{ N m}^{-2}, \tag{6.303}$$

$$\rho_a = 1.06 \text{ kg m}^{-3},$$

$$\Phi_a^+ = 2.875 \text{ kg s}^{-1}.$$

Thus

$$P_f = \frac{1}{0.6 \times 0.7} \frac{190}{1.06} 2.875 \text{ W} = 1220 \text{ W}.$$

7. CONSTRUCTIONAL PARTS
OF THE TRANSFORMER

7.1 Limb and yoke clamping structures

7.1.1 Limb clamping

Bonding

The edges of the sheets (laminations) in the limbs and in the bottom yoke of a fully stacked and provisionally clamped core are sprayed with adhesive epoxy-based varnish. By capillary action, the adhesive varnish penetrates the gaps between the sheets to a depth of about 20 to 25 mm. Connecting the sheet edges together reduces the magnetic noise of the core, binds the core together mechanically, prevents further rusting of cut edges of the sheets in the course of vacuum drying, and insulates the edges from each other, thereby reducing the risk of total or partial inter-lamination short-circuits in the core. The drawback of a bonded core is that it can no longer be dismantled, except by breaking it into pieces.

Wedging

Some factories follow the practice of combining core bonding with a method of wedging by means of channels made of hard aluminium alloy. As shown in Fig. 7.1, insulation is placed on the narrowest sheet in the yoke to carry the said hard aluminium channel, in which a wedge is laid. By driving the wedge between the core and cylinder isolating the L. V. winding from the core, a firm mechanical connection is established between the limb and the L. V. winding.

Banding

Glass-epoxy band

This band material is a glass-fibre tape impregnated with thermo-setting epoxy resin. Before baking the tape has plastic properties. Two different kind of band can be made of it.

(a) Wound-on band without contractive elements
The laminations, laid horizontally on the floor, are compressed by a hydraulic tool at a point adjacent to that where the band is to be applied. A pressboard strip and, over it, a few layers of plastic glass-epoxy tape is wound. Then, the layers are made to stick together by application of heat. Baking of the band takes place simultaneously with vacuum drying of the transformer. An advantage of the method

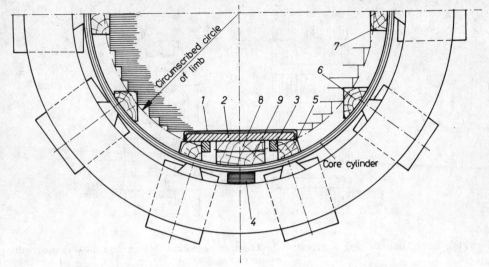

Fig. 7.1. Wedging by hard aluminium channel: *1* — insulating plate; *2* — clamping plate; *3* — cleat; *4* — strip; *5* to *7* — core fillet; *8* to *9* — wedging strip

is that no metallic clamping element is used. Its disadvantages are that the hydraulic equipment required for compressing the laminations and the banding machine are expensive; and since no subsequent corrective readjustment is possible, the banding has to be cut off and applied anew if found to be loose; finally, the result of baking cannot be checked, because any subsequent inspection is prevented by the presence of the windings. An arrangement of core clamping by banding wound on without using contractive elements is shown in Fig. 7.2.(a).

(b) Pre-baked band with contractive element

The bands are prepared in advance, independently of the core. From a glass-epoxy tape, still in its plastic condition, a circular hank is wound containing several layers and of a circumference corresponding to the periphery of the core. The contractive elements, made of plastic are pushed through the hank, and it is forced around a shaping drum of a diameter equal to that of the core by means of clamping bolts. The hank is then put into the baking oven. Through baking, the bands become bunched up into a flexible unit retaining its circular shape. It is so elastic that, when opened, it can be placed around the core. The plastic clamping elements are contracted by means of steel bolts (see Fig. 7.2(b)). An advantage is that its production is independent of core manufacturing, so it can be prepared and checked in advance. As long as the windings are not mounted on the core, it can be retightened at any time. Its drawback consists of being more complicated than banding without clamping elements, and because of the clamping elements, it is more costly.

As is commonly known, the tensile strength of glass is very high, $\sigma_B = 1$ to 1.2×10^9 N m^{-2}. Its elongation at rupture, however, is unusually low: $\varepsilon_B = 2$ to 3%.

537

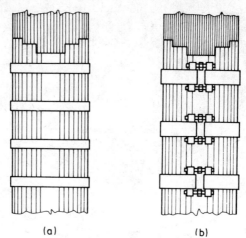

Fig. 7.2. (a) Limb clamping with band, without contracting element; (b) limb clamping with band, with contracting element

This means that a glass-epoxy band is unsuitable for the task of compressing the required parts, unless combined with some kind of elastic element. Even a very slight change in the longitudinal dimension leads to a total loosening of the banding, if no elastic element is coupled in series with the inelastic glass. In the case bands wound on the core the lamination itself acts as an elastic element, and in the type provided with contractive elements, the core and the clamping bolts fulfil this task.

The retightening of clamping elements cannot be avoided because of the presence of oil duct supporting pressboard spacers which are subject to shrinkage during drying. This brings about the necessity of drying of the core before it is mounted. The glass-epoxy banding may only be seated on some padding material—on wood or pressboard—because the edges of the steel core would cut it through.

Steel bands

(a) "Buckle of belt" type steel bands, (b) Adjustable steel bands

Some factories use the method termed "buckle of belt" type steel banding, as shown in Fig. 7.3, up to 10 MVA type rating. Before the band is mounted, its vicinity has to be compressed by a hydraulic tool. After mounting, the bent-back steel strip is secured by spot welding against opening. In order to avoid the formation of a closed turn around the core, the banding structure comprises a steel-reinforced insulating element provided with a cloth-base phenol coating.

Recently, for clamping the limbs of larger units, the adjustable steel bands shown in Fig. 7.4. have been introduced.

538

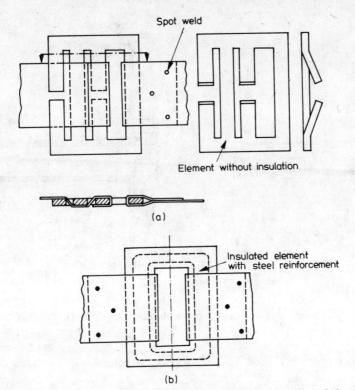

Fig. 7.3. Buckle of belt-type steel band limb clamping: (a) prefabricated part of belt; (b) insulated element interrupting continuity of steel band to prevent formation of a closed turn around the limb

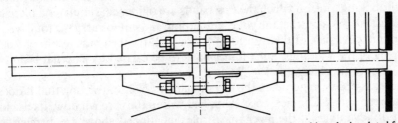

Fig. 7.4. Contracting element of steel band type of limb clamping. The band is to be insulated from the core lamination and an insulating element is to be inserted at one point. The steel band should be connected to the core at one point

Clamping by means of through-going studs

Through-going studs were formerly widely used for clamping purposes, but this method is now only applied for clamping the core packets of small gapped cores and the core limbs of furnace transformers. The through-going studs are either pushed through a phenol-based paper tube, or the same varnished paper is baked over the stud (see. Fig. 7.5).

539

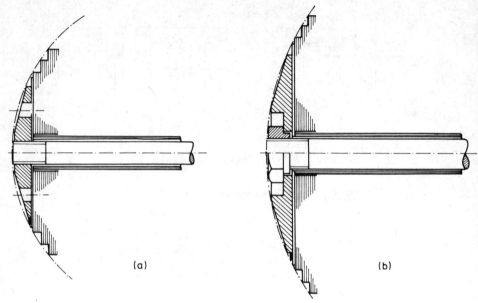

Fig. 7.5. Bolted clamps used for (a) smaller diameters; (b) larger diameters

7.1.2 Yoke contracting structures

Yoke contracting glass-epoxy bands

These are prepared in a similar way to pre-baked core bands, with the difference that the baking tool, on which the raw bands are tightened, simulates a piece of the yoke. Its task is to contract the yoke lamination, and not to carry the total weight of the entire active part of the transformer, nor to carry the loads resulting from the clamping of the windings. Such forces have to be carried by external beam structures or core tie plates (see Fig. 7.6).

The glass-epoxy band offers the advantage for yoke contracting that it does not increase the end spacer clearance, if applied within the core window. Its use brings about disadvantages if the mechanical rules mentioned above are disregarded. It has not come into general use.

Yoke contracting steel bands

When applied inside the core window, the end spacer distance has generally to be increased, and the banding must be carefully isolated from the core and yoke beam.

540

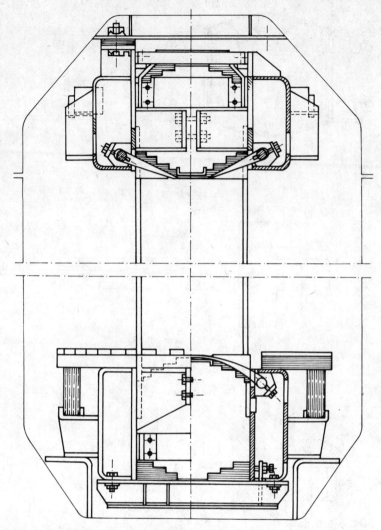

Fig. 7.6. Glass-epoxy band for contracting the yoke. The band may have the role of yoke contracting only

Yoke contracting external cylindrical bolt and flat steel

These are the most widely used contracting elements. With the use of flat steel contracting structures within the core window, the size of the end spacer is smaller than in the case of cylindrical bolts (see Fig. 7.7).

When applied outside the yoke they may be used, to a limited extent, for lifting. Their drawback is the need to increase the end spacer. Their advantage is simplicity and ease of their design. In H.V. transformers, the internal yoke bolts are also cylindrical (Fig. 7.8).

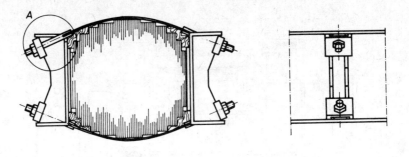

Section A

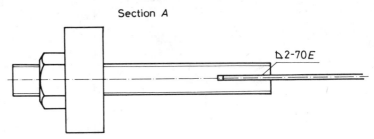

△2-70E

Fig. 7.7. Yoke contracting flat steel

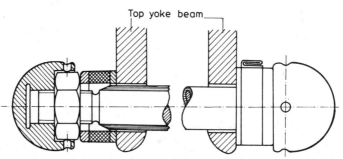

Top yoke beam

Fig. 7.8. Yoke contracting stud for contracting the yoke of large transformers, inside the core window

Yoke bolts

The force is applied to the yoke from outside by the yoke contracting structures described so far. This kind of bolt is pushed through the hole across the yoke laminations (see Fig. 7.9). It is excellent for contracting the yokes of framed cores. Its drawback is that the yoke of a non-framed core has to have holes which is undesirable in the case of cold-rolled steel cores.

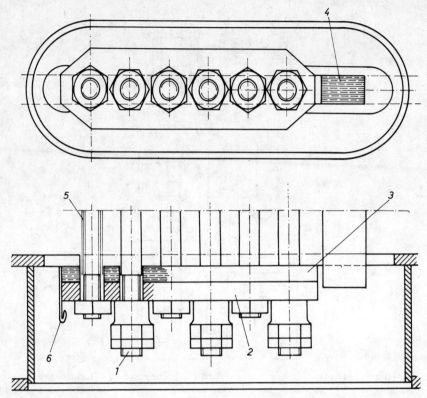

Fig. 7.9. Yoke contracting bolt clamping of framed core: *1* — stud; *2* — clamping plate; *3* — laminated wooden shim; *4* — laminated wooden spacer; *5* — baked-on insulation; *6* — earthing plate

Hammer-head flat steel yoke contracting clamping

The slots between frames of semi-framed or framed cores present themselves as places to mount a flat yoke contracting element. Such an internal contracting element is the only one having advantages with no disadvantages (see Fig. 7.10). It is simple, and the clamp can be re-tightened at any later time. Representing a further advantage, the yoke is compressed symmetrically through the frame beams.

7.1.3 Design of the limb-compressing glass-epoxy band

The core is clamped, at the places indicated in Fig. 7.11, by the bands through the tie plate and spacers. Under the effect of the clamping force a pressure p, $N\,m^{-2}$, arises in the core, (see Fig. 7. 12). Let us introduce the following notations:

A, m^2 is the cross-sectional area of the glass in the glass-epoxy band,

P, N   drawing force of the band,

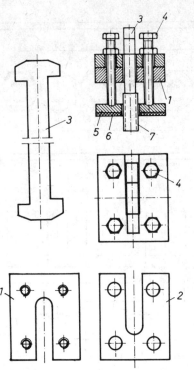

Fig. 7.10. Hammer-head yoke clamping structure: *1* and *2* — bearing plates; *3* — hammer-head flat steel; *4* — tightening bolt; *5* — yoke beam; *6* — yoke insulation; *7* — insulation of hammer-head flat steel

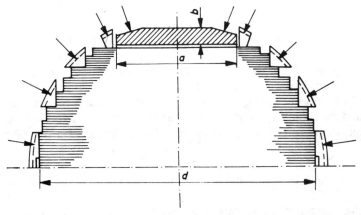

Fig. 7.11. A radial pressure on the core is produced by the band via the tie plate and wedging strips

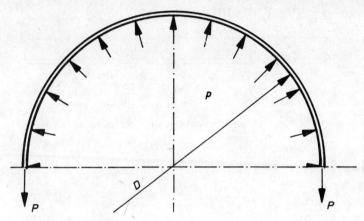

Fig. 7.12. Pressure p arises in the core under the effect of compression

d, m	width of widest sheet in the limb,
D, m	mean diameter of band,
p, Nm^{-2}	pressure exerted by the band,
p', Nm^{-1}	distributed load imposed on the tie plate,
l, m	spacing between two bands,
σ_g, Nm^{-2}	tensile stress arising in the glass,
σ_{ga}, Nm^{-2}	permissible tensile stress for glass,
σ_s, Nm^{-2}	bending stress arising in the tie plate,
σ_{sa}, Nm^{-2}	permissible tensile stress for the steel material of the tie plate,
a, m and b, m (see Fig. 7.11)	dimensions of the cross-sectional area of the tie plate,
M, Nm	moment imposed on the tie plate,
K_x, m^3	section modulus of the tie plate.

On the basis of Fig. 7.11,

$$pDl = 2P \tag{7.1}$$

and

$$P = A\sigma_g. \tag{7.2}$$

Thus

$$p = \frac{2A\sigma_g}{Dl} \tag{7.3}$$

and

$$\sigma_g = \frac{Dl}{2A}\, p. \tag{7.4}$$

The tie plate can be regarded as a multi-support beam rod carrying a distributed load of

$$p' = pa, \tag{7.5}$$

where the spacing of supports is l, equal to the spacing of bands (see Fig. 7.13).

35

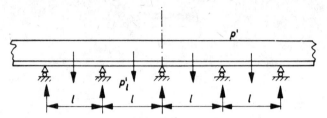

Fig. 7.13. Tie plate as continuous girder with distributed load

According to Clapeyron's equation, the moment acting on the beam (disregarding the sign) is

$$M = \frac{p'l^2}{12} = \frac{pal^2}{12}.$$ (7.6)

Taking the cross-sectional area of the tie plate—for the sake of simplicity—as a rectangle, the section modulus referred to the axis of bending is

$$K_x = \frac{ab^2}{6}.$$ (7.7)

The maximum bending stress is

$$\sigma_s = \frac{M}{K_x} = \frac{l^2}{2b^2} p.$$ (7.8)

Substituting the value of p from Eqn. (7.3) into Eqn. (7.8), and re-arranging the latter, with the permissible stresses for the materials concerned, the following formula is obtained for the spacing of bands

$$1 = \frac{\sigma_{sa}}{\sigma_{ga}} \frac{Db^2}{A}.$$ (7.9)

As seen from the above, the tie plate also has the task of core clamping, due to which a bending stress σ_s will arise in it. For the tie plate, the design stress is the sum resulting from compressive force imposed on the winding and from that due to the lifting of active parts. No overstressing occurs, if the stress caused by core clamping does not exceed 1/10 of the stress permissible for the material of the tie plate.

Example 7.1. Consider a glass-epoxy tape of dimensions 30×0.33 mm, and a band made of 14 parallel tapes. The cross-sectional area of the band is

$$A = 14 \times 30 \times 0.33 \times 10^{-6} \text{ m}^2 = 138.6 \times 10^{-6} \text{ m}^2.$$

Let us assume

$$\sigma_{sa} = \frac{137}{10} 10^6 \text{ Nm}^{-2} = 13.7 \times 10^6 \text{ Nm}^{-2}.$$

The tensile strength of glass is 1 to 1.2×10^9 Nm^{-2}. Let the permissible bending strength be:

$$\sigma_{ga} = 100 \times 10^6 \text{ Nm}^{-2}.$$

Let $b = 18 \times 10^{-3}$ m, $D = 0.5$ m.

546

Spacing between centre lines of bands is

$$l = \frac{\sigma_{sa}}{\sigma_{ga}} \frac{Db^2}{A} = \frac{13.7}{100} \frac{0.5 \times 18^2 \times 10^{-6}}{138.5 \times 10^{-6}} = 0.16 \text{ m}.$$

Let us now find the stresses arising in the band and in the tie plate. From Eqn. 7.4, calculating with a pressure $p = 1.5 \times 10^5 \text{ N m}^{-2}$,

$$\sigma_g = \frac{0.5 \times 0.16}{2 \times 138.6 \times 10^{-6}} 1.5 \times 10^5 \text{ Nm}^{-2} = 43.2 \times 10^6 \text{ Nm}^{-2}.$$

From Eqn. (7.8)

$$\sigma_s = \frac{0.16^2}{2 \times 18^2 \times 10^{-6}} 1.5 \times 10^5 \text{ Nm}^{-2} = 5.92 \times 10^6 \text{ Nm}^{-2}.$$

As seen from the above the stresses arising will be 43% of their permissible values in both the glass and in the steel. The compression force of the band is adjusted by tightening the bolts of the contracting element by means of a torque spanner.

7.2 Frame and clamping structures

The frame keeps the yoke parts of the core compressed. The top yoke beam keeps the winding in a pressed-down condition, while the bottom yoke beam supports the same winding from below and carries the core, the windings and inner leads when the inner part of the transformer is lifted out of the tank.

7.2.1 Designing

Permissible stresses

When designing the suspension girders, suspension bolts, and lifting bollards, the permissible tensile and bending stresses should not exceed 1/6 of the ultimate tensile strength of the material concerned.

The tie plates, cushion beams and their fixing elements are dimensioned to withstand the strains permissible in the static condition, yet the stresses resulting from lifting are higher by a dynamic factor of 1.3. Design is carried out with the following assumptions.

(a) The friction between core and yoke beam is left out of consideration.

(b) Lifting of the active part is effected through four bollards or suspension girders. One lifting point carries 1/3 of the weight of the active part. In order to avoid bending stresses in the upper yoke beam during lifting, the lifting bollards or suspension girders are located on the centre line of the outer limbs, above the tie plates. Thus, lifting will only give rise to stresses in four tie plates.

(c) The reaction force of coil clamping is evenly distributed among the six tie plates.

(d) Dimensioning of the suspension claws of tie plates is based on the sum of forces obtained from (b) and (c). Dimensions of tie plates of the middle limb are made

35*

equal to those of the two outer limbs, although the latter are not loaded during lifting.

(e) The cross beams above the top yoke are designed to carry the load, in addition to that caused by clamping together the yoke or imposed by holding it down. It is assumed that the load corresponding to 1/4 of the full core weight acts on the cross beams as a continuous load, evenly distributed along a length equal to the width of the widest yoke sheet, the cross beams sharing this load evenly. The cross beam is designed as a two-support beam stressed by tension and bending.
The tensile and bending stresses are then summarized.

(f) The cushion bars placed under the core are designed to carry the load resulting from the weight of the total active part, i.e. that of the winding above the yoke beam, and that caused by the part of the coil compression force falling above the yoke. It is assumed that this load is evenly distributed along the cushion bars.

The cushion bar, as a two-support beam is stressed by the load acting on it as a continuous load evenly distributed along a length corresponding to the step of the widest yoke sheet. The beam is designed for bending. If the yoke is also compressed by the cushion bar, then the latter is also to be designed for tension. The stresses resulting from bending and tension are then summarized.

Upper yoke beam

The force contracting the yoke acts on the upper yoke beam as a uniformly distributed continuous horizontal load, produced by the glass-epoxy bands or contracting studs placed within the core window, or by tightening the contracting bolts located outside the yoke. The coil clamping forces are in a vertical direction, balanced by the forces arising in the tie plates. The coil clamping forces act along the line of the clamping elements, whereas the reaction forces arise along the line of tie plates mounted on the front surface of the limb. During lifting of the inner part of the transformer, the lifting force acts on the upper yoke beam through the bollards and is transmitted by the tie plates.

For compressing the yoke laminations a surface pressure of $p = 1.5 \times 10^5 \ \mathrm{Nm}^{-2}$ is to be applied to the widest and longest yoke sheet. This compression force has to be reduced to the narrowest and shortest yoke sheet. Let us denote by p, $\mathrm{N\,m}^{-2}$, the pressure specified for the longest and widest yoke sheet of length h, m and width a, m, and by p_1, Nm^{-2}, the pressure acting on the shortest and narrowest sheet of length h_1, m and width a_1, m.
It can be written, from Fig. 7.14,

$$p_1 = \frac{ah}{a_1 h_1} p. \tag{7.10}$$

The compression force F_1, N acting on the front surface of the yoke is

$$F_1 = a_1 h_1 p_1. \tag{7.11}$$

This horizontal force is produced by the tension force F_2, N of clamping bolts of the upper yoke and by the tension force F_3, N of the bottom glass-epoxy band:

548

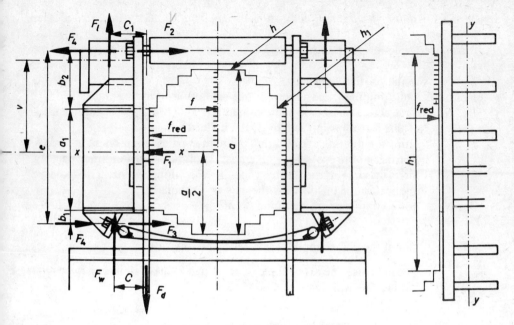

Fig. 7.14. Loads on the upper yoke beam

$$F_1 = F_2 + F_3 . \qquad (7.12)$$

Considering F_1 as a concentrated force acting at the middle height of the yoke sheet of width a_1, the forces concerned, using the notation in the diagram, are

$$F_2 = \frac{\dfrac{a_1}{2} + b_1}{e} F_1 , \qquad (7.13)$$

$$F_3 = \frac{\dfrac{a_1}{2} + b_2}{e} F_1 , \qquad (7.14)$$

where

$$e = a_1 + b_1 + b_2 . \qquad (7.15)$$

The bending moment acting in the vertical plane is

$$M_b = \left(\frac{a_1}{2} + b_2\right) F_2 = \left[\frac{a_1}{2}\left(\frac{a_1}{2} + b_1 + b_2\right) + b_1 b_2\right]\frac{1}{e} F_1 . \qquad (7.16)$$

The bending stress is

$$\sigma_b = \frac{M_b}{K_y} , \qquad (7.17)$$

$$\sigma_b \leqq \sigma_a . \qquad (7.18)$$

549

Stresses resulting from lifting and coil clamping, imposed on the upper yoke beam

Introduce the following notations (see Fig. 7.15):

F_m, N   weight of the inner part,

F_w, N   coil clamping force acting on one side of one winding stack,

F_d, N   force acting on one tie plate from lifting: $F_d = F_l + F_w$. The influence line of this force lies in the axis of the bollard,

F_l, N   lifting force acting at one lifting point, the vertical component of force F_r. For design purposes F_l is taken as 1/3 of the weight of inner part F_m,

F_r, N   force arising in the lifting rope at one lifting point. The point of application of this force is the axis of the bollard,

F_h, N   horizontal force at one lifting point, the horizontal component of force F_r,

$\dfrac{F_w}{2}$, N   half of the coil clamping force acting on one side of one winding stack.

The influence line of this force is the influence line of clamping forces over a quadrant winding circumference,

m, m   yoke spacing.

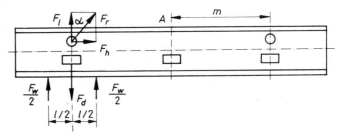

Fig. 7.15. Loads on the upper yoke beam resulting from lifting and coil clamping

The bending moment resulting from coil compression is

$$M_b = \frac{F_w l}{4}.\tag{7.19}$$

The bending stress is

$$\sigma_{h1} = \frac{M_b}{K_x}.\tag{7.20}$$

The moment produced by force F_w and, at lifting, the moment of force F_l is balanced by the moment of force F_4 (see Fig. 7.14). With the notation used there

$$F_w c + F_l c_1 = F_4 e.\tag{7.21}$$

Force F_4 is to be summed algebraically with force F_2 and F_3. Accordingly, the glass-epoxy band threaded through the core window has to be designed for $F_3 + F_4$. To avoid over-compression of the yoke sheets, the glass-epoxy band is tightened by

550

means of a torque spanner. The spindle outside the yoke is dimensioned to the greater force F_2. Forces F_w, F_l and F_4 impose a torque on the yoke beam, because their points of application do not fall in the same plane. With small transformers this torsional stress is neglected, but in the case of transformers of ratings 100 MVA and above, the frame is also checked for torsional loads.

When lifting the yoke beam is subjected to the pressure of force F_h. The compressive stress resulting from this pressure is

$$\sigma_{h2} = \frac{F_h}{A_h} + \frac{F_h v}{K_x}, \tag{7.22}$$

where A_h, m^2 is the cross-sectional area of the yoke beam, perpendicular to force F_h, v, m is the distance between the influence line of force F_h and the axis x–x of the yoke beam (Fig. 7.15).

The resultant of the vertical strains is

$$\sigma_h = \sigma_{h1} + \sigma_{h2} \tag{7.23}$$

is required to be

$$\sigma_h \leqq \sigma_a \tag{7.24}$$

in this case also.

If the points of lifting are on the centre line of the outer core limbs, then the bending stress arising during lifting of the active part can be neglected. If the lifting points do not coincide with the centre line of the outer limbs, then the stress resulting from bending should be added to the above expression for the vertical stress.

Stresses arising at the point of engagement of the tie plate claws

Area X in Fig. 7.16 is checked for surface pressure and cross-sectional area A for shearing stress. During lifting, the tie plate is loaded by the sum of forces F_l and F_w.

$$F_d = F_l + F_w. \tag{7.25}$$

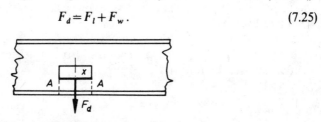

Fig. 7.16. Load at point of engagement of tie plate claw

Surface pressure on area X, m^2 is

$$\sigma_p = \frac{F_d}{X}. \tag{7.26}$$

It must be checked whether σ_p is lower then $\sigma_{pa} = 137 \times 10^6$ N m^{-2} permitted for A38 grade steel.

551

The shear stress brought about by force F_d is

$$\tau = \frac{F_d}{A}.$$ (7.27)

It must be checked whether the shear stress is lower than $\tau_a = 108 \times 10^6 \, \text{Nm}^2$ permitted for A38 grade steel.

Lower yoke beam

In the static condition the beam is subject to three different kinds of load. The first load is the compressive force acting on the yoke beam as a uniformly distributed load, caused by the glass-epoxy band located in the core window and by the clamping bolt accommodated outside the yoke. The second load is composed of the parts of the winding weight and winding clamping force falling above the bottom frame. The third load is the total core weight, and the part of the winding weight and winding clamping force falling on the bottom yoke. One part of the second load, viz. the winding clamping force is taken up by the tie plate through the frame. The other part of the second load, viz. the respective part of the winding weight is directly transferred by the frame to the longitudinal support of the tank. Finally, all the three forces of the third load are transmitted through the tension beam to the frame girder. Of these the force clamping the winding is counterbalanced by the tie plate, whereas the other two are in equilibrium with the reaction force provided by the longitudinal support of the tank. During lifting the distribution of forces differs from the static condition, in that all forces resulting from the weight of parts, i.e. the winding weight of the second load and both weights of the third load are taken up by the tie plate. Thus, the load acting on the tie plate is composed of the tensile force caused by coil compression and the tensile force due to the weight of the active part, as has been shown before. The dimensions of the lower yoke are identical with those of the upper one, hence force F_1 required for compressing the upper yoke and forces F_2 and F_3 producing the former are equal to those arising in the case of the upper yoke beam (see Fig. 7.17).

Let the part of the winding clamping force F_w, falling above the top frame be denoted by F'_w. The torque resulting from this force multiplied by c is counterbalanced by torque $F_4 e$. Thus,

$$F'_w c = F_4 e$$ (7.28)

from which

$$F_4 = F'_w \frac{c}{e}.$$ (7.29)

From the second stress, let the force produced by the weight of the winding be denoted by F_s and its line of action be taken to coincide with that of F_w in the section shown in Fig. 7.18. Further, let the weight of the total active part be denoted by F_m, and the force produced by this total weight less than the force exerted by the winding weight directly carried by the bottom be F'_m, i.e. let

$$F'_m = F_m - F_s.$$ (7.30)

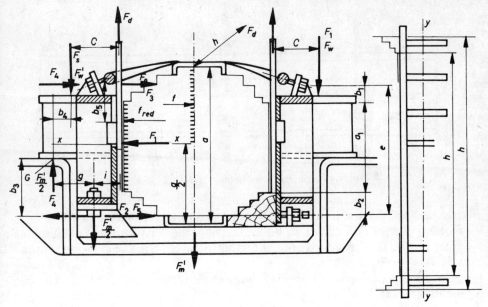

Fig. 7.17. Loads on lower yoke beam

From Fig. 7.17 the equilibrium of moments for point G may be written as

$$\frac{F'_m}{2}g + F_s b_4 = F_5 b_3,\tag{7.31}$$

whence

$$F_5 = \frac{\dfrac{F'_m}{2}g + F_s b_4}{b_3}.\tag{7.32}$$

Forces F_2, F_4 and F_5 have to be summed up algebraically. The cushion beam setting bolt is tightened by a torque spanner adjusted to this force. During lifting the weight of the total active part is carried by the tie plates. The total weight and the coil clamping force, except for forces F_s and F'_w, is transferred through the cushion beams to the frame, via the vertical bolts clamping together the cushion beams to the frame beam. Writing the equation for the equilibrium of moments

$$\left[\frac{F_m}{2} + \frac{F_w}{2} - (F_s + F'_w)\right]i + (F_s + F'_w)c = F_6 b_5,\tag{7.33}$$

from which

$$F_6 = \frac{\left[\dfrac{F_m}{2} + \dfrac{F_w}{2} - (F_s + F'_w)\right]i + (F_s + F'_w)c}{b_5}.\tag{7.34}$$

553

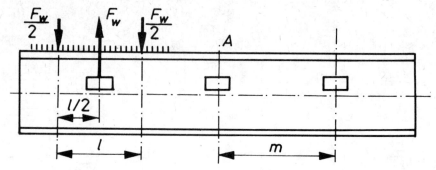

Fig. 7.18. Winding weight and coil clamping force acting on lower yoke beam

The glass-epoxy band is tightened by means of a torque spanner adjusted to force $F_3 + F_4$. During lifting force F_6 arises, therefore the bend is dimensioned to this greater force. The vertical load on the bottom frame beam is a bending load around axis x–x (see Fig. 7.18).

In the rest position, the bending moment caused by coil clamping, with the notation of Fig. 7.18, is

$$M_v = \frac{F_w l}{4}.$$
(7.35)

The bending stress is

$$\sigma_v = \frac{M_v}{K_x}.$$
(7.36)

The condition

$$\sigma_v \leqq \sigma_a$$

must be fulfilled.

In Eqn. (7.35) the coil clamping force has not been divided into the parts loading the bottom yoke and the frame beam, because the bottom beam is loaded by the former force through the cushion beam and the bolts connecting the frame beam with the cushion beam lie in almost the same line of action as the two $F_w/2$ forces.

The stress imposed on the point of engagement of the tie plate claws is the same as with the upper frame beam. This structural part is dimensioned in the same way as the claw engaged with the upper frame beam.

7.2.2 Design of the suspension bolt or suspension girder

For the case of four suspension bolts or suspension girders, the cross-sectional area of one bolt or one girder is

$$A = \frac{F_m}{3\sigma_a},$$
(7.37)

or, if the cross-sections are subject to a shearing stress, τ_a is to be written instead of σ_a.

554

In Eqn. 7.37, F_m is the force or weight resulting from the mass of the total inner part.

As stated in the introduction to this Chapter, σ_a must not be higher than 1/6 of the tensile strength of the material concerned.

The mass of the total inner part is composed of the following parts:

mass of core,
mass of frame,
mass of windings + insulating oil in them,
mass of major insulation with the oil in it and other insulation,
mass of inner leads.

7.2.3 Design of the cushion beam

The force resulting from the weight loading the cushion beams, as shown above, is F'_m, in N. This force acts as a load uniformly distributed over a length of b, m as shown in Fig. 7.19, on the z cushion beams, each of length l, m. The maximum moment is

$$M = \frac{F'_m}{2} \frac{2l-b}{4} \frac{1}{z}.$$ (7.38)

The highest stress resulting from bending is

$$\sigma'_b = \frac{M}{K}.$$

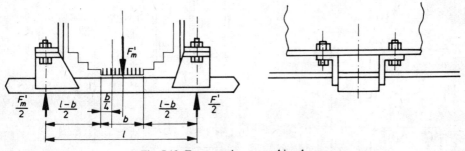

Fig. 7.19. Forces acting on cushion beam

The design stress is the product of this stress and a dynamic factor: $\sigma_b = 1.3\sigma'_b$. Quantity K in m³ is the section modulus of the cushion beam. It has been shown above that the cushion beam is loaded with forces $F_5 + F_3$ as well, so that the tensile stress calculated from these forces has to be added to the stress obtained from the equation. Here also the design stress should be less than the allowable stress.

7.2.4 Design of fixing bolts of the cushion beam

The cross-sectional area of one bolt, the number of fixing bolts for the cushion beam being n, is

$$A = \frac{F'_m}{nz\sigma_a} 1.3, \qquad (7.39)$$

where number 1.3 is the dynamic factor.

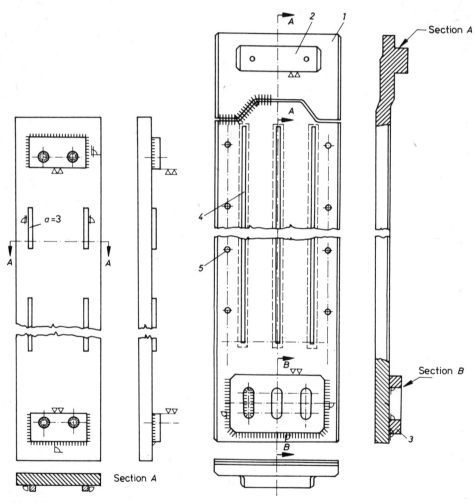

Fig. 7.20. Tie plate (a) tie plate of core wedged by hard aluminium trough: *1* — tie plate; *2* — claw; *3* — cleat; (b) tie plate of banded core: *1* — tie plate; *2* — upper claw; *3* — bottom claw; *4* — longitudinal split for eddy current reduction; *5* — bores for supporting rods

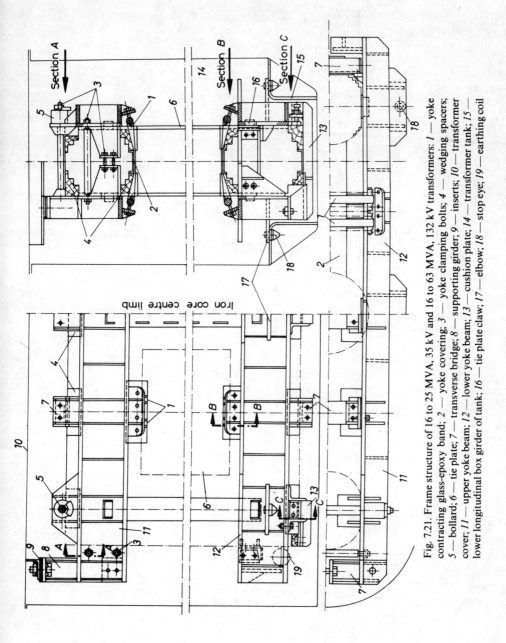

Fig. 7.21. Frame structure of 16 to 25 MVA, 35 kV and 16 to 63 MVA, 132 kV transformers: *1* — yoke contracting glass-epoxy band; *2* — yoke covering; *3* — yoke clamping bolts; *4* — wedging spacers; *5* — bollard; *6* — tie plate; *7* — transverse bridge; *8* — supporting girder; *9* — inserts; *10* — transformer cover; *11* — upper yoke beam; *12* — lower yoke beam; *13* — cushion plate; *14* — transformer tank; *15* — lower longitudinal box girder of tank; *16* — tie plate claw; *17* — elbow; *18* — stop eye; *19* — earthing coil

Iron core centre limb

557

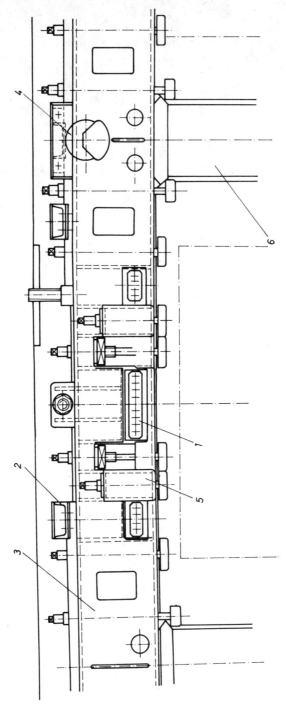

Fig. 7.22. Frame structure of semi-frame type 16 to 63 MVA, 132 kV transformers: *1* — yoke contracting bolt clamping of Fig. 7.9; *2* — transverse bridge; *3* — upper yoke beam; *4* — bollard; *5* — spring-type core clamping; *6* — tie plate

7.2.5 Design of the tie plate

As already mentioned, when lifting the inner part by means of 6 tie plates at four lifting points, one tie plate carries 1/3 of the weight of the inner part and 1/2 of the clamping force of one phase winding. For calculating the cross-sectional area a dynamic factor of 1.3 is also taken into account. With the notations used so far, the cross-sectional area of the tie plate and of the weld (see diagrams (a) and (b) of Fig. 7.20) is

$$A = \frac{\dfrac{F_m}{3} + \dfrac{F_w}{2}}{\sigma_a} 1.3 = \frac{F_l + \dfrac{F_w}{2}}{\sigma_a} 1.3. \qquad (7.40)$$

The contact area of the tie plate claw is calculated in the same way as for the top frame beam.

Frame constructions

The frame structures may be very different, showing a wide variety not only from the various manufacturers, but also depending on the type rating. Three frame constructions are presented as examples. The first is shown in Fig. 7.14, used for transformer ratings in the range 2.5 to 10 MVA and up to 33 kV.

The second construction is used for 16 to 25 MVA, 33 kV and for 16 to 63 MVA, 132 kV transformers (see Fig. 7.21).

The third frame structure has been adopted for 16 to 63 MVA, 132 kV semi-frame type core transformers, and is shown in Fig. 7.22.

7.3 Windings

7.3.1 Windings wound of rectangular section material
(Refer to Fig. 7.23)

The following types of windings are known:

(a) helical winding flat-wound on spacer;
(b) helical winding edge-wound on spacer;
(c) helical winding flat-wound on insulating cylinder;
(d) helical winding edge-wound on insulating cylinder;
(e) winding flat-wound, snapped into insulating cylinder;
(f) helical winding edge-wound, snapped into insulating cylinder;
(g) double-layer helical winding with interlayer cylinder;
(h) double-layer helical winding without inter-layer cylinder;
(i) double layer self-supporting helical winding;
(j) layer winding;
(k) spiral coil;
(l) winding made of continuously transposed conductors;

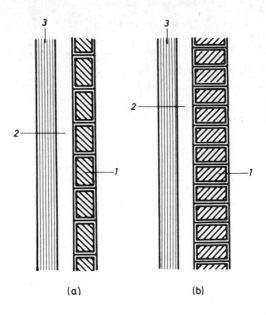

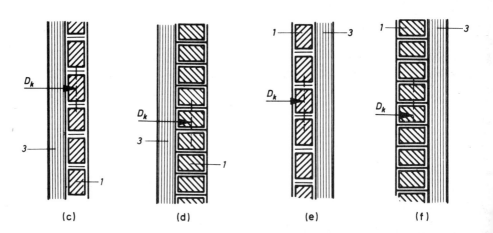

Fig. 7.23. (a) to (f): Helical windings:
1 — winding; *2* — strip; *3* — insulating cylinder; D_k — mean diameter

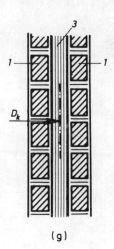

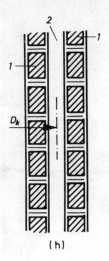

(g)   (h)

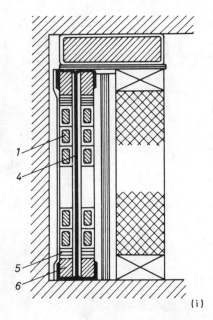

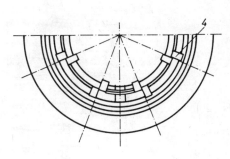

(i)

Fig. 7.23. (g) to (i): Helical windings:
1 — winding; *2* — strip; *3* — insulating cylinder; *4* — metal tape support; *5* — end spacer; *6* — steel end ring; D_k — mean diameter

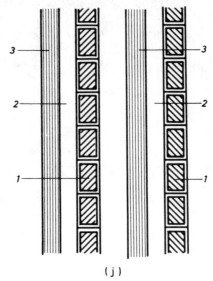

Fig. 7.23. (j) Layer winding: *1* — winding; *2* — strip; *3* — insulating cylinder

Arrangement of
parallel conductors
in Section *B*

6	5	4	3	2	1
1	6	5	4	3	2
2	1	6	5	4	3
3	2	1	6	5	4
4	3	2	1	6	5
5	4	3	2	1	6

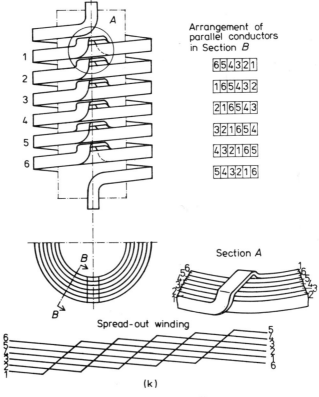

Section *A*

Spread-out winding

(k)

Fig. 7.23. (k): Spiral coil

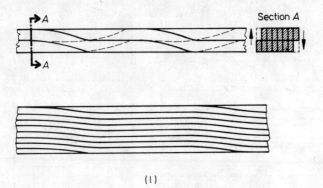

(l)

Fig. 7.23. (l): Winding made of continuously transposed conductors

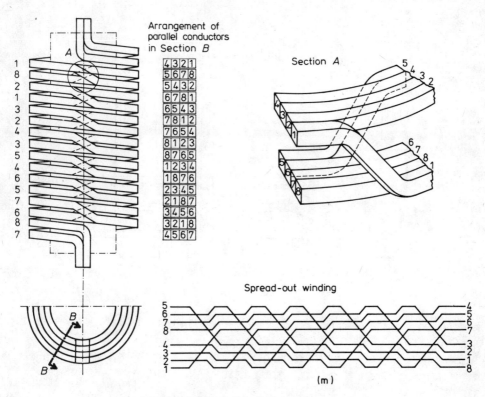

Arrangement of
parallel conductors
in Section B

Section A

Spread-out winding

(m)

Fig. 7.23. (m): Rolling spiral winding with two sections

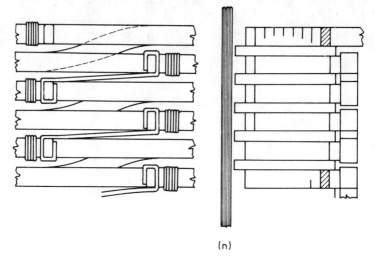

(n)

Fig. 7.23. (n): Two-section disc coil

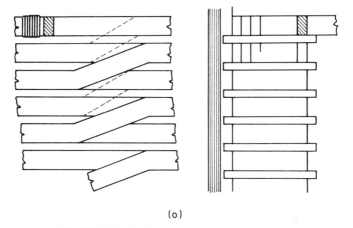

(o)

Fig. 7.23. (o): Continuously wound disc-type coil

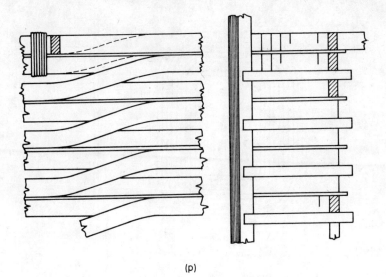

(p)

Fig. 7.23. (p): Continuously wound disc, consisting of double coils

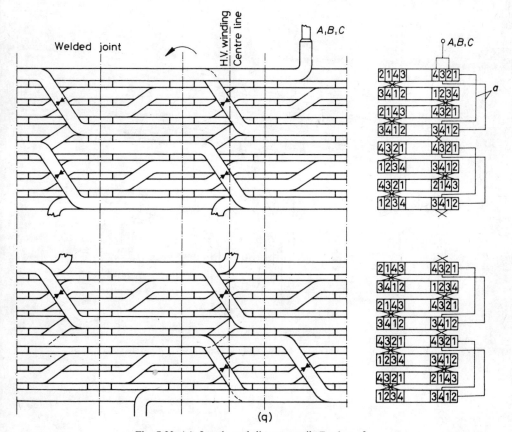

(q)

Fig. 7.23. (q): Interleaved disc-type coil, $E=4$; $p=2$.

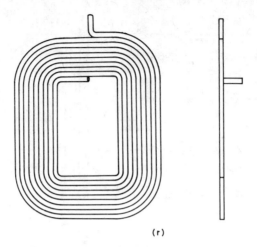

(r)

Fig. 7.23. (r): Pancake coil

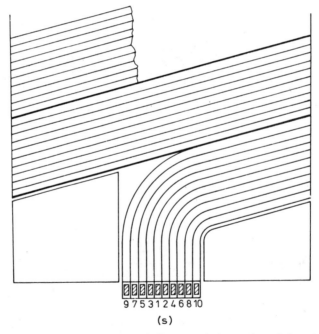

9 7 5 3 1 2 4 6 8 10

(s)

Fig. 7.23. (s): Single-layer tapping winding wound of several parallel conductors

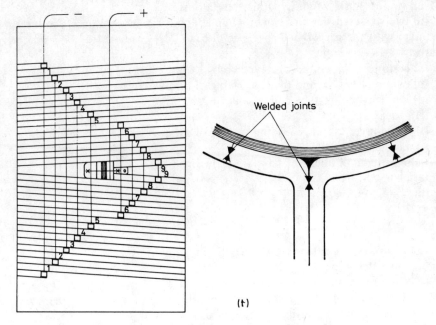

(t)

Fig. 7.23. (t): Single-layer, middle entry tapping winding

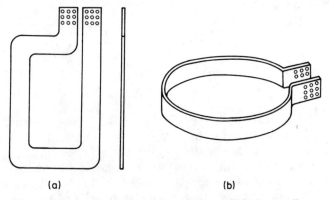

(a)   (b)

Fig. 7.24. Coils of plate or bar: (a) plate coil; (b) hoop coil

567

(m) rolling spiral winding with two sections;

(n) two-section disc which can be wound of one or more conductors connected in parallel with integer or half-turn per disc and/or with even or odd number of turns;

(o) continuously wound disc-type coil;

(p) continuously wound disc consisting of double coils;

(q) interleaved disc-type coil, $E = 4$; $p = 2$;

(r) pancake coil;

(s) single-layer tapping winding wound of several parallel conductors;

(t) middle entry coil.

Windings made of plates or bars are shown in Fig. 7.24.

(a) plate coil;

(b) hoop coil.

7.3.2. Fields of application of windings

On a core-type transformer the winding may be of

(a) concentric-type or rarely

(b) sandwich-type

The H.V. winding of a transformer provided with the concentric winding arrangement may be a layer winding, or a disc winding.

The L.V. winding, depending on the current rating, may either be of the layer type, or one of the types shown in Fig. 7.23 marked with (k) to (m) or (o) to (q). The windings in the sandwich arrangement are disc, pancake or plate coils.

The windings in the case of shell-type transformers may occasionally have

(a) a concentric arrangement, but in most cases they have

(b) sandwich windings.

7.3.3 Mechanical design of windings

7.3.3.1 Mechanical properties of copper and aluminium winding materials

In Fig. 7.25 the elongation of copper is plotted as a function of tensile stress. As will immediately be seen from the diagram, copper has no limit of linear elasticity, no yield point and no such tensile strength value as is normal in the case of steels. The relevant standards, e.g. BS 18: 1962, have introduced for such materials the concepts of "assumed limit of non-proportional elongation" and of the "yield point" pertaining to it. In Fig. 7.25, the 0.1% limit of non-proportional elongation and the pertinent yield point are shown. The limit of linear elasticity has also been entered in the diagram but this is no basis for design as applies in the case of steels. In the following short-circuit calculations the $\sigma_{0.1}$ yield point of copper and of aluminium behaving quite similarly to copper, is taken as the permissible tensile strength. This also means that the stress caused by a short-circuit reaches the permissible value, so the material will suffer permanent deformation, i.e. elongation.

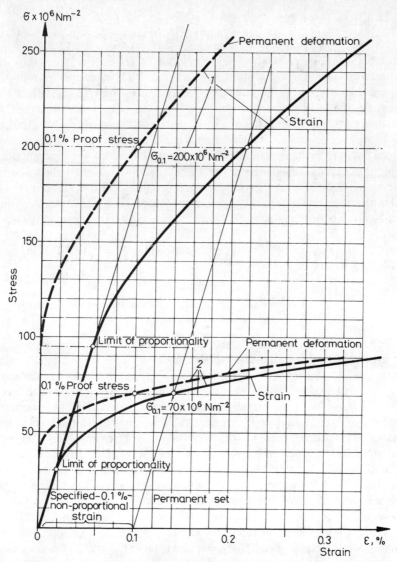

Fig. 7.25. Tensile stress and permanent deformation as a function of elongation. Continuous curves indicate the relation elongation vs. tensile stress, dotted-line curves indicate their relation permanent deformation vs. tensile stress: *1* — semi-hard copper; *2* — soft copper

When exposed to the electrodynamic effect of further short-circuits, this elongation will increase further.

Thus, in the following, the permissible tensile stresses are:
in fully annealed copper

$$\sigma_{0.1} = 70.632 \times 10^6 \ \text{Nm}^{-2}, \tag{7.41}$$

in semi-hard copper

$$\sigma_{0.1} = 196.2 \times 10^6 \ \mathrm{Nm^{-2}}. \tag{7.42}$$

The coils are made of semi-hard winding material. The allowable transient temperature rise during a short-circuit is $250\,^{\circ}\mathrm{C}$ in copper windings. Thus temperatures exceeding the annealing temperatures may occur, with consequent annealing of the material. Substantial reduction of $\sigma_{0.1}$ was observed of hard drawn material in the 200 to $250\,^{\circ}\mathrm{C}$ temperature range. Thus the winding will probably be unable to withstand the electrodynamic stresses arising at the next short-circuit. Fortunately, the temperature rise limit permitted by the standard is never reached, due to the lower current densities applied, shorter fault clearing times, and lower short-circuit currents. As is known, the strength of materials is increased by the cold metal working process. Thus, the value of $\sigma_{0.1}$ increases from $70 \times 10^6 \ \mathrm{N \ m^{-2}}$ to $156 \times 10^6 \ \mathrm{N \ m^{-2}}$, whereas electrical conductivity decreases by 1.25%.

The Young's modulus of elasticity for copper is $E_c = 110 \times 10^9 \ \mathrm{N \ m^{-2}}$, and for pressboard $E_{pr} = 0.25 \times 10^9 \ \mathrm{Nm^{-2}}$.

7.3.3.2 Resultant of axial forces, magnitude of winding compression force

Generally, the magnitude of axial forces cannot be calculated very accurately, because the stray flux pattern is disturbed by many external factors. Such external factors are the flux trap, the proximity of the yoke and limb, the dimensions of windings and leakage channel, etc. Calculation of axial forces consists of the following steps. With use of the computer program available for calculating the magnetic field, the radial stray flux density pertaining to the mean diameter of the windings is calculated as a function of winding height (see Fig. 7.26). The product of local stray flux density and the excitation of the winding disc is a force. When calculating such forces for each disc, a diagram identical in shape with that of Fig. 7.26(a) is obtained, since in the case of identical numbers of turns, current and axial size, the curve of radial flux density will be identical with that of the axial force, although this force will be of axial direction, unlike the direction of flux density. By separately adding the axial forces for the L.V. and H.V. windings, Fig. 7.26(b) is obtained. At the middle height of the winding stack, F_{wa} will be the maximum force by which the axial spacers located at the middle height will be compressed.

If the winding stacks are clamped with a force F_{wa} determined as described above, the uppermost discs will not work loose during short-circuits and no hammer-blow type shocks will have to be withstood by the windings; under such shocks the insulation will tend to be damaged and will eventually suffer internal interturn faults.

The compression force of the coil stack should be:

$$F_{wa} \geq p_w A_s, \tag{7.43}$$

where p_W, $\mathrm{Nm^{-2}}$ is the specific compression force, obtained from the dynamic forces calculation, A_s, $\mathrm{m^2}$ is the coil surface area covered by the axial spacers.

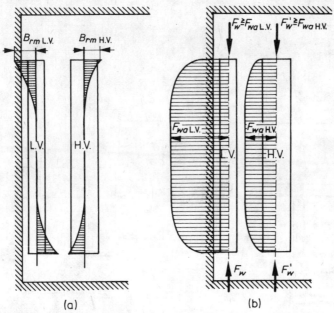

Fig. 7.26. (a) Radial leakage flux density pertaining to the mean coil diameter as a function of coil height; $B_{rm\,LV}$ is the maximum of maximal radial leakage flux density at the coil ends of the L.V. winding; $B_{rm\,HV}$ is the maximum of maximal radial leakage flux density at the coil ends of the H.V. winding; L.V. — low-voltage winding; H.V. — high-voltage winding. (b) Load acting on axial spacers as a function of coil height. $F_{wa\,LV}$ and $F_{wa\,HV}$ are the forces acting on the spacers located at middle height. If the coils are clamped with a force at least equal to that indicated, the top coils will not work loose under the effect of a short-circuit

7.3.4 Coil clamping

7.3.4.1 Screw- and spring-type clamping structures

The force for compressing the winding stacks is adjusted by appropriate clamping structures located in the top frame beam and hanging down from it. Two kinds of clamping structures are used, of screw-type and spring-type. Both types may be of two different arrangements: the shorter underlaid and the longer through-going variants (see Figs. 7.27 and 7.28).

As is shown in the diagrams, in the compressed state of the spring, the spring-type clamping acts as a screw-type clamp.

The axial spacers placed between the coils are made of hard (type Weidmann T IV) pressboard. Before being put into the winding, the spacers are pre-compressed in a tier-press under a pressure of 2.5 to 3×10^7 Nm^{-2}, and kept in a dried-out state until assembled. The ready-made coils are stored in a drying oven until placed over the limb, but before assembly they are compressed in a hydraulic press with the specified coil compression load and, if required, are set to the specified size. After vacuum drying, securing and re-tightening of bolts, the transformer is filled with oil

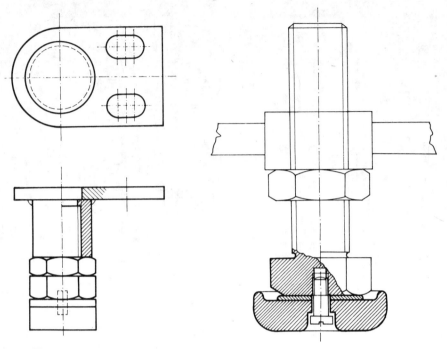

Fig. 7.27. (a) Underlaid screw-type clamping; (b) through-going screw-type clamping

under vacuum. The coil stack of the transformer, when so treated, will shrink by about 0.5% during the last vacuum drying. The size of the core window is filled out with insulating material, as given for the average case in Table 7.1.

During designing, the actual space factor of the window should be calculated. With the window height and percentage of shrinking known the reduction of axial length of the winding due to shrinkage can be computed. To attain a good clamping even during vacuum treatment the compressibility of the clamping spring should be twice the expected shrinkage. The clamping of coils is brought about by tightening the screws of the spring/screw-type clamping structures.

Table 7.1

Insulating material filling factor of the core window in coil stacks
of disc-type coils

Highest voltage of winding, kV	10	20	35	132	220	400	750
Filling factor of core window (%)	35	38	40	48	50	52	65

572

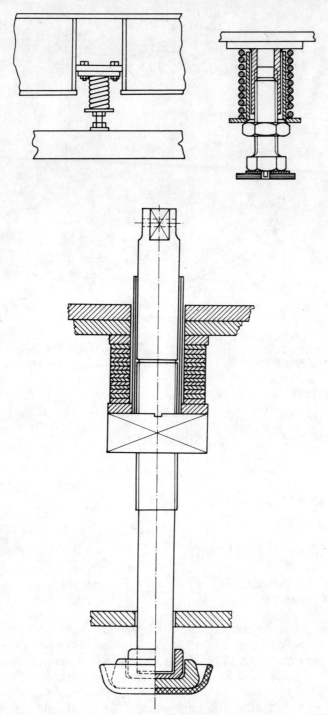

Fig. 7.28. (a) Underlaid spring-type clamping; (b) through-going spring-type clamping

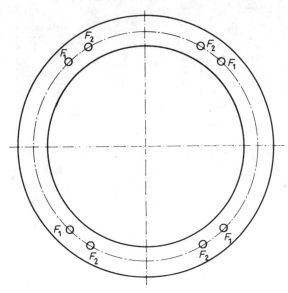

Fig. 7.29. Clamping ring. F_1, F_2 are the respective points of application of the coil clamps. The reaction force of spacers acts as a distributed load along the periphery pertaining to the mean diameter of the ring

The concentrated force provided by the coil clamping structure should be uniformly distributed among the radial spacers. A clamping ring is used for this purpose, as shown in Fig. 7.29.

7.4 Tank and cover

7.4.1 Tank types and requirements

Four types of oil tank are distinguished, as shown in Fig. 7.30. These are as follows.

1. Trough-type oil tank,
2. Bell-type tank,
3. Leaning-type tank,
4. Beaked-type tank.

The oil tank serves for enclosing the active parts and oil filling of the transformer. In order to make local drying of active parts and oil of the transformer possible, all transformers above 2.5 MVA have to withstand an internal overpressure of $5 \times 10^4 \, \mathrm{Nm^{-2}}$, moreover, if the voltage rating of any winding of the transformer exceeds 40.5 kV, then the tank has to be designed to withstand total vacuum. Instead of vacuum testing, the transformer is subject to an overpressure test performed with an internal overpressure of $1 \times 10^5 \, \mathrm{Nm^{-2}}$.

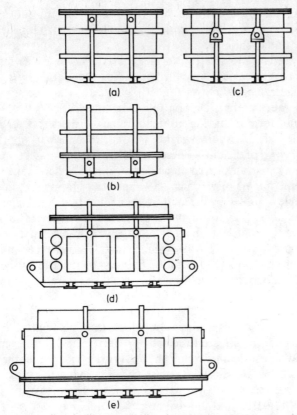

Fig. 7.30. (a) Trough-type oil tank; (b) bell-type oil tank; (c) leaning-type oil tank; (d) top-split beaked oil tank; (e) bottom-split beaked oil tank

The tank is required to withstand the stresses occurring during

1. transport,
2. lifting,
3. traction,
4. hoisting by hydraulic means.

The oil tank is an oil-tight welded steel construction, provided with ribs to carry the mechanical loads.

1. Trough-type oil tank

This is the most commonly used type. The active part may be lifted out of the tank, or replaced, only by means of a crane. The complete transformer including its tank is lifted at its bollards, mounted or welded on the sides of the tank. Lifting plates are welded to the bottom of the tank, for the application of hydraulic jacks. For lifting the transformer at its working site, e.g., for placing it on an elevated foundation, methods of stacking and hydraulic jacks are used. For moving the

575

transformer within the substation area, holes are provided at the bottom of the tank for drawhooks. Stiffening braces under the tank are located in the space between the tank bottom and the cushion beam.

No rollers are used for moving transformers of 10 MVA or higher ratings to their final position. The transformers are pulled onto their foundation over greased rails. When moving the transformer over longer distances within the substation, low-built roller-type trailer paths are employed.

With power transformers having no underframes, the brackets extending from the bottom girder which carry the weight of the entire inner part are supported by the longitudinal beams of the tank. These beams are triangular-section box girders.

The transformers are transported on platform cars or flats or, in the case of larger units, in cradle-type trailers or wagons. The transformer rating for which a trough-type tank is suitable is limited by the loading gauge. The platform height of platform vehicles lies around 0.85 m above the road surface or railhead. The lower the platform height of the vehicle available, the larger the transformers which may be provided with trough-type tanks. An advantage of the trough-type tank is that there is no need for an oil container to store the oil filling when the transformer is lifted out of its tank. The oil tank of the transformer may itself be used as the oil storage tank.

2. Bell-type tank

Its advantage is that, for inspecting or repairing the active part, the light-weight bell-shaped casing can be lifted off the active part by means of the small-capacity crane usually available in smaller repair shops. Its drawback is that the oil has to be drained off into a separate storage vessel before removing the tank. Moreover if there is even a small problem with the sealing of the tank tray, the oil has to be drained, the tank lifted off and the sealing newly prepared.

The beaked oil tank of very large units (of marginal rating) is often designed to have a bell-tank. In such cases the tank is designed to withstand the stresses occurring during transportation and to maintain its vacuum integrity. An advantage of this construction is that during repairs on site, a crane of capacity matched to the tank weight is sufficient.

3. Leaning-type tank

Its name comes from the way the unit is transported. During transportation, the transformer is suspended between the two longitudinal beams of a vehicle. The weight of the transformer is transmitted to these beams through the brackets located at the upper part of the tank. Owing to the lateral dimension of the longitudinal beams, the arrangement is suitable for transporting large transformers which are high but relatively narrow. The tank has to be designed for the required vacuum and loads occurring during transport. It is a preferred tank shape if no limitations are imposed during transportation. The leaning type oil tank may be either of trough or bell shape.

4. Beaked-type tank

This is the tank type suitable for the transport of the largest units. It may be of trough or bell type (bottom split). The transformer is transported together with its

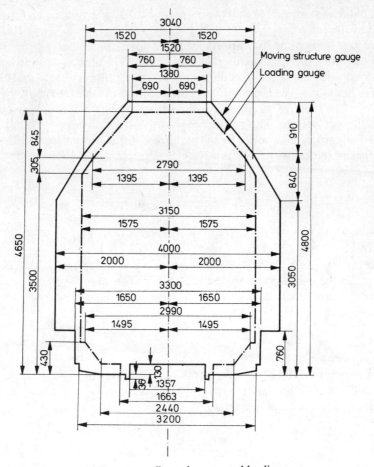

Fig. 7.31. Central European railway clearance and loading gauges

active part, so the tank has to be designed to withstand the vacuum and loads occurring during transport. In Central Europe the 525-ton vehicle of the Deutsches Bundesbahn is the largest capacity transport wagon. The tank is suspended on lugs from the beaks of the wagon.

7.4.2 Dimensions of the tank

The inside volume of the tank depends on the electrical performance data of the transformer. The required electrical clearances are determined by the designer of the electrical part. The inner surfaces of the tank and cover should be smooth, to reduce the flow resistance of the oil and to avoid spots where hot oil may get trapped. When vapour phase drying is used, drainage of precipitating kerosene collecting at the bottom of the tank should be ensured. The discharge valve located at the lowest point of the tank should permit the drainage of all the kerosene and, later, of the oil filling of the tank.

The first thing of major importance to be done in the course of designing is to check the tank size thoroughly. The platform height of the transportation means employed and the smallest loading gauge of the transport route must be known prior to designing. The Central European loading and clearance gauges are given in Fig. 7.31. The shape of the oil tank is also influenced by the load carrying capacity of the transport vehicle. The permissible width of a transformer equipped with a beaked oil tank is determined by the requirement that objects located along curved stretches of the railway track must not be injured.

7.4.3 Transport and lifting loads

Oil tanks of trough and bell shapes

During transport, the maximum acceleration to which the transformer may be exposed in the horizontal direction is $2\,\text{g ms}^{-2}$. The limit of vertical acceleration is $1\,\text{g ms}^{-2}$. Static loads have to be multiplied by 1.3 for the case of lifting.

Leaning type oil tank

Accelerations during transport are the same as with the trough shape tank. The dynamic factor for transport and lifting is 1.3. The load is assumed to be uniformly distributed on the leaning supports.

Beaked type oil tank

The loads expected to arise are shown in Fig. 7.32. Let the following notation be introduced:

F_1, N   is the load due to the weight of the transformer,
F_2, N   is the load due to the triangular shape bridge structure,
F, N   is the load due to the weight of the transformer and bridge structures

$$F = F_1 + 2F_2 , \qquad (7.44)$$

F_h, N is the horizontal load carried by the lower pivot,
F_v, N is the vertical load carried by the lower pivot,

$$F_v = \frac{F}{2} - F_2 = \frac{F_1}{2} . \qquad (7.45)$$

Writing the moments for point A, see Fig. 7.33(a):

$$\frac{F}{2} a - F_h b - F_2(a - e) = 0 \qquad (7.46)$$

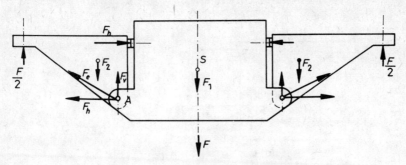

Fig. 7.32. Forces acting on beaked oil tank

from which

$$F_h = \frac{\frac{F_1}{2} a + F_2 e}{b}. \tag{7.47}$$

F_r, N is the resultant of loads carried by the lower pivot

$$F_r = \sqrt{F_h^2 + F_v^2}, \tag{7.48}$$

F_s, N   is the horizontal buffer impact imposed on the buffers of the transport vehicle; according to railway specifications

$$F_s = 1.96 \times 10^6 \, \text{N},$$

F_{s1}, N   buffer impact imposed on the lower carrying lug of the transformers,
F_{s2}, N   buffer impact imposed on the upper thrust plate of the transformer.
Force F_s is shared by the pivot and thrust plate in the ratio of the respective arms, see Fig. 7.33(b):

$$F_{s1} = \frac{F_s d}{b}, \tag{7.49}$$

$$F_{s2} = \frac{F_s c}{b}. \tag{7.50}$$

As is apparent from Fig. 7.33(c) the action of force F_{s1} is opposed to that of F_h acting on the lower pivot. Therefore, when calculating the load imposed on the pivot and on the lug, the value of F_h should be multiplied by a dynamic factor 1.3.

Let the force acting on the lug on one side of the transformer tank, i.e. on one pivot of the wagon be F_{st}.
According to the above

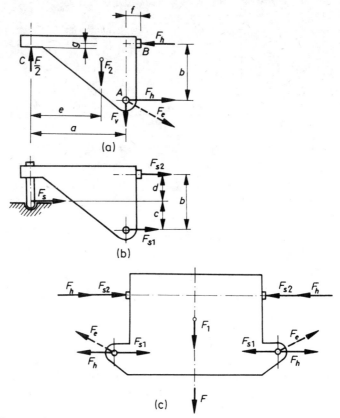

Fig. 7.33.(a) Forces acting on the triangular bridge structure; (b) buffer forces; (c) forces acting on the tank during transport

$$F_{st} = 1.3 \frac{F_e}{2}. \tag{7.51}$$

Let the compressive load acting on the thrust plate on one side of the transformer tank be F_p. As shown in Fig. 7.33(c), the upper thrust plate is loaded by the sum of forces F_h and F_{s2}. Thus, the load on the upper thrust plate is either

$$F_p = \frac{F_h + F_{s2}}{2}, \tag{7.52}$$

or

$$F_p = 1.3 \, F_h. \tag{7.53}$$

The larger compressive load is chosen. The tank is checked for capability of withstanding the load imposed by the effect of vacuum either by the diagonal fracture method, or by the nodal, finite-element method.

580

7.4.4 Cover types

Three different cover types are employed in practice, shown in Fig. 7.34. These are as follows:

1. plain cover
2. coffin shape cover
3. embossed cover

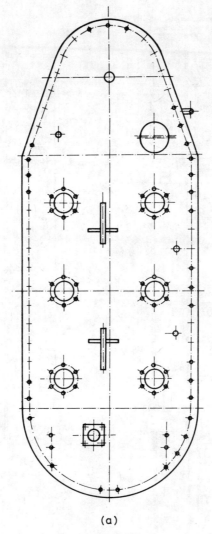

(a)

Fig. 7.34(a) Plain cover

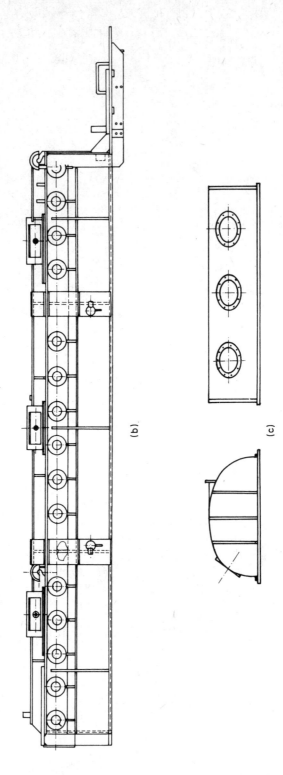

(b)

(c)

Fig. 7.34.(b) Coffin-shape cover; (c) embossed cover

7.5 Conservator

7.5.1 Guiding principles for design, operation, and air dryers

The density of transformer oil varies as a function of temperature. The volume of a given mass of oil changes by 8% under the effect of a temperature variation of 100 °C. The primary task of the conservator is to take up the volume increment of the oil present in the tank and cooling system and expanding with increasing temperature. The conservator is designed for a capacity corresponding to 10% of the volume of the oil filling at −20 °C. This corresponds to a temperature variation of 120 °C, i.e. at 100 °C oil would just fill the space of the conservator.

The secondary task of the conservator—hence its name—is to conserve the oil from the harmful consequences of moisture and oxidation.

Water absorption of the oil depends on the

1. extent of the oil–air interface,
2. relative humidity of the oil, the saturation moisture content being taken as 100%,
3. relative humidity of the air,
4. temperature of the oil and air, and on the difference between the two temperatures,
5. velocity of the air,
6. duration of exposure,
7. atmospheric pressure.

The rate of oxidation may be reduced by

1. reducing the oil–air interface,
2. decreasing the temperature at the boundary surface,
3. using an inert gas atmosphere instead of air, or applying hermetic sealing.

The oxidation of oil may be substantially reduced by using inhibitors. During operation of the transformer, air is being inhaled and exhaled by the conservator. Two methods are available for reducing the extent of air exchange in the conservator, and are generally employed in transformers above 132 kV. The first method, introduced by Dr. Csikós, is called "sparing-type service". This consists of switching on or off the pumps and fans of the compact coolers of forced oil flow transformers as a function of load, so as to keep the oil temperature approximately constant. Two temperatures are selected, one for the summer and one for the winter. The slow-rate breathing is thereby prevented or, at least, impeded. The quick-rate breathing caused by alternating sunshine and cloudy weather is attenuated by heat insulating the top part of the conservator, whereby the quick-rate breathing is considerably reduced.

The moisture content of the air entering the conservator is reduced by silica gel air dryers, as shown in in Fig. 7.35.

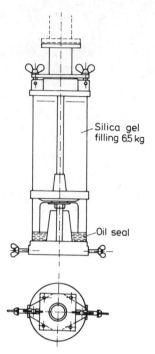

Fig. 7.35. Silica-gel air dryer

The effectiveness of the air dryer depends on the particle size of the filling, on how long it takes for the air to pass through the dryer, on the saturation of the filling, on the relative humidity of the air and on the temperature.

The moisture content of the air sucked through is only partly abstracted by the silica-gel air dryer. A modified version combined with a zeolite molecular filter is used in every transformer of 220 kV and higher voltage rating (see Fig. 7.36). This combined air dryer effectively absorbs the moisture content of the air.

The other method consists of removing the humidity of the air by freezing. The Drycol air dryer developed by GEC and shown in Fig. 7.37 operates on this principle. The most perfect moisture protection of conservators having their oil filling in contact with air has been developed by Dr. Béla Csikós. As will be seen, the equipment is also suitable for drying out oil. The equipment consists of a fan, a zeolite and silica-gel filled air dryer, and air heater and cooler heat exchangers (see Fig. 7.38).

Above the oil level in the conservator, dried hot air is kept flowing. The moisture leaving the oil is dragged along with the air into the cooler, where its relative humidity increases and the condensing moisture is collected in a condensate storage vessel.

The humid air gets rid of its moisture content while passing through the zeolite silica-gel air dryer. The dry air is driven by the fan into the heating heat exchanger, where it is warmed, and again blown into the coservator. Air sucked in from outside

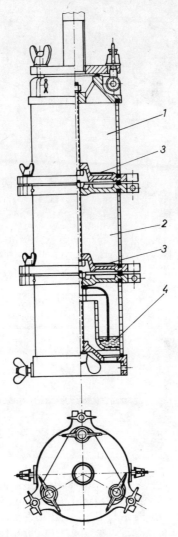

Fig. 7.36. Zeolite silica-gel air dryer: *1* — zeolite filling; *2* — silica gel filling; *3* — filter; *4* — oil seal

by breathing enters the system through the combined air-dryer. Apparently, the device is also capable of reducing the original moisture content of the oil. The 750 kV transformers installed at the Albertirsa substation are equipped with air-dryers of this type.

One method of protecting the oil from the moisture and oxygen content of the air is to fill the space above the oil level in the conservator with nitrogen. The nitrogen filling is driven into a gas-tight cuchion by the oil as it expands with rising temperature. The disadvantages of the method are that the nitrogen has to be kept under pressure and that it has to be supplemented from time to time. A further

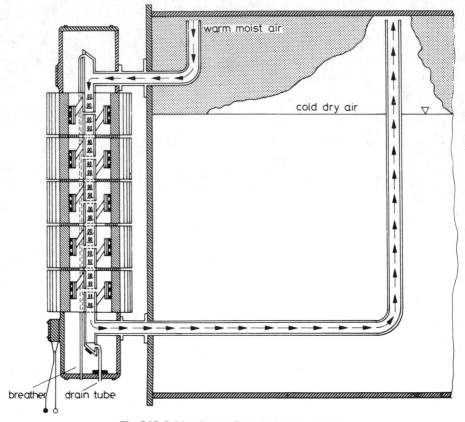

warm moist air

cold dry air

breather drain tube

Fig. 7.37. Peltier-element Drycol air dryer of GEC

disadvantage arises in saturation of the oil by the nitrogen. On reduction of external barometric pressure the gas segregates from the oil saturated with gas, causing the formation of gas bubbles which may then lead to breakdowns.

Aging of the oil under the effects of temperature and time is accompanied by the formation of water. Water may escape from oil whose surface is in contact with the air, but when sealed with nitrogen, water remains trapped in the oil. Other decomposition products may also leave the oil through its free surface, but cannot escape when the surface is sealed off by the nitrogen cushion. Another method of complete sealing is to make the conservator of two halves, with the plane of partition shut by a plastic diaphragm (see Fig. 7.39). This plastic diaphragm must be oil-resistant. The advantage of this method is its complete sealing, but all decomposition products remain trapped in the oil.

The conservators are designed for vacuum in a similar way to the tanks. During transport, the oil in the conservator surges and swings, giving rise to dynamic stresses and causing the oil of the transformer to foam. Surging may be prevented by internal baffle ribs.

586

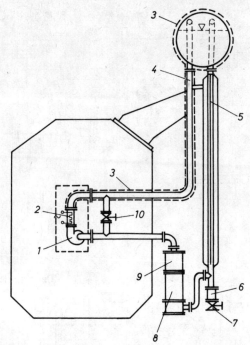

Fig. 7.38. Air dryer system of Csikós: *1* — fan; *2* — heater; *3* — heat insulation; *4* — hot air tube; *5* — cooler; *6* — condensate storage; *7* — water drain valve; *8* — silica gel; *9* — zeolite; *10* — by pass valve

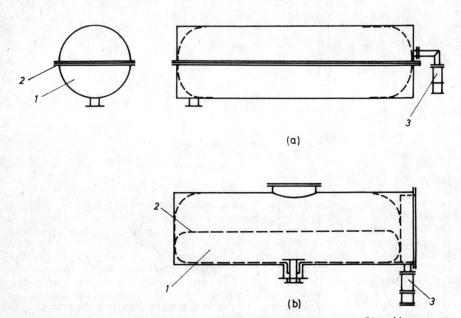

(a)

(b)

Fig. 7.39. Plastic diaphragm-type conservator (a) diaphragm-type conservator; (b) cushion-type conservator: *1* — conservator; *2* — diaphragm cushion; *3* — air-dryer

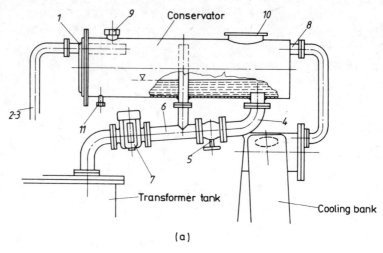

(a)

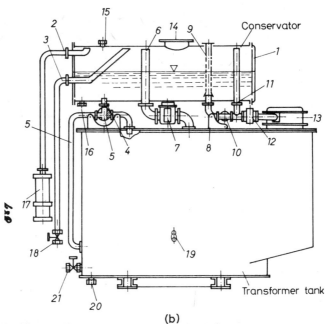

(b)

Fig. 7.40. Conservator accessories: (a) Bypass tube Buchholz connection *1* — oil level gauge; *2* — air dryer connecting tube; *3* — oil filling tube; *4* — bypass tube connection; *5* — stop valve of bypass tube; *6* — Buchholz relay connecting tube; *7* — Buchholz relay; *8* — vent valve connection, for radiators on separate foundations; *9* — oil filling hole; *10* — cleaning aperture; *11* — sludge drain; (b) Intercommunicating-tube type three-way cock arrangement: *1* — oil level gauge; *2* — air dryer connecting tube; *3* — oil filling tube; *4* — connecting tube of three-way cock; *5* — three-way cock; *6* — connecting tube of Buchholz relay; *7* — Buchholz relay; *8* — connecting tube of tap-changer filter; *9* — filter element; *10* — stop cock; *11* — bubble tube; *12* — protective relay of on-load tap changer; *13* — tap changer; *14* — cleaning aperture; *15* — oil filling hole; *16* — sludge drain; *17* — air dryer; *18* — oil filling hole and centrifuge connecting valve; *19* — oil sampling spherical cock; *20* — oil draining spherical cock; *21* — oil drain valve

The conservators of regulating transformers always consist of two parts. One part serves for taking up the expansion of the oil in the tank and cooling system, whereas the other is connected with the oil filling of the on-load tap changer.

7.5.2 Accessories of the conservator

The conservator of a transformer of medium rating is equipped with the following accessories shown in Fig. 7.40.

1. oil level gauge,
2. air dryer connecting pipe,
3. oil filling pipe connection,
4. buchholz and tap-changer protective relay,
5. connection of intercommunication or bypass pipes,
6. sludge drain valve,
7. cleaning apertures,
8. connection of vent pipes.

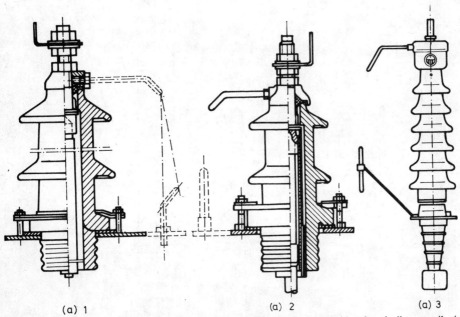

(a) 1   (a) 2   (a) 3

Fig. 7.41. Oil-filled bushings: (a)1 normal, open-bottom oil–air bushing; (a)2 reduced oil space oil–air bushing; (a)3 separate oil space oil–air bushing

589

7.6 Bushings

Bushings are required to lead out the winding ends from the inside of the transformer and to connect them to the outgoing line.

A distinction is made between

1. oil–air,
2. oil–oil,
3. oil–SF_6

bushings.

1. Types of oil–air bushings are as follows:
   (a) Oil-filled bushings, which may be of
   (a)1 normal, open-bottom oil–air bushing;
   (a)2 reduced oil space oil–air bushing;

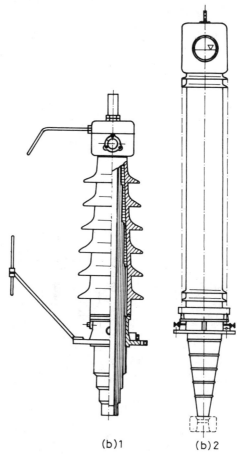

(b)1   (b)2

Fig. 7.42. Condenser bushings: (b)1 hard dielectric oil–air bushing; (b)2 soft paper dielectric oil–air bushing

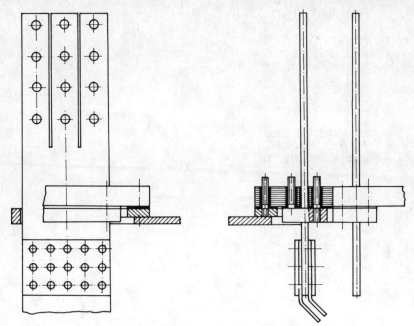

Fig. 7.43. Bushing with bar conductor

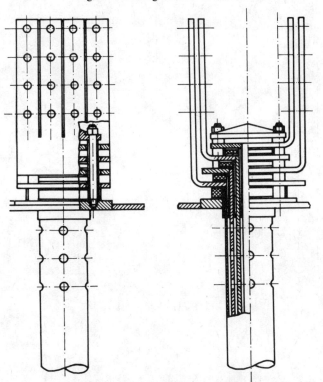

Fig. 7.44. Bushing with concentric conductor

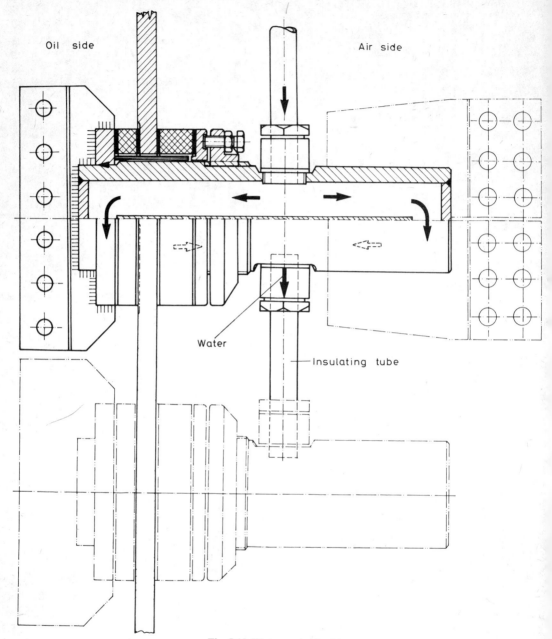

Oil side

Air side

Water

Insulating tube

Fig. 7.45. Water-cooled bushing

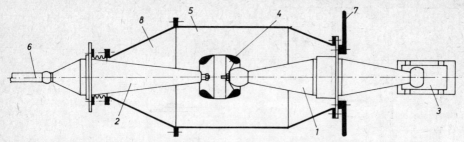

Fig. 7.46. Oil–oil condenser bushing: *1* — oil–oil condenser bushing; *2* — cable terminal; *3* — transformer-side screening fitting; *4* — screening fitting; *5* — disconnecting chamber; *6* — cable; *7* — transformer tank; *8* — transformer oil

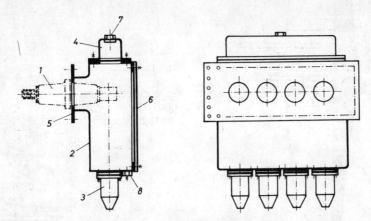

Fig. 7.47. Cable bushing. 6 kV 1000 A, three-phase neutral cable bushing: *1* — bushings; *2* — cable box; *3* — cable sheath connecting cup; *4* — conservator; *5* — fixing flange; *6* — closing cover of handling aperture; *7* — filling hole; *8* — drain hole

    (a)3 separate oil space oil–air bushing (see Fig. 7.41);
    (b) condenser bushings, which may be of;
    (b)1 hard dielectric oil–air bushing;
    (b)2 soft paper dielectric oil–air bushing (see Fig. 7.42);
    (c) bar conductor bushings (see Fig. 7.43);
    (d) concentric bushings (see Fig. 7.44);
    (e) water-cooled bushings (see Fig. 7.45).

2. Types of oil–oil bushings are as follows:
    (a) condenser bushings (see Fig. 7.46);
    (b) cable bushings (see Fig. 7.47).

3. Two kinds of oil–SF_6 bushings are known, both belonging to the condenser type bushings:
    (a) hard dielectric type,
    (b) soft paper dielectric type (see Fig. 7.48).

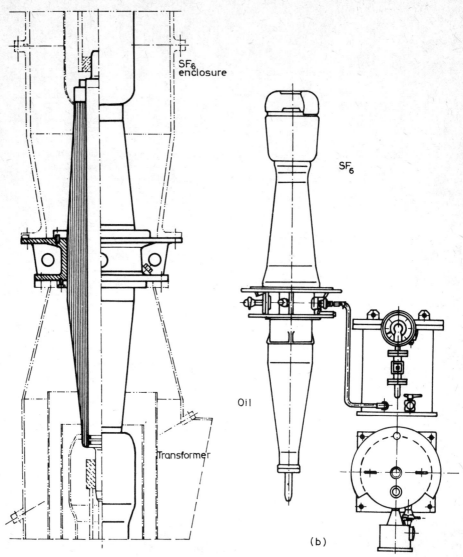

Fig. 7.48. Oil–SF₆ bushings: (a) hard dielectric oil–SF₆ condenser bushing DIELECTRA; (b) soft paper dielectric oil–SF₆ condenser bushing MICAFIL

7.7 Cooling equipment

ONAN cooling is used for approximately 25% of the total MVA rating of all transformers in operation. This cooling system appears on distribution transformers up to 2 MVA, but due to its simple maintenance requirements and low noise level many units of 75 to 100 MVA equipped with this type of cooling are encountered. The heat exchangers are either built up of circular or elliptical tubes, or are

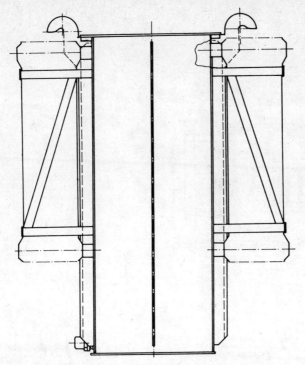

Fig. 7.49. Radiator groups welded to the tank side

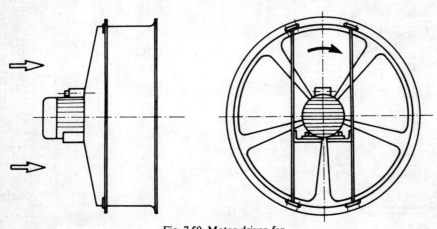

Fig. 7.50. Motor-driven fan

radiators. Up to a maximum of 25 kW cooler output, the heat exchangers are welded to the side of the tank as in Fig. 7.49. Up to a cooler output of 120 kW, the radiator groups built up of elements welded one behind another are connected to the tank side through butterfly valves. Up to maximum 500 kW cooler output, the radiator

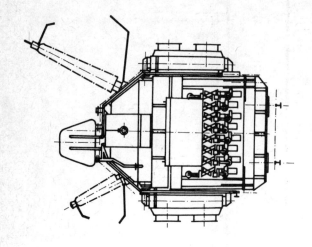

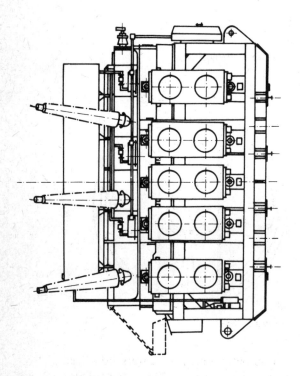

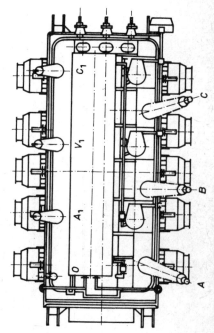

Fig. 7.51. Transformer with OFAF cooling

596

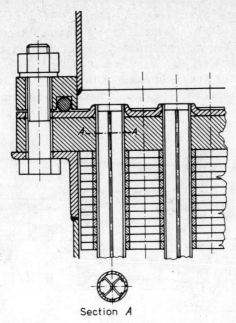

Section A

Fig. 7.52. 100 kW oil–air compact heat exchanger with connection of oil pipes into oil chamber

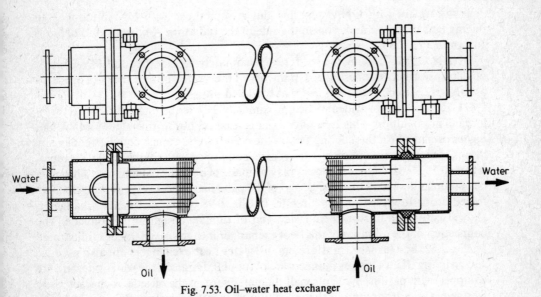

Water

Water

Oil

Oil

Fig. 7.53. Oil–water heat exchanger

groups are also connected through butterfly valves to the headers mounted on the side of the tank and communicate with the latter through gate valves. For cooler outputs above 500 kW, the coolers are mounted on separate foundations, but are connected to the tank through headers just as in the case of the 500 kW cooling units. The radiators of the radiator type cooling dealt with in Chapter 6 are manufactured in five different lengths. Diagrams of radiator elements are shown in Fig. 6.69, and their performance data given in Table 6.9. Up to 25 radiator elements are welded one behind another.

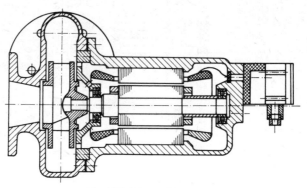

Fig. 7.54. Motor-driven pump

Transformers with ONAF cooling differ from those with ONAN cooling in having fans mounted under or on the side of the radiators. A normal kind of fan is shown in Fig. 7.50.

ONAN/ONAF cooling is used for approximately 33% of the total MVA rating of all transformers in operation. Up to 70% of their rated output they are operated as ONAN, beyond that load as ONAF cooled units.

OFAF cooling is provided with approximately 36% of the total rating of all transformers in operation. This type of cooling is used on transformers of 63 MVA and higher ratings. With this cooling system the oil is forced through compact coolers by means of motor driven pumps. A transformer equipped with OFAF cooling is represented in Fig. 7.51. Compact coolers may be made of copper or aluminium. They may be of tubular design on the oil side, and of the plate-fin surfaces i.e. plain, louvered, pin finned cross-flow type on the air side. In Fig. 6.98 a compact heat exchanger is shown with small fins on the air side and of tubular design on the oil side. The connection of the oil tubes of this heat exchanger into the oil chamber is illustrated in Fig. 7.52. For furnace transformers oil–water heat exchangers are also used, as shown in Fig. 7.53. These exchangers are of tubular design on the water side, and are equipped with pin-fins made of small-diameter wire on the oil side. A characteristic feature of the motor pumps is that no stuffing box is used in them. The squirrel-cage motor, which may be either a 4-pole or a 2-pole unit, runs immersed in transformer oil. The pump is a single-stage unit (see Fig. 7.54).

7.8 Noise suppression structures

As discussed in Chapter 1, the core is one of the sources of transformer noise, others being the cooling fans and, to a lesser extent, the oil pumps in the case of the OFAF, DOFAF and OFW systems of cooling.

Noise reduction of the core

Core noise is caused by the phenomenon of magnetostriction. The amplitude of magnetostriction as a function of core flux density is given in Fig. 7.55. As is apparent there, the peak-to-peak values in the HI-B lamination material hardly change over the range of flux densities 1.4 to 1.75 T. This means that it is not worth reducing the magnetic noise, not only for economic reasons, but because it would not be possible by choosing a reasonable lower value for the flux density of the core.

One method of reducing transformer noise is the bonding of sheet edges. This method has been described in Section 7.11. The other method is to provide a resilient support under the frame structure within the tank. By inserting a rubber strip 12 to 20 mm thick between beam 17 (Fig. 7.21) and longitudinal beam 15 of the tank, the vibration of the active part transmitted to the tank may be reduced considerably. The vibration originating in the core is transferred by the oil and the

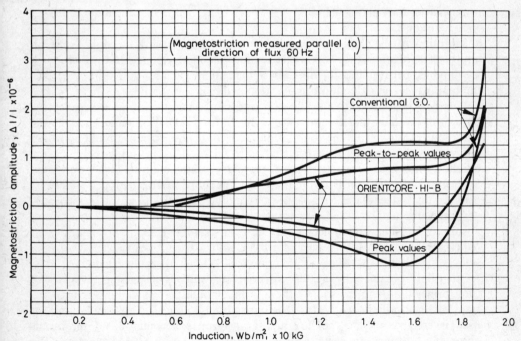

Fig. 7.55. Amplitude of magnetostriction as a function of flux density according to Cat. No. EXE 323 Sep. 1971 of Nippon Steel Corporation. It is apparent from the diagram that the magnetostriction of HI-B steel sheets is lower than that of the traditional grain-oriented cold-rolled transformer sheets

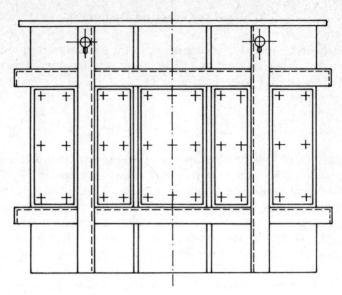

Fig. 7.56. Reinforced concrete elements mounted on the side of the oil tank to increase the mass for vibration damping

lower part of the tank to the tank wall, which thereby becomes the main radiating source of noise.

Vibration of the tank wall can be reduced by increasing its mass. Such an approach is shown in Fig. 7.56, where reinforced concrete panels 10 to 15 cm thick are mounted between the stiffening ribs of the tank by means of stud bolts welded to the tank wall. The reduction of noise obtainable is 6 to 8 dB.

Weather-resistant synthetic rubber pads 30 to 50 mm thick placed between the bottom of the tank and the rails mounted on the foundation mainly reduce the vibration transferred to the foundation, and contribute to noise reduction only indirectly. Owing to vibration transmitted to the foundation and, possibly, to the building, these structures may become secondary sources of noise. This is prevented by the resilient, vibration absorbing support. The rubber pads mentioned are placed between thick steel plates capable of carrying the dead weight of the transformer.

An effective method of reducing transformer noise is to place the tank in a sound-absorbing housing. This housing may be of brickwork or made of prefabricated panels. The acoustic absorptivity of a brick wall 12.5 cm thick plastered on both sides is 30 to 40 dB, that of a wall 36 cm thick is 35 to 40 dB. A noise reduction of 20 to 40 dB is obtained by a 12.5 cm concrete wall. The spacing between the transformer tank and the wall of the housing should permit walking around the transformer inside the wall. This spacing is about 75 cm. The cooling equipment itself is erected outside the housing. Transformers accommodated in sound-absorbing housings are usually provided with ONAN or ONAF cooling. The radiator groups are located on separate foundations. The cost of such a sound-absorbing housing is about 10% of that of the transformer concerned.

600

From among the auxiliaries, the cooling fans are to be treated as sources of noise. Especially high is the noise level of high-pressure fans employed in compact coolers. The noise level of fans can be made lower by speed reduction. If the fan speed is halved, i.e. when an 8-pole motor is used instead of a 4-pole driving unit, the motor output drops proportionally to the cube root of the ratio of speeds, the fan pressure drops proportionally to the square root of the ratio of speeds, and the flow of air delivered drops linearly with the ratio of speeds.

The noise level of the fan drops proportionally to the logarithm of the speed ratio. The reduction of noise level is

$$\Delta L = C \lg \frac{n_2}{n_1} \, dB, \qquad (7.54)$$

where the value C lies in the range of 45 to 55.

The noise level of the fans of transformers equipped with compact coolers can only be reduced by adopting suitable low-noise types. By installing the cooling unit in a housing the noise level may be considerably reduced, but filtering of the air in sandy areas may cause special problems. Varying the impeller diameter of the fan, leaving the speed unchanged, influences the noise level. Changing from a larger

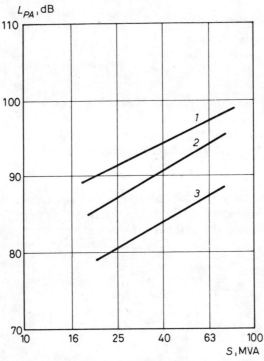

Fig. 7.57. Noise level of transformers *1* — DIN 42540/1966; *2* — normal design, VDI 2713; *3* — low-noise design VDI 2713

impeller diameter to a smaller one, the reduction of noise level will be

$$\Delta L = D \lg \frac{D_2}{D_1} \text{ dB},$$ (7.55)

where the value of factor D is between 60 and 70.

Permissible noise level of transformers

Data are given in DIN 42540/1966 up to 40 MVA (see Fig. 7.57). The values for larger transformers have been extrapolated in the diagram.

Noise levels according to VDI 2713 for transformers of normal designs and for those of low-noise designs are also shown in Fig. 7.57.

Permissible noise levels for residential areas

Generally, the noise levels permitted for residential areas in various countries differ. Nevertheless, the following values are most often adopted:

Permissible noise levels:

Residential areas: 35 dBA,
Loosely built-in areas: 40 dBA,
Densely built-in areas: 45 dBA.

Propagation of transformer noise

Let the acoustic power of the transformer be L_{PA}, dBA. At a distance of r, m away from the transformer, the noise level L_A, dBA will be

$$L_A = L_{PA} - 10 \lg (2\pi r^2).$$ (7.56)

The above formula considers the transformer as a source of sound standing on a sound reflecting plane and radiating into a hemispherical space. From the formula, it follows that doubling the distance reduces the noise level by 6 dB.

REFERENCES

[1] Abetti, P. A. (1953): Transformer models for the determination of transient voltages. *Trans., AIEE* **72**, III, 468–480.

[2] Abetti, P. A. (1960): Pseudo-final voltage distribution in impulsed coils and windings. *Trans., AIEE on PAS*, **79**, III, 87–91.

[3] Agarwal, Paul D. (1959): Eddy current losses in solid and laminated iron. *Trans. AIEE*, **78**, II, 169–179.

[4] Algbrant, A., Brierley, A. E., Hylten-Cavallius, N. and Ryder, D. H. (1966): Switching surge testing of transformers. *IEEE Trans. on PAS*, **85**, 54–61.

[5] Allan, D. J., Forrest, J. A. C., Howitt, E. L. and Petchall, A. T. (1974): Electric and acoustic location of discharges in transformers. Conference Toronto, Canada (Abstracts).

[6] Allen, N. G. H., Szpiro, O. and Campero, E. (1981): Thermal analysis of power transformer winding. *Electric Machines and Electromechanics*, **6**, 1–11. (Hemisphere Publishing Corporation).

[7] Arnold, E. and La Cour, J. L. (1904): *Die Transformatoren*. Springer, Berlin.

[8] Baatz, H. (1956): *Überspannungen in Energieversorgungsnetzen*. Springer, Berlin.

[9] Bickford, J. P. and Cornick, K. J. (1971): A summary of overvoltages on transmission systems. Paper presented at the Residential Symposium: *The Testing of Power Apparatus and Systems Operating in the Megavolt Range*. University of Manchester. Institute of Science and Technology.

[10] Binggeli, J., Froidevaux, J. and Kratzer, R. (1966): The treatment of transformers. Quality and completion criteria of the process. *CIGRE Report*, 110.

[11] Blume, L. F. G. and Boyajian, A. (1954): *Transformer Engineering*. John Wiley, New York.

[12] Boyajian, A. (1924): Theory of three-circuit transformers. *Trans. AIEE*, XVIII, 508–529.

[13] Brandes, D. (1977): Geräusche der Transformatoren des mittleren Leistungsbereiches. *Elektrizitätswirtschaft*, **76**, 195–199.

[14] Brechna, A. H. (1960): Erwärmung und Kühlung der Eisenkörper von Leistungtransformatoren. *Neue Technik, Zürich*, **2**, 36–42.

[15] Brinkmann, K. and Beyer, M. (1960): Über die Hochvakuum-Trocknung und Entgasung von Isolierölen für Höchstspannungskabeln. *ETZ* **81**, 744–749.

[16] BS 171. Specification for power transformers. 1970.

[17] BS 5800. Guide for the interpretation of the analysis of gases in transformers and other oil-filled electrical equipment in service. 1979.

[18] Carpenter, C. J. and Djurovic, M. (1975): Three-dimensional numerical solution of eddy currents in thin plates. *PROC. IEE*, **122**, 681–688.

[19] Cauer, W. (1934): Topologische Dualitätssätze und Reziprozitätstheoreme der Schaltungstheorie. *ZAMM*, **14**, 349.

[20] Cherry, E. C. (1949): The duality between electric and magnetic circuits and the formation of transformer equivalent circuits. *Proc. Phys. Soc.*, **62**, 101–111.

[21] Christl, H. (1959): Die Verteilung der Stromkräfte an Transformatorenwicklungen. *Archiv für Electrotechnik*, **44**, I, 1–11.

[22] Cliff, S. S. (1971): Philosophy of testing specifications. Paper presented at the Residential Symposium: *The Testing of Power Apparatus and Systems Operating in the Megavolt Range*. University of Manchester, Institute of Science and Technology.

[23] Colonias, S. (1976): Calculation of magnetic fields for engineering devices. *Trans. IEEE Mag.*, Vol. Mag-12, 1030–35.

[24] Csernátony-Hoffer, A. (1968): Experimental results with vacuum drying of thick soft paper insulation. *Trans. IEEE on Electrical Insulation*, Vol. El 3, No. 4. 96–105.

[25] Csernátony-Hoffer, A. (1968): Deterioration of paper insulation due to moisture absorption. (In Hungarian). *Electrotechnika*, **61**, 389–392.

[26] Djurovic, M. and Monson, I. E. (1977): 3-dimensional computation of the effect of the horizontal magnetic shunt on transformer leakage fields. *Trans. IEEE Mag.*, Vol. Mag-13, 1137–1139.

[27] Eckert, E. R. G. and Drake, R. M. jr. (1959): *Heat and Mass Transfer*. McGraw-Hill, New York.

[28] Edelmenn, H. (1959): Anschauliche Ermittlung von Transformator-Ersatzschaltbildern. *Archiv der elektrischen Übertragung*, **13**, 253–261.

[29] Fallou, B. (1970): Synthèse des travaux effectués au L.C.I.E. (Laboratoire Centrale des Industries Electriques) sur le complex papier-huile. *Revue générale de l'Electricité*, **79**, 645–661.

[30] Fehér, Gy. and Kerényi D. (1969): Die Berechnung der Stoss-Spannungsverteilung in Wicklungen und Wicklungsgruppen von Transformatoren mit Hilfe eines Digitalrechners. *Acta Tech. Acad. Sci. Hung.*, **65**, (1–2), 163–177.

[31] Fischer, E. (1952): Festigkeit der inneren Rohre von Transformator-Entwicklungen. *ETZ*, **73**, 121–33.

[32] Földiák, G. (1962): Determination of some significant non-standardized physical properties of transformer oils based on known standard characteristics. (In Hungarian). *Electrotechnika*, **65**, 160–164.

[33] Foschum, H. (1980): Erfahrungsbericht zur Gasanalyse an Öltransformatoren. *ELIN-Zeitschrift*, 1–2, 17–26.

[34] Froidevaux, J. (1956): Die Modelle, ein Hilfsmittel der Transformatorenberechnung. *Bulletin Sécheron*, **25**, 41.

[35] Fryxell, J., Agerman, E., Grundmark, B., Hessen, P. and Lampe, W. (1968): Réalisation d'essais de décharge partielles sur les transformateurs de puissance. *CIGRE Report*, 12–04.

[36] Gailhofer, G., Kury, H. and Rabus, W. (1968): Mesures de décharges partielles dans l'isolement des transformateurs de puissance. Principes et pratique. *CIGRE Report*, 12–15.

[37] Gallai, G. (1968) The new vacuum drying equipment of Ganz Electric Works. (In Hungarian). *Electrotechnika*, **61**, 433–441.

[38] Gert, R., Glavitsch, H., Tikhodeyev, N. N., Shur, S. S. and Thorén, B. (1972): Temporary overvoltages, their classification, magnitude, duration, shape and frequency of occurrence. *CIGRE Report*, 33–12.

[39] Goldstein, A. (1960): Neue Wege im Bau von Transformatorkernen. *ETZ*, **81**, 2, 53–59.

[40] Goss, N. (1953): New development in electrical strip steels characterized by fine grain structure approaching properties of a single christal. *Trans. Am. Soc. Met.*, **23**, 511–544.

[41] Gotter, G. (1963): *Erwärmung und Kühlung elektrischer Maschinen*. Springer, Berlin.

[42] Gröber, Erk. and Grigull, U. (1963): *Die Grundgesetze der Wärmeübertragung*. Springer, Berlin.

[43] Hadfield, R. (1889): On alloys of iron and silicon. *J. Iron Steel Inst.*, pp. 222–255.

[44] Hagenguth, J. H. (1944): Progress in impulse testing of transformers. *Trans. AIEE.*, **63**, 999–1005 and 1140–1445.

[45] Heller, B. (1965): Die Anfangsverteilung bei Spannungsstoss in einer Transformatorwicklung. *Acta Technika CSAV*, 3, 258.

[46] Heller, B. and Veverka, A. (1968): *Surge Phenomena in Electrical Machines*. Academia, Prague.

[47] High-voltage test techniques. Part 2: Test Procedures. *IEC Publications*, 60–2, 1973.

[48] Honey, C. C., Keil, C., Preston, L. L. and Yakov, S. (1966): Switching surge tests on conductor samples and transformer windings. *CIGRE Report*, 116.

[49] Hueter, E. and Buch, H. (1935): Über Transformatoren mit annähernd sinusformigen Magnetisierungsstrom. *ETZ*, 56, 933–937.

[50] Imre, L. (1969): Dimensioning of Cooling radiators of Naturally Cooled Oil Transformers and Investigation of their Cooling Process. (In Hungarian). Study, Budapest, Technical University, Faculty of Electrical Engineering, Department of Mechanics.

[51] Imre, L. (1976): Determination of steady-state temperature distribution of transformer windings by the heat flux network method. *Periodica Polytechnica, El. Engineering*, **20**, 261–471.

[52] Imre, L. et al. (1982): *Temperature rise and cooling of electric machines and appliances*. (In Hungarian). Műszaki Könyvkiadó, Budapest.

[53] Imre, L. and Bitai, A. (1979): A conception of simultaneous and hierarchic network modelling, for oil-cooled transformers. *Periodica Polytechnica*, **23**, 265.

[54] Imre, L. Szabó, I. and Bitai, A. (1978): Determination of the steady-state temperature field in naturally oil-cooled disc-type transformers. EC-20. Paper presented at the *Sixth International Heat Conference, Toronto, Canada*.

[55] Imre, L., Bitai, A. and Csényi P. (1979): Parameter sensitivity investigations of the warming of oil-transformers *Proc. of the 1st Int. Conf. Num. Meth. in Thermal Problems*. (Eds Lewis, R. W. and Morgan, K.), Pineridge Press, Swansea 850–858.

[56] Insulation co-ordination, *IEC Publication*, 71–1, 71–2, 1976.

[57] Irie, T. and Fukuda, B. (1975): Effect of insulating coating on domain structure in grain oriented 3% Si–Fe sheet as observed with a high voltage scanning electron microscope. *Amer. Inst. of Physics Conf. Proc. of the 21st Conf. on Magn. and Magn. Mat.* Philadelphia. (Abstracts).

[58] Johansen, O. S. (1968): Overvoltage problems in power transmission networks. (In Hungarian). *Electrotechnika*, **61**, 349–358.

[59] Karády, Gy. (1958): The stresses produced by chopped-wave tests in different transformer connections and their variation with wave duration. *CIGRE Report*, 114.

[60] Karsai, K. (1962): A transformer winding showing inhomogeneity in its last section. (In Hungarian). *Electrotechnika*, **55**, 538.

[61] Karsai, K. (1966): Calculation of pseudo-final voltage distribution in impulsed windings with a matrix method. *Acta Techn. Acad. Sci. Hung.*, **56**, (3–4); 309–318.

[62] Karsai, K. (1968): Some problems of drying high unit rating transformers. (In Hungarian). *Electrotechnika*, **61**, 427–432.

[63] Karsai, K. (1970): Are magnetic conditions of transformers truly reflected by their equivalent circuits? (In Hungarian). *Electrotechnika*, **63**, 297–301.

[64] Karsai, K. and Lovass-Nagy, V. (1963): Verteilung von Spannungen in inhomogenen Transformatorwicklungen beim Auftreten von Stosswellen. *ETZ*, **84**, 262.

[65] Karsai, K., Kerényi, D. and Ujházy, G. (1971): Einige Fragen des Stoss-spannungsschutzes von Transformatoren mit Scheibenwicklungen. *Periodica Polytechnica*, **15**, 1. 21–42.

[66] Karsai, K., Kerényi, D. and Kiss, L. (1971): Some aspects of design associated with the operational reliability of large and EHV transformers. *CIGRE Report*, 12–04.

[67] Kaufman, R. B. and Meador, J. R. (1968): Dielectric tests for EHV transformers. *Trans. IEEE on PAS*, **87**, 135–141.

[68] Kays, W. M. and London, A. L. (1955): *Compact Heat Exchangers*. McGraw-Hill, New York.

[69] Kézér, I. (1978): Transformer losses caused by the magnetic field at current carrying conductors. (In Hungarian). *Elektrotechnika*, **71**, 173–180.

[70] Kerr, H. W. and Palmer, S. (1964): Developments in the design of large power transformers. *Proc. IEEE*., Vol. 4. 823–832.

[71] Kiss, L. (1969): Thermal problems of transformers provided with forced directed oil cooling. (In Hungarian). *Electrotechnika*, **62**, 1–18.

[72] Kiss, L. (1974): Temperature rise of transformers under loads varying in time. (In Hungarian). *Electrotechnika*, **67**, 241–246.

[73] Kiss, L. (1976): Quasi-stationary warming of three-winding transformers. (In Hungarian). *Electrotechnika*, **69**, 241–254.

[74] Kiss, L. (1976): Quasi-stationary warming and cooling under conditions of free-flow heat exchange. (In Hungarian). *Electrotechnika*, **69**, 41–80.

[75] Kiss, L. (1980): *Nagrev i ochladidenie transformatorov*. Publishing House for Power Engineering, Moscow.

[76] Kiss, L., Szita, Z. and Ujházy, G. (1974): Some characteristic design problems of large power transformers. *CIGRE Report*, 12–07.

[77] Knaak, W. (1939): Zusätzliche Verluste durch Streufelder in den Wicklungen von Transformatoren. *Electrotechnik und Maschinenbau*, **57**, 7/8, 89–93.

[78] Knaak, W. and Schwaab, H. (1938): Zusätzliche Streuung bei Transformatoren. *ETZ*, **59**, 470–482.

[79] Kohn, S., Pamme, A., Pichon, A., Rogé, G. and Blumenthal, L. (1959): Les autotransformateurs du réseau à 380 kilovolts. *Revue générale de l'électricité*, 68, 44.

[80] Kovács, T. (1967): Digital computer methods for calculating short-circuit forces in transformer windings. (In Hungarian). *Electrotechnika*, **60**, 231–235.

[81] Kraussold, H. (1963): *VDI-Wärmeatlas*. Verein Deutscher Ingenieure, Düsseldorf.

[82] Kreuger, F. H. (1964): *Discharge Detection in High Voltage Equipment.* A Heywood Book, Temple Press Books Ltd., London.

[83] Küchler, R. (1956): *Die Transformatoren.* Springer, Berlin.

[84] Loading guide for oil-immersed transformers. *IEC Publication*, 354, 1972.

[85] Leschanz, A. (1971): Messverfahren und Messergebnisse bei der Isolationsüberwachung von Transformatoren. *Elektrotechnik und Maschinenbau*, 88, 410.

[86] Liska, J. (1962): *Electric machines I. Transformers.* (In Hungarian). 7th ed. Tankönyvkiadó, Budapest.

[87] Lovass-Nagy V. (1964): *Matrix calculus.* (In Hungarian). Tankönyvkiadó, Budapest.

[88] Lovass-Nagy, V. (1962): A matrix method of calculating the distribution of transient voltages in transformer windings. *The Institution of Electrical Engineers*, Monograph No. 5175. May.

[89] Márkon, S. and Szabó, L. (1977): Quasi-stationary magnetic field computing program and its application in special calculation of electric machines. *Ganz Electric Review*, No. 16. 35–46.

[90] Mihejev, M. A. (1963): *Fundamentals of practical heat transfer calculations.* (In Hungarian). 3rd ed. Tankönyvkiadó, Budapest.

[91] Mole, G. (1952): Design on performance of a portable discharge detector. *ERA-Rep.*, V/T 115.

[92] Morva, T. (1966): Verfahren zum Berechnen der elektrischen Feldstärke an Hochspannungselektroden. *ETZ*, **87**, 955.

[93] Moser, H. P. (1979): *Transformerboard.* (Sonderdruck der Zeitschrift Scientia Electrica) Birkhäuser AG, Basel.

[94] Moser, H. P. Dahinden, V., Friedrich, H., Lennarz, K. and Potocnik, O. (1981): Entwicklungstendenzen bei den Isolationssystemen für Transformatoren. *Bulletin des SEV*, 72, 674–679.

[95] Moser, H. P., Bahmann, E., Bahmann, E., Kiwaczynski, G. and Lennarz, K. (1964): Probleme der Koronamessung bei Leistungstransformatoren. Transformatorenwerk Lepper. *Technische Mitteilung*, 1, March.

[96] MSZ 9250–66. Insulations Coordination, Hungarian standard specification.

[97] MSZ 153/2–61 Insulating oils used in power transformers and circuit breakers. Hungarian standard specification.

[98] MSZ 9230/1–70. Transformers. Hungarian standard specification.

[99] Mulhall, V. R. (1980): The significance of the density of transformer oils. *Trans. IEEE on Electrical Insulation*, Vol. El-15 No. 6, 498–499.

[100] Musil, R. J., Preininger, G., Schopper, E. and Wenger, S. (1981): Voltage stresses produced by aperiodic and oscillating system overvoltages in transformer windings. *Trans. IEEE on PAS*, Vol. PAS-100, No. 1, January, 431–441.

[101] NEMA Publ., 107 (1964) Methods of measurements of radio, influence voltage (RIV) of right voltage apparatus.

[102] NEMA Publ. No. Tr. 1–1968: Transformers regulators and reactors.

[103] Norris, E. T. (1963): High-voltage power transformer insulation. *Proc. IEE*, 110, 428.

[104] Oesch, G. and Schatzl, H. (1976): Die Solventdampftrocknung von Leistungstransformatoren. (In zeitlich freier Folge erscheinendes Mitteilungsblatt für die Geschäftsfreunde der Micafil A. G., Zürich.) *Micafil Nachrichten*, August, MNV 46/1 d.

[105] Palkovics, Sz. (1961): *Work sheets for thermodynamical dimensions of heat exchangers.* (In Hungarian). Tankönyvkiadó, Budapest.

[106] Pass, F. (1971): Isolieröle. Allgemeine Grundlagen über chemischem Aufbau und Anwendungseigenschaften. *Elektrotechnik und Maschinenbau*, **88**, 290.

[107] Pattantyús, Á. G. (1942): *Fluid Mechanics.* (In Hungarian). Post-graduate Training Institute for Engineers and Managers, Budapest.

[108] Pfeiffer, G. (1965): Die Darstellung von Wanderwellen durch Exponentialfunktionen. *Wissenschaftliche Zeitschrift der Hochschule für Elektrotechnik, Ilmenau*, 11, 109.

[109] Power transformers. Part 3: Insulation levels and dielectric tests. *IEC Publication*, 76–3. 1980.

[110] Power transformers IEC, 76, 1976.

[111] Raznjevic, K. (1964):*Thermodynamic charts.* (In Hungarian). Müszaki Könyvkiadó, Budapest.

[112] Reiplinger, E. and Steler, H. (1977): Geräuschprobleme. *ETZ*, **98**, 3, 224–228.

[113] Rogowski, W. (1909): Über das Streufeld un den Streuinduktionskoeffizienten eines Transformators mit Scheibenwicklung und geteilten Endspulen. *Mitt. Forsch-Arb. VDI*, 71, 1–33.

[114] Rogowski, W. (1913): Über zusätzliche Kupferverluste, über die kritische Kupferhöhe einer Nut und über das kritische Widerstandsverhältnis einer Wechselstrommaschine. *Archiv für Elektrotechnik*, **2**, 3, 81–118.

[115] Rösch, V. (1957): Theorie des Rahmkernes für Drehstromtransformatoren. *L'Elettrotechnica*, **44**, 8, 476–486.

[116] Rüdenberg, A. (1962): *Elektrische Wanderwellen*. Springer, Berlin.

[117] Ryder, D. H. (1977): High voltage, high power transformers. *Electronics & Power*, April, 302–306.

[118] Sato, T., Kuroki K. and Tanaka, K. (1978): Approaches to the lowest core loss in grain-oriented 3% silicon steel with high permeability. *Trans. IEEE Mag.*, Vol. Mag-14, 350–352.

[119] Schlichting, H. (1958): *Grenzschicht-Theorie*. 3rd ed. Braun, Karlsruhe.

[120] Schmidt, J. and Fiedler, A. (1964): Vakuumtrocknung eines 400/231 kV Einphasenspartransformators mit 210 MVA Durchgangsleistung. *Elektrie*, 18, 91–94.

[121] Schober, J. (1971): *Inhibierte Izolieröle*. Bulletin SEV, 62, 1210.

[122] Specification for unused mineral insulating oils for transformers and switchgear. *IEC Publication*, 296, 1982.

[123] Stenkvist, E. (1958): Problems raised by short-circuits in large power transformers. *CIGRE Report*, 155.

[124] Switching surge test for oil-insulated power transformers. *IEE Committee Report. Trans. IEE on PAS*, **86**, 247, 1967.

[125] Stĕpina, J. (1961): Reduction of additional losses in compact iron by screening. (In Czech). *Elektrotecknický Obzor*, **50**, 5, 254–259.

[126] Stĕpina, J. (1962): Screening of transformer vessels with aluminium to reduce losses. (In Czech). *Elektrotechnický Obzor*, **51**, 7, 330–335.

[127] Szoldatics, E. (1971): Abnahme und periodische Überwachungsmessungen von Isolierölen. *Elektrotechnik und Maschinenbau*, 88, 300.

[128] Szűcs, L. (1961): *Theory of heat exchangers*. (In Hungarian). Post-graduate Training Institute for Engineers and Managers, Budapest.

[129] Takagi, T. a.o. (1978): Reliability improvement of 500 kV large capacity power transformer. *CIGRE Report*, 12–02.

[130] Lamierini magnetici a grano orientato. TERNI S.p.A Panetto and Petrelli, Spoleto, 1968.

[131] Thorén, B. (1974): Temporary overvoltages. Power System Overvoltages. Paper presented at the Residential Symposium: *The Testing of Power Apparatus and Systems Operating in the Megavolt Range*. The University of Manchester. Institute of Science and Technology.

[132] Timascheff, A. (1951): Über die Ersatzschaltung des Dreiwickeltransformators. *Elektrotechnik und Maschinenbau*, **68**, 277–288.

[133] Török, B. (1974): Impulse voltage transfer in transformer windings. (In Hungarian): *Elektrotechnika*, **67**, 117.

[134] Ujházy, G. (1966): Withstand Capability of Regulating Transformers to Voltage Surges. (In Hungarian). *Electrotechnika*, **59**, 107, 206.

[135] Vajda, Gy. (1964): *Deterioration of Insulants and its Detection*. Akadémiai Kiadó, Budapest.

[136] Veverka, A. and Chládek, J. (1961): Analyse der Dämpfung beim Modellieren der Stosserscheinungen in Transformatoren. In: *Probleme der Transformatoren*. Akademia, Prague. pp. 9–33.

[137] Wagner, K. W. (1915): Das Eindringen einer elektromagnetischen Welle in eine Spule mit Windungskapazität. *Elektrotechnik und Maschinenbau*, 33; 89, 105.

[138] Waldvogel, P. and Rouxel, R. (1956): Predetermination by calculation of the electrical stresses in a winding subjected to a surge voltage. *CIGRE Report*, 125.

[139] Waters, M. (1954): The measurement and calculation of axial electromagnetic forces in concentric transformer windings. *Proc. IEEE.*, **101**, II, 35–46.

[140] Weh, H. (1953): Die zweidimensionale Wärmeströmung im geschichteten Transformatorenkern. *Archiv für Elektrotechnik*, **41**, 2, 122–126.

[141] Weiland, K. (1979): Ein Verfahren zur Berechnung von Wirbelströmen in massiven, dreidimensionalen, beliebig geformten Eisenkörpern. *ETZ*, **100**, 9, 263–267.

[142] Yakov, S. (1977): Considerations about the impulse test procedure for power transformers. *Electra*, No. 55. December, 5–23.

[143] Zurmühl, R. (1964): *Matrizen und ihre technischen Anwendungen*. Springer, Berlin.

SUBJECT INDEX

614